Bioresource Technology: Concepts, Design and Applications

Bioresource Technology: Concepts, Design and Applications

Edited by **Elsa Cooper**

SYRAWOOD
PUBLISHING HOUSE
New York

Published by Syrawood Publishing House,
750 Third Avenue, 9th Floor,
New York, NY 10017, USA
www.syrawoodpublishinghouse.com

Bioresource Technology: Concepts, Design and Applications
Edited by Elsa Cooper

International Standard Book Number: 978-1-68286-226-1 (Hardback)

Printed in China.

Contents

Preface

As an academic discipline, the scope of bioresource technology is vast and it finds applications across many fields such as agriculture, etc. The aim of this book is to present researches that have transformed this discipline and aided its advancement. Biocatalysis, biotransformation, biosynthesis, bioenergy, etc. are some of the significant topics that have been covered within the book. It presents an overview of bioresource development, processing and utilization. The extensive content of the text provides the readers with a thorough understanding of the subject.

This book is the end result of constructive efforts and intensive research done by experts in this field. The aim of this book is to enlighten the readers with recent information in this area of research. The information provided in this profound book would serve as a valuable reference to students and researchers in this field.

At the end, I would like to thank all the authors for devoting their precious time and providing their valuable contributions to this book. I would also like to express my gratitude to my fellow colleagues who encouraged me throughout the process.

Editor

1

Optimization of upstream and downstream process parameters for cellulase-poor-thermo-solvent-stable xylanase production and extraction by *Aspergillus tubingensis* FDHN1

Dharmesh N Adhyaru[1], Nikhil S Bhatt[1*] and Hasmukh A Modi[2]

Abstract

Background: Xylanases are important members of the hemicellulolytic enzyme system. Xylanase plays a vital role in the hydrolysis of major hemicellulosic component xylan and converts it into xylooligosaccharides and ultimately yields xylose. Cellulase-lacking or cellulase-poor xylanase with high temperature and pH stability has gained special attention, especially in paper and pulp industries. Most of the available literature highlighted the fungal xylanase production by optimizing environmental and cultural parameters. However, the importance of enzyme recovery from fermented biomass still needs attention. In this study, upstream and downstream process parameters were studied for enhancing xylanase production and extraction by a newly isolated *Aspergillus tubingensis* FDHN1 under solid-state fermentation using low-cost agro-residues.

Results: In the present study, *A. tubingensis* FDHN1 was used for the xylanase, with very low level of cellulase, production under solid-state fermentation (SSF). Among various agro-residues, sorghum straw enhanced the xylanase production. Under optimized upstream conditions, the highest xylanase production 2,449 ± 23 U/g was observed. Upon characterization, crude xylanase showed stability over a broad range of pH 3.0 to 8.0 up to 24 h. The temperature stability revealed the nature of the xylanase to be thermostable. Native polyacrylamide gel electrophoresis (native PAGE) and zymogram analysis revealed the multiple forms of the xylanase. Due to the many industrially important characteristics of the xylanases, the study was elaborated for optimizing the downstream process parameters such as volume of extractant, extraction time, temperature and agitation speed to recover maximum xylanase from fermented sorghum straw. The highest amount of xylanase (4,105 ± 22 U/g) was recovered using 0.05 M sodium citrate buffer (pH 6.5) at 12:1 (*v/w*) extractant/solid ratio, 90-min extraction time, 150-rpm agitation speed and 40°C. Finally, detailed bioprocess optimization shows an overall 6.66-fold enhancement in the xylanase yield.

Conclusions: The present study consolidates the importance of upstream and downstream process optimization for the overall enhancement in the xylanase production. The xylanase from *A. tubingensis* FDHN1 shows the stability at different pH and temperature, and it was also active in the presence of organic solvents. These properties of xylanase are very much important from an industrial application point of view.

Keywords: Cellulase-poor xylanase; Upstream-downstream process optimization; *Aspergillus tubingensis* FDHN1; Solid-state fermentation; Thermo-solvent stable enzyme; Multiple xylanases

* Correspondence: bhattnikhil2114@gmail.com
[1]P. G. Department of Microbiology, Gujarat Vidyapeeth, M. D. Gramseva Mahavidhyalaya, Talim Kendra Campus, Sadra 382 320, Gujarat, India
Full list of author information is available at the end of the article

Background

Hemicelluloses are the second most abundant renewable resources, only exceeded by cellulose. Xylan constitutes a major component of hemicellulose, which is a heteropolysaccharide, having a chain of β-1,4-linked xylopyranose residues. The complete hydrolysis of xylan requires the combined action of various enzymes such as endo β-1,4-xylanase (E.C. 3.2.1.8), exoxylanase (β-D-xylan xylohydrolase), β-D-xylosidase (E.C. 3.2.1.37), etc. [1,2]. Recently, xylanases have attracted considerable attention due to its application in many industrial processes such as enzymatic bleaching of paper pulp, juice clarification, extraction of plant oils, texture improvement in bakery, bioconversion of agricultural waste, bioscouring in textiles and improvement of animal feed digestibility [3-5]. Due to such immense industrial applications, xylanases have a worldwide market of around 200 million dollars.

Filamentous fungi are good producers of xylanase because they are capable of producing high levels of extracellular enzymes and can be cultivated very easily. On an industrial scale, xylanases are mainly produced by *Aspergillus* and *Trichoderma* sp. in solid-state fermentation (SSF). The research has indicated that xylanase production differs with the strains and is regulated by physiological, nutritional and biochemical nature of the microbes employed [6]. The fermentation profile of an organism that influences metabolism-mediated production includes pH, temperature, carbon and nitrogen source, metal ion requirement, incubation time, inoculum size, etc. The production of an industrial enzyme by optimizing these growth parameters is of prime importance because an improper optimization of these factors may result in the lower enzyme production. In general, no defined medium has been established for the best production of any metabolite because the genetic diversity present in different microbial sources causes each organism or strain to have its own special conditions for maximum product yield [7]. For the commercialization of xylanase production, it is necessary to identify a microorganism that produces high levels of xylanase using cheaper carbon sources and having novel properties. As lignocellulosic materials are abundantly found in nature in the form of agricultural and industrial residues, it can be exploited as a potential substrate for growing the organisms.

Apart from medium optimization, recovery of enzyme is also an important aspect of SSF technology. An optimized recovery process would generate concentrated enzyme extract, prevent enzyme loss, reduce the number of the downstream process step and thus gain better commercial viability. A wealth of information is available on medium optimization for xylanase production [8-11], but studies on xylanase recovery from fermented substrate are scarce [11]. Therefore, it is highly imperative to optimize production and extraction parameters which further facilitate economic design of the full-scale operation system for newly isolated microbial strains.

In the present study, we have highlighted the optimization of upstream process for xylanase production using newly isolated *Aspergillus tubingensis* FDHN1 under SSF. Produced xylanase was characterized for important properties such as pH, temperature and solvent tolerance. Looking towards the important features of this xylanase and thrust for its higher yield, downstream process parameters for xylanase recovery were also optimized.

Methods

Agro-residue preparation, chemicals and medium components

Different agro-residues such as caster straw, wheat straw, maize straw, rice straw, sorghum straw, barley straw, sugarcane bagasse, wood chips and groundnut shell were collected locally from agricultural farms. To remove the adhered surface dust particles, agro-residues were washed in distilled water, dried at 50°C and chopped into small pieces. These were ground in mixture to achieve average particle size of 0.5 to 2 mm.

Chemicals such as oat spelt xylan, carboxymethyl cellulose (CMC), bovine serum albumin (BSA) and dinitrosalicylic acid (DNSA) were purchased from HiMedia Laboratories Ltd (Mumbai, India). All the other chemicals, media, salts and reagents used were of analytical grade (Sigma-Aldrich, St. Louis, MO, USA; HiMedia, Mumbai, India; and Merck & Co., Inc., Whitehouse Station, NJ, USA).

Fungal isolation, identification and inoculum preparation

The novel cellulase-poor xylanase-producing fungal strain FDHN1 was isolated in our laboratory from active compost pit (approximately 10-cm depth). The strain was identified at the National Fungal Culture Collection of India (NFCCI), Pune, India, on the basis of morphological and molecular study. The strain was identified as *A. tubingensis*. Morphologically, the strain showed white velvety colonies which later turns to greyish-black on potato dextrose agar (PDA) plates. Under stereoscopic observation, the conidial heads were found to be globose, radiating, split into columns, greyish metallic in colour and measured up to 260 × 260 μm in size. The conidiophores were found to be hyaline to subhyaline which later turned into brown when old, smooth-walled and up to 625 × 5 μm in size. The vesicles were globose to subglobose, fertile all over and up to 39.0 × 45.5 μm in size. 18S rDNA gene sequence of fungus has been submitted to NCBI (Accession No. KF971693).

Spore inoculum was prepared from 5-day grown culture on PDA slants at 35°C. Suspension was prepared by lightly brushing the slant containing fungal mycelia with sterile distilled water to give a final spore count of approximately 1×10^6 spores/mL.

Xylanase production using agro-residues and upstream process optimization

Caster straw, wheat straw, maize straw, rice straw, sorghum straw, barley straw, groundnut shell, sugarcane bagasse and wood chips were used as substrates to produce xylanase under SSF. Erlenmeyer flasks (250 mL) containing 5 g of each substrates and 5 mL of mineral salt medium (g L^{-1}: KH_2PO_4, 3.0; $NaNO_3$, 2.5; $MgSO_4 \cdot 7H_2O$, 1.0; $CaCl_2$, 0.05; $FeSO_4 \cdot 7H_2O$, 7.5; $MnSO_4$, 2.5; $ZnSO_4 \cdot 7H_2O$, 3.6; $CoCl_2 \cdot 6H_2O$, 3.0 and 0.1% Tween-80 ($v \cdot v^{-1}$), pH 5.0) were autoclaved at 121°C for 15 min, cooled, inoculated with 1 mL of spore suspension and incubated at 35°C for 6 days.

SSF conditions were optimized by changing one factor at a time while keeping the other factors constant. To improve xylanase production, various upstream process parameters such as sorghum straw concentration (3.0 to 15 g), incubation period (2 to 8 days), substrate to moisture ratio (1:1 to 1:7 *w*/*v*), medium pH (3.0 to 9.0) and temperature (25 to 55°C) were optimized. Furthermore, effect of various carbon sources (glucose, fructose, sucrose, xylose, lactose, maltose and mannitol at 0.5% concentration), organic nitrogen sources (peptone, beef extract, urea, yeast extract, tryptone, malt extract, meat extract, gelatine and skimmed milk powder at 0.05 g 'N' equivalent in 50 mL medium) and inorganic nitrogen sources (ammonium hydrogen phosphate, ammonium dihydrogen phosphate ammonium sulphate, ammonium per sulphate, ammonium nitrate, ammonium chloride and sodium nitrate at 0.05 g 'N' equivalent in 50 mL medium) were evaluated for xylanase production. The xylanase production was also assessed in the presence of different metal additives ($CuCl_2$, $MgCl_2$, $SnCl_2$, $CaCl_2$, $HgCl_2$, $CoCl_2$, $MnCl_2$, $NiCl_2$ and $FeCl_3$, 0.2%) and modulators (Tween-40, Tween-80, Triton X-100, glycine, glycerol, EDTA and SDS, 0.1%).

Enzyme assay and protein estimation

Xylanase activity was determined by adopting the modified method of Bailey et al. [12]. The reaction mixture containing 450 μL of oat spelt xylan (1%) prepared in 0.05 M citrate buffer (pH 5.0) and 50 μL of enzyme was incubated at 55°C for 10 min. The reaction was stopped by adding 2 mL of DNSA. The mixture was then heated in boiling water bath for 10 min and cooled down to room temperature, and then, 3 mL of water was added to the system. CMCase activity was assayed according to the method of Ghose [13]. The reaction system containing 1,000 μL of CMC (1%) dissolved in 0.05 M citrate buffer (pH 5.0) and 500 μL of enzyme was incubated at 60°C for 60 min. After incubation, 3 mL of DNSA was added to the system followed by 10-min incubation in a boiling water bath. The absorbance of the colour developed during xylanase and cellulase assay was measured at 540 nm as described by Miller [14]. One unit of xylanase/cellulase activity was defined as the amount of enzyme required to liberate 1 μmol of xylose/glucose equivalent per minute under the specified conditions. The unit 'U/g' denotes the xylanase/cellulase activity in international unit per gram dry substrate.

Soluble protein was determined by Folin's method using bovine serum albumin as standard [15].

Enzymatic properties of crude xylanase

The pH optima for xylanase activity was determined by measuring its activity in various buffers (0.05 M) such as sodium citrate (pH 3.0 to 6.0), sodium phosphate (pH 6.0 to 8.0) and glycine-NaOH (pH 8.0 to 10). The pH stability of xylanase was studied by incubating it in the above buffers for 24 h, and its residual activity was determined at a regular interval of 3 h.

The optimal temperature for xylanase activity was determined by assaying its activity at different temperatures (30 to 70°C). The thermostability of xylanase was determined at 30 to 70°C by measuring its activity at a regular interval of 30 min up to 120 min.

The solvent stability of xylanase was assessed by incubating the enzyme in the presence of various organic solvents (methanol, ethanol, propanol, butanol, acetone and benzene, at 10 to 30% concentration) for 30 min, and at the end of the reaction, residual activity was measured.

Native PAGE and xylan zymography

Native polyacrylamide gel electrophoresis (native PAGE) of the crude enzyme extract was performed at room temperature using 0.05 M citrate buffer (pH 5.0). The electrophoresed gel was stained with Coomassie Brilliant Blue G-250 for molecular weight determination [16]. A replica gel containing xylan (0.5%) was incubated at 50°C for 5 min in buffer and then stained in 0.1% Congo Red solution. After removal of dye using 1 M NaCl, gel was transferred to 5% acetic acid solution. Clear bands against a dark background indicated xylanase activity.

Enzyme extraction and downstream process optimization

Initially, the xylanase extraction was performed using 0.05 M citrate buffer (pH 5.0) at an extractant/solid ratio of 10:1 (*v*/*w*). Next, the suspension was agitated at 120 rpm for 60 min (35°C), centrifuged at 10,000×*g* for 15 min (4°C) and finally the clear supernatant was used as a crude enzyme source.

In order to find the most suitable extractant for xylanase recovery from fermented sorghum straw, seven different solvents (0.05 M citrate buffer, pH 5.0; 0.05 M sodium citrate buffer, pH 6.5; double-distilled water, pH 7.0; 0.05 M phosphate buffer, pH 7.5; 0.05 M Tris-HCl buffer, pH 8.0; 0.1% Tween-80 and 1.0% NaCl) at an extractant/solid ratio of 10:1 (*v*/*w*) were used. After selecting extractant type, different recovery parameters such as extractant/solid ratio (4:1 to 16:1 (*v*/*w*)), extraction time (15 to 120 min), agitation speed (50 to 300 rpm) and extraction temperature (25 to 50°C) were also optimized.

After optimizing extraction conditions, the percentage efficiency of xylanase recovery was studied under unoptimized and optimized extraction conditions. The percentage efficiency of xylanase recovery was calculated as follows:

$$\%\ \text{Xylanase recovery} = \frac{\text{Xylanase activity extracted}}{\text{Total xylanase activity in fermented substrate}}$$

Effect of optimized upstream and downstream conditions on total xylanase yield

At the end of detailed upstream and downstream process optimization, additional set of experiment for total xylanase productivity was carried out under the optimized conditions and obtained results were compared with earlier-reported xylanase-producing *Aspergillus* strains.

Results and discussion

Selection of agro-residues for xylanase production

Due to its high cost, pure xylan is not affordable for large-scale production of xylanase. Utilization of agricultural residues for enzyme production may serve a dual purpose: on one side, it provides a valuable metabolite, and on the other side, it helps in solving environmental solid waste disposal problem. Therefore, various low-cost agricultural residues were explored for the xylanase production. After 6 days of cultivation, among the tested agro-residues, sorghum straw supported maximum xylanase activity (615.5 ± 19.21 U/g) followed by wheat straw, sugarcane bagasse and rice straw (Table 1). Till date studies on xylanase production using sorghum straw as a substrate under SSF are scarce. It is believed that the substrate from graminaceous plants containing arabinoxylans supports high xylanase production [17]. The high xylanase titre when sorghum straw was used may be attributed to the nature of hemicelluloses, presence of some activators, surface pore size and favourable degradability of the carbon source. Thus, sorghum straw was used as a substrate throughout the optimization of upstream and downstream process parameters for xylanase production by *A. tubingensis* FDHN1 under SSF. Previously, use of sorghum straw has been reported as a potent inducer for xylanase production from *Thermomyces lanuginosus* under SSF by Sonia et al. [18]. However, xylanase production under SSF using wheat bran has also been reported by *Aspergillus niger* [11], *A. niger* and *Aspergillus niveus* [19].

In the present study, a variation in xylanase production was observed with the change in sorghum straw concentration from 3.0 to 15 g. Maximum xylanase production (1,329 ± 29 U/g) with very poor cellulase activity (0.42 ± 0.02 U/g) was obtained using 9 g sorghum straw under SSF (Figure S1 in Additional file 1).

Evaluation of upstream bioprocess parameters for xylanase production

Effect of incubation period and substrate to moisture ratio

The fermentation time for the maximum enzyme production is dependent upon the organism type and cultural/environmental conditions. Xylanase production at different time intervals during SSF showed that it was initiated from the second day and reached the maximum (1,217 ± 14 U/g) on the fifth day with specific xylanase activity of 53.54 ± 1.88 U/mg (Figure 1). Xylanase activity was found to decrease sharply with prolonged incubation. This result suggested that the fermentation endpoint should be controlled carefully because non-specific proteases secreted by the fungus may degrade the synthesized xylanase. The decrease in xylanase production may also be due to the accumulation of the end products during fermentation. SSF has also been carried out by Pandya and Gupte [20] for 10 days to produce xylanase from *A.*

Table 1 Effect of various agro-residues on xylanase production from *A. tubingensis* FDHN1 under SSF

Agricultural residue[a]	Xylanase activity (U/g)	Protein (mg/g)	Specific activity (U/mg)	Cellulase activity (U/g)
Caster waste	284 ± 10.00	31.40 ± 1.24	9.04 ± 0.36	1.26 ± 0.03
Sugarcane bagasse	459.7 ± 8.21	33.75 ± 1.0	13.62 ± 0.62	1.04 ± 0.02
Wood chips	78.1 ± 3.47	21.22 ± 0.69	3.68 ± 0.15	2.81 ± 0.08
Wheat straw	495.5 ± 6.46	37.5 ± 1.65	13.21 ± 0.31	1.5 ± 0.04
Maize straw	344.1 ± 14.26	32.19 ± 1.11	10.68 ± 0.48	0.73 ± 0.01
Rice straw	453.2 ± 12.36	33.85 ± 0.56	13.38 ± 0.71	2.0 ± 0.1
Sorghum straw	615.5 ± 19.21	32.87 ± 1.41	18.72 ± 0.91	0.6 ± 0.02
Groundnut shell	252.9 ± 8.29	26.84 ± 0.9	9.42 ± 0.33	0.71 ± 0.06
Barley straw	185 ± 4.34	35.23 ± 1.21	5.25 ± 0.21	1.05 ± 0.04

[a]SSF was carried out using 5 g of substrate moistened with 5 mL mineral salt medium pH 5.0 at 35°C for 6 days. The data presented are the mean values of three replicates with standard deviations.

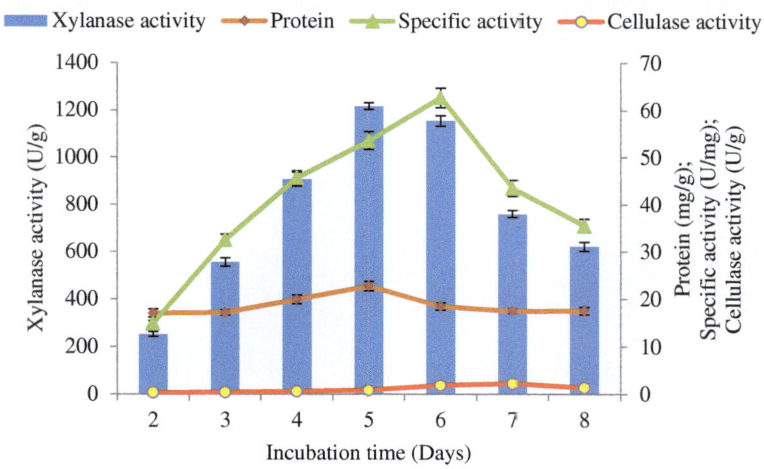

Figure 1 Effect of incubation time on the xylanase production. SSF was carried out using 9 g sorghum straw for respective days (2 to 8 days) at 35°C, and samples were analysed at a regular interval of 24 h.

tubingensis JP-1 using wheat straw. They have noticed optimum xylanase yield on the eighth day of incubation. However, an increase in the xylanase production was observed by Goyal et al. [21] up to 17 days of fermentation using various agro-residues under SSF by *Trichoderma viride.* The above results indicated that the fermentation time for enzyme production under SSF depends on the growth rate of the microorganism and its pattern for enzyme production. The strain *A. tubingensis* FDHN1 used in the present study showed relatively high xylanase production within a short incubation period which could be attractive for large-scale xylanase production.

Apart from incubation time, moisture content plays a major role during SSF in the regulation of microbial metabolism known as water activity [22]. In the present investigation, influence of moisture content was studied at different solid/liquid ratios of 1:1 to 1:7 (*w*/*v*). The highest xylanase production (1,659 ± 17 U/g) and specific xylanase activity (81.60 ± 1.18 U/mg) were obtained at 1:5 (*w*/*v*) moisture ratio after 5 days of incubation, while maximum cellulase production (2.03 ± 0.11 U/g) was achieved at 1:6 (*w*/*v*) moisture ratio. In some fungi, high xylanase production has been shown to be firmly associated with cellulase production [23], but *A. tubingensis* FDHN1 did not produce much cellulase at different time intervals and moisture ratios. At 1:6 and 1:7 (*w*/*v*) moisture ratios, the xylanase production decreased to 1,463 ± 6 and 1,397 ± 6 U/g, respectively (Table 2). The obtained results indicated that the deviations from the optimum level moisture content lead to reduced xylanase production. In SSF, moisture content is an important factor that determines the overall success of a process by altering the physical properties of the solid substratum. The lower xylanase production at higher moisture levels could attribute to the decrease in porosity, alteration in particle structure or lower oxygen transfer, whereas the lower moisture content leads to a reduction in the diffusion of the nutrients in the substrate, lower degree of swelling and higher water tension [24]. Previously, optimum moisture level of 1:5 (*w*/*v*) has been shown for the xylanase production after 10 days under SSF by Pandya and Gupte [20].

Effect of pH and temperature

The enzyme production by microbial strains strongly depends on the initial pH of the medium as it influences many enzymatic processes and transport of various components across the cell membrane [25]. In the present study, xylanase production was noticed in the wide range of pH 3.0 to 9.0 (Table 2) and high xylanase production (2,012 ± 18 U/g) with poor cellulase activity (1.64 ± 0.12 U/g) was achieved at pH 6.0. However, substantial xylanase production was also observed at pH 5.0 (1,606 ± 7 U/g), 7.0 (1,924 ± 15 U/g), 8.0 (1,745 ± 13 U/g) and 9.0 (1,565 ± 19 U/g). These results indicated that *A. tubingensis* FDHN1 was not only grown successfully in acidic and alkaline pH but also generated considerable xylanase titre. However, xylanase production by *A. tubingensis* JP-1 has been observed to decrease drastically under acidic and alkaline pH ranges [20]. Liao et al. [26] have reported higher xylanase production by *Penicillium oxalicum* GZ-2 at pH 5.0. The highest xylanase has been produced by *A. niveus* RS2 at a neutral to alkaline pH than at an acidic and high alkaline pH [27].

In the present study, optimum xylanase production (1,998 ± 12 U/g) and specific activity (71.35 ± 2.97 U/mg) were noticed at 40°C. At the same temperature, cellulase activity was increased to 2.87 ± 0.16 U/g. The xylanase and cellulase activity data suggested that the temperature is an important parameter which regulates the enzyme production by *A. tubingensis* FDHN1. At higher temperatures, xylanase production declined sharply and it was 50% at 55°C (Table 2). Although, the physiological changes

Table 2 Evaluation of physiological parameters for xylanase production from *A. tubingensis* FDHN1 under SSF

Variable	Substrate/moisture ratio[a] (*w/v*)						
	1:1	1:2	1:3	1:4	1:5	1:6	1:7
Xylanase activity (U/g)	1,064 ± 12	1,257 ± 8	1,356 ± 9	1,430 ± 13	1,659 ± 17	1,463 ± 6	1,397 ± 6
Protein (mg/g)	16.34 ± 0.63	18.20 ± 0.46	16.69 ± 0.79	20.04 ± 0.87	20.33 ± 0.29	19.53 ± 0.67	18.29 ± 0.75
Specific activity (U/mg)	65.11 ± 2.47	69.06 ± 3.21	81.24 ± 3.66	71.35 ± 2.59	81.60 ± 1.18	74.91 ± 3.34	76.38 ± 3.09
Cellulase activity (U/g)	0.21 ± 0.04	0.53 ± 0.02	0.64 ± 0.01	1.53 ± 0.06	2.03 ± 0.11	2.59 ± 0.10	2.07 ± 0.15
Variable	**pH[b]**						
	3	4	5	6	7	8	9
Xylanase activity (U/g)	978 ± 5	1,331 ± 12	1,606 ± 7	2,012 ± 18	1,924 ± 15	1,745 ± 13	1,565 ± 19
Protein (mg/g)	16.87 ± 0.57	21.97 ± 0.92	25.01 ± 1.03	24.44 ± 0.87	26.27 ± 1.00	26.24 ± 0.81	24.03 ± 0.78
Specific activity (U/mg)	57.97 ± 2.88	60.58 ± 1.24	64.21 ± 2.55	82.32 ± 2.31	73.25 ± 3.09	66.50 ± 2.99	65.12 ± 3.18
Cellulase activity (U/g)	ND	ND	0.78 ± 0.08	1.64 ± 0.12	2.33 ± 0.07	3.17 ± 0.13	2.56 ± 0.13
Variable	**Temperature[c] (°C)**						
	25	30	35	40	45	50	55
Xylanase activity (U/g)	1,452 ± 19	1,808 ± 15	1,860 ± 15	1,998 ± 12	1,623 ± 10	1,331 ± 15	1,056 ± 10
Protein (mg/g)	25.29 ± 0.98	26.66 ± 1.21	27.93 ± 0.77	28.00 ± 0.61	24.33 ± 1.03	24.34 ± 1.14	23.33 ± 0.62
Specific activity (U/mg)	57.41 ± 2.43	67.81 ± 2.08	66.59 ± 1.26	71.35 ± 2.97	66.70 ± 3.04	54.68 ± 1.33	45.26 ± 1.97
Cellulase activity (U/g)	ND	0.64 ± 0.10	2.15 ± 0.08	2.87 ± 0.16	2.06 ± 0.13	1.31 ± 0.04	0.80 ± 0.06

[a]SSF was carried out using 9 g of sorghum straw moistened with mineral salt medium (pH 5.0) at 35°C for 5 days; [b]SSF was carried out using 9 g of sorghum straw moistened with mineral salt medium of various pH values at 1:5 (*w/v*) moisture ratio at 35°C for 5 days; [c]SSF was carried out using 9 g of sorghum straw moistened with mineral salt medium (pH 6.0) for 5 days at various temperatures. ND, not determined. The data presented are the mean values of three replicates with the standard deviations.

induced by high temperature during enzyme production are not completely understood. It has been suggested that at higher temperatures, microorganisms may synthesize only a reduced number of proteins essential for growth and other physiological processes [8]. It has been shown in a previous study that *A. tubingensis* JP-1 was unable to produce a considerable xylanase yield above 30°C [20]. The results of incubation time, pH and temperature study indicate that *A. tubingensis* FDHN1 has far better xylanase production ability across the pH and temperature range within short fermentation time as compared to earlier reported *A. tubingensis* JP-1.

Influence of carbon and organic/inorganic nitrogen sources

Carbon sources are very important for the growth and metabolic processes of microorganisms. The xylanase and cellulase production in the presence of different carbon sources are shown in Table 3. Addition of 0.5% xylose in the fermentation medium enhanced the xylanase production (2,267 ± 12 U/g) with low cellulase activity (1.85 ± 0.09 U/g). However, fructose, mannitol, lactose, maltose, glucose and sucrose were affected adversely to the xylanase production. The level of cellulase production was reached to 3.68 ± 0.07 U/g in the presence of glucose, but the xylanase production was affected negatively. Xylanase and cellulase are known to be inducible in fungi and are affected by the nature of the substrate used in the study [23]. To enhance the xylanase production, xylose was selected as the best carbon source and its concentration in medium was varied from 0.1 to 1.1%. The results indicated that 0.3% xylose gave high xylanase yield (2,349 ± 17 U/g) (Figure S2 in Additional file 1). It has been suggested that xylose is not only a carbon source but also an effective inducer for the xylanase production [28]. In the present study, xylose concentration above 0.3% reduced the xylanase production gradually. Such carbon source concentration-dependent enzyme production regulation has been noticed in different xylanase-producing microbial strains by various research groups [29,30].

Formation of extracellular enzymes is greatly influenced by the availability of nitrogen source as it is the ultimate precursor of protein synthesis. Furthermore, nitrogen source can significantly affect the pH of the medium during the course of fermentation, and hence, the enzyme activity and stability may get influenced. Among the different organic nitrogen sources tested, gelatine was found to support the maximum xylanase production (2,453 ± 11 U/g). In this study, all the tested nitrogen sources enhanced the xylanase production as compared to control (Table 4). During the study with different inorganic nitrogen sources, sodium nitrate induced the xylanase production (2,589 ± 22 U/g). The xylanase production from *Trichoderma viride* in the presence of $NaNO_3$ as an inorganic nitrogen source has also been reported by Goyal et al. [21]. In the present study, combination of gelatine and sodium nitrate further improved the

Table 3 Effect of carbon sources on xylanase production from *A. tubingensis* FDHN1 under SSF

Carbon source[a]	Xylanase activity U/g	Protein mg/g	Specific activity U/mg	Cellulase activity U/g
Control[b]	1,875 ± 13	26.84 ± 0.33	69.85 ± 2.86	2.31 ± 0.11
Glucose	1,601 ± 14	28.06 ± 0.79	57.05 ± 1.93	3.68 ± 0.07
Fructose	1,871 ± 11	27.70 ± 1.13	67.54 ± 2.37	2.06 ± 0.14
Sucrose	1,644 ± 10	25.51 ± 0.94	64.44 ± 2.11	2.57 ± 0.06
Xylose	2,267 ± 12	28.61 ± 1.22	79.23 ± 3.40	1.85 ± 0.09
Lactose	1,267 ± 19	24.36 ± 1.05	52.01 ± 1.23	1.63 ± 0.13
Maltose	1,463 ± 12	23.87 ± 0.98	61.29 ± 2.74	3.11 ± 0.13
Mannitol	1,206 ± 19	20.96 ± 0.87	57.53 ± 2.66	2.49 ± 0.08

[a]SSF was carried using 9 g sorghum straw under optimized physiological conditions (at 5-day incubation, 1:5 (*w/v*) substrate/moisture ratio, pH 6.0 and 40°C); [b]the control medium used was devoid of any carbon source. The data presented are the mean values of three replicates with the standard deviations.

xylanase (2,697 ± 18 U/g) and cellulase (3.86 ± 0.06 U/g) yield.

Influence of metal additives and modulators

The choice of an appropriate additive is of great importance for the successful production of xylanase. Among tested metal compounds, $MgCl_2$ exhibited the clear stimulation in xylanase yield (2,446 ± 27 U/g) followed by $CuCl_2$, $CaCl_2$, $MnCl_2$ and $FeCl_3$. However, $HgCl_2$, $CoCl_2$, $SnCl_2$ and $NiCl_2$ responded negatively towards xylanase production (Table S1 in Additional file 1). Similarly, addition of various modulators to the culture medium exerts a range of effects on the enzyme secretion. In the presence of Tween-80, the levels of cellulase and xylanase activities were increased to 4.21 ± 0.13 and 2,449 ± 23 U/g, respectively, which could be due to its favourable effect on cell permeability and thus it affects the secretion of certain proteins [31]. In addition,

Table 4 Effect of nitrogen sources on xylanase production from *A. tubingensis* FDHN1 under SSF

Nitrogen source[a]	Xylanase activity U/g	Protein mg/g	Specific activity U/mg	Cellulase activity U/g
Control[b]	1,986 ± 13	25.84 ± 0.56	76.85 ± 3.45	1.63 ± 0.08
Organic				
Peptone	2,071 ± 11	25.92 ± 1.48	79.89 ± 2.64	3.02 ± 0.18
Beef extract	2,189 ± 10	24.35 ± 1.34	89.89 ± 3.89	2.54 ± 0.09
Urea	2,432 ± 14	25.1 ± 0.89	96.89 ± 1.26	2.81 ± 0.13
Yeast extract	2,401 ± 11	26.63 ± 1.08	90.16 ± 4.15	1.67 ± 0.11
Tryptone	2,004 ± 12	24.77 ± 1.07	80.90 ± 3.86	3.72 ± 0.16
Malt extract	2,348 ± 15	24.18 ± 0.96	97.10 ± 2.41	3.09 ± 0.16
Meat extract	2,262 ± 14	25.34 ± 1.21	89.26 ± 3.02	2.84 ± 0.06
Gelatine	2,453 ± 11	25.09 ± 1.14	97.76 ± 3.37	3.28 ± 0.12
Skim milk powder	2,347 ± 11	24.69 ± 0.88	95.05 ± 3.98	1.52 ± 0.15
Inorganic				
$(NH_4)_2HPO_4$	2,148 ± 20	25.61 ± 1.05	83.87 ± 2.46	2.44 ± 0.07
$(NH_4)_2H_2PO_4$	2,367 ± 14	4.89 ± 0.46	95.09 ± 1.84	3.81 ± 0.14
$(NH_4)_2SO_4$	2,272 ± 21	23.5 ± 1.12	96.68 ± 2.92	3.33 ± 0.11
$(NH_4)_2S_2O_8$	2,430 ± 20	24.31 ± 0.98	99.95 ± 3.36	2.16 ± 0.14
NH_4NO_3	2,464 ± 15	25.17 ± 1.26	97.89 ± 2.71	2.62 ± 0.08
NH_4Cl	2,435 ± 16	22.96 ± 1.11	106.05 ± 1.50	1.38 ± 0.12
$NaNO_3$	2,589 ± 22	23.84 ± 1.03	108.59 ± 2.08	3.59 ± 0.17
Gelatine + $NaNO_3$	2,697 ± 18	25.98 ± 1.02	103.50 ± 3.57	3.86 ± 0.06

[a]SSF was carried using 9 g sorghum straw under optimized physiological conditions (at 5-day incubation, 1:5 (*w/v*) substrate/moisture ratio, pH 6.0 and 40°C); [b]the control medium used was devoid of any organic or inorganic nitrogen source. The data presented are the mean values of three replicates with the standard deviations.

Tween-40, glycine and glycerol also showed its stimulatory effects on the xylanase synthesis.

Enzymatic properties of crude xylanase

Effect of pH on the activity and stability

Xylanase activity was studied in the wide range of pH 3.0 to 10 using different buffers to find out the best pH for the enzyme. The crude enzyme extract exhibited the maximum xylanase activity at pH 5.0 and retained more than 80% of its activity between pH 3. to 7.0 (Figure 2A). The xylanase activity was found to be affected slightly by variation in the pH outside its optimum pH. However, it decreased drastically at pH 9.0 and 10. Betini et al. [19] have measured xylanase activity at different pH 4.0 to 8.0. They have observed the optimum pH 5.0 to 5.5 for *A. niveus*, pH 5.5 to 6.0 for *A. niger* and pH 5.0 for *Aspergillus ochraceus* xylanase activities from crude enzyme extracts. During pH stability studies, *A. tubingensis* FDHN1 xylanase showed high stability towards a broad range of pH 3.0 to 9.0 (Figure 2B). The enzyme retained its 50% activity (t_{50}) after 9, 15 and 15 h of incubation in the acidic range pH 3.0, 4.0 and 5.0, respectively. At pH 6.0 and 7.0, xylanase showed good stability even after 24 h. However, at pH 7.0, 8.0 and 9.0, the half xylanase activities were noticed after 15, 9 and 6 h, respectively. These results indicate that xylanase from *A. tubingensis* FDHN1 is highly active and stable within a broad pH range as compared to the reported xylanases from several other *Aspergillus* sp. [19,32].

Effect of temperature on the activity and stability

The effect of different temperatures on xylanase activity is shown in Figure 3A. The highest xylanase activity was observed at 50°C. In the present study, approximately 30 and 70% reduction in the xylanase activity were observed at 55 and 60°C, respectively. Enzyme activity reduced

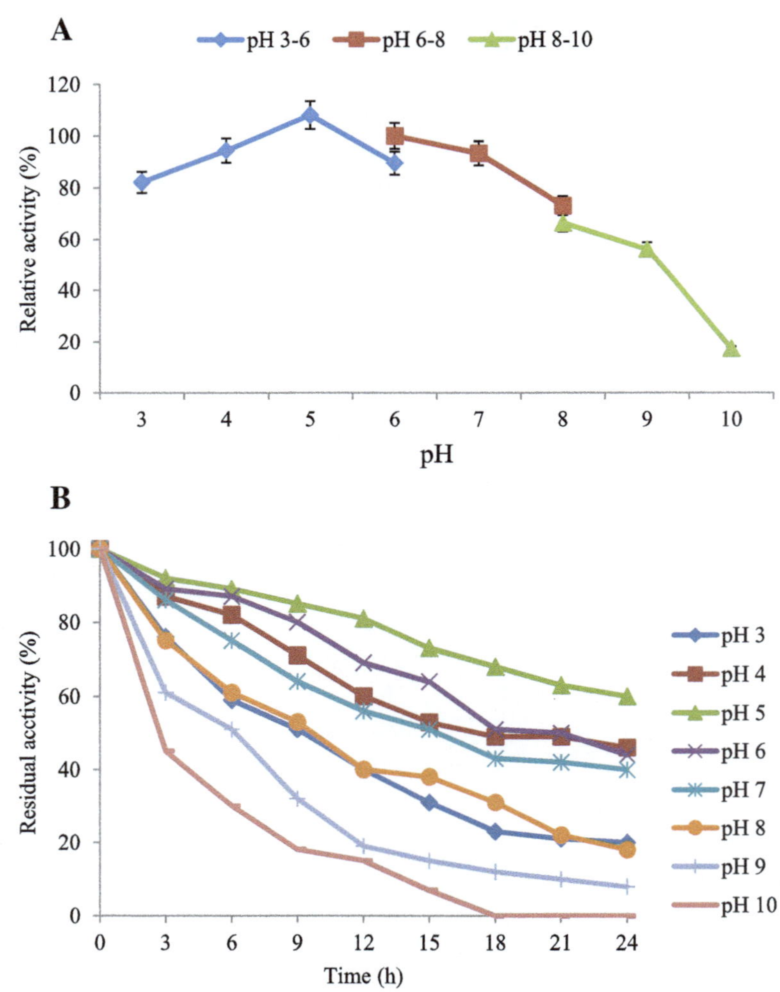

Figure 2 Optimal pH (A) and pH stability (B) of the xylanase from *A. tubingensis* FDHN1. The influence of pH on xylanase activity was determined using 0.05 M sodium citrate (pH 3.0 to 6.0), sodium phosphate (pH 6.0 to 8.0) and glycine-NaOH (pH 8.0 to 10) buffers. The pH stability was studied at a regular interval of 3 up to 24 h.

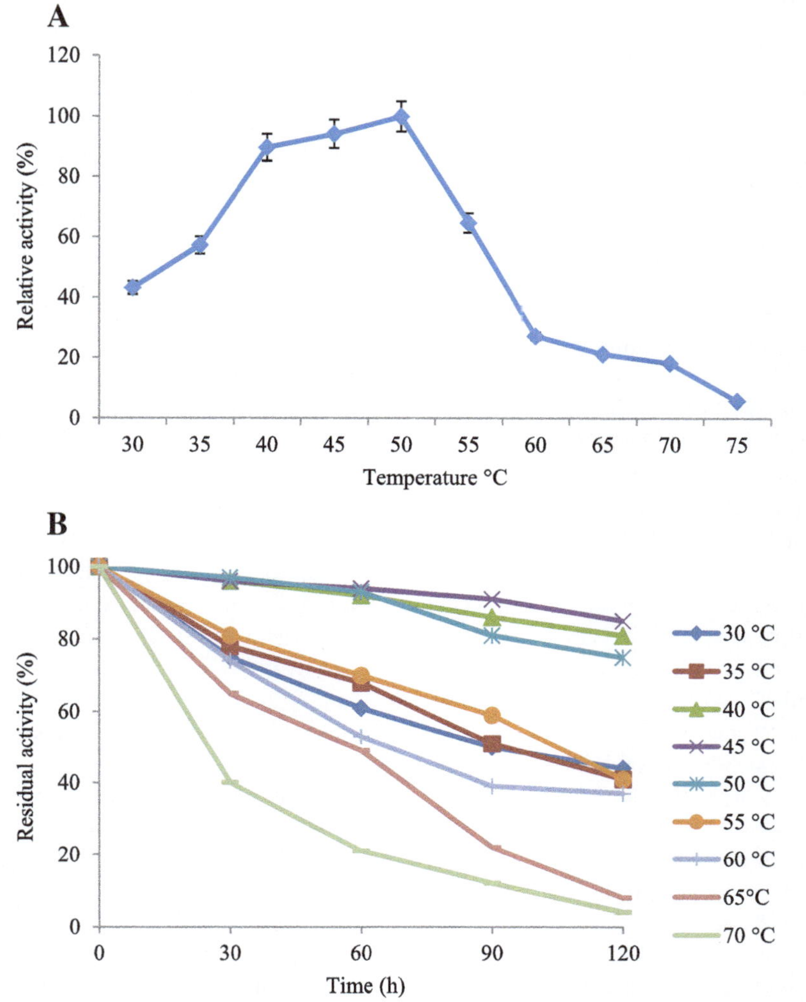

Figure 3 Optimal temperature (A) and thermal stability (B) of the xylanase from *A. tubingensis* FDHN1. Xylanase activity was measured at different temperatures to determine its optimal range. For the determination of thermal stability, the residual activity of the xylanase was measured after 30, 60, 90 and 120 min pre-incubation at different temperatures.

drastically with further increase in the temperature. The present xylanase exhibited more than 70% thermal stability at temperature range of 35 to 55°C after 60 min (Figure 3B), and it showed maximum stability at 45°C by retaining 85% residual activity after 120 min. The xylanase t_{50} was observed after 60 and 90 min at 30 and 35°C, respectively. At 40, 45 and 50°C, t_{50} was noticed after 195, 180 and 155 min, respectively. The xylanase retained 51, 53 and 50% activity even at temperature 55°C (90 min), 60°C (60 min) and 65°C (50 min), respectively. These results show that the activity and stability of xylanase at various temperatures were high as compared to the existing *Aspergillus* sp. xylanases [19,32,33].

Effect of organic solvents on the activity

Published studies of the mechanism of adaptation of enzymes for functioning into the organic solvent are relatively few. In the present study, effects of various organic solvents and its concentration on the xylanase activity were examined (Figure S3 in Additional file 1). The xylanase retained more than 85% of its initial activity with all the solvents at 10% concentration except benzene. In the presence of 30% acetone, butanol and propanol, xylanase retained more than 70% of the initial activity after 30 min. This result is interesting with an aim of using the present xylanase in an application requiring the incubation of the enzyme with organic solvents. The mechanism of organic solvent tolerance has been investigated by Ogino et al. [34] in a *Pseudomonas aeruginosa* PST-01 protease by site-directed and random mutagenesis. They have reported that the disulfide bonds and amino acid residues located on the surface of the molecule play an important role in the organic solvent stability of the enzymes. Gupta et al. [35] have proposed that the presence of hydrophobic clusters on the protein surface and disulfide bonds was responsible for the solvent-stable nature of enzymes. Previously, Ines et al. [36] have studied the effect of various solvents up to 40% concentration on the activity of xylanase produced by *Talaromyces thermophilus* and observed maximum activity in the presence of 5% ethanol. However, reduced xylanase activity has been observed in the presence of methanol when its concentration was increased from 20 to 40%.

Native PAGE and xylan zymography

The crude enzyme from *A. tubingensis* FDHN1 was analysed by native PAGE and zymography. Figure 4 shows the presence of three clear zones upon xylan zymography, suggesting the presence of at least three xylanases. The approximate molecular weight of three xylanases, namely, Xyl-D1, Xyl-D2 and Xyl-D3 was found to be approximately 94, 30 and 18.60 kDa, respectively. This study for the first time highlights multiplicity of the xylanase from *A. tubingensis*. It is believed that the protein modification (e.g. post-translational cleavage) such as glycosylation, proteolysis or both could be responsible for the genesis of multiple enzymes [37]. Apart from post-translational modifications, multiplicity might also be due to the differential mRNA processing, post-secretional modification and the presence of different alleles of the same gene [3]. In 2010, two thermostable xylanases with a molecular weight of 25.2 and 30 kDa were identified by Sharma et al. from *Malbranchea flava* which were active under the alkaline conditions [38].

Thus, cellulase-poor xylanase from *A. tubingensis* FDHN1 seems a novel enzyme, being active and stable at broad pH and temperature ranges and in the presence of organic solvents. The multiplicity is also an interesting feature of this xylanase. Due to such desirable properties of xylanase, the present study was elaborated to achieve its better recovery from fermented sorghum straw by optimizing downstream process parameters.

Evaluation of downstream process parameters for maximum xylanase recovery

Selection of an appropriate extraction solvent

Recovery of the enzyme from solid support is another important aspect of SSF. An ideal solvent extracts the enzyme selectively and completely in minimum contact time. During initial solvent screening experiments, 0.05 M sodium citrate buffer (pH 6.5) was found suitable for efficient xylanase recovery (3,573 ± 22 U/g) with cellulase activity of 4.02 ± 0.08 U/g from the fermented sorghum straw. Extractions with all other solvents gave lower enzyme recoveries (Table 5). Interestingly, in this study, the xylanase was highly active and stable within the acidic range of pH and its optimum recovery was also found in the presence of an acidic extractant. This could be due to the fact that the use of suitable buffer solutions as a solvent helps in maintaining the pH-dependent enzyme stability. Earlier, distilled water and sodium citrate buffer 50 mM, pH 6.5 were used by Pal and Khanum [11] and Sonia et al. [18] as a solvent for the effective recovery of xylanase from cultivated biomass, respectively. Fernendez-Lahore et al. [39] have pointed that the hydrophobic or hydrophilic nature of fungal mycelia, ionic bonds, hydrogen bonds and van der Waal forces determines the efficiency of extractant for enzyme recovery.

Effect of extractant volume and extraction time

In this study, volumes of extractant and extraction time were optimized to improve the xylanase recovery. An extractant/solid ratio was kept in the range of 4:1 to 16:1 (*v*/*w*) while extraction time was varied from 15 to

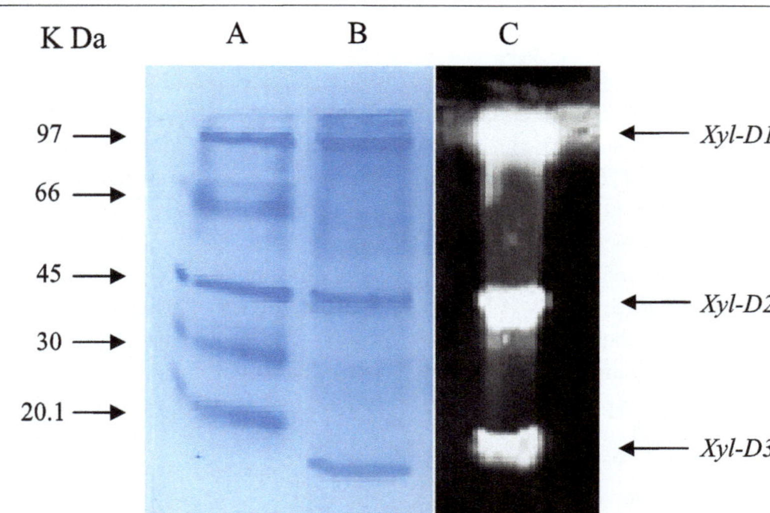

Figure 4 Native PAGE and zymogram analysis of crude enzyme extract from *A. tubingensis* FDHN1. (A) Molecular weight marker, **(B)** crude enzyme extract, and **(C)** xylan zymography.

Table 5 Effect of various extraction solvents on xylanase recovery from the fermented sorghum straw

Name of extractant[a]	Xylanase activity (U/g)	Protein (mg/g)	Specific activity (U/mg)	Cellulase activity (U/g)
Citrate buffer (0.05 M, pH 5.0)	2,591 ± 50	24.86 ± 0.99	104.22 ± 3.55	3.26 ± 0.11
Na-citrate buffer (0.05 M, pH 6.5)	3,573 ± 22	27.96 ± 1.08	127.79 ± 2.13	4.02 ± 0.08
Double-distilled water (pH 7.0)	2,675 ± 26	26.88 ± 1.19	99.51 ± 4.67	3.83 ± 0.05
Phosphate buffer (0.05 M, pH 7.5)	3,256 ± 23	25.79 ± 0.79	126.25 ± 4.06	3.54 ± 0.14
Tris buffer (0.05 M, pH 8.0)	2,859 ± 13	23.96 ± 0.96	119.32 ± 3.24	4.87 ± 0.17
Tween-80 (0.2% *v/v*)	3,166 ± 10	25.47 ± 0.67	124.30 ± 3.00	3.09 ± 0.06
NaCl (1.0% *w/v*)	2,416 ± 18	26.88 ± 1.07	91.93 ± 2.67	4.00 ± 0.13

[a]SSF was carried using 9 g sorghum straw under optimized physiological (at 5-day incubation, 1:5 *w/v*) substrate/moisture ratio, pH 6.0 and 40°C) and nutritional conditions (xylose 0.3%, gelatine and $NaNO_3$ both at 0.05 g N equivalent in 50 mL medium). For xylanase recovery, the fermented sorghum straw was mixed with the solvents separately at an extractant/solid ratio of 10:1 (*v/w*) and agitated at 120 rpm for 60 min at 37°C. The data presented are the mean values of three replicates with the standard deviations.

120 min. During the study, the best xylanase extraction (3,769 ± 22 U/g) was obtained at an extractant/solid ratio of 12:1 (*v*/*w*) and the extraction was further improved (3,951 ± 17 U/g) for an extraction time of 90 min. Such increase in the xylanase extraction could be due to the maximum solubilization of the enzyme from the fermented sorghum straw (Figure 5A,B). At 12:1 (*v*/*w*) extractant/solid ratio, the cellulase activity was 4.89 ± 0.14 U/g which was reached to 5.17 ± 0.11 U/g at 90 min extraction. The improved xylanase recovery under the above optimized conditions is extremely important because concentrated crude extract ease downstream processes mainly the purification step, reducing the time and cost of enzyme recovery. The use of small volumes of solvent for xylanase recovery greatly reduces the energy requirements, equipment size and pollution, but the volumes that are too low leads to unsatisfactory recoveries, since a significant fraction of the xylanase remains in the fermented solid [40]. Previous studies have shown that 10 to 11 mL g^{-1} [11] and 17 mL g^{-1} [41] solvent/substrate ratio effectively recovers maximum xylanase that was produced by *Aspergillus niger* DFR-5 and *Trichoderma longibrachiatum*, respectively.

Effect of agitation speed and extraction temperature

The xylanase extraction was also studied at different agitation speed and temperature values. During the study, 150 rpm was found suitable for the xylanase recovery

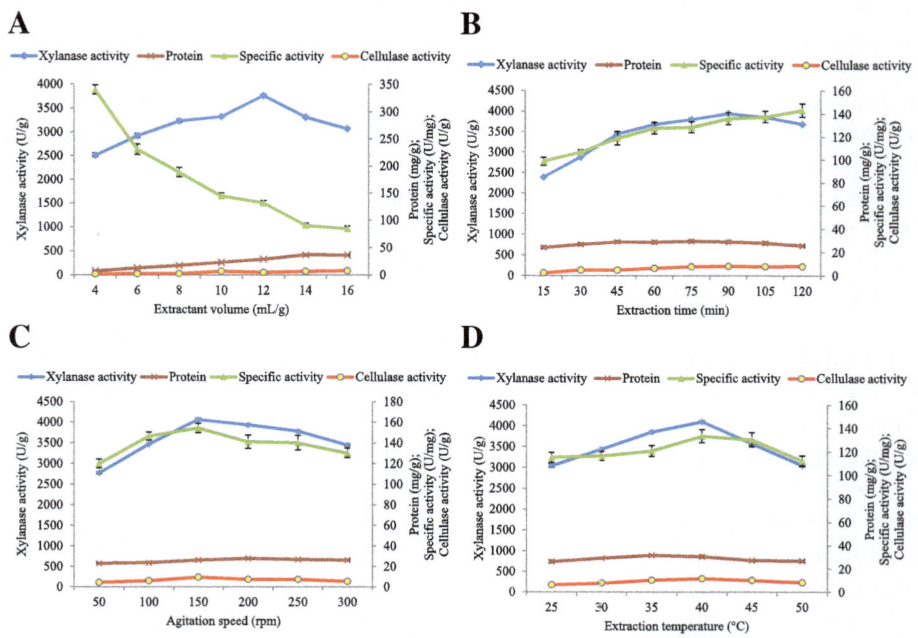

Figure 5 Optimization of xylanase recovery parameters using 0.05 M sodium citrate buffer pH 6.5. (A) Xylanase recovery was performed using different volumes of extractant at 120 rpm for 60 min at 37°C, **(B)** xylanase recovery was performed at different time intervals using 12:1 (*v/w*) extractant/solid ratio at 120 rpm and 37°C, **(C)** xylanase recovery was performed at different agitation rates using 12:1 (*v/w*) extractant/solid ratio for 90 min at 37°C, and **(D)** xylanase recovery was performed at different temperatures using 12:1 (*v/w*) extractant/solid ratio at 150 rpm for 90 min.

Table 6 Comparison of *A. tubingensis* FDHN1 with different *Aspergillus* sp. for xylanase yield and productivity

Microorganism	Substrate	Xylanase activity (U/g)	Productivity (U/g/day)	Reference
Aspergillus tubingensis FDHN1[a]	Sorghum straw	4,105	821.00	This study
Aspergillus tubingensis JP-1	Wheat straw	6,887	859.90	[20]
Aspergillus niger DFR-5	Wheat bran	4,465	744.16	[11]
Aspergillus fischeri Fxn 1	Wheat bran	1,024	341.33	[37]
Aspergillus fumigates	Wheat bran	1,055	211.00	[38]
Aspergillus fumigates	Sugarcane bagasse/wheat bran (1:1)	821.0	164.20	[38]
Aspergillus fumigates	Soybean bran	558.3	111.66	[38]
Aspergillus niger	Wheat bran	1,285	257.00	[38]
Aspergillus niger	Soybean bran	484.2	96.84	[38]
Aspergillus niger	Sugarcane bagasse/wheat bran (1:1)	1,050	210.00	[38]

[a]Highest xylanase yield, 4,105 U/g, was obtained when SSF was carried out using 9 g sorghum straw under optimized physiological (at 5-day incubation, 1:5 (*w/v*) substrate/moisture ratio, pH 6.0 and 40°C) and nutritional conditions (xylose 0.3%, gelatine and $NaNO_3$ both at 0.05 g N equivalent in 50 mL medium) along with optimized recovery conditions (at an extractant/solid ratio of 12:1 (*v/w*), 150 rpm, 90 min and 40°C).

(4,067 ± 27 U/g) which was further improved at 40°C (4,105 ± 22 U/g). The maximum cellulase activity under above optimized conditions was 5.36 ± 0.17 U/g (Figure 5C,D). In the present study, xylanase was most stable at 45°C and its optimum recovery was observed at 40°C suggesting that the enzyme is optimally active at moderately high temperatures. Ghildyal et al. [40] have reported that the prolonged higher rate of agitation causes a loss of enzyme activity. The maximum xylanase extraction has been observed at the highest rate of agitation (200 rpm) with an extraction time of 60 min at 37°C by Pal and Khanum [11]. However, 25°C and 200 rpm agitation speed for xylanase recovery has also been reported by Azin et al. [41].

Effect of un-optimized and optimized recovery conditions on xylanase recovery efficiency

After studying the detailed downstream process parameters, the efficiency of un-optimized and optimized recovery conditions for xylanase recovery were studied at different time intervals. The results indicated that xylanase recovery efficiency was 89.4% under the optimized conditions after 90 min, whereas it was only 53.3% under un-optimized conditions after 60 min (Figure S4 in Additional file 1). After 90 min, the recovery efficiency remained almost constant indicating that the maximum xylanase from the fermented sorghum straw was leached into the extraction solvent during this period. In the present study, the maximum xylanase recovery within a short time indicated the importance of detailed downstream process optimization.

The existing novel strain of *A. tubingensis* FDHN1 has been found to show ability to produce higher cellulase-poor xylanase yield under controlled optimized conditions. The overall productivity of xylanase by *A. tubingensis* FDHN1 has been compared with the existing xylanases produced by different *Aspergillus* sp. under SSF (Table 6).

Conclusions

The present study has revealed the potential of a newly isolated *A. tubingensis* FDHN1 to produce xylanase under SSF. The prominent induction of xylanase has been observed in the presence of the least highlighted agricultural residue sorghum straw. The optimization of upstream and downstream process parameters enhanced the xylanase yield from 615.5 ± 19.21 to 4,105 ± 22 U/g, suggesting the importance of the detailed bioprocess profiling. Apart from that cellulase-poor nature, high temperature, pH and solvent stability are suitable features for its application in paper, juice, feed improvement and in bioconversion of lignocelluloses to the solvents. The multiplicity in xylanase from *A. tubingensis* FDHN1 has been shown for the first time in the present study. Further scaling-up and purification of the xylanase still deserve more attention to reach the commercial feasibility and to find the novel properties of xylanase from this novel strain of *A. tubingensis*.

Additional file

Additional file 1: Table S1 and Figures S1 to S4. Table S1. Effect of metals and modulators on xylanase production from *A. tubingensis* FDHN1 under SSF. **Figure S1.** Effect of sorghum straw concentration (3 to 15 g) on xylanase production. SSF was carried out by moistening sorghum straw with mineral salt medium pH 5.0 at 35°C for 6 days. **Figure S2.** Effect of xylose concentration (0.1 to 1.1%) on xylanase production. SSF was carried using 9 g sorghum straw under optimized physiological conditions (at 5 days incubation, 1:5 (w:v) substrate: moisture ratio, pH 6.0 and 40°C). **Figure S3.** Effect of organic solvents (10 to 30%) on the xylanase activity. The crude xylanase was incubated in the presence of various organic solvents concentration for 30 min and at the end of reaction residual activities were measured. **Figure S4.** Comparison of the un-optimized and optimized extraction conditions on the xylanase recovery efficiency (%). Optimized extraction conditions were extractant/ solid ratio 12: 1 (v:w), 150 rpm agitation speed and 40°C. The data presented are the mean values of three replicates with the standard deviations.

Abbreviations

SSF: solid-state fermentation; PDA: potato dextrose agar; *w/v*: weight by volume; M: molarity; U/g: xylanase/cellulase activity as unit per gram dry substrate; U/mg: specific xylanase activity per milligram of protein; Native PAGE: native polyacrylamide gel electrophoresis; *v/w*: volume by weight; kDa: kilodalton.

Competing interests

The authors declare that they have no competing interests.

Authors' contributions

DNA designed experiments, carried out the laboratory work and wrote the manuscript. NSB provided supervision and research direction of the experimental work, as well as editing the manuscript. HAM critically observed the results and finalized the manuscript draft. All authors contributed intellectually via scientific discussions during the study and read and approved the final manuscript.

Authors' information

DNA is registered as a Ph.D. Fellow and is also DST-INSPIRE-JRF at P. G. Department of Microbiology, Gujarat Vidyapeeth, Sadra- 382 320, Gujarat, India. NSB is a Professor at P. G. Department of Microbiology, Gujarat Vidyapeeth, Sadra- 382 320, Gujarat, India. HAM is a Reader at the Department of Life Sciences, School of Sciences, Gujarat University, Ahmadabad, Gujarat, India.

Acknowledgements

One of the authors (DNA) gratefully acknowledges the financial assistance from the Department of Science and Technology (DST), Ministry of Science and Technology, Govt. of India, for the award of Junior Research Fellowship under the INSPIRE Program during the course of the investigation. The authors would like to thank Mr. Mahesh Bhatt, Gujarat Arts & Science College, Ahmedabad, for the language editing.

Author details

[1]P. G. Department of Microbiology, Gujarat Vidyapeeth, M. D. Gramseva Mahavidhyalaya, Talim Kendra Campus, Sadra 382 320, Gujarat, India. [2]Department of Life Sciences, School of Sciences, Gujarat University, Navrangpura, Ahmedabad, Ahmedabad 380009, Gujarat, India.

References

1. Tseng MJ, Yap MN, Ratanakhanokchai K, Kyu KL, Chen ST (2002) Purification and characterization of two cellulase-free xylanases from an alkaliphilic *Bacillus firmus*. Enzyme Microb Technol 30:590–595
2. Collins T, Gerday C, Fellre G (2005) Xylanases, xylanase families and extremophilic xylanases. FEMS Microbiol Rev 29:3–23
3. Polizeli MLTM, Rizzatti ACS, Monti R, Terenzi HF, Jorge JA, Amori DS (2005) Xylanase from fungi: properties and industrial applications. Appl Microbiol Biotechnol 67:577–591
4. Butt MS, Tahir-Nadeem M, Ahmad Z, Sultan MT (2008) Xylanases and their applications in baking industry. Food Technol Biotechnol 46:22–31
5. Nagar S, Mittal A, Kumar D, Kumar L, Gupta VK (2012) Immobilization of xylanase on glutaraldehyde activated aluminium oxide pellets for increasing digestibility of poultry feed. Process Biochem 47:1402–1410
6. Coughlan MP, Hazlewood GP (1993) β-1, 4-D xylan-degrading enzyme systems: biochemistry, molecular biology and applications. Appl Biochem 17:259–289
7. Rao CS, Sathish T, Laxmi MM, Laxmi GS, Rao RS, Prakasham RS (2008) Modeling and optimization of fermentation factors for enhancement of alkaline protease production by isolated *Bacillus circulans* using feed-forward neural network and genetic algorithm. J Appl Microbiol 104:889–898
8. Gawande PV, Kamat MY (1999) Production of *Aspergillus* xylanase by lignocellulosic waste fermentation and its application. J Appl Microbiol 87:511–519
9. Park Y, Kang S, Lee J, Hong S, Kim S (2002) Xylanase production in solid state fermentation by *Aspergillus niger* mutant using statistical experimental designs. Appl Microbiol Biotechnol 58:761–766
10. Maciel GM, Vandenberghe LPS, Haminiuk CWI, Fendrich RC, Bianca BED, Brandalize TQS, Pandey A, Soccol CR (2008) Xylanase production by *Aspergillus niger* LPB 326 in solid state fermentation using experimental designs. Food Technol Biotechnol 46(2):183–189
11. Pal A, Khanum F (2010) Production and extraction optimization of xylanase from *Aspergillus niger* DFR5 through solid state fermentation. Bioresour Technol 101:7563–7569
12. Bailey MJ, Biely P, Poutanen K (1992) Interlaboratory testing of methods for assay of xylanase activity. J Biotechnol 23:257–270
13. Ghose TK (1987) Measurement of cellulase activities. Pure Appl Chem 59:257–268
14. Miller LG (1959) Use of dinitrosalicylic acid reagent for determination of reducing sugars. Anal Chem 31:426–428
15. Lowry OH, Rosebrough NJ, Farr AL, Randall RJ (1951) Protein measurement with Folin phenol reagent. J Biol Chem 31:426–428
16. Laemmli UK (1970) Cleavage of structural proteins during the assembly of the head of bacteriophage T4. Nature 227(5259):680–685
17. Puls J, Suhuseil J (1993) Chemistry of hemicelluloses: relationship between hemicellulose structure and enzyme required for hydrolysis, in hemicellulose and hemicellulases. Portland Press, London, pp 103–127
18. Sonia KG, Chadha BS, Saini HS (2005) Sorghum straw for xylanase hyper production by *Thermomyces lanuginosus* (D2W3) under solid state fermentation. Bioresour Technol 96:1561–1569
19. Betini JHA, Michelin M, Peixoto-Nogueira SC, Jorge JA, Terenzi HF, Polizeli MLTM (2009) Xylanase from *Aspergillus niger*, *Aspergillus niveus* and *Aspergillus ochraceus* produced under solid-state fermentation and their application in cellulose pulp bleaching. Biprocess Biosyst Eng 32:819–824
20. Pandya JJ, Gupte A (2012) Production of xylanase under solid state fermentation by *Aspergillus tubingensis* JP-1 and its application. Bioprocess Biosyst Eng 35:769–779
21. Goyal M, Kalra KL, Sareen VK, Soni G (2008) Xylanase production with xylan rich lignocellulosic wastes by a local soil isolate of *Trichoderma viride*. Braz J Microbiol 39:535–541
22. Biswas R, Sahai V, Mishra S, Bisaria VS (2010) Bioprocess strategies for enhanced production of xylanase by *Melanocarpus albomyces* IITD3A on agro-residual extract. J Biosci Bioeng 110(6):702–708
23. Kang SW, Park YS, Lee JS, Hong SI, Kim SW (2004) Production of cellulases and hemicellulases by *Aspergillus niger* KK2 from lignocellulosic biomass. Bioresour Technol 91:153–156
24. Hasseltine CW (1972) Solid state fermentations. Biotechnol Bioeng 14:517–532
25. Prakasham RS, SubbaRao C, Rao RS, Rajesham S, Sarma PN (2005) Optimization of alkaline protease production by *Bacillus* sp. using Taguchi methodology. Appl Biochem Biotechnol 120:133–144
26. Liao H, Xu C, Tan S, Wei Z, Ling N, Yu G, Raza W, Zhang R, Xu QSY (2012) Production and characterization of acidophilic xylanolytic enzymes from *Penicillium oxalicum* GZ-2. Bioresour Technol 123:117–124
27. Sudan R, Bajaj BK (2007) Production and biochemical characterization of xylanase from an alkalitolerant novel species *Aspergillus niveus* RS2. World J Microbiol Biotechnol 23:491–500
28. Priem B, Dobberstein J, Emies CC (1991) Production of β-1,4- xylanase in continuous culture by *Aureobasidium pullulans* CBS 58475. Biotechnol Lett 13:149–154
29. Lakshmi GS, Rao CS, Rao RS, Hobbs PJ, Prakasham RS (2009) Enhanced production of xylanase by newly isolated *Aspergillus terreus* under solid state fermentation using palm industrial waste: a statistical optimization. Biochem Eng J 48:51–57
30. Oliveira LA, Porto ALF, Tambourgi EB (2006) Production of xylanase and protease by *Penicillium janthinellum* CRC 87-M- 115 from different agricultural wastes. Bioresour Technol 97:862–867
31. Bakri Y, Mohammed J, Mohammed IEA (2008) Improvement of xylanase production by *Cochliobolus sativus* in submerged culture. Food Tehnol Biotechnol 46(1):116–118
32. Delabona PS, Pirota RDPB, Codima CA, Tremacoldi CR, Rodrigues A, Farinas CS (2013) Effect of initial moisture content on two rainforest *Aspergillus* strains cultivated on agro-industrial residues: biomass-degrading enzymes production and characterization. Ind Crop Prod 42:236–242
33. Lu FX, Lu M, Lu ZX, Bie XM, Zhao HZ, Wang Y (2008) Purification and characterization of xylanase from *Aspergillus ficcum* AF-98. Biresour Technol 99:5938–5941
34. Ogino H, Uchiho T, Yokoo J, Kobayashi R, Ichise R, Ishikawa H (2001) Role of intermolecular disulfide bonds of the organic solvent-stable PST-01 protease in its organic solvent stability. Appl Environ Microbiol 67:942–947

35. Gupta A, Ray S, Kapoor S, Khare SK (2008) Solvent-stable *Pseudomonas aeruginosa* PseA protease gene: identification, molecular characterization, phylogenetic and bioinformatic analysis to study reasons for solvent stability. J Mol Microbiol Biotechnol 15:234–243
36. Ines MA, Mohamed G, Ines BBR, Ali G, Hafedh B (2012) The effect of *Talaromyces thermophilus* cellulase-free xylanase and commercial laccase on lignocellulosic components during the bleaching of kraft pulp. Int Biodet Biodeg 75:43–48
37. Li XL, Ljungdahl LG (1994) Cloning, sequencing and regulation of a xylanase gene from the fungus *Aureobasidium pullulans* Y-2311-1. Appl Environ Microbiol 60:3161
38. Sharma M, Chadha BS, Saini HS (2010) Purification and characterization of two thermostable xylanases from *Malbranchea flava* under alkaline conditions. Bioresour Technol 101:8834–8842
39. Fernendez-Lahore HM, Fraile ER, Cascone O (1998) Acid protease recovery from a solid state fermentation system. J Biotechnol 62:83–93
40. Ghildyal NP, Ramakrishna M, Lonsane BK, Karanth NG (1991) Efficient and simple extraction of mouldy bran in a pulsed column extractor for recovery of amyloglucosidase in concentrated form. Process Biochem 26:235–241
41. Azin M, Moravej R, Zareh D (2007) Production of xylanase by *Trichoderma longibrachiatum* on a mixture of wheat bran and wheat straw: optimization of culture conditions by Taguchi method. Enzyme Microb Technol 40:801–805

Biosurfactant, polythene, plastic, and diesel biodegradation activity of endophytic *Nocardiopsis* sp. mrinalini9 isolated from *Hibiscus rosasinensis* leaves

Mrinalini J Singh* and Padmavathy Sedhuraman

Abstract

Background: Endophytic non-streptomycetes group of actinomycetes from *Hibiscus rosasinensis* leaves were screened for biosurfactant, polythene, plastic, and diesel biodegradation activities. Biosurfactant activity was evaluated by hemolysis, drop-collapsing test, lipase production, oil-spreading technique, bacterial adhesion to hydrocarbon (BATH) assay, penetration assay, and emulsification assay. Similarly, polythene, plastic, and diesel biodegradation activity were also carried out.

Findings: Among the five non-streptomycetes isolates, only one newly found actinomycete isolate named as A9 exhibited significant biosurfactant and biodegradation activity. Based on Bergey's manual of systematic bacteriology, the strain A9 was similar to *Nocardiopsis* sp. Molecular characterization and phylogenetic analysis support the classification of the isolate as a new strain which was named as *Nocardiopsis* sp.mrinalini9. The isolate was closely related to the type strain of *Nocardiopsis synnemataformans* sharing a 16S rRNA gene sequence similarity of 99%. The partial 16S rRNA sequence of the isolate (1061 bp) was deposited in Genebank under the accession number KF909126. Still, DNA-DNA hybridizations, phenotypic comparisons, and chemotaxonomic analysis need to be performed to confirm its novelty.

Conclusion: It is evident from the study that rare endphytes associated with medicinal plants has got excellent hydrocarbon biodegradation capacity.

Keywords: Biosurfactant; Biodegradation; Bergey's manual of systematic bacteriology; *Nocardiopsis* sp.mrinalini9

Findings

Introduction

Surfactants like wetting agents or petroleum derivatives lower the surface tension between two liquids or between a liquid and a solid. But, they are non-degradable leading to toxic effects. To solve this difficulty, eco-friendly microbes were explored for biosurfactant property with high solubility, high biodegradability, and low toxicity with diverse applications.

Though plastic and polythene are durable and strong with wide applications, they seem to be non-degradable leading to severe pollution issues by disturbing the whole ecosystem. Even their recycling are not followed widely due to economic viability.

Similar to plastic issues, even the hydrocarbons like petrol and diesel spills seem to be the global environmental pollutant causing deleterious pollution effects in spite of being the principle source of energy. Successful application of bioremediation technology solely depends on in situ application of a microbial consortium with high biodegradation capacity.

Yousaf et al. [1] isolated hydrocarbon degraders *Pseudomonas*, *Arthrobacter*, *Enterobacter*, and *Pantoea* species from the root and stem tissues of Italian ryegrass and birds foot trefoil vegetated in hydrocarbon-contaminated soil. Similarly, Siciliano et al. [2] found that endophytic hydrocarbon degraders in tall fescue (*Festuca arundinacea*) and rose clover (*Trifolium fragiferum*) at an aged-hydrocarbon-contaminated site with up to 4% of culturable endophytes possessing genes involved in hydrocarbon degradation.

* Correspondence: mrinalini.singh50@gmail.com
Department of Botany, Nirmala college for Women, Red fields, Coimbatore 641018, Tamilnadu, India

There is increasing interest in the role of endophytes in phytoremediation application as they can serve as vector delivering biodegradative capacities inside the plant [3]. However, it is not fully understood, whether potentially degrading endophytes are active once inside the plant and thereby contribute to the detoxification of the pollutant and better survival of the plant under toxic conditions.

Hence, the screening of effective microbes among the diverse marine and terrestrial microbial communities has been carried out worldwide for the last few decades. Our search especially focused on least-explored endophytic non-streptomycetes actinobacteria isolated from less-explored medicinal plants since the search for rare isolates proved to be the treasure house of bioactivity with varied applications. Endophytic actinomycetes were usually explored for antimicrobial activities and plant growth promotion studies. This might be the first report which led to the discovery of potent endophytic *Nocardiopsis* sp. mrinalini9, a new endophytic strain with good biosurfactant and biodegradation property which could address the serious environmental issues.

Materials and methods

Sample collection

Healthy leaf samples of common medicinal plant *Hibiscus rosasinensis* were gathered from the garden of Nirmala College for Women, Coimbatore (11.0183° N, 76.9725° E) surrounded by Western Ghats of Southern India during the period of December 2011 to January 2012. The collected plant materials were taken to the laboratory, preserved at 4°C in sealed plastic bags, and subjected to isolation work within 96 h.

Isolation of endophytic actinobacteria

Healthy leaf samples were cut into small pieces (2 × 2 cm) and washed by running tap water for 1–2 min to remove the soil particles completely. The resultant were subjected to a five-step surface sterilization procedure as per the method of Sheng *et al.* [4]: a 4- to 10-min wash in 5% sodium hypochlorite, followed by a 10-min wash in 2.5% sodium sulphite, a 5-min wash in 75% ethanol, a wash in sterile water, and a final rinse in 10% sodium bicarbonate for 10 min to disrupt the plant tissues and inhibit the fungal growth. At this point, the final washed solution was spread onto ISP 2 (yeast extract malt extract agar) and observed for microbial growth to validate the surface-sterilized protocol and the result was negative. After the sterility check, the surface-sterilized tissues were subjected to continuous drying at 100°C for 15 min. The surface-treated samples were aseptically transferred on humic acid vitamin agar and incubated at 28°C for 2 to 8 weeks [4]. Then, they were purified on yeast extract malt extract agar and incubated at room temperature (28 ± 2°C) for 30 days.

Screening methods for biosurfactant assay

Biosurfactant assay was determined for the pure culture of actinobacteria by seven different methods, namely, (i) hemolysis, (ii) drop-collapsing test, (iii) lipase test, (iv) oil-spreading technique, (v) bacterial adhesion to hydrocarbon (BATH) assay, (vi) penetration assay, and (vii) emulsification assay.

Hemolysis activity Biosurfactant assays of the isolates were evaluated by the hemolysis test on blood agar plate. The isolates were streaked on the blood agar and the plates were incubated at 28°C for 7 days. The plates were then observed for zone of clearance around the colonies as discussed elsewhere [5,6].

Drop-collapsing test Two microliters of mineral oil was added to a 96-well microtitre plate and equilibrated for 1 h at 37°C. Five microliters of the culture supernatant was added to the surface of the oil in well. The shape of drop on the oil surface was noted after 1 min. The culture supernatant that collapsed the oil drop was indicated as positive, and the culture supernatant which failed to collapse the oil drop was indicated as negative. Distilled water was used as negative control. The procedure is discussed elsewhere with slight modifications [6,7].

Oil-spreading technique In oil-spreading assay, as discussed by Morikawa *et al.* [8], 10 μL of crude oil was added to the surface of 40 mL of distilled water in a Petri dish to form a thin oil layer. Then, 10 μL of culture or culture supernatant were gently placed on the center of the oil layer. The presence of biosurfactant would displace the oil and a clear zone would form. The diameter of the clearing zone on the oil surface would be visualized under visible light and measured after 30 s, which correlates to the surfactant activity, also known as oil displacement activity.

Bacterial adhesion to hydrocarbons The hydrophobicity of the cells can be measured by BATH assay as discussed by Rosenberg *et al.* [9]. Actinobacterial cells were washed twice with phosphate buffer salt solution and were suspended in it to give an optical density of ~0.5 at 600 nm. One hundred microliters of crude oil (diesel) was added to 2 mL of cell suspension and was vortex-shaken for 3 min in test tubes. After shaking, crude oil and aqueous phase were allowed to separate for 1 h. Optical density of the aqueous phase was then measured at 600 nm in a spectrophotometer. Hydrophobicity is expressed as the percentage of cell adherence to crude oil and was calculated as follows:

100 * (1-optical density of the aqueous phase/ optical density of the initial suspension)

Three independent determinations were made and the mean values were calculated.

Penetration assay This assay relies on the contact of two insoluble phases which leads to a color change as described by Maczek *et al.* [10].

In this assay, the cavities of a 96-well microplate were filled with 150 μL of a hydrophobic paste made up of oil and silica gel. This separated paste was covered with 20 μL of oil. Then, 10 μL of a red-staining solution (safranin) was added to 90 μL of the supernatant. That colored supernatant was then placed on the surface of the oil-covered paste.

Emulsification assay Emulsification assay was carried out using petroleum as described by Cooper *et al.* [11].

Two milliliters of hydrocarbon, i.e., petrol was taken in a test tube to which 1 mL of cell-free supernatant obtained after centrifugation of the culture was added and was vortexed for 2 min to ensure homogenous mixing of both the liquids. The emulsification activity was observed after 24 h and it was calculated by using this formula:

Total height of the emulsion/Height of the aqueous layer * 100

Plastic degradation

Plastic degradation steps were followed as per the standard procedure with slight modifications, as discussed by Kathiresan [12]. The fresh polythene bags and disposable plastic cups were collected and heat-dried at 40°C in hot air oven for 14 days and then cut into small discs of 1-cm diameter. Pre-weighed polythene and plastic discs of 100 mg were aseptically transferred to conical flasks as separate batches containing 50 mL of sterile 0.3% yeast extract broth, inoculated with actinomycetes culture. Control was maintained with polythene and plastic discs in the microbe-free medium. All these flasks were left in a shaker at 37°C for bacterial cultures and 28°C for actinomycetes cultures for 2 months maintaining 180 revolutions/min. After 2 months of incubation, the polythene and plastic discs were washed thoroughly with distilled water and dried over night at 45°C, and then final weight was measured. From the data collected, weight loss of the plastic strip was calculated.

Diesel degradation

Diesel degradation steps were followed as per the standard procedure with slight modifications, as discussed by Smita *et al.* [13]. For confirming the diesel biodegradation ability of the test isolates, Bacto Bushnell Haas broth was used. The pure culture isolates were introduced into 50 mL of Bacto Bushnell Haas broth incorporated with sterile diesel (1% *v*/*v*). Control was maintained with only 1% sterile diesel in the microbe-free medium. All these flasks were left in a shaker at 28°C for actinomycetes cultures for 7 days maintaining 180 revolutions/min. The aliquots in the flask were monitored daily for color change from deep blue to pink and then colorless. On a daily basis, 5 mL of the aliquots were collected from each flask and the absorbance was noted on spectrophotometer at 600-nm wavelength.

Vapor phase transfer diesel utilization test was carried out for the confirmation of actual diesel utilizing microbes with the isolates that showed faster color change in the primary step as per the standard procedure [13,14]. The isolates that showed faster color change were streaked on mineral salt agar media. The inner side of the Petri dish was covered with a sterile filter paper (Whatman no. 1) saturated with filter-sterilized diesel. Through vapor phase transfer, diesel was used as sole carbon and energy source for the growth of microbes. All the plates were kept inverted and incubated at 28°C for 7 to 14 days for the growth of diesel utilizers.

Identification of actinomycetes

The isolates were identified according to morphological criteria, including characteristics of colonies on plate, morphology of substrate and aerial hyphae, morphology of spores, and pigment production as discussed elsewhere [15,16]. These isolates were preserved in 20% (*v*/*v*) glycerol for subsequent investigation.

Molecular identification

DNA from the potent strain was isolated employing Chromous DNA isolation kit (Chromous Biotech, Bangalore, India) according to the manufacturer protocol. The 16S rRNA gene fragment was amplified by using universal primers corresponding to positions 8 to 27 for the forward primer and 1492 to 1510 for the reverse primer mentioned below (Table 1).

Forward primer: 5′-AGAGTTTGATCMTGGCTCAG-3′

Reverse primer: 5′-TACGGYTACCTTGTTACGACTT-3′

Sequencing was performed by using Big Dye terminator cycle sequencing kit (Applied BioSystems, Waltham, MA, USA). Sequencing products were resolved on an Applied Biosystems model 3500XL automated DNA sequencing system (Applied BioSystems, USA). Nucleotide sequence

Table 1 PCR amplification protocol

94°C	94°C	55°C	72°C	72°C
5 min	30 s	30 s	1 min	5 min
Initial denaturation	Denaturation	Annealing	Extension	Final Extension
	35 cycles			

of 16S rRNA gene from actinomycetes strain was determined and compared for similarity level with the reference species present in genomic database bank using NCBI BLAST program multiple sequence alignment, and the phylogenetic tree was constructed using MEGA software as described by Tamura *et al.* [17].

Results and discussion

Biosurfactant assay

Yeast extract malt extract agar plates spread with the last water from the washing of plant samples failed to grow colonies of microorganisms after 2 weeks of incubation. This indicated that the five-step surface sterilization protocol was effective at killing the epiphytic microorganisms. Thus, the subsequent isolates can be considered true endophytes.

Aseptical transfer of surface-treated *H. rosasinensis* leaf samples and subsequent plating on ISP agar resulted in isolation of five non streptomycetes colonies. Only one non-streptomycetes isolate named as A9 showed distinct biosurfactant, diesel, polythene, and plastic degradation.

On blood agar, it showed a clear zone around the colonies causing lysis of blood (Figure 1).

In the drop collapsing test, a drop of surfactant producing culture was flattened on oil surface leading to positive reaction. A9 provided complete spreading due to the reduction in force or interfacial tension between the liquid drop and the hydrophobic surface. The stability of drops is dependent on surfactant concentration and correlates with surface and interfacial tension.

In oil-spreading technique, when the supernatant of the given strain was added to the plates containing oil, it displaced the oil showing a zone of displacement of 4.8 cm.

In BATH assay, after mixing aqueous suspension of washed microbial cells with a distinct volume of a hydrocarbon, petrol for 2 min, the two phases were allowed to separate. Hydrophobic cells were bound to hydrocarbon droplets and had risen with the hydrocarbon. They were removed from the aqueous phase. The turbidity of the aqueous phase was measured. The decrease in the turbidity of the aqueous phase correlates to the hydrophobicity of the cells. The percentage of cells bound to the hydrophobic phase (H) was found to be 54%.

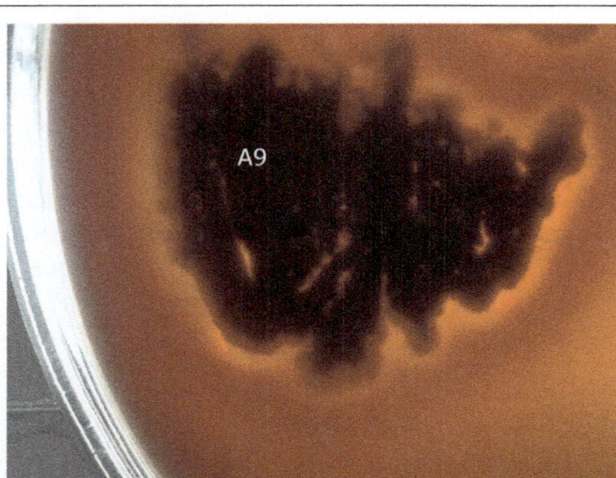

Figure 1 Biosurfactant assay. Hemolysis activity of Isolate A9.

In penetration assay, the presence of biosurfactant was confirmed since the hydrophilic liquid break through the oil film barrier into the paste. The silica is entering the hydrophilic phase and the upper phase changed from clear red to cloudy white within 15 min. The described effect relies on the phenomenon that silica gel is entering the hydrophilic phase from the hydrophobic paste much more quickly if biosurfactants are present. Biosurfactant-free supernatant will turn cloudy but stay red.

In emulsification assay, screening was done by adding 2 mL of diesel fuel to 1 mL of culture supernatant and kept overnight. Around 60% emulsification was observed after 24 h.

Polythene and plastic biodegradation

Polythene degradation is noticed by 22% weight loss whereas plastic degradation is noticed by 10% weight loss in A9 isolate after 2 months compared to negative control (Figure 2).

The weight loss in the polythene and plastic after 2 months implies that the organism uses polythene/plastic for their partial degradation. They colonize to form biofilm due to their hydrophobicity which aids in polymer biodegradation activity as discussed by Usha *et al.* [18]. In the primary degradation, the main chain cleaves leading to the formation of low-molecular weight fragments (oligomers), dimers, or monomers as discussed elsewhere [18,19]. The degradation is due to the extra cellular enzyme secreted by the organism. These low molecular weight compounds are further utilized by the microbes as carbon and energy sources. The resultant-broken down fragments must be completely used by the microorganisms; otherwise, there will be potential disruption in environmental cycle and resultant health consequences as discussed elsewhere [19,20].

Diesel biodegradation

During the primary step of diesel biodegradation, the strain A9 showed the fastest color change in Bacto Bushnell Haas broth medium resulting in decrease of absorbance leading to highest biodegradation (Figure 3). The same isolate was further tested for vapor phase transfer method. The actinomycetes after 6 days showed better growth confirming to be a better hydrocarbon degrader. The color change in Bacto Bushnell

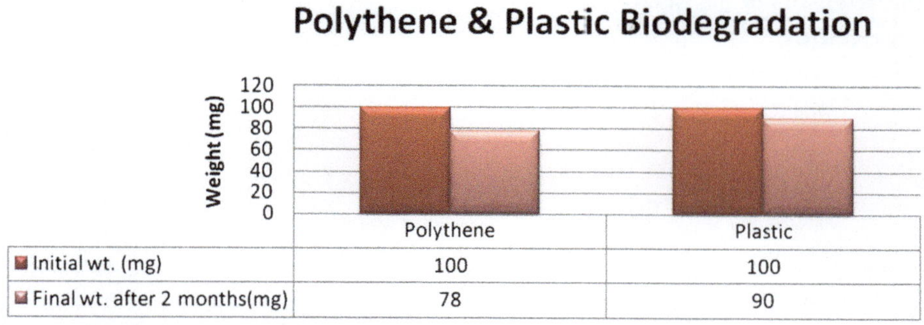

Figure 2 Polythene and plastic biodegradation pattern exhibited by A9 isolate.

Haas broth medium during the primary step of diesel biodegradation is mainly due to the reduction of the indicator by the oxidized products of hydrocarbon degradation as discussed by Smita *et al.* [13]. Hence, the given isolate was considered to be potential hydrocarbon oxidizers.

The strain A9 which showed effective biosurfactant and biodegradation properties was identified both in morphological and molecular level.

The isolate produced irregular, raised, grey white colonies with light yellow substrate mycelium on ISP2 agar surface. Dark brown melanin pigment was produced on ISP7 agar medium and no soluble pigments were produced by the isolate. By viewing under ×1,000 magnification, gram-positive rods with rounded ends, branched hyphae with non-motile spores were found. Therefore, based on actinomycetes identification key as discussed elsewhere [15,16], the isolate was assigned to the genus *Nocardiopsis* (Table 2).

Though it was assigned to the genus *Nocardiopsis*, the strain did not match with any of the given species. Hence, molecular characterization by 16S rRNA ribotyping and phylogeny analysis was carried out in order to identify isolate A9 up to species level.

Phylogenetic analysis and species identification

Using BLAST search in the NCBI data bank, sequences homologous to 16S rRNA gene of the isolate A9 were collected. The DNA sequences were aligned and phylogenetic tree was constructed using MEGA software (bootstrap method) as described by Tamura *et al.* [15] (Figure 4). The evolutionary history was inferred using the neighbor-joining method. The percentage of replicate trees in which the associated taxa clustered together in the bootstrap test (1,000 replicates) was shown next to the branches. Comparison of 16S rRNA nucleotide gene sequence of the isolated strain A9 with corresponding *Nocardiopsis* sequences clearly showed that the organism form a distinct phyletic line with *Nocardiopsis* genera. The isolate was closely related to the type strain of *Nocardiopsis synnemataformans* sharing a 16S rRNA gene sequence similarity of 99%. Hence, the molecular results support the classification of the isolate as a new strain which

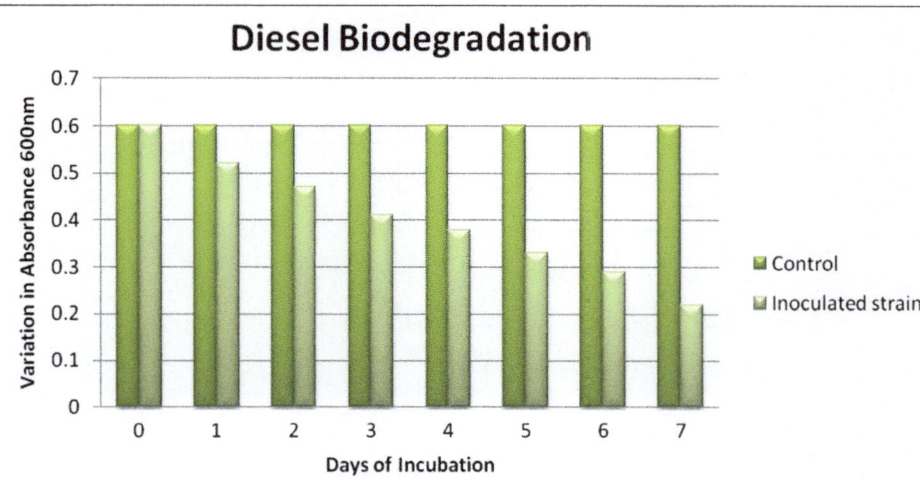

Figure 3 Diesel biodegradation pattern exhibited by A9 isolate.

Table 2 Cultural characteristics of the isolate A9

Medium	Colony characters					
	Growth	Aerial mycelium	Basal mycelium	Color on reverse side	Soluble pigment	Melanin production
ISP-1	++	White	Colorless	Reddish brown	Reddish brown	−
ISP-2	+++	White	Cheese like	Reddish brown	Reddish brown	−
ISP-4	+++	White	Cheese like	Reddish brown	Reddish brown	−
ISP-5	++	Dull White	Colorless	Reddish brown	−	−
ISP-6	++	White	Cheese like	Reddish brown	−	−
ISP-7	+	White	Cheese like	Brown	Brownish	+
SCA	+++	White	Cheese like	Light olive green	Light green	−
NA	+	White	Cheese like	Reddish brown	Reddish brown	−
SDA	+++	White	Cheese like	Reddish brown	Reddish brown	−
Microscopy	Gram-positive rods with rounded ends; branched hyphae with spores; non-motile					

+, less; ++, moderate; +++, good, +/−, doubtful.

was named as *Nocardiopsis* sp. mrinalini9. The partial 16S rRNA sequence of isolate A9 (1,061 bp) was deposited in Genbank under the accession number KJ909126. Still, DNA-DNA hybridizations, phenotypic comparisons, and chemotaxonomic analysis need to be performed to confirm its novelty.

Conclusion

In this paper, we have reported the isolation and identification of *Nocardiopsis* sp. mrinalini9 with remarkable biosurfactant and biodegradation capacity. It was also noted that the same isolate remains endophytic to Neem leaves also as described by Mrinalini *et al.* [21]. Compared to bacteria and fungi, actinomycetes possess greater potential to degrade hydrocarbons [19,22,23]. Especially, *Nocardiopsis* sp. seems to have higher biosurfactant and biodegradation value [19,24] compared to *Streptomyces* sp. which showcases better efficacy of rare actinomycetes. To the best of our knowledge, the novelty lies in the fact that not much research has been carried out on rare actinomycetes present in medicinal plants of Western Ghats of India with regard to its biosurfactant and biodegradation

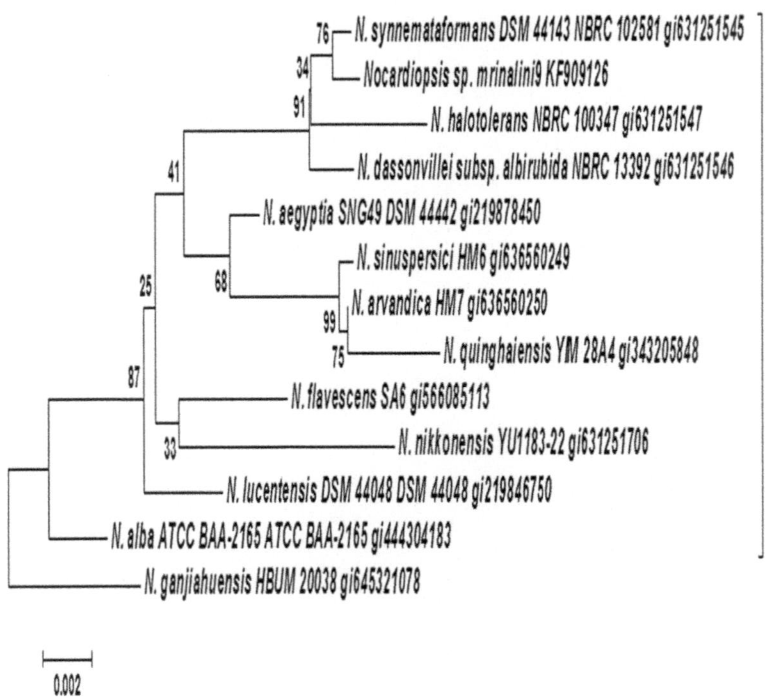

Figure 4 Phylogram depicting the taxonomic position of *Nocardiopsis* sp. mrinalini9.

activity. Further studies are in progress with respect to the extraction and purification of biosurfactant from *Nocardiopsis* sp. mrinalini9 which will have wider applications in industrial process and bioremediation. Microorganisms residing in the floral environment are diverse and unique. Intensive studies must be carried out to unravel its unexhausted reserve of bio-metabolites.

Competing interests

The authors declare that there is no conflict of interests regarding the publication of this article.

Authors' contribution

MJS designed the whole study and carried out the experiment while PS guided the first author and helped in manuscript preparation. Both authors read and approved the final manuscript.

Acknowledgement

We are thankful to the authorities of Nirmala College for Women, Coimbatore, Tamil Nadu, for providing the necessary facilities. The first author is sincerely grateful to University Grants Commission, Government of India, for providing the award of Junior Research Fellowship in Science to pursue the research work. We are also thankful to Yaazh Xenomics, Chennai, for 16S rRNA ribotyping service.

References

1. Youssef NH, Duncan KE, Nagle DP, Savage KN, Knapp RM, McInerney MJ (2004) Comparison of methods to detect biosurfactant production by diverse microorganism. J Microbiol Meth 56:339–347
2. Siciliano SD, Fortin N, Mihoc A, Wisse G, Labelle S, Beaumier D, Ouellette D, Roy R, Whyte LG, Banks MK, Schwab P, Lee K, Greer CW (2001) Selection of specific endophytic bacterial genotypes by plants in response to soil contamination. Appl Environ Microbiol 67:2469–2475
3. Barac T, Taghavi S, Borremans B, Provoost A, Oeyen L, Colpaert JV, Vangronsveld J, Van der Lelie D (2004) Engineered endophytic bacteria improve phytoremediation of watersoluble, volatile, organic pollutants. Nat Biotechnol 22:583–588
4. Qin S, Li J, Chen HH, Zhao GZ, Zhu WY, Jiang CL, Xu LH, Li WJ (2009) Isolation, diversity, and antimicrobial activity of rare actinobacteria from medicinal plants of tropical rain forests in Xishuangbanna. China Appl Environ Microbiol 75:6176–6186
5. Carillo P, Mardarz C, Pitta-Alvarez S (1996) Isolation and selection of biosurfactant producing bacteria. World J Microbiol Biotechnol 12:82–84
6. Deepika L, Kannabiran K (2010) Biosurfactant and heavy metal resistance activity of *Streptomyces* spp. isolated from saltpan soil. Br J Pharmacol Toxicol 1(1):33–39
7. Yousaf S, Andria V, Reichenauer TG, Smalla K, Sessitsch A (2010) Phylogenetic and functional diversity of alkane degrading bacteria associated with Italian ryegrass (*Lolium multiflorum*) and Birdsfoot trefoil (*Lotus corniculatus*) in a petroleum oil-contaminated environment. J Hazard Mater 184:523–532
8. Morikawa M, Ito M, Imanaka T (1992) Isolation of a new surfactin producer *Bacillus pumilus* A- 1, and cloning and nucleotide sequence of the regulator gene, psf-1. J Ferment Bioeng 74:255–261
9. Rosenberg M, Gutnick DL, Rosenberg E (1980) Adherence of bacteria to hydrocarbons: a simple method for measuring cell-surface hydrophobicity. FEMS Microbiol Lett 9:29–33
10. Maczek J, Junne S, Götz P (2007) Examining biosurfactant producing bacteria - an example for an automated search for natural compounds. In: Application Note CyBio AG
11. Cooper DG, Goldenberg BG (1987) Surface active agents from two *Bacillus* species. Appl Environ Microbiol 53:224–229
12. Kathiresan K (2003) Polythene and plastics - degrading microbes from the mangrove soil. Rev Biol Trop 51(3):629–634
13. Chaudhry S, Luhach J, Sharma V, Sharma C (2012) Assessment of diesel degrading potential of fungal isolates from sludge contaminated soil of petroleum refinery, Haryana. Res J Microbiol 7(3):182–190
14. Thijsse GJ, der Linden V (1961) Iso alkane oxidation by a *Pseudomonas*. Antonie Van Leeuwenhoek 27:171–179
15. Shirling EB, Gottlieb D (1966) Methods for characterization of *Streptomyces* species. Int J Syst Bacteriol 16:313–340
16. Williams ST, Sharpe ME, Holt JG (1989) Bergey's manual of systematic bacteriology, vol 4. Williams and Wilkins, Baltimore
17. Tamura K Stecher G, Peterson D, Filipski A, Kumar S (2013) MEGA6: Molecular Evolutionary Genetics Analysis Version 6.0. Mol Biol Evol 30:2725–2729
18. Vasile, C (1993). Degradation and decomposition, in Handbook of polyolefins synthesis and properties, edited by C Vasile and R B Seymour (Marcel Dekker Inc, New York) pp: 479–509.
19. Usha R, Sangeetha T, Palaniswamy M (2011) Screening of polyethylene degrading microorganisms from garbage soil. Libyan Agr Res Center J Int 2(4):200–204
20. Narayan R (2006) Biobased and biodegradable polymer materials: rationale, drivers and technology exemplars. Am Chem Soc Symp 939(18):282
21. Singh MJ, Padmavathy S (2014) Isolation, screening and characterization of endophytic PGPR actinomycetes present commonly in Neem and Tulsi leaves - *in vitro* study (Tomato). Int J Recent Sci Res 5(3):574–579
22. Kannahi M, Sudha P (2013) Screening of polythene and plastic degrading microbes from Muthupet mangrove soil. J Chem Pharm Res 5(8):122–127
23. Idemudia MI, Nosagie OA, Omorede O (2014) Comparative assessment of degradation potentials of bacteria and actinomycetes in soil contaminated with motorcycle spent oil. Asian J Sci Tech 5(8):482–487
24. Gandhimathi RK, Seghal TA, Hema S, Joseph RRT, Shanmughapriya PS (2009) Production and characterization of lipopeptide biosurfactant by a sponge-associated marine actinomycetes *Nocardiopsis alba* MSA10. Bioproc Biosys Eng 32(6):825–835

Optimization of drying parameters for mango seed kernels using central composite design

Franck Junior Anta Akouan Ekorong[1], Gaston Zomegni[2], Steve Carly Zangué Desobgo[3*] and Robert Ndjouenkeu[4]

Abstract

Background: The combined effect of drying temperature and time was evaluated on residual water content, yield of oil extraction, total phenolic compounds and antioxidant activity of seed kernel from a Cameroonian local variety of mango (*Local Ngaoundere*). Response surface methodology (RSM) using central composite design (CCD) as tool, was used to develop, validate and optimize statistical models in order to establish the impact of the drying parameters (temperature and time) either alone or in combination.

Results: It was shown that drying temperature individually in its first order (X_1) contributed 30.81, 21.11, 41.28 and 33.24% while drying time individually in its first order (X_2) contributed 39.91, 15.12, 29.92 and 25.87% for residual water content, yield of oil extraction, total phenolic components and antioxidant activity respectively. The increase of drying temperature increased antioxidant activity while the other physicochemical characteristics such as water content, yield of oil extraction and total phenolic components decreased. Concerning drying time, only water content was reduced with an increase of that factor. The synergetic effect of drying temperature and time was effective only for antioxidant activity. A compromise for optimization were then fixed for water content ≤ 10% w/w; oil content ≥ 9% w/w; total polyphenols ≥ 1 mg/g and antioxidant activity ≥ 1000 mg AAE/100 g DM. A simulation for optimization gave, for 60 H and 60°C for drying time and temperature respectively permitted to obtain 4.10% w/w, 9.53% w/w, 1340.28 mg AAE/100 g DM and 1.16 mg/g for water content, oil content, antioxidant activity and total polyphenols respectively.

Conclusions: The physicochemical characteristics studied was globally influenced by the chosen factors (drying time and temperature).

Keywords: Mango seed kernels; Drying; CCD; RSM; Modelling; Optimization

Background

Consumption and industrial exploitation of mango generate significant waste, mainly made of the peels and the seed kernel of the fruit, which account for 7% to 22% of the weight of the whole fruit [1]. In Cameroon, a rapid estimation of this waste based on the national mango production, evaluated to 539,000 t in 2012 [2], showed that nearly 38,500 to 121,000 t of mango waste are produced. Environmental, hygienic, and public health problems result in unorganized management of this waste [3,4]. Valorization of mango seed kernels through production of butter and biofunctional flour are the main solutions technologically proposed, since different studies have shown that mango seed kernels contain various phenolic compounds. Mango seed kernel butter is used in cosmetics for its non-saponifiable matter content and its antioxidant activity potential [5-7]. The mango seed kernel flour displays interesting antioxidant and antibacterial properties [8-13]. These properties have also been found in mango peels [3]. Valorizing this biowaste as a potential source of non-conventional oil and natural antioxidants represents an opportunity to improve mango producer's income, particularly in regions where poverty is current. In this respect, drying of the seed kernels is one of the main technological steps both for efficient extraction of the functional components and for inactivation of enzymatic degradation of the raw material. Drying of mango seed kernels before extraction contributes then to stabilize the product and to increase the yield of extraction.

* Correspondence: desobgo.zangue@gmail.com
[3]Department of Food Processing and Quality Control, University Institute of Technology (UIT) of The University of Ngaoundere, P.O. Box 455, Ngaoundere, Cameroon
Full list of author information is available at the end of the article

Since the temperature and the time of drying may affect the activity and the stability of bioactive compounds, due to chemical and enzymatic degradation, low evaporation, and/or thermal degradation, a badly carried out drying can lead to physicochemical reactions which can lead to losses of the textural and nutritional values [14] and thus bring damage in the quality of the product. Optimizing the drying of seed kernels would then enable to make sure that the product obtained has desired quality.

The present paper aims at determining the optimal drying parameters (temperature and time) for mango seed kernels in order to improve the butter extraction with preservation of its total phenolic compounds and antioxidant capacity. Response surface methodology is used in this respect, the optimization procedure consisting in determining the drying temperature and time which minimize moisture and maximize total polyphenol content, oil content, and antioxidant activity.

Methods

Material

Biological material

Mango (*Local Ngaoundere*) was collected from a farm in Ngaoundere (Adamawa region, Cameroon) at harvest stage maturity acceptable by consumers.

Chemicals

Acetone, n-hexane, tannic acid, methanol, trichloroacetic acid, and ascorbic acid were obtained from Sigma-Aldrich Chemie GmbH, Munich, Germany. Sodium carbonate and phosphate buffer solution was obtained from SERVA Electrophoresis GmbH, Germany. Folin-Ciocalteu, potassium hexacyanoferrate, and ferric chloride were from Fisher Scientific UK Ltd., Bishop Meadow Road, Loughborough, UK.

Sample preparation

The mango flesh was removed, the seed sundried, and the seed kernels extracted manually using a stainless steel knife to open the shell. The seed kernels were then open in two cotyledons of about 9.7 ± 2.4 mm thickness for drying.

Drying process

The drying process parameters considered for the study were drying temperature and drying time. Preliminary tests were allowed to fix the limits of these factors. The seed kernel cotyledons were laid on a tray and dried using fixed temperatures and times according to the central composite design (CCD), in a tropical oven dryer [15]. The air velocity in the drying chamber was constant (0.55 ms^{-1}), and the relative humidity of the chamber was 75 ± 3% (Hanna Instruments HI 8564 Thermo hygrometer, Hanna Instruments, Woonsocket, RI, USA). The dried seed kernels were reduced in powder form with knife mills (Retsch GM 200 GmbH, Retsch-Allee 1-542781 Haan, Germany) which are particularly suitable for grinding and homogenizing soft to medium-hard, elastic, fibrous, dry, or wet materials. It was assisted by a Retsch sieving shaker AS 300 permitted to obtain a particle size less than 1 mm. That powder was preserved in a plastic bag at 4°C in darkness until use.

Table 1 Central composite design: coded variables, real variables, and responses

Number	Coded variables		Real variables		Responses											
					Moisture (%)			Oil content (%)			Total polyphenols (% DM)			Antioxidant activity (eq g of VitC/100 g DM)		
	x_1	x_2	X_1	X_2	Exp	Cal	Res	Exp	Cal	Res	Exp	Cal	Res	Exp	Cal	Res
1	−1.00	−1.00	44.85	13.64	16.59	17.86	−1.27	9.14	9.05	0.09	0.99	1.11	−0.12	477.12	581.02	−103.90
2	1.00	−1.00	75.15	13.64	10.36	10.30	0.05	8.78	8.68	0.11	0.37	0.41	−0.04	781.64	924.62	−142.98
3	−1.00	1.00	44.85	61.36	7.72	8.11	−0.39	9.16	9.32	−0.16	1.30	1.30	0.00	892.26	723.22	169.04
4	1.00	1.00	75.15	61.36	1.72	0.79	0.93	8.79	8.94	−0.15	1.05	0.97	0.08	2,326.22	2,196.42	129.80
5	0.00	0.00	60.00	37.50	4.80	5.79	−0.99	9.45	9.56	−0.11	0.86	0.94	−0.08	1,144.65	1,148.92	−4.27
6	0.00	0.00	60.00	37.50	6.31	5.79	0.52	9.42	9.56	−0.14	0.91	0.94	−0.03	1,164.00	1,148.92	15.08
7	0.00	0.00	60.00	37.50	4.91	5.79	−0.88	9.49	9.56	−0.07	0.81	0.94	−0.13	1,150.18	1,148.92	1.26
8	−1.32	0.00	40.00	37.50	12.35	11.16	1.20	9.16	9.12	0.04	1.29	1.21	0.08	708.29	753.06256	−44.78
9	1.32	0.00	80.00	37.50	0.53	1.34	−0.81	8.65	8.63	0.02	0.49	0.53	−0.04	1,966.91	1,952.15056	14.76
10	0.00	−1.32	60.00	6.00	18.61	17.75	0.86	8.92	9.08	−0.16	0.89	0.78	0.11	596.20	404.3872	191.81
11	0.00	1.32	60.00	69.00	4.56	5.03	−0.47	9.66	9.44	0.22	1.20	1.27	−0.07	1,116.10	1,337.6272	−221.53
12	0.00	0.00	60.00	37.50	4.77	5.79	−1.02	9.39	9.56	−0.16	1.07	0.94	0.13	1,145.52	1,148.92	−3.40
13	0.00	0.00	60.00	37.50	6.41	5.79	0.62	9.83	9.56	0.27	0.92	0.94	−0.02	1,149.64	1,148.92	0.72
14	0.00	0.00	60.00	37.50	7.45	5.79	1.66	9.75	9.56	0.19	1.05	0.94	0.11	1,147.22	1,148.92	−1.70

Experimental design, modelling, validation of the model, and optimization

Response surface methodology (RSM) with CCD was used to carry out the experiments in order to model and optimize the following responses: residual water content, yield of oil extraction, total phenolic components, and antioxidant activity of the dried seed kernels. The independent variables (factors) were drying temperature (x_1) and drying time (x_2). The intervals of these factors were respectively 40°C to 80°C and 6 to 69 h (Table 1). The interval values of the factors were chosen considering the heat-sensitive effect of seed kernel components on oil [14] and polyphenols [16,17].

From the coded variables, many equations were used to transform them into real values to realize experiments in the laboratory. Those equations were as follows:

$$X_i = X_{0i} + x_i \times \Delta X_i \quad (1)$$

$$N = k^2 + 2k + k_0 \quad (2)$$

With the two factors, CCD had given a total of 14 experiments (with six replicates at the central point) as shown in Table 1. The value of α was calculated in order to respect the orthogonality criterion [18] using the formula:

$$\alpha = \left(\frac{2^k \left(\sqrt{2^k + 2k + n_0} + \sqrt{2^k} \right)^2}{4} \right)^{\frac{1}{4}} \quad (3)$$

Mathematical models describing the relationships among the process-dependent variable and the independent variables in a second-order equation were developed [19]. Design-based experimental data were matched according to the following second-order polynomial equation:

$$Y = \beta_0 + \sum_{i=1}^{k} \beta_i x_i + \sum_{i=1}^{k} \beta_{ii} x_i^2 + \sum\sum_{i<j}^{k} \beta_{ij} x_i x_j + \varepsilon \quad (4)$$

where Y is the response, x_i and x_j are the variables, β_0 is the constant, β_i is the coefficient of the linear terms, β_{ii} is the coefficient of the quadratic terms, and β_{ij} is the coefficient of the interaction terms.

The coefficients of the models and statistical analysis (ANOVA) were obtained using the Minitab version 16 software (Minitab, Ltd., Brandon Court, Unit E1-E2 Progress Way, Coventry, CV3 2TE, UK), and the curves were plotted using Sigmaplot version 12.1 (Systat Software, Inc., 1735 Technology Drive, Suite 430, San Jose, CA 95110, USA).

Validating the models was obtained by calculating the absolute average deviation (AAD), the bias factor (B_f), and the accuracy factor (A_f) [20,21] which were expressed as follows:

$$\text{AAD} = \frac{\left[\sum_{i=1}^{N} \left(\frac{\left| Y_{i,\exp} - Y_{i,\text{cal}} \right|}{Y_{i,\exp}} \right) \right]}{N} \quad (5)$$

$$B_f = 10^{\frac{1}{N} \sum_{i=1}^{N} \log\left(\frac{Y_{i,\text{cal}}}{Y_{i,\exp}} \right)} \quad (6)$$

$$A_{f1} = 10^{\frac{1}{N} \sum_{i=1}^{N} \left| \log\left(\frac{Y_{i,\text{cal}}}{Y_{i,\exp}} \right) \right|}, \quad (7)$$

where $Y_{i,\exp}$ and $Y_{i,\text{cal}}$ are respectively experimental and calculated responses and N is the number of experiments used in the calculation.

Each linear, interaction, and quadratic contribution of each factor were obtained as follows:

For linear terms,

$$\text{Contribution (\%)} = \frac{|\beta_i|}{\sum_{i=1}^{k} |\beta_i| + \sum_{i=1}^{k} |\beta_{ii}| + \sum\sum_{i<j}^{k} |\beta_{ij}|} \quad (8)$$

For quadratic terms,

$$\text{Contribution (\%)} = \frac{|\beta_{ii}|}{\sum_{i=1}^{k} |\beta_i| + \sum_{i=1}^{k} |\beta_{ii}| + \sum\sum_{i<j}^{k} |\beta_{ij}|} \quad (9)$$

For interaction terms,

$$\text{Contribution (\%)} = \frac{|\beta_{ij}|}{\sum_{i=1}^{k} |\beta_i| + \sum_{i=1}^{k} |\beta_{ii}| + \sum\sum_{i<j}^{k} |\beta_{ij}|} \quad (10)$$

Lastly, optimization was done using the software Mathcad 15.0 (build 15.0.0.436 Parametric Technology Corporation, 140 Kendrick Street, Needham, MA 02494, USA). The conditions fixed were to minimize moisture and total polyphenols and maximize oil content and antioxidant activity. The use of Sigmaplot version 12.1 (Systat Software, Inc., 1735 Technology Drive, Suite 430, San Jose, CA 95110, USA) permitted to draw the contour plots and superimpose the graphs in order to determine the optimal zone.

Analysis

Moisture content

The determination of the moisture content was done using a standard method [22] and an isothermal oven (Heraeus, Type: T6, manufacturing no. 20001046, Kendro Laboratory Products, Langenselbold, Germany) for drying a known mass of the sample (using an electronic balance Gibertini no. 125186, made in Milan, Italy) at 105°C for

24 h. After cooling in a desiccator, the samples were reweighed (using an electronic balance Gibertini no. 125186, made in Milan, Italy). The water content (TE) representing the ratio of mass before and after heating in the oven is determined as a percentage.

Lipid content

The extraction of the lipid from seed kernels was made by the Soxhlet extraction method [23], using n-hexane as solvent. In practice, 600 g of dried mango seed kernel powder was used. The oil-hexane mixture obtained after extraction was separated on a rotary evaporator (Heidolph Salvis Electronic W 60) at 70°C to recover hexane. The oily extract, previously dried in an oven at 105°C for 20 min to evaporate residual n-hexane solvent, was then weighed. The lipid content was determined in g/100 g of dry matter (DM).

Determination of total phenolic compound

These analyses were based on the oxidation/reduction principle and employed Folin-Ciocalteu reagent [24]. Briefly, 0.5 g of each sample was weighed in a glass beaker, and 10 mL of 70% acetone was added. The whole was stirred for 20 min at room temperature (25°C). The extract was centrifuged at 3,000 G for 10 min at 4°C. The supernatant was recovered and stored at 4°C. Volumes of 0, 10, 20, 30, 40, 50, 60, 70, 80, 90, and 100 µL of a tannic acid standard solution (0.1 mg/mL) for calibration curve and a suitable volume of extract were introduced into test tubes. The volumes were made up to 500 µL with distilled water. To these solutions, 250 µL of Folin-Ciocalteu reagent (1 N) and 1.25 mL of sodium carbonate (20%) were added. The mixtures were agitated and incubated at room temperature in the dark. After 40 min, the absorbance was read at 725 nm against the blank. The amount of total phenolic compound expressed by weight of tannic acid was then determined.

The antioxidant activity

To obtain the extract, 250 mg of powder was mixed in 25 mL of methanol at room temperature (25°C) for 2 h and centrifuged for 10 min at 4,000 G. The residue was extracted again with 25 mL of methanol and centrifuged (4,000 G) for 2 h again. The mixture of the two volumes of extract was then evaporated using a rotary evaporator to reduce the sample to 25 mL. This extract was recovered in sealed tubes and stored at 4°C until use. The antioxidant capacity of the different powder samples was assessed by determining their ability to reduce iron (III) to iron (II) [25].

In a test tube, 1 mL of each extract was mixed with 2.5 mL of a phosphate buffer solution (0.2 M, pH 6.6) and 2.5 mL of 1% potassium hexacyanoferrate [$K_3Fe(CN)_6$]. The whole was incubated for 30 min at 50°C in a water bath. Then, 2.5 mL of 10% trichloroacetic acid was added and the mixture was centrifuged for 10 min using a Heraeus Biofuge Primor, Kendro Laboratory Products, Langenselbold, Germany. After that, 2.5 mL of the supernatant was taken and mixed with 2.5 mL of distilled water and 0.5 mL of an aqueous solution of 0.1% $FeCl_3$. The absorbance was read at 700 nm using a spectrophotometer Rayleigh VIS-723N (Beijing Beifen-Ruili Analytical Instrument (Group) Co., Ltd., Beijing, China). A calibration curve was done using ascorbic acid as reference at different concentrations. The total reducing power was expressed as equivalent of ascorbic acid (AAE).

Results and discussion

Mathematical modelling

Modelling of the action of drying temperature and time on four key drying parameters, residual water content, yield of oil extraction, total phenolic components, and antioxidant activity was carried out by modelling the experimental design required for laboratory purposes (Table 1). The mathematical models obtained were as follows, respectively:

$$Y_{\text{water}}(x_1, x_2) = 5.793 - 3.718x_1 - 4.816x_2 + 0.060x_1x_2 + 0.261x_1^2 + 3.212x_2^2 \quad (11)$$

$$Y_{\text{oil}}(x_1, x_2) = 9.558 - 0.187x_1 + 0.134x_2 - 0.002x_1x_2 - 0.391x_1^2 - 0.172x_2^2 \quad (12)$$

$$Y_{\text{polyphenols}}(x_1, x_2) = 0.939 - 0.258x_1 + 0.187x_2 + 0.092x_1x_2 - 0.040x_1^2 + 0.048x_2^2 \quad (13)$$

$$Y_{\text{antioxidant}}(x_1, x_2) = 1{,}148.92 + 454.2x_1 + 353.5x_2 + 282.4x_1x_2 + 116.9x_1^2 - 159.5x_2^2 \quad (14)$$

with $Y_{\text{water}}(x_1, x_2)$ representing the mathematical model for water content; $Y_{\text{oil}}(x_1, x_2)$, the mathematical model for oil content; $Y_{\text{polyphenols}}(x_1, x_2)$, the mathematical model for total polyphenols; $Y_{\text{antioxidant}}(x_1, x_2)$, the mathematical model for antioxidant activity; x_1, the drying temperature; and x_2, the drying time.

The mathematical models were polynomials and validated according to the method described by Ross [20] as shown in Table 2. The factors of the models were of first

Table 2 Model validation data

Models	R^2	AAD	B_f	A_f
$Y_{\text{water}}(x_1, x_2)$	0.966	0.246	1.023	1.245
$Y_{\text{oil}}(x_1, x_2)$	0.824	0.014	1.000	1.015
$Y_{\text{polyphenols}}(x_1, x_2)$	0.890	0.083	1.008	1.086
$Y_{\text{antioxidant}}(x_1, x_2)$	0.948	0.090	0.996	1.097

Table 3 Estimated coefficient impact and contributions to moisture, oil content, total polyphenols, and antioxidant activity

Source	Moisture			Oil content			Total polyphenols			Antioxidant activity		
	Coefficients	*P*	Contribution (%)	Coefficients	*P*	Contribution (%)	Coefficients	*P*	Contribution (%)	Coefficients	*P*	Contribution (%)
A:x_1	−3.718	0.000	30.81	−0.187	0.033	21.11	−0.258	0.000	41.28	454.200	0.000	33.24
B:x_2	−4.816	0.000	39.91	0.134	0.104	15.12	0.187	0.002	29.92	353.500	0.000	25.87
AA	0.261	0.612	2.16	−0.391	0.001	44.13	−0.040	0.397	6.40	116.900	0.080	8.55
BB	3.212	0.000	26.62	−0.172	0.067	19.41	0.048	0.321	7.68	−159.500	0.026	11.67
AB	0.060	0.924	0.50	−0.002	0.986	0.23	0.092	0.137	14.72	282.400	0.004	20.67

degree (x_1 and x_2), of second degree (x_1^2 and x_2^2), and of interaction (x_1x_2) form. They were statistically significant or not if the probability (*P*) was ≤0.05 or ≥0.05, respectively (Table 3).

Effect of drying temperature

The impact of drying temperature (x_1) on physicochemical characteristics of *Local Ngaoundere* mango seed kernels was significant on the decrease of water content ($P = 0.000$), oil content ($P = 0.033$), and total polyphenols ($P = 0.000$), respectively (Table 3). Moreover, it had significant impact on the increase of antioxidant activity with $P = 0.000$ (Table 3).

The effect of drying temperature on the moisture of mango seed kernels is shown in Figure 1a. Moisture decreased from 23.2% *w*/*w* (at 40°C) with increasing drying temperature to attain a minimum level of 13.1% *w*/*w* at 80°C. In this case, it could be linked to effective moisture diffusivity which increased with a decrease in moisture content. This may indicate that as the moisture content decreased, the permeability to vapor increased, and the pore structure remained open. The temperature of the seed kernels raised rapidly in the initial stages of drying, due to more absorption of heat during drying, as the seed kernels could have a high loss factor at higher moisture content. This increases the water vapor pressure inside the pores inducing the opening of seed kernel pores so that, in the first stage of drying, liquid diffusion of moisture could be the main mechanism of moisture transport. As drying progressed further, vapor diffusion could have been the dominant mode of moisture diffusion in the later part of drying as reported by the literature [26-30].

The effect of drying temperature on the oil content of mango seed kernels is shown in Figure 2. The oil content started from 8.64% *w*/*w* at 40°C, then increased to 9.10% *w*/*w* at 56°C slightly (with a gap of 0.46% *w*/*w* obtained), and decreased significantly to 8.15% *w*/*w* at 80°C (with a gap of 0.95% *w*/*w* obtained) with a probability $P = 0.033$ (Table 3). This could be explained by lipid autoxidation and photo-oxidation. In fact, due to the increase of drying temperature and oxygen, the oil oxidation autoxidation and photo-oxidation was promoted. This could have been virtually inevitable since mango seed kernel oil content polyunsaturated triglycerides could play a role in that oxidation [31]. Other factors that could affect oxidation included moisture content, presence of metals, enzyme activity, UV light, protein content, and other chemical reactions [32,33].

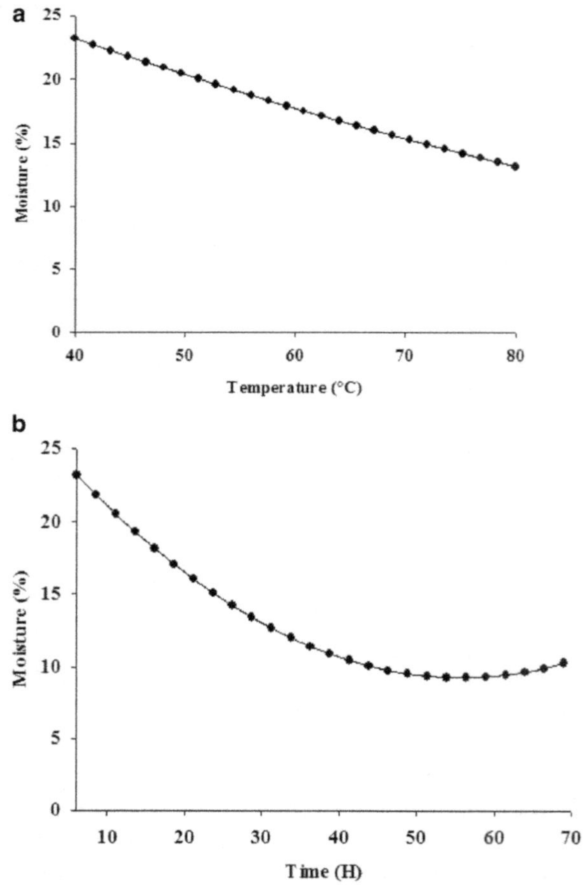

Figure 1 Effect of drying temperature and drying time on moisture. (a) Effect of drying temperature as the sole factor on the moisture of mango seed kernels (drying time fixed at 6 h). **(b)** Effect of drying time as the sole factor on the moisture of mango seed kernels (drying temperature fixed at 40°C).

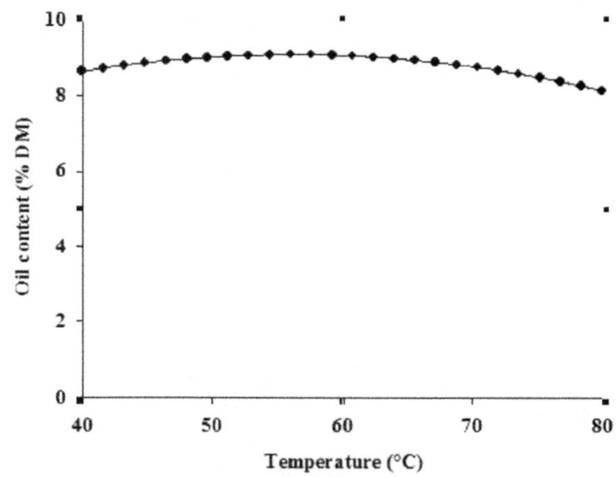

Figure 2 Effect of drying temperature on oil content. Effect of drying temperature as the sole factor on the oil content of mango seed kernels (drying time fixed at 6 h).

The effect of drying temperature on total polyphenols of mango seed kernels is shown in Figure 3a. Total polyphenols decreased from 1.20 mg/g (at 40°C) with increasing drying temperature to attain a minimum level of 0.20 mg/g at 80°C. The reduced levels of the polyphenol compounds obtained from oven-dried mango seed kernels resulted from the degradation of phenolic compounds at high temperatures, due to chemical, enzymatic, or thermal decomposition [34].

The effect of drying temperature on antioxidant activity of mango seed kernels is shown in Figure 4a. The antioxidant activity started from 500.59 mg AAE/100 g DM at 40°C, then decreased to 390.21 mg AAE/100 g DM at 54.72°C (with a gap of 110.38 mg AAE/100 g DM obtained), and increased significantly to 715.57 mg AAE/100 g DM at 80°C (with a gap of 325.36 mg AAE/100 g DM obtained) with a probability $P = 0.000$ (Table 3). This could be explained by considering that high temperatures promote the inactivation of oxidative enzymes

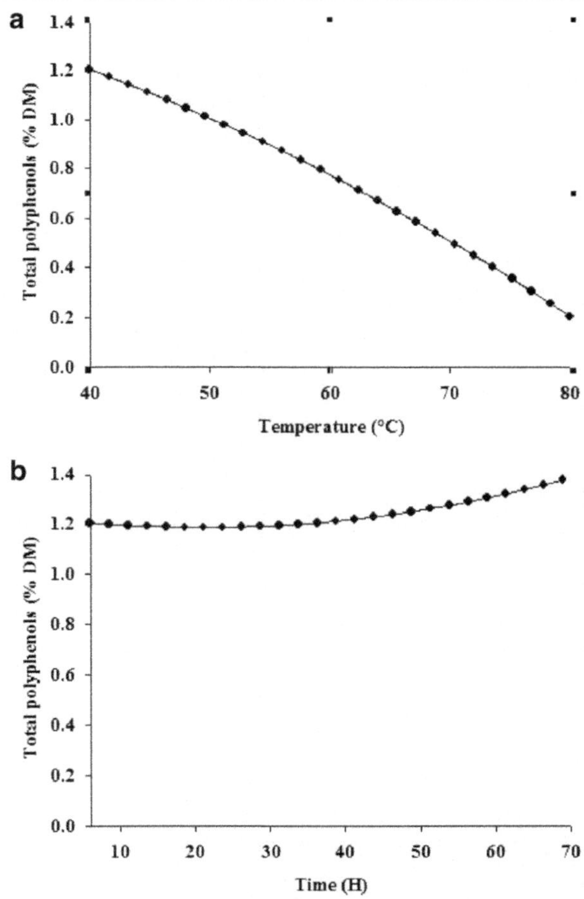

Figure 3 Effect of drying temperature and drying time on total polyphenol content. (a) Effect of drying temperature as the sole factor on total polyphenol content of mango seed kernels (drying time fixed at 6 h). **(b)** Effect of drying time as the sole factor on total polyphenol content of mango seed kernels (drying temperature fixed at 40°C).

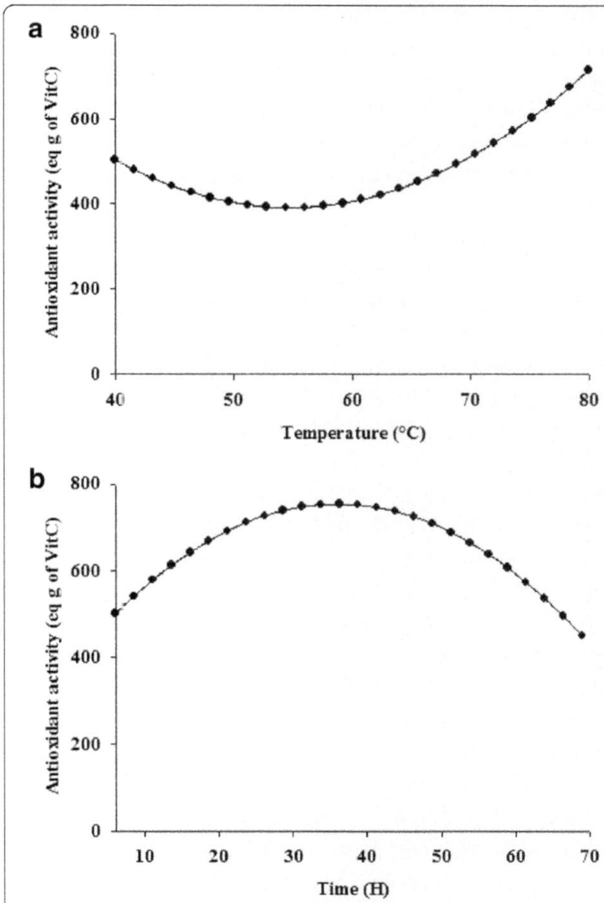

Figure 4 Effect of drying temperature and drying time on antioxidant activity. (a) Effect of drying temperature as the sole factor on antioxidant activity of mango seed kernels (drying time fixed at 6 h). **(b)** Effect of drying time as the sole factor on antioxidant activity of mango seed kernels (drying temperature fixed at 40°C).

[35], avoiding the degradation of antioxidants. Furthermore, at high temperatures, the generation and accumulation of Maillard-derived melanoidins with a varying degree of antioxidant activity could also enhance the antioxidant properties of extracts [36].

Effect of drying time

The drying time (x_2) as a sole factor on some physicochemical characteristics of *Local Ngaoundere* mango seed kernels had a significant impact on the decrease of water content ($P = 0.000$) and had a significant impact on the increase of total polyphenols ($P = 0.002$) and antioxidant activity ($P = 0.000$), respectively (Table 3), while it had no significant impact on the increase or decrease of oil content with $P = 0.104$ (Table 3).

The effect of drying time on the moisture of mango seed kernels is shown in Figure 1b. It started from 23.2% *w*/*w* at 6 h and then decreased to a minimum value of 9.3% *w*/*w* at 55.68 h. After that, a slight increase until 10.3% *w*/*w* was observed at 69 h. This could be explained by the fact that when air held the maximum possible amount of vapor, the vapor exerted a saturation vapor pressure, and since the water vapor present was less than the maximum, the air took up more moisture. This evaporation which took place from the surface of the seed kernels induced moisture decrease, and it could be attained using time [37].

Although phenolic compounds are considered as heat-sensitive compounds [38], the increase in the total polyphenols as the drying time lengthened was observed (Figure 3b). In fact, it started from 1.20 mg/g at 6 h and then increased to 1.38 mg/g at 69 h. As far as this aspect is concerned, the literature shows contradictory results. But it could be explained by the sum of the content of the individual phenolic acids in the free fraction which significantly increased as the drying time lengthened as observed also for mandarin pomace [39].

The effect of drying time on antioxidant activity of mango seed kernels is shown in Figure 4b. It started from 500.58 mg AAE/100 g DM at 6 h and then increased to a maximum value of 753.64 mg AAE/100 g DM at 36.05 h. After that, it decreased until 449.71 mg AAE/100 g DM at 69 h. This result highlights the relationship between the previously observed enhancement of antioxidant potential and the increase in the content of some individual polyphenols. In fact, this could be explained by the fact that Maillard reaction products, which can be formed as a consequence of heat treatment or time, generally exhibit strong antioxidant properties [40-43].

Effect of interaction drying temperature/drying time

The impact of interaction of drying temperature/drying time (x_1x_2) on some physicochemical characteristics of

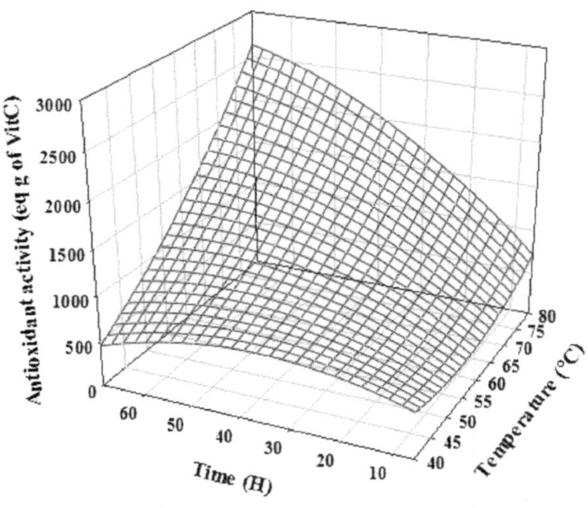

Figure 5 Effect of interaction (drying temperature/drying time) on antioxidant activity of mango seed kernels.

Local Ngaoundere mango seed kernels was not significant on water content ($P = 0.924$), oil content ($P = 0.986$), and total polyphenols ($P = 0.137$) (Table 3), while it was significant on the increase of antioxidant activity ($P = 0.004$) (Table 3).

The effect of interaction of drying temperature/drying time (x_1x_2) on the increase of antioxidant activity of mango seed kernels is shown in Figure 5. This could be explained as before by the fact that an increase of temperature and time results in more Maillard reaction products, permitting to exhibit strong antioxidant activities as mentioned in the literature [40-43].

Optimization

The results obtained for the action of drying parameters (time and temperature) on water content, fats content, total polyphenols, and antioxidant activity on the basis of the models were optimized to determine satisfactory domains of compromise. These domains were obtained for four key physicochemical characteristics of mango seed kernels, by fixing them at water content ≤10% *w*/*w* [44], oil content ≥9% *w*/*w*, total polyphenols ≥1 mg/g, and antioxidant activity ≥1,000 mg AAE/100 g DM.

A drying parameter couple of 60 h and 60°C respectively for drying time and temperature permitted to obtain 4.10% *w*/*w*, 9.53% *w*/*w*, 1,340.28 mg AAE/100 g DM, and 1.16 mg/g respectively for water content, oil content, antioxidant activity, and total polyphenols. Also, a drying parameter of 60 h and 65°C respectively for drying time and temperature permitted to obtain 2.92% *w*/*w*, 9.43% *w*/*w*, 1,590.49 mg AAE/100 g DM, and

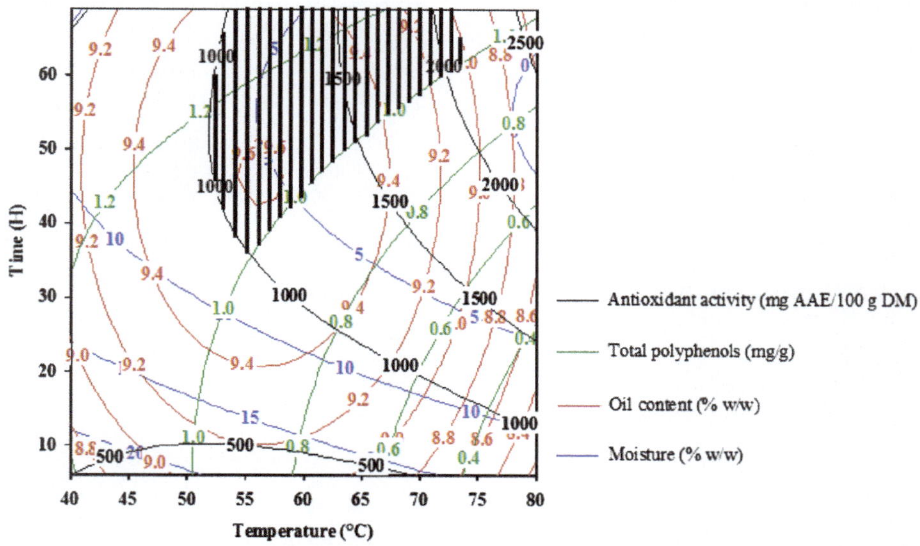

Figure 6 Response surface curves for drying temperature and time combinations. Response surface curves for drying temperature and time combinations providing for compromised physicochemical properties for mango seed kernels (*Local Ngaoundere*).

1.10 mg/g respectively for water content, oil content, antioxidant activity, and total polyphenols. The results were thus confirming that the areas exploitable (Figure 6) for efficient drying respecting the conditions fixed before were valid.

Conclusions

The effects of drying parameters (drying temperature and time) on some physicochemical characteristics of mango seed kernels were studied. Drying temperature was impacting more than the drying time. The interaction was only significant for antioxidant activity, meaning that there was no synergetic action of drying temperature and time for water, oil, and total polyphenol content. The study showed that satisfactory properties could be achieved when acting on drying parameters. Optimization of physicochemical properties of mango seed kernels showed that compromise could permit to obtain physicochemical characteristics of importance in order to assess the exploitability of the results for the valorization of mango seed kernels.

Abbreviations

A_f: accuracy factor; ANOVA: analysis of variance; AAE: ascorbic acid equivalent; AAD: average absolute deviation; B_f: bias factor; X_{0i}: center of variable; CCD: central composite design; R^2: coefficient of determination; β_{ij}: coefficient of the interactions terms; β_i: coefficient of the linear terms; β_{ii}: coefficient of the quadratic terms; x_i: coded variables given by the Doehlert table; β_0: constant term; DM: dry matter; $Y_{i,exp}$: experimental response; ΔX_i: increment; k_0: number of center points; *N*: number of experiments; *k*: number of variables; *P*: probability level; X_i: real variables; Res: residue; *Y*: response; RSM: response surface methodology; $Y_{i,cal}$: theoretical response.

Competing interests

The authors declare that they have no competing interests.

Authors' contributions

All authors have directly participated in the planning, execution, or analysis of this study. All authors of this paper have read and approved the final version submitted.

Acknowledgements

This work was undergone in the framework of the research group TQI2A (Technologie, Qualité et Innovations Agro Alimentaires) funded by AIRD (Agence Inter Etablissements de Recherche pour le Développement).

Author details

[1]Department of Process Engineering, National School of Agro-Industrial Sciences (ENSAI), University of Ngaoundere, P.O. Box 455, Ngaoundere, Cameroon. [2]Department of Textile and Leather Engineering, The Higher Institute of the Sahel, University of Maroua (ISS), P.O. Box 46 Maroua, Cameroon. [3]Department of Food Processing and Quality Control, University Institute of Technology (UIT) of The University of Ngaoundere, P.O. Box 455, Ngaoundere, Cameroon. [4]Department of Food Science and Nutrition, National School of Agro-Industrial Sciences (ENSAI) of The University of Ngaoundere, P.O. Box 455, Ngaoundere, Cameroon.

References

1. Bafode BS (1988) Projet de transformation et de conditionnement des mangues à Boundiali en Côte d'Ivoire. SIARC (Section des ingénieurs alimentaires/ région chaude), Montpellier
2. FAOSTAT (2014) Mangoes gross production value (Cameroon). FAO Statistics Division. http://faostat.fao.org/site/613/DesktopDefault.aspx?#ancor. Accessed 26 June 2014
3. Ashoush IS, Gadallah MGE (2011) Utilization of mango peels and seed kernels powders as sources of photochemicals in biscuit. World J Dairy & Food Sciences 6(1):35–42
4. Dorta E, Gloria ML (2012) Using drying treatments to stabilize mango peel and seed: effect on antioxidant activity. LWT Food Sci Technol 45:261–268
5. Abdalla AEM, Darwish SM, Ayad EHE, El-Hamahmy RM (2007) Egyptian mango by-product 1: compositional quality of mango seed kernel. Food Chem 103:1134–1140
6. Nzikou JM, Kimbonguila A, Matos L, Loumouamou B, Pambou-Tobi NPG, Ndangui CB, Abena AA, Silou TH, Scher J, Desobry S (2010) Extraction and characteristics of seed kernel oil from mango (*Mangifera indica*). Res J Environmental and Earth Sci 2(1):31–35

7. Solis-Fuentes JA, Duran-de-Bazua MC (2004) Mango seed uses: thermal behavior of mango seed almond fat and its mixtures with cocoa butter. Bioresour Technol 92:71–78
8. Abdalla AEM, Darwish SM, Ayad EHE, El-Hamahmy RM (2007) Egyptian mango by-product 2: antioxidant and antimicrobial activities of extract and oil from mango seed kernel. Food Chem 103:1141–1152
9. Arogba SS (1997) Physical, chemical and functional properties of Nigerian mango (*Mangifera indica*) kernel and its processed flour. J Sci Food Agric 73:321–328
10. Ebiloma UG, Arogba SS, Aminu OR (2011) Some activities of peroxydase from mango (*Mangifera indica* L. Var. Mapulehu) kernel. Int J Biol Chem 5:200–206
11. Fayeye TR, Joseph K (2004) Effects of dietary dehulled, sundried mango seed kernel meal on growth and carcass characteristics of fryer rabbit. J Agriculture, Res Dev 3:129–139
12. Joseph JK, Aboladji J (1997) Effect of replacing maize with graded levels of cooked Nigerian mango seed kernels (*Mangifera indica*) in the performance, carcass yield and meat quality of broiler chickens. Bioressource Technology 61:99–102
13. Maisuthisakul P, Harnsilawat T (2011) Characterization and stabilization of the extract from mango seed kernel in a cosmetic emulsion. Kasetsart J Nat Sci 45(3):521–529
14. Turkan A, Refik P (2007) Changes in the drying characteristics and water activity values of selected pistachio cultivars during hot air drying. J Food Process Eng 30:607–624
15. Kuitche A, Kouam J, Edoun M (2006) Modélisation du profil de température dans un séchoir construit dans l'environnement tropical. J Food Eng 76:605–610
16. Larrauri JA, Ruperez P, Saura-Calixto F (1997) Effect of drying temperature on the stability of polyphenols and antioxidant activity of red grape pomace peels. J Agric Food Chem 45(4):1390–1393
17. Podsedek A (2007) Natural antioxidants and antioxidant capacity of Brassica vegetables: a review. LWT - Food Sci Technology 40:1–11
18. Louvet F, Delplanque L (2005) Les Plans d'Expériences : une approche pragmatique et illustrée. Expérimentique Orléans, France
19. Giovanni M (1983) Response surface methodology and product optimization. J Food Technol 37:41–45
20. Ross T (1996) Indices for performance evaluation of predictive models in food microbiology. J Appl Bacteriol 81:501–508
21. Baranyi J, Pin C, Ross T (1999) Validating and comparing predictive models. Int J Food Microbiol 48:159–166
22. AFNOR (1982) Recueil des normes françaises des produits dérivés des fruits et légumes, 1st edn. Association Française de Normalisation, Paris, France
23. IUPAC (1979) Standard methods for the analysis of oils, fats and derivatives. International Union of Pure and Applied Chemistry, 6th edn. Pergamon Press Ltd., Great Britain
24. Makkar HPS, Blummel M, Borowy NK, Becker K (1993) Gravimetric determination of tannins and their correlations with chemical and protein precipitation methods. J Sci Food and Agriculture 61:161–165
25. Oyaizu M (1986) Studies on products of browning reaction prepared from glucosamine. Jpn J Nutr 44:307–314
26. Reyes A, Ceron S, Zuniga R, Moyano P (2007) A comparative study of microwave-assisted air drying of potato slices. Biosys Eng 98:310–318
27. Pickles CA (2003) Drying kinetics of nickeliferous limonitic laterite ores. Miner Eng 16:1327–1338
28. Sharma GP, Prasad S (2004) Effective moisture diffusivity of garlic cloves undergoing microwave-convective drying. J Food Eng 65:609–617
29. Celma AR, Cuadros F, Lopez-Rodriguez F (2012) Convective drying characteristics of sludge from treatment plants in tomato processing industries. Food Bioprod Process 90:224–234
30. Sharma GP, Verma RC, Pathare PB (2005) Thin-layer infrared radiation drying of onion slices. J Food Eng 67:361–366
31. Tallman KA, Roschek B, Porter NA (2004) Factors influencing the autoxidation of fatty acids: effect of olefin geometry of the nonconjugated diene. J Am Chem Soc 126:9240–9247
32. Mujumdar AS (2007) Handbook of industrial drying, 3rd edn. Taylor & Francis, UK
33. Frankel EN (1985) Chemistry of autoxidation: mechanism, products and flavor significance. In: Min DB, Smouse TH (eds) Flavor chemistry of fats and oils. AOCS Press, Champaign, IL, p 1
34. Nicoli MC, Anese M, Parpinel M (1999) Influence of processing on the antioxidant properties of fruit and vegetables. Trends Food Sci Technol 10(3):94–100
35. Sanjuán N, Benedito J, Clemente G, Mulet A (2000) The influence of pretreatments on the quality of dehydrated broccoli stems. Food Sci Technol Int 8(3):227–234
36. Que F, Mao L, Fang X, Wu T (2008) Comparison of hot air drying and freeze drying in the physicochemical properties and antioxidant activities of pumpkin (*Cucurbita moschata* Duch) flours. Int J Food Sci Technol 43(7):1195–1201
37. Potter NN, Hotchkiss JH (1998) Food science, 5th edn. Springer. New York
38. Erbay Z, Icier F (2009) Optimization of hot air drying of olive leaves using response surface methodology. J Food Eng 91:533–541
39. Hayat K, Zhang X, Farooq U, Abbas S, Xia S, Jia C, Zhong F, Zhang J (2010) Effect of microwave treatment on phenolic content and antioxidant activity of citrus mandarin pomace. Food Chem 123:423–429
40. Eichner K (1981) Antioxidative effect of Maillard reaction intermediates. Progr Food Nutr 5:441–451
41. Anese M, Pittia P, Nicoli MC (1993) Oxygen consuming properties of heated glucose-glycine aqueous solutions. Ital J Food Sci 5:75–79
42. Nicoli MC, Anese M, Manzocco L, Lerici CR (1997) Antioxidant properties of coffee brews in relation to the roasting degree. Lebensm-Wiss u-Technol 30:292–297
43. Nicoli MC, Anese M, Parpinel MT, Franceschi S, Lerici CR (1977) Study on loss and/or formation of antioxidants during processing and storage. Cancer Lett 114:71–74
44. Gupta MK (1993) Processing to improve soybean oil quality. Inform 4(11):1267–1272

4

Current status and future perspectives of kinetic modeling for the cell metabolism with incorporation of the metabolic regulation mechanism

Yu Matsuoka[1] and Kazuyuki Shimizu[1,2*]

Abstract

It becomes more and more important to develop appropriate models for the efficient design of the cell factory for microbial biofuels and biochemical productions, since the appropriate model can predict the effect of culture environment and/or the specific pathway genes knockout on the growth characteristics. Among various modeling approaches, kinetic modeling is promising in the sense of realizing the essential feature of metabolic regulation. A brief overview is given for the current status of the kinetic modeling of the cell metabolism from the point of view of metabolic regulation focusing on *Escherichia coli* (but not limited to *E. coli*). For the proper modeling, it is important to realize the systems behavior by integrating different levels of information to understand and unravel the underlying principles of the living organisms, namely, it is important to properly understand how the environmental stimuli are detected by the cell, how those are transduced, and how the cell metabolism is regulated, and to express these into the model. In particular, it is important to incorporate the enzymatic regulations of Pyk, Pfk, and Ppc by fructose-1,6-bisphosphate (FBP), phosphoenol pyruvate (PEP), and acetyl-coenzyme A (AcCoA) to realize the flux-sensing and homeostatic behavior. The proper modeling for phosphotransferase system (PTS) and the transcriptional regulation by cAMP-Crp and Cra is also important to simulate the main metabolism in relation to catabolite regulation. The coordinated regulation between catabolic and anabolic (nitrogen source-assimilation) metabolisms may be simulated by the behavior of keto acid such as α-ketoglutarate (αKG). The metabolism under micro-aerobic conditions may be made by incorporating the global regulators such as ArcA/B and Fnr. It is quite important to develop quantitative kinetic models, which incorporate enzyme level and gene level regulations from the biological science and metabolic engineering points of view.

Keywords: Virtual microbe; Systems biology; Kinetic model; Metabolic regulation; Metabolic engineering; Dynamics; Metabolism; *Escherichia coli*

Introduction

Microbial production of biofuels and biochemicals from renewable resources or biomass has been paid recent attention from global sustainability and environmental protection points of view, and many attempts have been made for the cell design by metabolic engineering approach. However, the practical application is limited in many cases, and more innovative design of cell factories is desired [1]. On the other hand, significant progress has been made on molecular biology from the reductionist point of view. However, the molecular knowledge alone is in many cases not sufficient to understand the cell system's behavior, where the system's behavior emerges from the interactions between the characterized molecules [2]. Thus, the systems biology approach has been paid recent attention in the post genome era. The ultimate goal of systems biology is to reconstruct a cell system into the computer which can predict observable phenotypes. If this could be attained, the effects of culture environment and/or the specific genetic mutation on the metabolism can be

* Correspondence: shimi@bio.kyutech.ac.jp
[1]Kyushu Institute of Technology, Iizuka, Fukuoka 820-8502, Japan
[2]Institute of Advanced Biosciences, Keio University, Tsuruoka, Yamagata 997-0017, Japan

predicted without conducting many exhaustive experiments, and metabolic engineering may be made more efficient with verification by the experiment for the selected mutants in the optimized condition based on the computer simulation. Thus, the appropriate model can contribute for the efficient design of cell factories from the practical application point of view.

It is quite important to quantitatively understand the complex and highly interrelated cellular behavior from biological science and metabolic engineering points of view. This may be attained with the help of informatics and systems biology by integrating different levels of ever increasing data with deep insight into the available data by biological knowledge [2-4].

In living organisms, metabolic network, defined as the set and topology of metabolic biochemical reactions in a cell, plays an essential role for the cell to survive, where it is under well-organized control. Thousands of different biochemical reactions as well as transport processes are linked together to break down organic compounds to generate energy (catabolism) and to synthesize macromolecular compounds (anabolism) for the cell synthesis. Similarly, complex signaling networks interconvert signals or stimuli that are important for the cellular function and interactions with the environment. This implies the importance of the transfer of information in signal transduction pathways and cascades designed to maximize the efficiency for cellular responses to environmental perturbations.

In order to understand the cell system in response to culture environment, the coupling between the recognition or sensing of the environmental condition and adjustment of the metabolic system must be properly incorporated into the model. In particular, it is important to incorporate the coupling between enzymatic reactions and the transcriptional regulation [5]. Moreover, although local regulation mechanisms are known to exist, it is not clear how those local regulation systems are coordinated on the systems level, where this may be made by 'distributed sensing of the intracellular metabolic fluxes' [5].

In the present article, current status of kinetic modeling is overviewed from the point of view of proper modeling with incorporation of metabolic regulation mechanism. Metabolic regulation analysis is critical for the proper modeling and has to be made in evaluating the performance of the designed cell as well as for reengineering the cell factories [6]. In bacterial adaptation to the culture environment, the global regulators detect the change in culture environment and control the metabolic pathway genes [7-9]. Here, the modeling of the metabolic regulation is considered focusing on *Escherichia coli* (but not limited to *E. coli*) based on the kinetic modeling approach with consideration of metabolic regulation.

Basic modeling approach

A variety of models have been proposed in the past, where they are discriminated from others depending on the underlying assumptions for the modeling, the data they require, and the accuracy of the model prediction [10]. The types of modeling formalism depend on such characteristics [11].

The model development may start with considering the network structure with kinetic rate expressions, model structure, parameter identification, and model validation, which may differ depending on the purpose of using a model [12]. It must be careful that the determination of kinetic rate expression is not straightforward due to the difficulty in identifying the mechanisms of enzymes and transporters [13], and therefore, some appropriate model simplification may be considered. Although parameter identification, sensitivity analysis, identifiability, experimental design, and optimization are important for the modeling in practice [12,14,15], here, we rather focus on the kinetic modeling with consideration of metabolic regulation.

Metabolic flux analysis

Among different levels of information, the metabolic fluxes are located on top of those, and it is the most important information from the phenotypic fermentation point of view [16-18], and it can be used for the analysis of the specific pathway gene knockout on the metabolism [19,20]. ^{13}C-metabolic flux analysis (^{13}C-MFA) has been shown to be useful for the metabolic regulation analysis [16,19-23]. However, this is essentially the analysis method for the physiological state of the organism based on mass balance together with isotopomer balance, and it does not have the predictability. It is highly desirable and useful to be able to predict the cell growth characteristics and the metabolic changes in response to the change in culture environment and/or the specific pathway mutation.

Flux balance analysis and its extensions

Flux balance analysis (FBA) and its extension to genome scale has made significant progress as it requires only basic knowledge of the metabolic reaction stoichiometry and has a reasonably accurate predictability. Significant efforts have also been made to integrate gene level regulation and metabolic networks to reveal the regulation mechanism [24,25]. In such approach, however, some appropriate objective functions such as the maximization of the cell growth rate, the specific substrate consumption rate, and/or the metabolite production rate must be introduced due to excess degrees of freedom. It was, however, shown that no single objective function can accurately represent the flux data for the different culture condition [26]. Rather, a vector-valued objective function or multiple objective functions must be considered, resulting in the Pareto optimal set to represent the metabolic fluxes [27], where the influential objective function may be the maximum ATP yield, maximum biomass yield, and minimum sum of absolute fluxes (which corresponds to minimum enzyme investment).

The formulation may be as follows:

$$\max J = [j_1, j_2, \ldots, j_k]^T \tag{1a}$$

Subject to:

$$\sum_{j=1}^{n} S_{ij} v_j = 0 \quad (i = 1, 2, \ldots, m) \tag{1b}$$

where j_i is the ith objective function, S_{ij} is the stoichiometric coefficient, and v_j is the flux. However, it is not easy to determine the optimal point on the Pareto optimal set in practice depending on the variety of culture conditions.

FBA approach together with MFA information may be considered for the metabolic engineering purpose such as OptKnock [28], a bi-level programming framework for identifying gene knockout for the strain improvement. This has been extended as OptReg to consider not only knockouts but also overexpression and downregulations of various reactions in the network [29]. Another extension has also been made as OptForce [30], OptFlux [31], and differential bees FBA (DFFBA) with OptKnock to identify the optimal gene knockout strategies for maximizing the yield of the desired phenotypes while sustaining the growth rate [32]. Further extension has been made as OptStrain aiming at guiding pathway modifications, through reaction additions and deletions, of microbial networks for the overproduction of targeted compounds based on stoichiometrically balanced approach imposing maximum product yield requirements, pinpointing the optimal substrates, and evaluating the different microbial hosts such as *Helicobacter pylori, E. coli, S. cerevisiae,* and other microorganisms [33].

Stoichiometry-based strain design algorithms are often formulated as bi-level mixed integer linear programming problems [28-30,34,35], where outer level optimizes the objective function(s), while the inner level optimizes the cellular system that counteract any externally imposed genetic or environmental perturbations [36,37]. Different fitness functions may be considered [38,39].

The linear property of stoichiometric equations underlying FBA is the computational advantage and allows for genome-scale extension. However, it is not easy to confirm the designed cell metabolism in view of enzymatic reactions with intracellular metabolite concentrations.

The problem in FBA and its extension to genome-scale is the difficulty for the dynamic analysis as compared to kinetic modeling approach. Some extension has been made by incorporating kinetic expressions of multiple carbon sources and other nutrients into the quasi steady-state [40-42]. The dynamic multi-species metabolic modeling (DMMM) approach has been considered by incorporating the metabolite uptake kinetics into stoichiometric models of a microbial consortium [43,44]. On the other hand, steady-state flux distributions obtained from FBA and stoichiometric information have been used to parameterize genome-scale kinetic models applicable for small perturbations [45-48]. Lin-log kinetic expression and thermodynamics may be incorporated to constrain FBA simulation [49].

Although some attempts have been made by the hybrid type of stoichiometric/kinetics-based modeling approach [45,50,51], its potential may not be fully investigated. The dynamic flux balance analysis (dFBA) may be considered for diauxic growth of *E. coli* consuming glucose and acetate by taking into account the constraints that govern the cell growth at different phases in the batch culture [52]. Moreover, dFBA may be used for the co-culture with multiple sugars for the cellulosic biofuels production [53-55]. Recently, OptForce formalism has been extended as k-OptForce by bridging the gap between stoichiometric approach and kinetics-based approach, where the procedure seamlessly integrates the mechanistic detail given by kinetic models within a constraint-optimization framework tractable for genome-scale models [56].

The proper formulation for the interaction between the metabolism and gene expression by applying the principle of growth optimization enables the accurate prediction of multi-scale phenotypes [25], where constitutively expressed genes show growth-rate-dependent expression trends [57,58]. This implies the economic ways of the cell system that is regulated in response to global change in metabolic efficiency [59]. Moreover, such optimality model may be used for the adaptive laboratory evolution [60].

The construction of a virtual microbe will be an ambitious but realistic target that builds a novel resource that can provide significant benefits in the variety of practical applications. As an extension of the constraint-based genome-scale models [61], a whole cell computational model was developed for *Mycoplasma genitalium,* a urogenital parasite adored by synthetic biologists for its reduced genome [62]. This model constitutes 28 processes of the cell's operation, where these include processes that track exchanges with the extracellular medium, all the metabolic fluxes, the state of the supercoiled chromosome, transcription of all active genes, processing of all mRNAs, translation of all proteins, formation of all macromolecular complexes including RNA polymerases and ribosomes, and progresses of cytokinesis and FtsZ polymerization [63]. This may be the dawn of virtual cell biology [64], and this might even go beyond the previous attempt of the so-called 'a grand challenge of the twenty-first century' [65].

Once again, although powerful and attractive for the possible extension to the whole cell modeling or the so-called virtual microbes, the main drawback of the above approach is the difficulty in incorporating 'explicitly' the metabolic regulation mechanism.

Kinetic modeling and incorporation of metabolic regulation

Kinetic models for the metabolism require quantitative expressions that connect fluxes and metabolite concentrations. Those can be obtained with respect to time by solving a set of ordinary differential equations (ODEs) such as:

$$\frac{dx_i}{dt} = \sum_{j=1}^{n_i} S_{ij} v_j \left(v_j{}^{\max}, x, p\right) \quad (\mathrm{i} = 1, \ldots, \mathrm{m}) \tag{2a}$$

where the typical mechanistic expression may be the Michaelis-Menten type such as:

$$v_j = \frac{v_j{}^{\max} S}{K_s + S} \tag{2b}$$

or Hill type expression may be considered, where $v_j^{\max}$ and K_s are the model parameters, and S is the substrate concentration for the corresponding pathway reaction. These expressions require detailed enzyme reaction mechanism and characterization [66,67]. In order to reduce the computational burden in association with kinetic modeling, various approximate kinetic forms such as lin-log [68-70] and log-lin [71] kinetics, power-law kinetic expressions such as S-system [72], generalized mass action [45], and others [73-75] have been proposed.

As mentioned in the previous section, although the stoichiometric constraint-based genome-scale metabolic models have been developed for a variety of organisms [76], it is not easy to incorporate or express the effects of intracellular metabolites and enzyme activities appropriately with such approach. Although some attempts have been made for incorporating transcriptional regulation into FBA framework in the form of Boolean rules [77-80], the regulatory rules are not based on the metabolic regulation mechanisms, but on the basis of the available data which may be the manifestation of part or snapshot of the real regulation mechanism [81].

In contrast to the stoichiometric models, the kinetic modeling approach is attractive in the sense that such mechanism can be incorporated into the model appropriately. The primary attempts of incorporating the regulation mechanisms into kinetic models have been made by cybernetic modeling approach, where the organisms are considered to utilize the available nutrient sources with the maximum efficiency by the optimal strategy [82]. This approach has been extended to more structural models that contain detailed pathways [83]. More recently, this approach has been considered for the potential applications to metabolic engineering [84,85]. In such modeling approach, an elementary mode was considered as a metabolic subunit to model cellular regulatory processes, where the elementary modes are a set of metabolic pathways by which the cellular metabolic routes can be completely described, and any feasible fluxes can be represented by their combinations at steady state [86]. The elementary modes consist of a minimal set of reactions that function at steady state, which implies that the elementary mode cannot be a functional unit if any reaction is removed [86]. The hybrid type modeling has also been developed by assuming quasi-steady-state for the intracellular metabolites [87,88], where several applications were made for *E. coli* [88] and for yeast [81].

In the kinetic modeling approach, it is critical to identify kinetic parameter values and kinetic rate laws applicable to a variety of genetic and/or environmental perturbations. Moreover, the large-scale extension may be limited by considering unambiguous kinetic model parameterization [89]. Several attempts have been made towards postulating a generalized uniform kinetic expression such as approximate enzyme kinetic equations [68,73,90-93], S-system formalism [94,95], or a combination of *in vitro*-based lumped and approximate rate equations [96,97]. However, the predictability may not be the satisfied level [68,75].

Recently, the ensemble modeling (EM) has been considered to cope with large-scale kinetic modeling by successively reducing the size of parameter space based on the available experimental data such as fluxes and/or intracellular metabolite concentrations together with thermodynamic constraints for the direction of the net fluxes [98]. In EM approach, any type of pathway reaction mechanism can be considered as well as already known mechanism, where each reaction is decomposed into elementary reaction steps with mass action kinetics [99] such as:

$$A + E \underset{k_{-1}}{\overset{k_1}{\Leftrightarrow}} AE \underset{k_{-2}}{\overset{k_2}{\Leftrightarrow}} BE \underset{k_{-3}}{\overset{k_3}{\Leftrightarrow}} B + E \tag{3}$$

where A and B are the metabolites and E is the enzyme.

The EM procedure starts with initially assumed kinetic models that predict the experimentally observed phenotypic characteristics, and the additional data such as those of the strain under environmental and/or genetic perturbations are used to screen the models until a minimal set of kinetic models are obtained [99]. This modeling approach has been successfully applied for lysine production [100], fatty acid production [101], aromatic production [98], robustness analysis for engineered non-native pathways [102], and modeling cancer cells [103]. Moreover, this approach has been applied for the modeling of *E. coli* that reasonably predicts the fluxes and intracellular metabolite concentrations of wild type and its single gene knockout mutants [104] based on the available multi-omics data [105].

Modeling of the main metabolism for catabolite regulation

The metabolic reactions of the central metabolism play important roles for energy generation and the production of the precursors for biosynthesis, and those form the hub on which all nearly catabolic and anabolic processes are built. Metabolic regulation of the central metabolism plays a key role in the adaptation of organisms to changes in their environment. The overall structure of central metabolic pathways is remarkably well conserved in the living organisms. Thus, the metabolic model of the central metabolism will provide a platform for further extension to peripheral metabolism and incorporation.

An attempt has been made for the modeling of the main metabolic pathways to simulate the dynamic behavior of *Saccharomyces cerevisiae* in response to the pulse addition of the carbon limited growth condition and measurement by fast sampling system [106,107]. The kinetic model equations for the glycolysis and pentose phosphate (PP) pathway have been developed for *E. coli* to simulate the transient data obtained by the fast sampling system [108]. The kinetic models for the tricarboxylic acid (TCA) cycle and anaplerotic pathways as well as glycolysis and PP pathway were also considered to simulate the typical batch and continuous cultures with some rule-based approach, where the cell growth rate was estimated based on the specific ATP production rate computed from the fluxes [109]. The kinetic modeling for the main metabolism of *E. coli* has also been made based on fluxomics and metabolomics data [110].

A wealth of information is available on genetic regulation, biochemistry, and physiology of cellular metabolism in response to culture environment, and some attempt has been made for the modeling and simulation, where it is important to make modeling based on the integrated information from gene level to flux level by incorporating the roles of transcription factors [5,111,112]. The important steps are how to incorporate (i) the effect of culture environment on global regulators, (ii) the effects of global regulators on the metabolic pathway genes, (iii) the effects of metabolic pathway genes on the corresponding enzyme activities, as well as (iv) the effects of enzyme level regulations (Figure 1).

Importance of the modeling for the main metabolic pathways

Although the modeling of the restricted metabolic pathways such as glycolysis only or glycolysis plus PP pathway, etc. may be useful depending on the purpose of using the model such as short-time transient responses against pulse addition of substrate, it is by far important to model the whole main metabolic pathways such as glycolysis, TCA cycle, PP pathway, together with anaplerotic and gluconeogenic pathways. This enables us to simulate the typical batch culture, where the metabolic transition occurs from glucose-rich (glycolysis) condition to acetate-rich (gluconeogenic) condition in *E. coli* and others.

In relation to the model development of the main metabolism, the accurate estimation of the cell growth rate is critical for the practical application point of view. In general, the cell growth rate may be expressed as a function of substrate such as Monod type model:

$$\frac{dX}{dt} = \mu(S)X \tag{4a}$$

with:

$$\mu(S) = \frac{\mu_m S}{K_s + S} \tag{4b}$$

where X and S are the cell and substrate concentrations, and μ is the specific cell growth rate. However, the saturation constant K_s is small, and the dynamics depend on the maximum specific growth rate parameter μ_m which is usually constant, resulting in the difficulty in estimating the reasonable cell growth rate by Monod type model and its modification. The importance of accurate estimation of the cell growth rate is more eminent for the dynamic simulation of the specific gene knockout mutants and for the effect of culture condition. Thus, it is desirable to be able to accurately predict the cell growth rate in general.

The cell growth rate is determined by the catabolic reactions such as ATP production as well as anabolic reactions under typical growth conditions. Once the main metabolic pathways could be appropriately modeled, the ATP production rate can be estimated. The model equations are established by the mass balance with kinetic equations for the main metabolic pathways (Figure 2). The solution to such ODEs can be used to compute the fluxes with respect to time. This enables us to compute the specific ATP production rate with respect to time such as:

$$\nu_{ATP} = OP_{NADH} + OP_{FADH_2} + \nu_{Pgk} + \nu_{Pyk} + \nu_{Ack} + \nu_{SCS} - \nu_{Pfk} - \nu_{Pck} - \nu_{Acs} \tag{5}$$

where ν is the reaction rate of the pathway (Figure 2). OP_{NADH} and OP_{FADH_2} are the specific ATP production

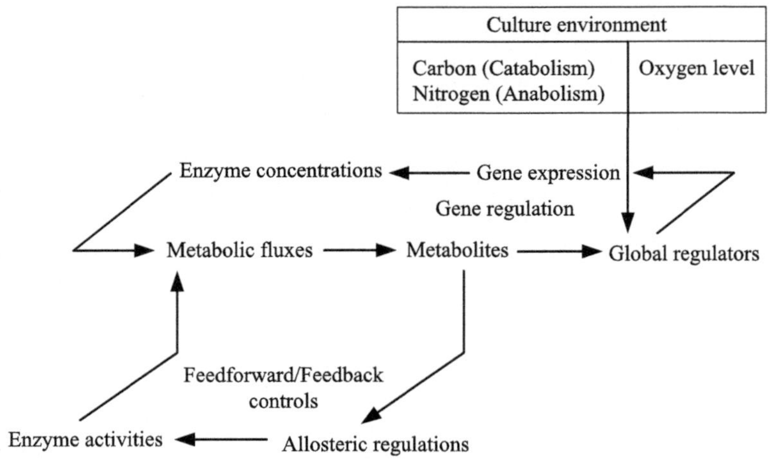

Figure 1 Overall metabolic regulation scheme.

rate via oxidative phosphorylation by NADH and $FADH_2$, respectively, and those may be expressed as:

$$OP_{NADH} = (v_{GAPDH} + v_{PDH} + v_{KGDH}(+v_{ICDH}) + v_{MDH}) \times (P/O) \quad (6a)$$

$$OP_{FADH_2} = v_{SDH} \times (P/O)' \quad (6b)$$

where (P/O) and $(P/O)'$ are the P/O ratios for NADH and $FADH_2$, respectively, and those are most likely to be 2.5 and 1.5, respectively, under typical aerobic condition. Those may be considered as model parameters and can be adjusted by the experimental data [109].

Now, ^{13}C-MFA shows the correlation between the specific ATP production rate and the specific cell growth rate [109,113-115]. This indicates that the above v_{ATP} can be used to estimate the specific growth rate, and in fact, it was shown that this approach allows us to estimate the cell growth rate and fluxes of the specific gene knockout mutant for *E. coli* to some extent [109,112].

In particular, in the case of anaerobic fermentation, NADH re-oxidation and substrate level phosphorylation for ATP generation are important, and ATP generation by acetate kinase (Ack) pathway is critical for survival in the case of using only xylose as a carbon source [116]. This may be simulated by the model with the cell growth rate taking into account the effect of ATP production rate as mentioned above.

Moreover, if the main metabolism was appropriately modeled, the specific CO_2 production rate can be also estimated by:

$$v_{CO_2} = v_{PGDH} + v_{PDH} + v_{ICDH} + v_{KGDH} + v_{Mez} + v_{Pck} - v_{Ppc} \quad (7)$$

where this can be used to estimate CO_2 evolution rate (CER), and thus the cell yield may be estimated together with other metabolite production rates and the cell growth rate. Since CER can be measured in practice, this may be also validated by the experimental data.

In relation to NADH production as mentioned above, the specific NADPH production rate can be also estimated as:

$$v_{NADPH} = v_{G6PDH} + v_{PGDH}(+v_{ICDH}) + v_{Mez} \quad (8)$$

where NADH is produced in many bacteria, while some bacteria such as *E. coli* produce NADPH at isocitrate dehydrogenase (ICDH). It has also been observed that the specific NADPH production rate is linearly correlated with the specific growth rate from the point of view of anabolism. This means that the flux from G6P to the oxidative PP pathway can be estimated as far as the oxidative PP pathway is dominant for NADPH production, once the specific growth rate was determined from the catabolic ATP production rate [109].

Metabolic regulation mechanisms to be incorporated in the kinetic model

As mentioned in the 'Kinetic modeling and incorporation of metabolic regulation' section, several efforts have been made for the appropriate kinetic models which can describe the metabolic regulation in response to genetic and/or environmental perturbations. Here, we consider the metabolic regulation mechanisms that have to be incorporated into the kinetic models, where the enzyme level regulation such as allosteric regulation may be incorporated into the kinetic rate expression, while the transcriptional regulation may be expressed as functions of transcription factors (TFs), where the activities of TFs may be considered to be functions of intracellular metabolites as will be mentioned next.

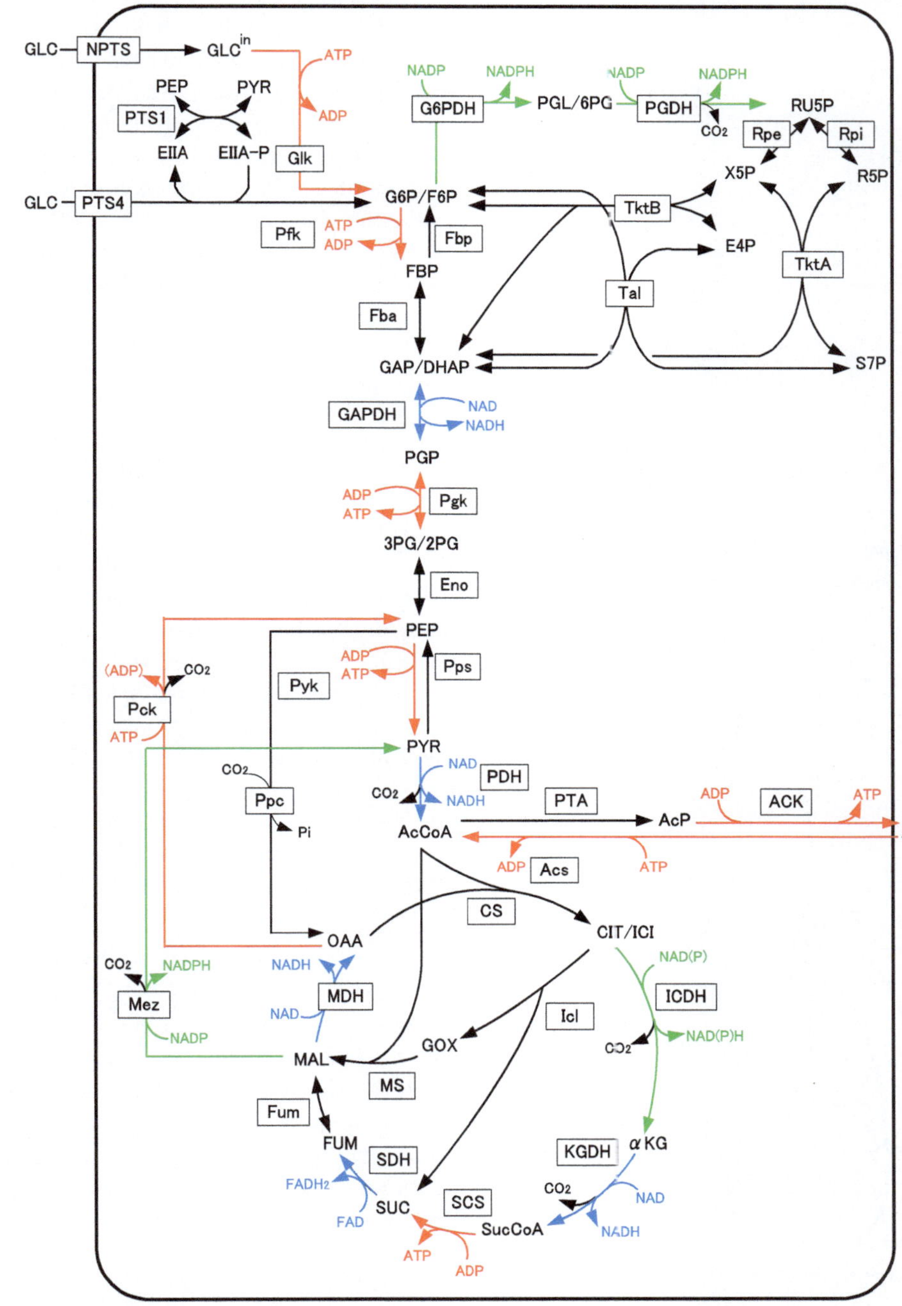

Figure 2 Main metabolic pathways.

The cell system achieves the coupling between recognition and adjustment through TFs, whose activities respond to the culture environment, and regulate the expression of the associated genes. This combined recognition and adjustment forms the reaction networks that overarch the metabolic and genetic layers [5]. In general, fast action is made by the enzyme level regulation such as the feed-forward activation of pyruvate kinase (Pyk) by fructose 1,6-bisphosphate (FBP), and the feedback inhibition of phosphofruct kinase (Pfk) by phosphoenol pyruvate (PEP), a motif that enables a high level of the upstream metabolite to lower the level of the downstream metabolite [117] (Figure 3a). The slow action is made through the transcriptional regulation, where cAMP-Crp activates the expression of TCA cycle genes, while (FBP-inhibited) catabolite repressor/activator, Cra activates the expression of gluconeogenic pathway genes as well as some of the TCA cycle genes and the glyoxylate

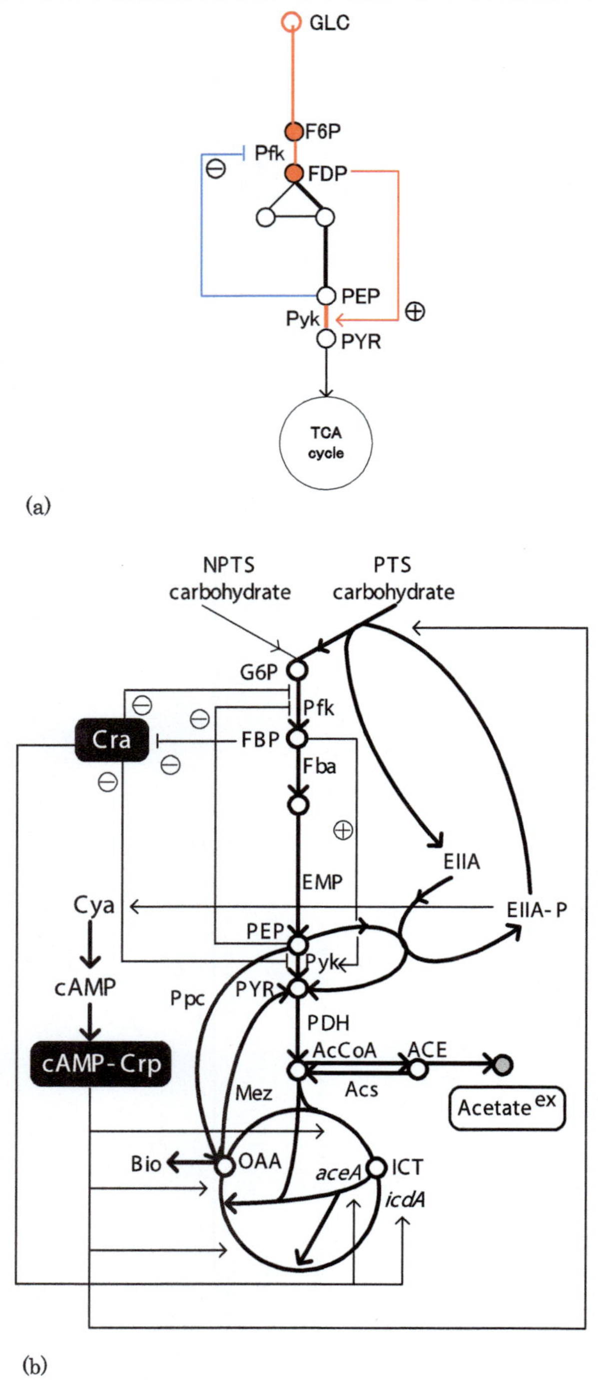

Figure 3 Metabolic regulation of the main metabolism. (a) Enzyme level regulation of glycolysis. **(b)** Transcriptional and enzyme level regulations of the main metabolic pathways.

pathway genes, and represses the expression of the glycolysis genes in the case of *E. coli* (Figure 3b), which will be explained later.

The levels of the flux-signaling metabolites become coupled, enabling a robust, coherent response of the TFs. The coherent behavior of the overall system is not established by a common transcriptional master regulator, but arises from the molecular interactions within the system itself [5]. It may be considered that the system of reactions of the lower glycolysis and the feed-forward activation of Pyk by FBP translate flux information into the concentration of FBP, and that this feed-forward activation affects the linearization of the glycolytic kinetics, where the glycolysis from FBP to PEP may be expressed as the reversible Michaelis-Menten (MM) equation, while Pyk may be expressed as the irreversible MM equation [118] such as:

$$v \rightarrow \text{FBP} \underset{k_{-1}}{\overset{k_1}{\Leftrightarrow}} \text{PEP} \xrightarrow{\text{Pyk}} \text{PYR} \tag{9}$$

where feed-forward activation of FBP on Pyk may be expressed by Monod-Wyman-Changeux (MWC) kinetics [118]. In fact, feed-forward regulation has been known to ensure the structural robustness against perturbations [117]. This mechanism may be conserved in many organisms. For example, in *S. cerevisiae*, sugar uptake rate is well correlated with the respiratory and fermentative pathways, or the specific ethanol production rate, and the similar relationship may be seen between the glycolytic flux and FBP [119,120], where Pyk is also feed-forward activation by FBP in *S. cerevisiae* [121].

In order to realize the above mechanism, the kinetic expression of Pyk (and also phosphoenolpyruvate carboxylase (Ppc)) must be a positive function of FBP such as:

$$v_{\text{Pyk}} = v_{\text{Pyk}}\Big([\text{PEP}], [\text{PYR}], [\text{FBP}], p_{\text{Pyk}}\Big) \tag{10}$$

where [$\cdot$] denotes the concentration, and PEP and pyruvate (PYR) are the substrate and product of Pyk reaction, respectively. FBP is the allosteric activator, and p_{Pyk} is the kinetic parameter vector for Pyk. The feed-forward activation of Pyk by FBP may be enhanced by the feedback inhibition of Pfk by PEP, where the kinetic expression for Pfk must be a negative function with respect to PEP:

$$v_{\text{Pfk}} = v_{\text{PFk}}([\text{F6P}], [\text{FBP}], [\text{PEP}], p_{\text{Pfk}}) \tag{11}$$

where F6P and FBP are the substrate and product of Pfk reaction, respectively, while PEP is the allosteric inhibitor, and p_{Pfk} is the parameter vector for Pfk reaction, namely, if the sugar uptake rate or the upper glycolytic flux were increased, FBP increases, and in turn allosterically activates Pyk, which decreases PEP concentration, and the feedback inhibition of Pfk by PEP is relaxed and causes further increase in the upper glycolytic flux (Figure 3a). In the nominal growth condition, the feedback inhibition of Pfk by PEP may not be important, but this may cause oscillatory behavior under certain condition [122].

Moreover, the increased pyruvate goes down through pyruvate dehydrogenase (PDH) to acetyl-coenzyme A (AcCoA), where AcCoA becomes homeostatic, namely, the increase in AcCoA activates Ppc reaction [123,124], thus reducing the upcoming Pyk-PDH fluxes, increases oxaloacetate (OAA), and activates citrate synthase (CS) reaction, and therefore activating the outgoing flux from AcCoA. In this way, AcCoA concentration decreases, forming the feed-back regulation against the initial increase in AcCoA (Figure 4). This mechanism can be realized by expressing the Ppc activity as a positive function of AcCoA as well as FBP such as:

$$v_{\mathrm{Ppc}} = v_{\mathrm{Ppc}}\left([\mathrm{PEP}], [\mathrm{OAA}], [\mathrm{FBP}], [\mathrm{AcCoA}], p_{\mathrm{Ppc}}\right) \tag{12}$$

where PEP and OAA are the substrate and product of Ppc reaction, respectively, and FBP and AcCoA are the allosteric activators. In another view, this phenomenon may be considered as the feed-forward regulation in the sense that the repression of TCA cycle activity is detected by the increase in AcCoA, which causes the activation of the anaplerotic Ppc pathway, and backs up the precursor metabolite such as OAA since it is expected to be decreased due to deactivated TCA cycle.

As the glucose uptake rate increases, the TCA cycle flux tends to increase by the increased OAA and AcCoA, and then NADH is overproduced. The accumulated NADH inhibits CS and ICDH allosterically [125], forming feedback regulation, and thus results in AcCoA accumulation, which in turn causes acetate overflow metabolism. This enzyme level regulation by NADH in the TCA cycle can be verified by incorporating NADH oxidase (NOX) [126] or nicotinic acid [125], whereby activating TCA cycle. This effect is more enhanced under *arcA* mutant [127]. In the long run, the expression of TCA cycle genes is eventually repressed by the transcriptional regulation by cAMP-Crp toward steady state as will be explained later. The inhibitory effect of NADH on CS and ICDH may be expressed explicitly in the rate equation, but the problem is that the estimation of $NADH/NAD^+$ pool is difficult without detailed proper modeling of the respiratory chain, which is not easy at this stage.

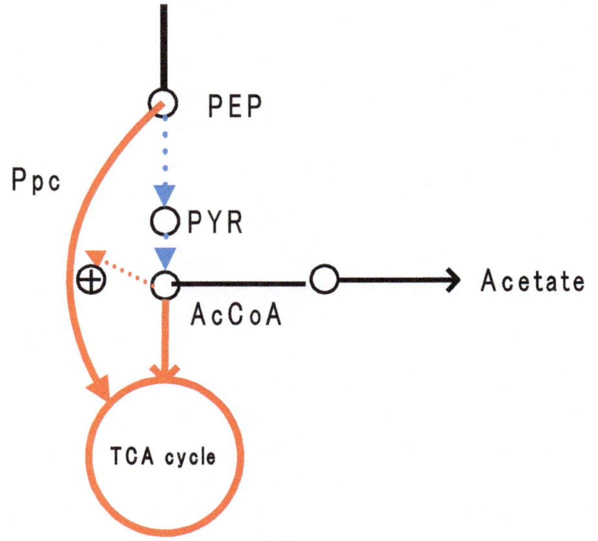

Figure 4 Homeostasis of AcCoA by the activation of Ppc.

The typical growth condition changes from glucose-rich to acetate-rich in the batch culture. This requires a significant reorganization of the central metabolism from glycolysis to gluconeogenesis. Although the molecular mechanism underlying the metabolic transition from glucose to acetate has been extensively investigated in *E. coli* [128], its dynamics have been poorly understood. Since it is critical for the cell to efficiently and quickly reprogram the metabolism under the changing environmental condition, the cell must have the elaborate managing system.

The expression of the reaction rate for Ppc is the function of both FBP and AcCoA as mentioned above, which then enables us to simulate the ultrasensitive regulation of anaplerosis [129], namely, after glucose depletion, FBP concentration decreases accordingly, where Ppc and Pyk activities decrease in turn by the allosteric regulation, and PEP consumption is almost completely turned off. These make PEP concentration to be increased, and this buildup of PEP is kept during certain period [114], and this may serve to quickly uptake the glucose by PTS if it becomes available again [129]. This mechanism is important for the fed-batch culture compensated by DO-stat or pH-stat, where carbon limitation often occurs periodically, and the uptake of carbon source can be made quickly and efficiently without delay. Such phenomenon can be simulated by the model as mentioned above as compared to the case without feed-forward regulation mechanism. This feed-forward regulation mechanism is also important for the modeling and simulation of lactic acid bacteria, where lactate dehydrogenase (LDH) as well as Pyk is also activated by FBP, thus producing lactate quickly and lowering the pH around the cell as soon as the glucose is available [130]. Although the kinetic model for lactic acid bacteria has been developed previously [131], the above mechanism is not incorporated, and thus the simulation result does not properly reflect the real characteristics.

Moreover, after glucose depletion, FBP level drops, and thus Ppc activity decreases, while PEP carboxykinase (Pck) activity is activated by the activated Cra caused by the decreased FBP. This reveals the mechanism of avoiding the futile cycling caused by Ppc and Pck during gluconeogenic phase [129], where ATP generation becomes important. During the active glycolysis with enough sugars available, this futile cycling occurs, and loses ATP without efficient use for the compensation of the flexible metabolic fluxes and the metabolic regulation [123]. This may be simulated by the appropriate models taking into account both enzymatic and transcriptional regulations.

Now, the enzyme level regulation in the glycolysis made by Pyk and Pfk as well as FBP and PEP as mentioned above keeps increasing the substrate uptake rate, where this makes the system unstable. This is counterbalanced by the transcriptional regulation by cAMP-Crp, where cAMP level decreases due to the lower level of phosphorylated EIIA (EIIA-P), and lower activity of adenylate cyclase (Cya) at higher glucose consumption rate. Since *ptsG* which encodes EIIBC of PTS is under control of cAMP-Crp, the glucose uptake rate is repressed by the lower level of cAMP-Crp (Figure 3b). Thus, the transcriptional repression of PTS by cAMP-Crp must be incorporated into the model to realize such feedback regulation for the glucose uptake rate. The molecular mechanism for catabolite regulation has been illustrated by several researchers [132-136], and the PTS and catabolic regulation have been modeled by several researchers [5,109,111,137].

Moreover, the TCA cycle is transcriptionally repressed by the lower level of cAMP-Crp as the glucose consumption rate was increased. In the continuous culture (Chemostat), the effect of the cell growth rate or the dilution rate on the metabolism can be appropriately simulated for *E. coli* [112]. Namely, as the specific growth rate was increased, FBP increases due to the increase in the fluxes of the upper glycolysis and Cra decreases. Moreover, PEP concentration decreases as the PTS flux increases, since PEP is the co-substrate of PTS, and in turn EIIA-P decreases, resulting in the decrease in Cya activity and cAMP-Crp level. The decrease in cAMP-Crp causes the repression of the TCA cycle, and acetate overflow metabolism occurs as the specific growth rate increases. The trend of the simulation result by reflecting the above mechanism [112] is the similar as the experimental data [105] (Figure 5).

In *E. coli*, acetate is formed from AcCoA by Pta-Ack and from pyruvate by pyruvate oxidase, Pox [128]. Acetate can be metabolized to AcCoA either by the reversed reactions of Pta-Ack or by acetyl-coenzyme A synthetase (Acs). Acetate formation has been known to be due to metabolic imbalance, where the rate of AcCoA formation via glycolysis surpasses the capacity of the TCA cycle in *E. coli* [138]. Pox and Acs may be expressed as functions of the sigma factor (σ^{38}) RpoS, but it may be difficult to predict the behavior of RpoS, while Acs may be expressed as a function of cAMP-Crp, where Acs is activated by cAMP-Crp during gluconeogenic phase [112].

Among intracellular metabolites, α-keto acid such as αKG turns to be a master regulator for catabolite regulation and co-ordination of different regulations [139]. Namely, when favored carbon source such as glucose was depleted, αKG level fall, and cAMP increases to stimulate other carbon catabolic machinery. Namely, when preferred carbon source such as glucose is abundant, the cell growth rate becomes higher with lower cAMP level, while if it is scarce, the cell growth rate declines with higher cAMP level. In particular, under nitrogen (N)-limitation, αKG accumulates due to decreased activity of glutamate dehydrogenase (GDH) and inhibits carbon assimilation, where there is less need for carbon-catabolic enzymes, and more demand for those involved in such nutrient assimilation. On the other hand, when anabolic nutrient such as ammonia is in excess, αKG concentration decreases due to activated GDH, producing glutamate (Glu) from αKG, cAMP level increases, and carbon catabolic enzymes increases to accelerate carbon assimilation. Namely, αKG coordinates the catabolic (C)-regulation and N-regulation by inhibiting EI of PTS [140] or cAMP via Cya [58,141]. Moreover, the physiological function of cAMP signaling goes beyond simply enabling hierarchical utilization of carbon sources as will be mentioned later but also controls the function of the proteome [139,142]. In order to model such phenomenon, EI of PTS has to be expressed as the inhibition by αKG, or Cya has to be expressed as the inhibition by keto acids such as OAA and PYR as well as αKG, where the modeling for nitrogen regulation will be mentioned later.

In the case of biofuels production from cellulosic biomass, the hydrolyzed biomass contains multiple sugars, and those sugars are selectively assimilated with catabolite repression depending on the type of microorganism used [143,144]. The metabolic regulation differs depending on the carbon sources used.

Glycerol has been paid recent attention for the production of biofuels and biochemicals, since it is a by-product of the biodiesel production [145-149]. In *E. coli*, glycerol is transported and phosphorylated to produce dihydroxy acetone phosphate (DHAP) of the central metabolism via the pathway encoded by *glpF, glpK*, and *glpD*, where ATP (or in certain cases PEP) is used for the phosphorylation at glycerol kinase (GlpK) reaction, while NADH is produced at glycerol 3-phosphate dehydrogenase (G3PDH) reaction under aerobic condition (Figure 6). These genes are under catabolic regulation by cAMP-Crp, so that glycerol is assimilated after glucose was depleted if glucose co-exists. NADH production at G3PDH becomes important for the biofuels production under anaerobic condition affecting $NADH/NAD^+$ balance for dehydrogenase reactions. In the case of using glycerol as a single carbon source, cAMP-Crp increases

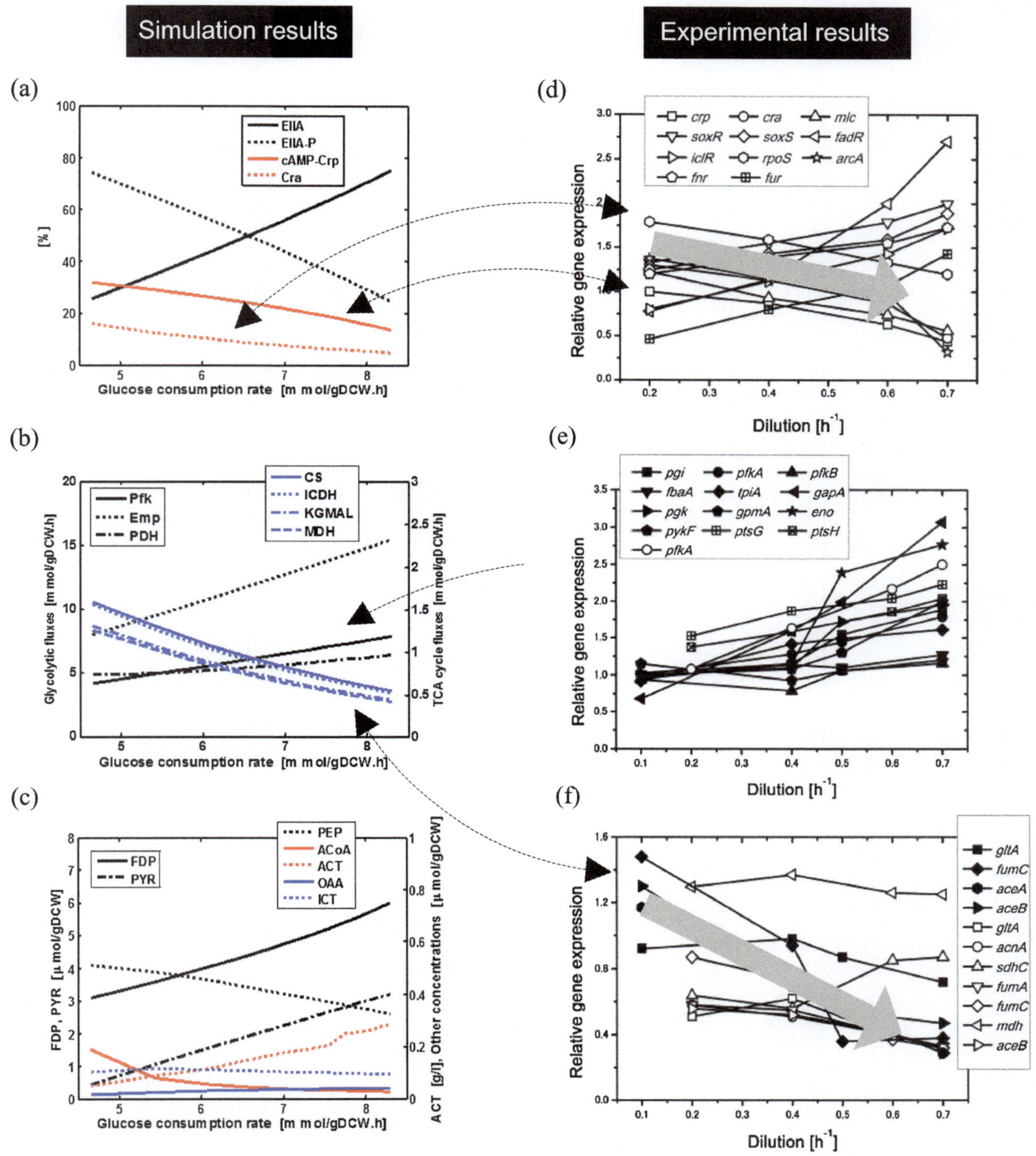

Figure 5 Comparison of the simulation results (a-c) [112] and the experimental data (d-f) [105,113].

due to the increase in the phosphorylated $EIIA^{Glc}$, where cAMP-Crp induces *glpFKD* genes. Since FBP concentration decreases in the case of using glycerol as a carbon source, Cra is activated, and this together with upregulation of cAMP-Crp causes *pckA* gene as well as TCA cycle genes to be upregulated [150,151]. The kinetic expressions for the glycerol uptake pathways have been proposed [152]. This together with the inclusion of PTS and the transcriptional regulation as mentioned above enables the simulation for the case of using multiple carbon sources such as glucose and glycerol. Moreover, the enhancement of the TCA cycle caused by the increase in cAMP-Crp can be also simulated in the case of using glycerol as a single carbon source.

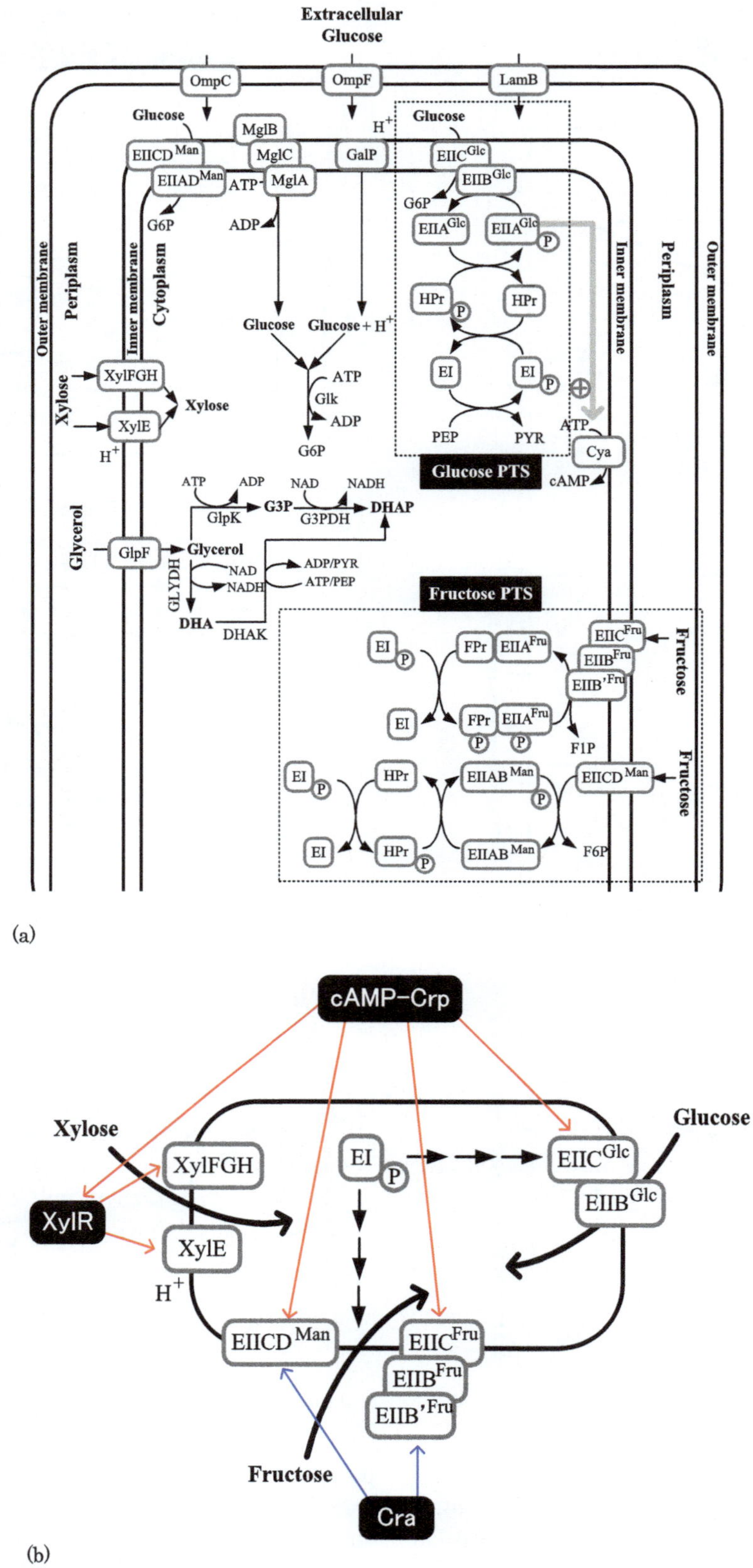

Figure 6 Glycerol-, fructose-, and xylose-assimilating pathways (a,b).

In the case of using fructose, it is transported by fructose-PTS, which has its own HPr-like protein domain called FPr. Namely, the phosphate of PEP is first transferred to EI (as EI-P), but then this phosphate is transferred to FPr instead of HPr, and in turn the phosphate is transferred via fructose specific $EIIA^{Fru}$ and $EIIBC^{Fru}$, and phosphorylates fructose, where phosphorylated fructose becomes fructose 1-phosphate (F1P) [153]. The *fruBKA* operon is under control of cAMP-Crp, and thus glucose is preferentially consumed by glucose PTS when glucose co-exists. On the other hand, this operon is repressed by Cra [154]. Because of this, *cra* gene knockout enables co-consumption of glucose and fructose with fructose to be consumed faster as compared to glucose [155], where activated FruB in *cra* mutant competes with HPr (for glucose phosphorylation) for the phosphate of EI-P. Since phosphorylation of $EIIA^{Glc}$ via HPr becomes lower [156], the glucose uptake rate decreases as compared to the wild-type strain [155]. This phenomenon may be also simulated by the similar expression as glucose-PTS but compete the phosphate of EI with glucose-PTS (Figure 6).

In the case of using xylose as a carbon source, it is transported either by an ATP-dependent high affinity ABC transporter encoded by *xylFGH* or ATP-independent low affinity proton symporter encoded by *xylE* [157,158]. In the case of xylose utilization, the transcription factor XylR regulates *xylAB*/*xylFGH* [159], where *xylR* is under control of cAMP-Crp, and then catabolite repression occurs when glucose co-exists, where glucose is preferentially consumed first. The kinetic model for xylose uptake pathways as well as Entner-Doudoroff (ED) pathways has been proposed for *Zymomonas mobilis* [160], and thus it is necessary to incorporate the activation of the transporter by cAMP-Crp, and this can be made for the catabolite repression when co-exist with glucose [112] (Figure 6).

Modeling for the peripheral metabolism

Although it is critical to consider the main metabolism for the metabolic regulation as well as for the cell growth rate, the peripheral metabolic pathways become important for the practical applications such as amino acids fermentation.

The kinetic model for lysine synthetic pathways from OAA in the TCA cycle has been proposed [161], and this can be used to apply sensitivity analysis such as metabolic control analysis (MCA) to identify the limiting pathways in *Corynebacterium glutamicum* [162]. This investigation revealed that lysine production is primarily controlled by aspartokinase (Ask) and lysine permease. This was verified by the experiment using the recombinant strain overexpressing Ask, resulting in the significant increase in lysine production, although that flux did not increase as much as would be expected by MCA [162].

Shikimic acid production and aromatic amino acids production may be also simulated based on the formation of the precursor metabolites such as erythrose 4-phosphate (E4P) and PEP in the central metabolism [111]. Other amino acid fermentation may be also simulated using dynamic metabolic models [111,137,163].

Modeling for the metabolism under oxygen limitation

Most of the biofuels and biochemical productions by microbes is made by the fermentation under anaerobic condition, and thus it is important to properly model such fermentation as well as under aerobic condition, where the latter is often employed for the enhancement of the cell growth rate before anaerobic condition to improve the productivity of the target metabolites.

Although the modeling and computer simulation of a microbial cell cultivated under anaerobic condition such as lactate fermentation [131], and acetone-butanol-ethanol fermentation [164-166] has been proposed by several researchers, the regulatory mechanisms are rarely incorporated. Moreover, cofactor balance such as NADH balance becomes important under anaerobic condition, and thus it may be better to appropriately incorporate in the model equations. However, this is not so easy without proper modeling of the respiratory pathways under different oxygen concentration.

In order to properly model the metabolic transition from aerobic to anaerobic conditions, the roles of global regulators such as ArcA/B and Fnr must be properly incorporated, where the effect of dissolved oxygen concentration on the activation of such TFs has been reported [167,168], and this may be taken into account for the simulation under microaerobic conditions. In particular, it is important to properly simulate the behavior at the branch point of PYR, where the reaction rate through PDH, v_{PDH} must be the negative function of ArcA (or phosphorylated ArcA, ArcA-P), while the reaction rate through pyruvate formate lyase (Pfl), v_{Pfl} is the positive function with respect to ArcA and Fnr, where PYR is converted to formate (FOR) and AcCoA (Figure 7). Moreover, ethanol-forming pathway from AcCoA, alcohol dehydrogenase (ADH) must be activated where NADH is required for this reaction. As for TCA cycle, α-ketoglutarate dehydrogenase (KGDH) may be repressed by ArcA, while formate reductase (Frd) is activated by Fnr, thus the TCA cycle will be branched under anaerobic conditions. Some attempt has been made to estimate the fluxes in relation to such global regulators [169].

Modeling for the nitrogen regulation

Next to carbon (C) catabolite regulation, the nitrogen (N) regulation is also important [170], and the silicon-cell models have been developed based on experimental kinetic data for the enzymes, involved that predict the flux of assimilation of extracellular ammonia into glutamate in *E. coli*.

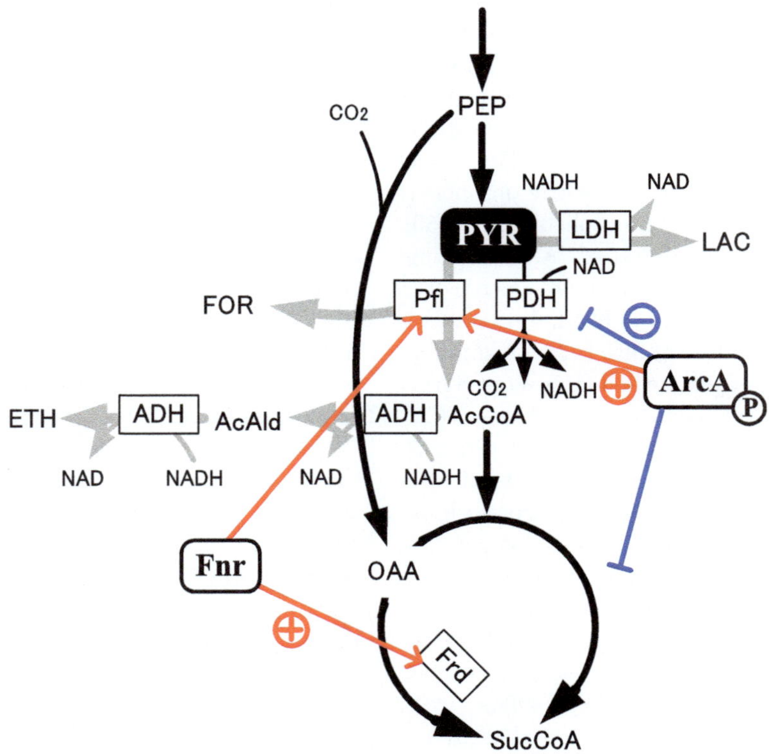

Figure 7 Anoxic regulation of the metabolic pathways by ArcA and Fnr.

Glutamate (Glu) and glutamine play key roles in cellular metabolism and serve as precursors of protein synthesis. Glutamate can be synthesized from two different pathways such as by one simple step reaction catalyzed by glutamate dehydrogenase (GDH) from αKG, and by glutamine synthetase (GS) and glutamate synthase (GOGAT) [171]. GS is active during low ammonium concentration while GDH is active at higher ammonium concentrations, where GS has a higher affinity than GDH for ammonia ($K_m = 0.1$ and 1.1 mM, respectively) [172,173]. The activity of GS was controlled by P_{II} which acts in response to the concentration of glutamine and αKG [174,175]. AmtB is used for NH_3 for transport when the ammonia concentration is lower than 0.05 μM [176], while the AmtB will be blocked when it was higher than 50 μM.

The activation (adenylylation) and inactivation (deadenylylation) of GS depends on C/N ratio such as αKG (C)-to-glutamine (N) ratio. In the case where C/N ratio was higher, GS will be adenylylated, otherwise, GS is deadenylylated. Sensing of the C/N ratio for GS adenylylation involves the protein P_{II}, which is in two forms such as urydylylated PI-UMP and deurydylylated P_{II}. Urydylylation/deurydylylation of P_{II} catalyzed by UT/UR enzymes are promoted by glutamine (Figure 8).

The model developed by Bruggeman et al. [173] combined metabolic regulation with signal transduction through the covalent modification of P_{II} and GS by urydylyl transferase (UTase) and adenylyl transferase (ATase). It shows that the regulation is distributed between the two modes of regulation. However, the model may be incomplete in the sense that αKG pool size was assumed to be constant, while it changes significantly during nitrogen perturbations, where it is not only the substrate for ammonia assimilation but also a regulator of the GS covalent modification cascade [177]. Moreover, it is important to capture the interdependence of metabolite pools and growth, where metabolite pool size of αKG affects the glucose uptake by inhibiting EI of PTS [140].

In order to see the effect of ammonium assimilation, the main metabolic pathway must be considered. This model involves GDH, GS, and GOGAT pathways together with nitrogen regulation mechanism. At present, several kinetic models have been proposed for ammonium assimilation [173,176,178], but little has been analyzed for the relationship between cell growth rate and NADPH production rate in relation to ammonium assimilation. Moreover, it is strongly desirable to combine the models for catabolite regulation and nitrogen regulation in order to simulate the coordinated regulation between C- and N-regulations via the dynamic behavior of intracellular metabolite αKG.

Conclusions

Completeness of the model may not be necessary for it to improve predictions or rationalizations. The uncertainty

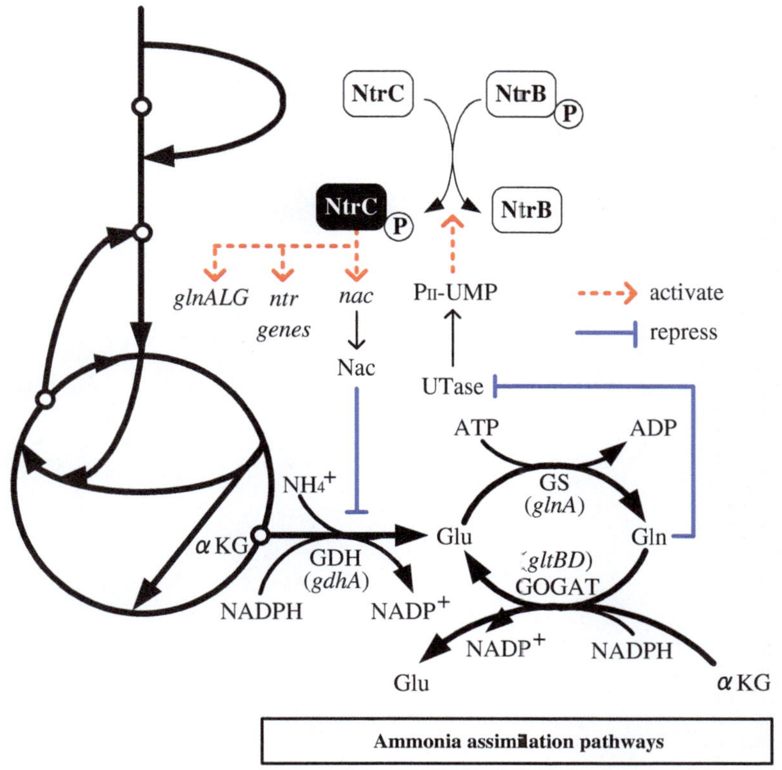

Figure 8 Ammonia assimilation pathways.

of the model or the simulation result comes either from uncertainty of the model parameters, uncertainty of the model structure due to ambiguity in the selection of rate laws, or uncertainty caused by neglecting the regulation mechanism or neglecting cofactor balances etc. The importance of the above factors depends on the strains and culture conditions. It is useful in understanding cellular mechanism, and the process towards the development of the whole cell metabolism could well pay off. The appropriate model will be of immense value to the following:

- Biotechnologists aiming to improve fermentation performances such as the yield and productivity of the target metabolite,
- Microbial engineers aiming to design novel microbes able to capture available carbon and produce bio-fuels and biochemicals,
- Basic scientists aiming to understand the metabolic regulation system in the living organisms, which can be used for the synthetic biology, and
- Systems biologists aiming to advance the science of modeling.

The modeling approach will greatly exceed the importance of the microbial genome sequencing projects, as it will be much closer to understanding biological function and will have widespread practical application.

In the present article, it is stressed the importance of incorporating the enzyme level and transcriptional regulations appropriately in the kinetic model to predict the cell's growth characteristics under environmental and/or genetic perturbations. The drawback of the kinetic modeling is the increase in the kinetic model parameters as the system becomes large, and thus it may be difficult to expand to genome-scale. The reasonable idea may be to consider the kinetic modeling only for the main metabolism, and the simplified model may be considered for the peripheral metabolisms.

Moreover, it is quite important to combine the catabolic regulation model with nitrogen regulation model for the coordination between C- and N-regulations, where the intracellular pool sizes of α-keto acids play important roles affecting PTS and cAMP level.

The simulation result based on the model developed must be verified by experiments, or the simulation result may give hint for additional experimental design. In this way, modeling approach together with experimental works contributes to the innovation for the efficient design of the cell factories for biofuels and biochemical production.

Abbreviations

6PG: 6-phosphogluconate; FBA: Flux balance analysis; MFA: Metabolic flux analysis; ED pathway: Entner-Doudoroff pathway; TCA cycle: Tricarboxylic acid cycle; PTS: Phosphotransferase system; Crp: cAMP receptor protein; Cya: Adenylate cyclase; EI: Enzyme I; EII: Enzyme II; $EIIA^{Glc}$-P: Phosphorylated $EIIA^{Glc}$; HPr: Histidine-phosphorylatable protein; Cra: Catabolite repressor/activator; Arc system: Anoxic respiration control system; Fnr: Fumarate and nitrate reduction; Ack: Acetate kinase; Acs: Acetyl coenzyme A synthetase; ADH: Alcohol dehydrogenase; Ask: Aspartokinase; CS: Citrate synthase; Eno: Enolase; Fbp: Fructose bisphosphatase; Frd: Fumarate reductase; Fum: Fumarase; GAPDH: Glyceraldehyde-3-phosphate dehydrogenase; GDH: Glutamate dehydrogenase; Glk: Glucokinase; GlpK: Glycerol kinase; GOGAT: Glutamate synthase; G3PDH: Glycerol 3-phophate dehydrogenase; G6PDH: Glucose-6-phosphate dehydrogenase; GS: Glutamine synthetase; ICDH: Isocitrate dehydrogenase; Icl: Isocitrate lyase; KGDH: Ketoglutarate dehydrogenase; LDH: Lactate dehydrogenase; MDH: Malate dehydrogenase; Mez: Malic enzyme; MS: Malate synthase; Pck: Phosphoenolpyruvate carboxykinase; PDHc: Pyruvate dehydrogenase complex; Pfk: Phosphofructokinase; Pfl: Pyruvate formate-lyase; 6PGDH: 6-phosphogluconate dehydrogenase; Pgi: Phosphoglucose isomerase; Pgk: phosphoglycerate kinase; Pox: Pyruvate oxidase; Ppc: Phosphoenolpyruvate carboxylase; Pps: Phosphoenolpyruvate synthase; Pta: Phosphotransacetylase; Pyk: Pyruvate kinase; Rpe: Ribulose phosphate epimerase; Rpi: Ribose phosphate isomerase; SCS: Succinyl-CoA synthetase; SDH: Succinate dehydrogenase; Tal: Transaldolase; TktA: Transketolase I; TktB: Transketolase II; AcCoA: Acetyl-coenzyme A; ACE: Acetate; AcP: Acetyl phosphate; CIT: Citrate; DHAP: Dihydroxy acetone phosphate; E4P: Erythrose-4-phosphate; ETH: Ethanol; F1P: Fructose-1-phosphate; F6P: Fructose-6-phosphate; FBP: Fructose-1,6-bisphosphate; FOR: Formate; G6P: Glucose-6-phosphate; GAP: Glycelaldehyde-3-phosphate; GLC: Glucose; GLU: Glutamate; GOX: Glyoxylate; ICT: Isocitrate; LAC: Lactate; MAL: Malate; OAA: Oxaloacetate; PEP: Phosphoenol pyruvate; PYR: Pyruvate; R5P: Ribose-5-phosphate; RU5P: Ribulose-5-phosphate; S7P: Sedoheptulose-7-phosphate; SUC: Succinate; X5P: Xylulose-5-phosphate; αKG: α-ketoglutarate.

Competing interests

The authors declare that they have no competing interests.

Authors' contributions

YM and KS both contributed for the preparation of the manuscript. Both authors read and approved the final manuscript.

Acknowledgements

The discussion on the virtual microbe is also made with Prof. H. Westerhoff of the University of Manchester, Prof. J. McFadden of the Surrey University in UK, and the UK-Japan systems biology project funded by JST (Japan) and BBSRC (UK).

References

1. Shimizu K (2014) Microbial production of biofuels and biochemicals from biomass. NOVA publ, Co, New York
2. Kitano H (2002) Systems biology: a brief overview. Science 295:1662–1664
3. Kitano H (2002) Computational systems biology. Nature 420:206–210
4. Stelling J (2004) Mathematical models in microbial systems biology. Curr Opin Microbiol 7:513–518
5. Kotte O, Zaugg JB, Heinemann M (2010) Bacterial adaptation through distributed sensing of metabolic fluxes. Mol Sys Biol 6:355
6. Vemuri GN, Aristidou A (2005) Metabolic engineering in the -omics era: elucidating and modulating regulatory networks. Microbiol Mol Biol Rev 69:197–216
7. Shimizu K (2014) Regulation systems of bacteria such as *Escherichia coli* in response to nutrient limitation and environmental stresses. Metabolites 4:1–35
8. Matsuoka Y, Shimizu K (2011) Metabolic regulation in *Escherichia coli* in response to culture environments via global regulators. Biotechnol J 6:1330–1341
9. Chuvukov V, Gerosa L, Kochanowski K, Sauer U (2014) Coordination of microbial metabolism. Nat Rev 12:327–340
10. Selinger DW, Wright MA, Church GM (2003) On the complete determination of biological systems. Trends Biotechnol 21:251–254
11. Machado D, Costa R, Rocha M, Ferreira E, Tidor B, Rocha I (2011) Modeling formalisms in systems biology. AMP Expre 1:1–34
12. Almquist J, Cvijovic M, Hatzimanikatis V, Nielsen J, Jirstrand M (2014) Kinetic models in industrial biotechnology-improving cell factory performance. Metabolic Eng 24:38–60
13. Costa RS, Machado D, Rocha I, Pereira EC (2011) Critical perspective on the consequences of the limited availability of kinetic data in metabolic dynamic modeling. IET Syst Biol 5:157–163
14. Ashyraliyev M, Fomekong-Nanfack Y, Kaandorp JA, Blom JG (2009) Systems biology: parameter estimation for biochemical models. FEBS J 276:886–902
15. Cvijovic M, Bordel S, Nielsen J (2011) Mathematical models of cell factories: moving towards the core of industrial biotechnology. Microb Biotechnol 4:572–584
16. Sauer U (2006) Metabolic networks in motion: ^{13}C-based flux analysis. Mol Syst Anal 2:62
17. Long CP, Antoniewicz MR (2014) Metabolic flux analysis of *Escherichia coli* knockouts: lessons from the Keio collection and future outlook. Curr Opin Biotechnol 28:127–133
18. Quek LE, Nielsen LK (2014) Steady-State 13C Fluxomics Using OpenFLUX. In: Krömer JO, Nielsen LK, Blank LM (eds) Metabolic flux analysis: methods and protocols, vol. 1191, Springer, New York, 209-224
19. Shimizu K (2004) Metabolic flux analysis based on ^{13}C-labeling experiments and integration of the information with gene and protein expression patterns. Adv Biochem Eng Biotechnol 91:1–49
20. Shimizu K (2013) Bacterial cellular metabolic systems. Woodhead Publ Ltd., Oxford
21. Matsuoka Y, Shimizu K (2014) ^{13}C-Metabolic flux analysis for *Escherichia coli*. In: Krömer JO, Nielsen LK, Blank LM (eds) Metabolic flux analysis: methods and protocols, vol. 1191, Springer, New York, 261-289
22. Shimizu K (2009) Toward systematic metabolic engineering based on the analysis of metabolic regulation by the integration of different levels of information. Biochem Eng J 46:235–251
23. Wittman C (2007) Fluxome analysis using GC-MS. Microb Cell Fact 6:6
24. Herrgard MJ, Lee B-S, Portnoy V, Palsson BO (2006) Integrated analysis of regulatory and metabolic networks reveals novel regulatory mechanisms in *Saccharomyces cerevisiae*. Genome Res 16:627–635
25. O'Brien EJ, Lerman JA, Chang RL, Hyduke DR, Palsson BO (2013) Genome-scale models of metabolism and gene expression extend and refine growth phenotype prediction. Mol Sys Biol 9:693
26. Schuetz R, Kuepfer SU (2007) Systematic evaluation of objective functions forpredicting intracellular fluxes in *Escherichia coli*. Mol Syst Biol 3:119
27. Schuetz R, Zamboni N, Zampieri M, Heinemann M, Sauer U (2012) Multidimensional optimality of microbial metabolism. Science 336:601–604
28. Burgard AP, Pharkya P, Maranas CD (2003) Optknock: a bilevel programming framework for identifying gene knockout strategies for microbial strain optimization. Biotechnol Bioeng 84:647–657
29. Pharkya P, Maranas CD (2006) An optimization framework for identifying reaction activation/inhibition or elimination candidates for overproduction in microbial systems. Metab Eng 8:1–13
30. Ranganathan S, Suthers PF, Maranas CD (2010) OptForce: an optimization procedure for identifying all genetic manipulations leading to targeted overproductions. Plos Comput Biol 6:e1000744
31. Rocha I, Maia P, Evangelista P, Vilaca P, Soares S, Pinto JP, Nielsen J, Patil KR, Ferreira EC, Rocha M (2010) OptFlux: an open-source software platform for *in silico* metabolic engineering. BMC Syst Biol 4:45
32. Choon YW, Mohamad MS, Deris S, Illias RM, Chong CK, Chai LE, Omatu S, Corchado JM (2014) Differential bees flux balance analysis with OptKnock for *in silico* microbial strains optimization. PLoS One 9:e102744
33. Pharkya P, Burgard AP, Maranas CD (2014) OptStrain: a computational framework for redesign of microbial production systems. Genom Res 14:2367–2376
34. Yang L, Cluett WR, Mahadevan R (2011) EMILiO: a fast algorithm for genome-scale strain design. Metab Eng 13:272–281
35. Cotten C, Reed JL (2013) Constraint-based strain design using continuous modifications (CosMos) of flux bounds finds new strategies for metabolic engineering. Biotechnol J 8:595–604
36. Ibarra RU, Edwards JS, Palsson BO (2002) *Escherichia coli* K-12 undergoes adaptive evolution to achieve *in silico* predicted optimal growth. Nature 420:186–189

37. Segrè D, Vitkup D, Church GM (2002) Analysis of optimality in natural and perturbed metabolic networks. Proc Natl Acad Sci U S A 99:15112–15117
38. Rark JM, Kim TY, Lee SY (2009) Constraints-based genome-scale metabolic simulation for systems metabolic engineering. Biotechnol Adv 27:979–988
39. Zomorrodi AR, Suthers PF, Ranganathan S, Maranas CD (2012) Mathematical optimization applications in metabolic networks. Metab Eng 14:672–686
40. Covert MW, Xiao N, Chen TJ, Karr JR (2008) Integrating metabolic, transcriptional regulatory and signal transduction models in *Escherichia coli*. Bioinformatics 24:2044–2050
41. Meadows AL, Karnik R, Lam H, Forestell S, Snedecor B (2010) Application of dynamic flux balance analysis to an industrial *Escherichia coli* fermentation. Metab Eng 12:150–160
42. Feng X, Xu Y, Chen Y, Tang YJ (2012) MicrobesFlux: a web platform for drafting metabolic models from the KEGG database. BMC Syst Biol 6:94
43. Zhuang K, Izallalen M, Mouser P, Richter H, Risso C, Mahadevan R, Lovley DR (2011) Genome-scale dynamic modeling of the competition between Rhodoferax and Geobacter in anoxic subsurface environments. Isme J 5:305–316
44. Salimi F, Zhuang K, Mahadevan R (2010) Genome-scale metabolic modeling of a clostridial co-culture for consolidated bioprocessing. Biotechnol J 5:726–738
45. Jamshidi N, Palsson BØ (2008) Formulating genome-scale kinetic models in the post-genome era. Mol Syst Biol 4:171
46. Jamshidi N, Palsson BØ (2010) Mass action stoichiometric simulation models: incorporating kinetics and regulation into stoichiometric models. Biophys J 98:175–185
47. Smallbone K, Simeonidis E, Broomhead DS, Kell DB (2007) Something from nothing - bridging the gap between constraint-based and kinetic modelling. FEBS J 274:5576–5585
48. Smallbone K, Simeonidis E, Swainston N, Mendes P (2010) Towards a genome-scale kinetic model of cellular metabolism. BMC Syst Biol 4:6
49. Fleming RM, Thiele I, Provan G, Nasheuer HP (2010) Integrated stoichiometric, thermodynamic and kinetic modelling of steady state metabolism. J Theor Biol 264:683–692
50. Antoniewicz MR (2013) Dynamic metabolic flux analysis-tools for probing transient states of metabolic networks. Curr Opin Biotechnol 24:973–978
51. Hoffner K, Harwood SM, Barton PI (2013) A reliable simulator for dynamic flux balance analysis. Biotechnol Bioeng 110:792–802
52. Mahadevan R, Edwards JS, Doyle FJ (2002) Dynamic flux balance analysis of diauxic growth in *Escherichia coli*. Biophys J 83:1331–1340
53. Hanly TJ, Henson MA (2011) Dynamic flux balance modeling of microbial co-cultures for efficient batch fermentation of glucose and xylose mixtures. Biotechnol Bioeng 108:376–385
54. Hanly TJ, Urello M, Henson MA (2012) Dynamic flux balance modeling of *S. cerevisiae* and *E. coli* co-cultures for efficient consumption of glucose/xylose mixtures. Appl Microbiol Biotechnol 93:2529–2541
55. Hanly TJ, Henson MA (2013) Dynamic metabolic modeling of a microaerobic yeast co-culture: predicting and optimizing ethanol production from glucose/xylose mixtures. Biotechnol Biofuels 6:44
56. Chowdhury A, Zomorrodi AR, Maranas CD (2014) k-OptForce: integrating kinetics with flux balance analysis for strain design. PLoS Comput Biol 10:e1003487
57. Klumpp S, Hwa T (2008) Growth-rate dependent partitioning of RNA polymerases in bacteria. PNAS USA 105:20245–20250
58. Klumpp S, Zhang Z, Hwa T (2009) Growth-rate dependent global effects on gene expression in bacteria. Cell 139:1366–1375
59. Valgepea K, Adamberg K, Seiman A, Vilu R (2013) *Escherichia coli* achieves faster growth by increasing catalytic and translational rates of proteins. Mol Biosyst 9:2344–2358
60. Harcomb WR, Delaney NF, Leiby N, Klitgord N, Marx CJ (2013) The ability of flux balance analysis to predict evolution of central metabolism scales with the initial distance to the optimum. PLoS Comput Biol 9:e1003091
61. Edwards JS, Covert MW, Palsson BØ (2002) Metabolic modelling of microbes: the flux-balance approach. Environ Microbiol 4:133–140
62. Karr JR, Sanghvi JC, Macklin DN, Gutschow MW, Jacobs JM, Bolival B Jr, Assad-Garcia N, Glass JI, Covert MW (2012) A whole-cell computational model predicts phenotype from genotype. Cell 150:389–401
63. Gunawardera J (2012) Silicon dreams of cells into symbols. Nature 30:838–840
64. Freddolino PL, Tavazoie S (2012) The dawn of virtual cell biology. Cell 150:248–250
65. Tomita M (2001) Whole-cell simulation: a grand challenge of the 21st century. Trends in Biotech 19:205–210
66. Heinrich R, Rapoport TA (1974) A linear steady-state treatment of enzymatic chains. General properties, control and effector strength. Eur J Biochem 42:89–95
67. van Riel NA (2006) Dynamic modelling and analysis of biochemical networks: mechanism-based models and model-based experiments. Brief Bioinform 7:364–374
68. Heijnen JJ (2005) Approximative kinetic formats used in metabolic network modeling. Biotechnol Bioeng 91:534–545
69. Wu L, Wang WM, van Winden WA, van Gulik WM, Heijnen JJ (2004) A new framework for the estimation of control parameters in metabolic pathways using lin-log kinetics. Eur J Biochem 271:3348–3359
70. del Rosario RCH, Mendoza E, Voit EO (2008) Challenges in lin-log modelling of glycolysis in *Lactococcus lactis*. Iet Syst Biol 2:136–149
71. Hatzimanikatis V, Emmerling M, Sauer U, Bailey JE (1998) Application of mathematical tools for metabolic design of microbial ethanol production. Biotechnol Bioeng 58:154–161
72. Wang L, Hatzimanikatis V (2006) Metabolic engineering under uncertainty-II: analysis of yeast metabolism. Metab Eng 8:142–159
73. Pozo C, Marín-Sanguino A, Alves R, Guillén-Gosálbez G, Jiménez L, Sorribas A (2011) Steady-state global optimization of metabolic non-linear dynamic models through recasting into power-law canonical models. BMC Syst Biol 5:137
74. Sorribas A, Hernandez-Bermejo B, Vilaprinyo E, Alves R (2007) Cooperativity and saturation in biochemical networks: a saturable formalism using Taylor series approximations. Biotechnol Bioeng 97:1259–1277
75. Liebermeister W, Klipp E (2006) Bringing metabolic networks to life: convenience rate law and thermodynamic constraints. Theor Biol Med Model 3:41
76. Kim JI, Song HS, Sunkara SR, Lali A, Ramkrishna D (2012) Exacting predictions by cybernetic model confirmed experimentally: steady state multiplicity in the chemostat. Biotechnol Prog 28:1160–1166
77. Covert MW, Palsson BØ (2002) Transcriptional regulation in constraints-based metabolic models of *Escherichia coli*. J Biol Chem 277:28058–28064
78. Covert MW, Palsson BØ (2003) Constraints-based models: regulation of gene expression reduces the steady-state solution space. J Theor Biol 221:309–325
79. Covert MW, Schilling CH, Famili I, Edwards JS, Goryanin II, Selkov E, Palsson BØ (2001) Metabolic modeling of microbial strains *in silico*. Trends Biochem Sci 26:179–186
80. Herrgård MJ, Fong SS, Palsson BØ (2006) Identification of genome-scale metabolic network models using experimentally measured flux profiles. Plos Comput Biol 2:676–686
81. Song HS, Morgan JA, Ramkrishna D (2009) Systematic development of hybrid cybernetic models: application to recombinant yeast co-consuming glucose and xylose. Biotechnol Bioeng 103:984–1002
82. Ramkrishna D, Kompala DS, Tsao GT (1987) Are microbes optimal strategists. Biotechnol Progr 3:121–126
83. Varner J, Ramkrishna D (1999) Metabolic engineering from a cybernetic perspective. 1. Theoretical preliminaries. Biotechnol Prog 15:407–425
84. Young JD (2005) A system-level mathematical description of metabolic regulation combining aspects of elementary mode analysis with cybernetic control laws. PhD thesis, Purdue University
85. Young JD, Henne KL, Morgan JA, Konopka AE, Ramkrishna D (2008) Integrating cybernetic modeling with pathway analysis provides a dynamic, systems-level description of metabolic control. Biotechnol Bioeng 100:542–559
86. Schuster S, Fell DA, Dandekar T (2000) A general definition of metabolic pathways useful for systematic organization and analysis of complex metabolic networks. Nat Biotechnol 18:326–332
87. Kim JW, Dang CV (2005) Multifaceted roles of glycolytic enzymes. Trends Biocem Sci 30:142–150
88. Kim JI, Varner JD, Ramkrishna D (2008) A hybrid model of anaerobic *E. coli* GJT001: combination of elementary flux modes and cybernetic variables. Biotechnol Prog 24:993–1006
89. Teusink B, Passarge J, Reijenga CA, Esgalhado E, van der Weijden CC, Schepper M, Walsh MC, Bakker BM, van Dam K, Westerhoff HV, Snoep JL (2000) Can yeast glycolysis be understood in terms of *in vitro* kinetics of the constituent enzymes? Testing biochemistry. Eur J Biochem 267:5313–5329
90. Chakrabarti A, Miskovic L, Soh KC, Hatzimanikatis V (2013) Towards kinetic modeling of genome-scale metabolic networks without sacrificing stoichiometric, thermodynamic and physiological constraints. Biotechnol J 8:1043–1057
91. Hatzimanikatis V, Bailey JE (1996) MCA has more to say. J Theor Biol 182:233–242
92. Smallbone K, Messiha HL, Carroll KM, Winder CL, Malys N, Dunn WB, Murabito E, Swainston N, Dada JO, Khan F, Pir P, Simeonidis E, Spasić I,

Wishart J, Weichart D, Hayes NW, Jameson D, Broomhead DS, Oliver SG, Gaskell SJ, McCarthy JE, Paton NW, Westerhoff HV, Kell DB, Mendes P (2013) A model of yeast glycolysis based on a consistent kinetic characterisation of all its enzymes. FEBS Lett 587:2832–2841
93. Stanford NJ, Lubitz T, Smallbone K, Klipp E, Mendes P, Liebermeister W (2013) Systematic construction of kinetic models from genome-scale metabolic networks. PLoS One 8:e79195
94. Savageau MA (1970) Biochemical systems analysis. 3. Dynamic solutions using a power-law approximation. J Theor Biol 26:215–226
95. Voit Eberhard O (2013) Biochemical systems theory: a review. ISRN Biomathematics 2013:897658
96. Dräger A, Kronfeld M, Ziller MJ, Supper J, Planatscher H, Magnus JB, Oldiges M, Kohlbacher O, Zell A (2009) Modeling metabolic networks in *C. glutamicum*: a comparison of rate laws in combination with various parameter optimization strategies. BMC Syst Biol 3:5
97. Costa RS, Machado D, Rocha I, Ferreira EC (2010) Hybrid dynamic modeling of *Escherichia coli* central metabolic network combining Michaelis-Menten and approximate kinetic equations. Biosystems 100:150–157
98. Rizk ML, Liao JC (2009) Ensemble modeling for aromatic production in *Escherichia coli*. PLoS One 4:e6903
99. Tan YK, Liao JC (2012) Metabolic ensemble modeling for strain engineers. Biotechnol J 7:343–353
100. Contador CA, Rizk ML, Asenjo JA, Liao JC (2009) Ensemble modeling for strain development of L-lysine-producing *Escherichia coli*. Metab Eng 11(4–5):221–233
101. Dean JT, Rizk ML, TanY DKM, Liao JC (2010) Ensemble modeling of hepatic fatty acid metabolism with a synthetic glyoxylate shunt. Biophys J 98:1385–1395
102. Lee Y, Lafontaine Rivera JG, Liao JC (2014) Ensemble modeling for robustness analysis in engineering non-native metabolic pathways. Metab Eng 25:63–71
103. Khazaei T, McGuigan A, Mahadevan R (2012) Ensemble modeling of cancer metabolism. Front Physiol 3:135
104. Khodayari A, Zomorrodi AR, Liao JC, Maranas CD (2014) A kinetic model of *Escherichia coli* core metabolism satisfying multiple sets of mutant flux data. Metab Eng 25:50–62
105. Ishii N, Nakahigashi K, Baba T, Robert M, Soga T, Kanai A, Hirasawa T, Naba M, Hirai K, Hoque A, Ho PY, Kakazu Y, Sugawara K, Igarashi S, Harada S, Masuda T, Sugiyama N, Togashi T, Hasegawa M, Takai Y, Yugi K, Arakawa K, Iwata N, Toya Y, Nakayama Y, Nishioka T, Shimizu K, Mori H, Tomita M (2007) Multiple high-throughput analyses monitor the response of *E. coli* to perturbations. Science 316:593–597
106. Rizzi M, Baltes M, Theobald U, Reuss M (1997) *In vivo* analysis of metabolic dynamic in *Saccharomyces cerevisiae*: II. Mathematical model. Biotechnol Bioeng 55:592–608
107. Theobald U, Mailinger W, Baltes M, Rizzi M, Reuss M (1997) *In vivo* analysis of metabolic dynamic in *Saccharomyces cerevisiae*: I. Experimental observations. Biotechnol Bioeng 55:305–316
108. Chassagnole C, Noisommit-Rizzi N, Schmid JW, Mauch K, Reuss M (2002) Dynamic modeling of the central carbon metabolism of *Escherichia coli*. Biotechnol Bioeng 79:53–73
109. Kadir TA, Mannan AA, Kierzek AM, McFadden J, Shimizu K (2010) Modeling and simulation of the main metabolism in *Escherichia coli* and its several single-gene knockout mutants with experimental verification. Microb Cell Fact 9:88
110. Peskov K, Mogilevskaya E, Demin O (2012) Kinetic modelling of central carbon metabolism in *Escherichia coli*. FEBS J 279:3374–3385
111. Usuda Y, Nishio Y, Iwatani S, Van Dien SJ, Imaizumi A, Shimbo K, Kageyama N, Iwahata D, Miyano H, Matsui K (2010) Dynamic modeling of *Escherichia coli* metabolic and regulatory systems for amino-acid production. J Biotechnol 147:17–30
112. Matsuoka Y, Shimizu K (2013) Catabolite regulation analysis of *Escherichia coli* for acetate overflow mechanism and co-consumption of multiple sugars based on systems biology approach using computer simulation. J Biotechnol 168:155–173
113. Yao R, Hirose Y, Sarkar D, Nakahigashi K, Ye Q, Shimizu K (2011) Catabolic regulation analysis of *Escherichia coli* and its *crp, mlc, mgsA, pgi* and *ptsG* mutants. Microb Cell Fact 10:67
114. Toya Y, Ishii N, Nakahigashi K, Hirasawa T, Soga T, Tomita T, Shimizu K (2010) ^{13}C-metabolic flux analysis for batch culture of *Escherichia coli* and its Pyk and Pgi gene knockout mutants based on mass isotopomer distribution of intracellular metabolites. Biotechnol Prog 26:975–992
115. Toya Y, Nakahigashi K, Tomita M, Shimizu K (2012) Metabolic regulation analysis of wild-type and *arcA* mutant *Escherichia coli* under nitrate conditions using different levels of omics data. Mol Biosyst 8:2593–2604
116. Hasona A, Kim Y, Healy FG, Ingram LO, Shanmugam KT (2004) Pyruvate formate lyase and acetate kinase are essential for anaerobic growth of *Escherichia coli* on xylose. J Bacteriol 186:7593–7600
117. Kremling A, Bettenbrock K, Gilles ED (2008) A feed-forward loop guarantees robust behavior in *Escherichia coli* carbohydrate uptake. Bioinformatics 24:704–710
118. Kochanowski K, Volkmer B, Gerosa L, Haverkorn van Rijsewijk BR, Schmidt A, Heinemann M (2013) Functioning of a metabolic flux sensor in *Escherichia coli*. Proc Natl Acad Sci U S A 110:1130–1135
119. Huberts DH, Niebel B, Heinemann M (2012) A flux-sensing mechanism could regulate the switch between respiration and fermentation. FEMS Yeast Res 12:118–128
120. Christen S, Sauer U (2011) Intracellular characterization of aerobic glucose metabolism in seven yeast species by 13C flux analysis and metabolomics. FEMS Yeast Res 11:263–272
121. Boels E, Hollenberg CP (1997) The molecular genetics of hexose transport in yeasts. FEMS Microbiol Rev 21:85–111
122. Ricci JCD (1996) Influence of phosphenolpyruvate on the dynamic behavior of phosphofructokinase of *Escherichia coli*. J Theor Biol 178:145–150
123. Yang C, Hua Q, Baba T, Mori H, Shimizu K (2003) Analysis of *Escherichia coli* anaprelotic metabolism and its regulation mechanisms from the metabolic responses to altered dilution rates and phosphoenolpyruvate carboxykinase knockout. Biotechnol Bioeng 84:129–144
124. Lee B, Yen J, Yang L, Liao JC (1999) Incorporating qualitative knowledge in enzyme kinetic models using fuzzy logic. Biotechnol Bioeng 63:722–729
125. Nizam SA, Zhu JF, Ho PY, Shimizu K (2009) Effects of *arcA* and *arcB* genes knockout on the metabolism in *Escherichia coli* under aerobic condition. Biochem Eng J 44:240–250
126. Vemuri GN, Eiteman MA, Altman E (2006) Increased recombinant protein production in *Escherichia coli* strains with overexpressed water-forming NADH oxidase and a deleted ArcA regulatory protein. Biotechnol Bioeng 94:538–542
127. Vemuri GN, Altman E, Sangurdekar DP, Khodursky AB, Eiteman MA (2006) Overflow metabolism in *Escherichia coli* during steady-state growth: transcriptional regulation and effect of the redox ratio. Appl Environ Microbiol 72:3653–3661
128. Wolfe AJ (2005) The acetate switch. Microbiol Mol Biol Rev 69:12–50
129. Xu YF, Amador-Noguez D, Reaves ML, Feng XJ, Rabinowitz JD (2012) Ultrasensitive regulation of anapleurosis via allosteric activation of PEP carboxylase. Nat Chem Biol 8:562–568
130. Voit E, Neves AR, Santos H (2006) The intricate side of systems biology. Proc Natl Acad Sci U S A 103:9452–9457
131. Hoefnagel MHN, Starrenburg MJC, Martens DE, Hugenholtz J, Kleerebezem M, Van Swam II, Bongers R, Westerhoff HV, Snoep JL (2002) Metabolic engineering of lactic acid bacteria, the combined approach: kinetic modeling, metabolic control and experimental analysis. Microbiol 148:1003–1013
132. Kremling A, Jahreis K, Lengeler JW, Gilles ED (2000) The organization of metabolic reaction networks: a signal-oriented approach to cellular models. Metab Eng 2:190–200
133. Kremling A, Gilles ED (2001) The organization of metabolic reaction networks. II. Signal processing in hierarchical structured functional units. Metab Eng 3:138–150
134. Kremlng A, Fischer S, Sauter T, Bettenbrock K, Gilles ED (2004) Time hierarchies in the *Escherichia coli* carbohydrate uptake and metabolism. BioSystems 73:57–71
135. Sauter T, Gilles ED (2004) Modeling and experimental validation of the signal transduction via the *Escherichia coli* sucrose phospho transferase system. J Biotech 110:181–199
136. Bettenbrock K, Fischer S, Kremling A, Jahreis K, Sauter T, Gilles ED (2006) A quantitative approach to catabolite repression in *Escherichia coli*. J Biol Chem 281:2578–2584
137. Nishio Y, Usuda Y, Matsui K, Kurata H (2008) Computer-aided rational design of the phosphotransferase system for enhanced glucose uptake in *Escherichia coli*. Mol Syst Biol 4:160
138. Majewski RA, Domach MM (1990) Simple constrained-optimization view of acetate overflow in *Escherichia coli*. Biotech Bioeng 35:732–738
139. Rabinowitz J, Silhavy TJ (2012) Metabolite turns master regulator. Nature 500:283–284

140. Doucette CD, Schwab DJ, Wingreen NS, Rabinowitz JD (2011) Alpha-ketoglutarate coordinates carbon and nitrogen utilization via enzyme I inhibition. Nat Chem Biol 7:894–901
141. Scott M, Gunderson CW, Mateescu EM, Zhang Z, Hwa T (2010) Interdependence of cell growth and gene expression: origins and consequences. Science 330:1099–1102
142. You C, Okano H, Hui S, Zhang Z, Kim M, Gunderson CW, Wang Y-P, Lenz P, Yan D, Hwa T (2013) Coordination of bacterial proteome with metabolism by cyclic AMP signaling. Nature 500:301–306
143. Vinuselvi P, Kim MK, Lee SK, Ghim C-M (2012) Rewiring carbon catabolite repression for microbial cell factory. BMB Rep 45(2):59–70
144. Gorke B, Stulke J (2008) Carbon catabolite repression in bacteria: many ways to make the most out of nutrients. Nature Rev Microbiol 6:613–24
145. Vasudevan P, Briggs M (2008) Biodiesel production-current state of the art and challenges. J Ind Microbiol Biotechnol 35:421–430
146. Dharmadi Y, Murarka A, Gonzalez R (2006) Anaerobic fermentation of glycerol by *Escherichia coli*: a new platform for metabolic engineering. Biotechnol Bioeng 94:821–829
147. Clomburg JM, Gonzalez R (2013) Anaerobic fermentation of glycerol: a platform for renewable fuels and chemicals. Trends Biotechnol 31:20–28
148. Almeida JRM, Fávaro LCL, Betania F, Quirino BF (2012) Biodiesel biorefinery: opportunities and challenges for microbial production of fuels and chemicals from glycerol waste. Biotechnol for Biofuels 5:48
149. Martínez-Gómez K, Flores N, Castañeda HM, Martínez-Batallar G, Hernández-Chávez G, Ramírez OT, Gosset G, Encarnación S, Bolivar F (2012) New insights into *Escherichia coli* metabolism: carbon scavenging, acetate metabolism and carbon recycling responses during growth on glycerol. Micob Cell Fact 11:46
150. Oh MK, Liao JC (2000) Gene expression profiling by DNA microarrays and metabolic fluxes in *Escherichia coli*. Biotechnol Prog 16:278–286
151. Peng L, Shimizu K (2003) Global metabolic regulation analysis for *Escherichia coli* K12 based on protein expression by 2-dimensional electrophoresis and enzyme activity measurement. Appl Microbiol Biotechnol 61:163–178
152. Cintolesi A, Clomburg JM, Rigou V, Zygourakis K, Gonzalez R (2012) Quantitative analysis of the fermentative metabolism of glycerol in *Escherichia coli*. Biotechnol Bioeng 109:187–198
153. Saier MH, Ramseier TM (1996) The catabolite repressor/activator (Cra) protein of enteric bacteria. Journal of Bacteriology 178:3411–3417
154. Kornberg HL (2001) Routes for fructose utilization by *Escherichia coli*. J Mol Microbiol Biotechnol 3:355–359
155. Yao R, Shimizu K (2013) Recent progress in metabolic engineering for the production of biofuels and biochemicals from renewable sources with particular emphasis on catabolite regulation and its modulation. Process Biochem 48:1409–1417
156. Crasnier-Mednansky M, Park MC, Studley WK, Saier MH Jr (1997) Cra-mediated regulations of *Escherichia coli* adenylate cyclase. Microbiology 143:785–792
157. Griffith JK, Baker ME, Rouch DA, Page MG, Skurray RA, Paulsen IT, Chater KF, Baldwin SA, Henderson PJ (1992) Membrane transport proteins: implications of sequence comparisons. Curr Opin Cell Biol 4:684–695
158. Sumiya M, Davis EO, Packman LC, McDonald TP, Henderson PJ (1995) Molecular genetics of a receptor protein for d-xylose, encoded by the gene *xylF*, in *Escherichia coli*. Receptors Channels 3:117–128
159. Song S, Park C (1997) Organization and regulation of the d-xylose operons in *Escherichia coli* K-12: XylR acts as a transcriptional activator. J Bacteriol 179:7025–7032
160. Altintas MM, Eddy CK, Zhang M, McMillan JD, Kompala DS (2006) Kinetic modeling to optimize pentose fermentation in *Zymomonas mobilis*. Biotechnol Bioeng 94:273–295
161. Yang C, Hua Q, Shimizu K (1999) Development of a kinetic model for L-lysine biosynthesis in *Corynebacterium glutamicum* and its application to metabolic control analysis. J Biosci Bioeng 88:393–403
162. Hua Q, Yang C, Shimizu K (2000) Metabolic control analysis for lysine synthesis using *Corynebacterium glutamicum* and experimental verification. J Biosci Bioeng 90:184–192
163. Nishio Y, Ogishima S, Ichikawa M, Yamada Y, Usuda Y, Masuda T, Tanaka H (2013) Analysis of L-glutamic acid fermentation by using a dynamic metabolic simulation model of *Escherichia coli*. BMC Sys Biol 7:92
164. Li R-D, Li Y-Y, Lu L-Y, Ren C, Li Y-X, Liu L (2011) An improved kinetic model for the acetone-butanol-etahnol pathway of *Clostridium acetobutyricum* and model-based perturbation analysis. BMC Sys Biol 5:S12
165. Shinto H, Tashiro Y, Kobayashi G, Sekiguchi T, Hanai T, Kuriya Y, Okamoto M, Sonomoto K (2007) Kinetic modeling and sensitivity analysis of acetone-butanol-ethanol production. J Biotechnol 131:45–56
166. Shinto H, Tashiro Y, Kobayashi G, Sekiguchi T, Hanai T, Kuriya Y, Okamoto M, Sonomoto K (2008) Kinetic study of substrate dependency for higher butanol production in acetone-butanol-ethanol fermentation. Proc Biochem 43:1452–1461
167. Alexeeva S, Hellingwerf KJ, de Mattos JT (2003) Requirement of ArcA for redox regulation in *Escherichia coli* under microaerobic but not anaerobic or aerobic conditions. J Bacteriol 185:204–209
168. Shalel-Levanon S, San K-Y, Bennett GN (2005) Effect of oxygen, and ArcA and FNR regulators on the expression of genes related to the electron transfer chain and the TCA cycle in *Escherichia coli*. Metab Eng 7:364–374
169. Cox SJ, Levanon SS, Bennett GN, San K-Y (2005) Genetically constrained metabolic flux analysis. Metab Eng 7:445–456
170. van Heeswijk WC, Westerhoff HV, Boogerd FC (2013) Nitrogen assimilation in *Escherichia coli*: putting molecular data into a systems perspective. Microbio Mol Biol Rev 77:628–695
171. Rhee SG, Chock PB, Stadtman ER (1985) Glutamine synthetase from *Escherichia coli*. Methods Enzymol 113:213–241
172. Sakamoto N, Kotre AM, Savageau MA (1975) Glutamate dehydrogenase from *Escherichia coli*: purification and properties. J Bacteriol 124:775–783
173. Bruggeman FJ, Boogerd FC, Westerhoff HV (2005) The multifarious short-term regulation of ammonium assimilation of *Ecsherichia coli*: dissection using an *in silico* replica. FEBS J 272:1965–1985
174. Atkinson MR, Blauwkamp TA, Bondarenko V, Studitsky V, Ninfa AJ (2002) Activation of the *glnA*, *glnK*, and *nac* promoters as *Escherichia coli* undergoes the transition from nitrogen excess growth to nitrogen starvation. J Bacteriol 184:5358–5363
175. Reitzer L (2003) Nitrogen assimilation and global regulation in *Escherichia coli*. Annu Rev Microbiol 57:155–176
176. Ma H, Boogerd FC, Goryanin I (2009) Modelling nitrogen assimilation of *Escherichia coli* at low ammonium concentration. J Biotechnol 144:175–183
177. Yuan J, Doucette CD, Fowler WU, Feng XJ, Piazza M, Rabitz HA, Wingreen NS, Rabinowitz JD (2009) Metabolomics-driven quantitative analysis of ammonia assimilation in *E. coli*. Mol Syst Biol 5:302
178. Lodeiro A, Melgarejo A (2008) Robustness in *Escherichia coli* glutamate and glutamine synthesis studied by a kinetic mode. J Biol Phys 34:91–106

Enhanced production of flavonoids by methyl jasmonate elicitation in cell suspension culture of *Hypericum perforatum*

Jing Wang[1,2], Jie Qian[3], Lingyun Yao[1,2*] and Yanhua Lu[1,2*]

Abstract

Background: Flavonoids of *Hypericum perforatum* are important secondary metabolites which have been widely utilized in medicine for a range of purposes. The use of methyl jasmonate (MeJA) elicitation for the enhancement of flavonoid production in cell suspension culture of *H. perforatum* would be an efficient alternative method for the flavonoid production.

Results: MeJA influenced the cells growth and flavonoid production. The optimal elicitation strategy was treatment of the cell cultures with 100 μmol/L MeJA on day 15, which resulted in the highest flavonoid production (280 mg/L) and 2.7 times of control cultures. The activities of catalase (CAT) were inhibited after MeJA treatment in the cell cultures, while the activities of phenylalanine ammonia lyase (PAL) increased, which led to the enhancement of flavonoid production.

Conclusion: MeJA elicitation is a useful method for the enhancement of flavonoid production in cell suspension culture of *H. perforatum*.

Keywords: MeJA; *Hypericum perforatum*; Suspension culture; Flavonoids

Background

Hypericum perforatum, commonly known as St. John's wort, is a perennial herb native to Europe and also a traditional medicinal plant which has been utilized in Chinese folk medicine for a range of purposes [1,2]. The medicinal applications of *H. perforatum*, including skin wounds, eczema, burns, diseases of the alimentary tract, and psychological disorders, have been related to the phenolic composition of the plant, particularly to hypericins, hyperforins, and flavonoids [3,4]. The efficacy of medical constituents of *H. perforatum* is based on the whole secondary metabolites, rather than the presence of single chemical compound [5]. Flavonoids, specifically quercetin and its glycoside derivatives, comprise the major group of biologically active metabolites in *H. perforatum* and are important biochemical markers in authenticating the herbal plant materials [4,6].

The consumption of *H. perforatum*-derived products as pharmaceutical preparations or food additives has increased dramatically, and it is presently one of the most consumed medicinal plants in the world [4,7]. The extracts of *H. perforatum* currently used in foods and pharmaceutics, mostly composed of flavonoids, are mainly obtained from the top aerial parts collected in the flowering stage [6,8]. However, the quality of the flavonoid-rich extracts derived from field-grown plants may be affected by many environmental factors as well as biological processes [5]. Furthermore, field cultivation of *H. perforatum* requires a long growth period and plant management, which is a slow and laborious process [6]. Therefore, an alternative method for more efficient and controllable production of flavonoids from *H. perforatum* is urgently required.

Plant cell, tissue, and organ cultivation technology has been successfully applied to the production of pharmaceutically valuable compounds and other fine chemicals of commercial interest in recent years [9]. For instance, cell suspension cultures of *Panax ginseng* have been widely used for the production of saponin and other metabolites [10]. Plant-mediated green biomimetic synthesis of silver

* Correspondence: Lyyao1982@qq.com; luyanhua@ecust.edu.cn
[1]State Key Laboratory of Bioreactor Engineering, Department of Bioengineering, East China University of Science and Technology, 130 Meilong Rd., Shanghai 200237, China
Full list of author information is available at the end of the article

nanoparticle was considered a widely acceptable technology for rapid production of silver nanoparticles for successfully meeting the excessive need and current market demand and resulting in a reduction in the employment or generation of hazardous substances to human health and the environment [11]. Till now, a number of efforts have been made for the establishment of cell and/or adventitious root cultures of *H. perforatum* for the production of hypericins, phenolics, and xanthones [12-15]. The information on the factors affecting biomass and aforementioned metabolites of *H. perforatum* was also investigated through elicitation and culture optimization [1,16-18]. However, there are few reports on the production of flavonoids from cell suspension culture of *H. perforatum*, and little is known about the strategies for hyperproduction of biomass and flavonoids.

Many cell cultures have been established from plants, but they seldom produce sufficient amounts of the required secondary metabolites. Secondary metabolite biosynthesis in plants depends on environmental stresses; their accumulation can be stimulated by precursors and elicitors [19]. Precursors are intermediate compounds of the secondary metabolite biosynthesis cycles, which would be toxic to the culture if not used at an appropriate stage or concentration [20]. Elicitors are biological (components of microbial cells and poly- and oligosaccharides), chemical (heavy metals, pesticides, and the signaling compounds in plant defense responses), or physical (cold shock, UV, hyperosmotic stress, ultrasound, and pulsed electric field) factors that induce enzymatic activity against stress [19-21]. Jasmonic acid (JA) and its methyl ester, methyl jasmonate (MeJA), have been proposed to be important signaling compounds in the process of elicitation leading to the hyperproduction of various secondary metabolites [14]. They have also been reported to play a key role in signal transduction processes that regulate defense responses in plants and shown effective to enhance the production of secondary metabolites in cell cultures [14,19].

In the present study, the effect of MeJA on cell growth and flavonoid biosynthesis in the *H. perforatum* cell suspension culture was investigated in small batches. Some parameters, such as elicitation time and MeJA concentration, on biomass and flavonoid production were studied in detail. The activities of key enzymes (catalase and phenylalanine ammonia lyase) related to plant stress responses and secondary metabolite biosynthesis were investigated as well. To the best of our knowledge, this is the first report on the induction of flavonoid production by MeJA in cell suspension culture of *H. perforatum.*

Materials and methods

Chemicals

MeJA was supplied by Drug Institute of East China University of Science and Technology (Shanghai, China). Salicylic acid (SA), ammonium metavanadate (NH_4VO_3), and nickel sulfate ($NiSO_4$) were purchased from Shanghai Zhong Lan Chemical Company (Shanghai, China). Acetonitrile of HPLC grade was purchased from TEDIA Company, USA. All other chemicals (analytical grade) were purchased from Shanghai Chemical Co., Ltd. (Shanghai, China).

Plant material and callus induction

H. perforatum plants were provided by Prof. Jie Qian (Tongji University, China) and were cultivated in the greenhouse. The fresh stem explants of *H. perforatum* were washed thoroughly with tap water, surface-sterilized with 75% ethanol for 60 s, and then soaked in 10% sodium hypochlorite solution for 30 s and rinsed seven times in sterile deionized water. The clean and sterilized stem explants were cut into 5-mm segments and inoculated on solid Murashige and Skoog (MS) [22] medium supplemented with 1.0 mg/L 2,4-D, 0.2 mg/L 6-benzyladenine (BA), and 25 g/L sucrose to induce callus formation. After 1 month, the successfully induced callus was separated from the explants and cultured separately until used for the establishment of cell suspensions.

Suspension cultures of *H. perforatum* cells

Cell suspension cultures were initiated from the friable callus and maintained in the liquid MS medium (pH 5.8) supplemented with 1.0 mg/L 2,4-D, 0.2 mg/L 6-benzyladenine (BA), and 25 g/L sucrose. A sterile standard sieve with aperture of 300 μm was used to filter the initial established suspension cells to obtain homogeneous cultures. The pre-weighed cells were cultured in a 250-mL Erlenmeyer flask containing a 50-mL medium; cultures were placed on a rotary shaker shaking at 120 rpm at 25 ± 2°C under continuous illumination. Every 20 days, cells were subcultured to fresh media with 5.0-g fresh weight (FW) in 250-mL flasks.

Elicitation of cell suspension culture

Four elicitors including MeJA, SA, NH_4VO_3, and $NiSO_4$ were used to study the elicitation effect on flavonoid production by the cell suspension cultures of *H. perforatum.* MeJA and SA were dissolved in ethanol, NH_4VO_3 and $NiSO_4$ were dissolved in water, and the four elicitors were filter-sterilized before adding into the suspension cultures. Based on the results of preliminary experiments, the feeding concentration of MeJA, SA, NH_4VO_3, and $NiSO_4$ was 50, 100, 50, and 15 μM, respectively. As a control, filter-sterilized ethanol was added to the cell suspension culture with a final concentration of 0.2%.

According to the results of the above experiments, MeJA was selected for further study. The MeJA induction time and feeding concentration, known to be crucial to plant cell culture, were investigated by single-factor experimental design by varying a single factor at a time and keeping other factors at a constant value. All experiments were performed in triplicate, and data are expressed as the mean of three samples with standard deviation.

Extraction and HPLC analysis of secondary metabolites

H. perforatum extracts were prepared by soaking 0.5 g of the dried cells in 10 mL of methanol and treated with 1 h of ultrasonic extraction for two times. The extracts were then centrifuged at 10,000 rpm at 4°C. The supernatant was combined and used for HPLC analysis and flavonoid determination.

The HPLC system of Shimadzu LC-10Avp Plus with a PDA detector (SPD-M20A) and a C_{18} column (4.6 mm × 250 mm, 5 μm, Eclipse XDB C18) was used for the qualitative and quantitative analyses of flavonoids and other metabolites in the methanol extracts. The mobile phase was acetonitrile (A) and 0.1% trifluoroacetic acid (B). A gradient method was used for the separation of the extracted samples: 0-25 min, 15%-40% A; 25-40 min, 40%-100% A. The elution flow rate and detection wavelength were set at 1.0 mL/min and 254 nm, respectively. The identity of peaks separated by HPLC was confirmed by the injection of standard, and UV spectral analysis was carried out to confirm compound identification.

Determination of biomass and flavonoid content

The plant cell biomass is expressed as the gram fresh cell weight and/or dry cell weight per liter. Cell suspension cultures were filtered and washed by deionized water, then cells were collected and weighed to get the FW. The collected cells were dried in the vacuum drying oven at 45°C until constant weight (DW) was attained.

The flavonoid content was assayed using UV colorimetric method as reported elsewhere [5,13]. Briefly, 0.5 mL of the methanolic plant cell extract was mixed with 2 mL of distilled water and subsequently with 0.15 mL of a 5% $NaNO_2$ solution. After 6 min, 0.15 mL of a 10% $AlCl_3$ solution was added and the mixture was allowed to stand for a further 6 min before 2 mL of 4% NaOH solution was added. Absorbance of the mixed solution was measured at 510 nm using a UV-Vis spectrophotometer (UV-1650PC, Shimadzu, Japan). Rutin was used as standard compound for the quantification of total flavonoid. Results were expressed as milligram of rutin equivalents per gram of dry cells. Data were expressed as means ± SD for three replications.

Enzyme activity analysis

Suspension culture cells were harvested for evaluation of the catalase (CAT) activity level as described by Georgiew et al [23]. The decomposition of H_2O_2 was followed by measuring the decrease in absorbance at 240 nm. One unit of CAT is the amount that decomposes 1 μmol of H_2O_2 for 1 min at pH 7.0 and 25°C. Protein was estimated by the procedure of Lowry et al. [24], using crystalline bovine albumin as standard. Phenylalanine ammonia lyase (PAL) was extracted from fresh *H. perforatum* cells with borate buffer (pH 8.8). The cells were ground in the buffer (0.15 g/mL) for 2 min with a pestle and mortar on ice and then centrifuged at 10,000 rpm and 4°C for 20 min to obtain a solid-free extract. The PAL activity was determined based on the conversion of L-phenylalanine to cinnamic acid as described by Wu and Lin [21].

Results and discussion

Callus induction and cell suspension culture establishment

In the preliminary work, calyx, leaf, petal, and stem segments of *H. perforatum* were used as explants for callus induction. It revealed that callus induction was significantly affected by the type of explants and the stem was the best explant for the initiation of a friable callus (Figure 1A,B,C). The stem-derived callus was incubated at 25 ± 2°C under continuous illumination on a solid MS medium and subcultured every 4 weeks. Cell suspension culture was initiated once the stable cell lines were obtained on the solid MS medium. Given the potential for deriving useful secondary metabolites from plant resource, cell suspension cultures of *H. perforatum* were established to produce hypericin, naphtodianthrones, and phenylpropanoids [12,14,18]. In addition, some efforts have been made for the establishment of adventitious root cultures of *H. perforatum* for the production of flavonoids, phenolics, and xanthones [5,15-17]. In the present study, cell suspension culture was successfully established (Figure 1D) from the stem-derived calli of *H. perforatum* for further investigation of flavonoids' production performance.

To understand the growth and flavonoid accumulation in *H. perforatum* cells for determining the optimal cultivation time, cultures were analyzed throughout a 30-day period (Figure 2A). The cell growth exhibited a lag phase or slow growth period in the first 10 days; subsequently, the cells entered their exponential growth phase with the maximum DW of 8.2 g/L reached on day 20. As shown in Figure 2B, the flavonoid production is related to cell growth and its content reached maximum with a value of 16 mg/g DW during stationary or declining phase between day 20–25. The content of total flavonoids in cell suspension culture of *H. perforatum* (16 mg/g DW) was lower than that of the adventitious root cultures (42.7 mg/g DW) as reported earlier [5]. However, the culture period for the suspension cells of *H. perforatum* (3 weeks) was much shorter than that of the adventitious root cultures (6 weeks). In addition, the flavonoid production in cell suspension culture of *H. perforatum* can be further improved by the supplementation of effective elicitors

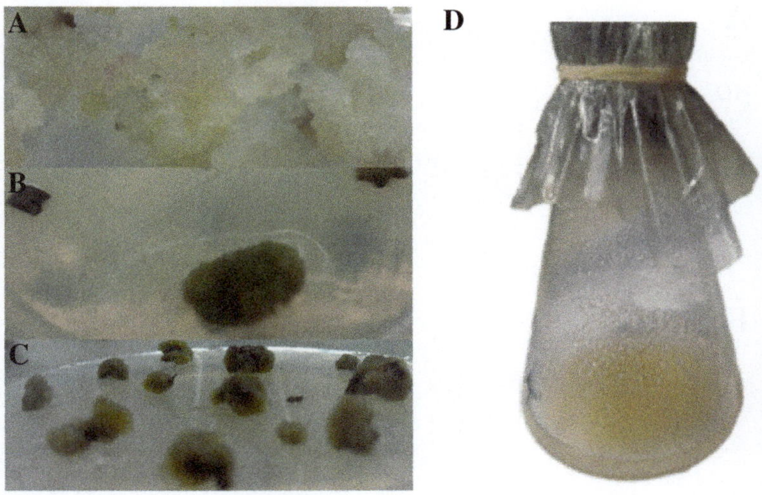

Figure 1 Callus induced from different explants of *H. perforatum* and established cell suspension cultures. (A) Callus induced from stem explants; **(B)** callus induced from leaf explants; **(C)** callus induced from calyx explants; **(D)** cell suspension culture.

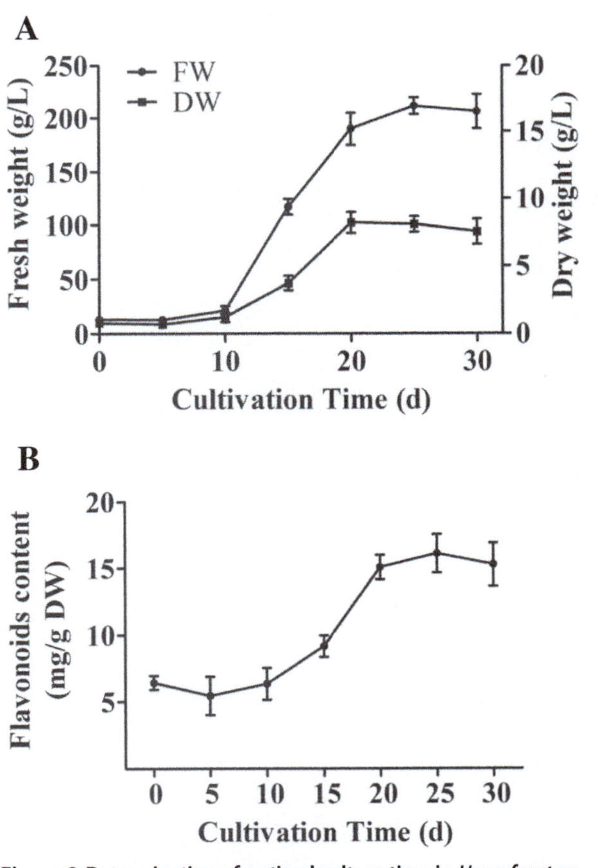

Figure 2 Determination of optimal culture time in *H. perforatum* cells. Kinetic profiles of cell growth **(A)** and content of flavonoids **(B)** in *H. perforatum* cell suspension cultures.

[18,19]. Therefore, it is crucial to screen potential elicitors for the hyperproduction of flavonoids in a shortened culture period by cell suspension culture of *H. perforatum*.

HPLC was used to recognize the flavonoids biosynthesized in the cell cultures on the basis of the retention times and UV spectras with those of reference standards. As shown in Figure 3A,B,C,D, hyperin and quercetin were the main flavonoids obtained from the established cell suspension cultures, which was in accordance with a previous study [6]. In addition, the medium was examined, and no flavonoids were detected.

Selection of MeJA as the elicitor in cell suspension culture of *H. perforatum*

Some known abiotic elicitors for plant secondary metabolites such as signaling molecules (MeJA and SA) and inorganic salts (NH_4VO_3 and $NiSO_4$) were tested for their effect on cell growth and flavonoid production by the *H. perforatum* cell suspension culture. Four treatments, MeJA (50 μM), salicylic acid (100 μM), NH_4VO_3 (50 μM), and $NiSO_4$ (15 μM) were applied to the cell cultures on day 15 based on our previous lab results. As shown in Figure 4, the growth of cell cultures was inhibited by the addition of the four elicitors as compared to the control cultures. Similar findings were also observed by Dong and Zhong [25], in which the application of elicitation in suspension cell culture severely inhibited the growth of *Taxus chinensis* cells. However, elicitation treatments may have varied effects on different cell lines. Some researchers have reported that the growth of cell cultures was not affected by elicitors [26]. The effects of the elicitors on flavonoid production are also given in Figure 4. Flavonoid content was promoted by MeJA and

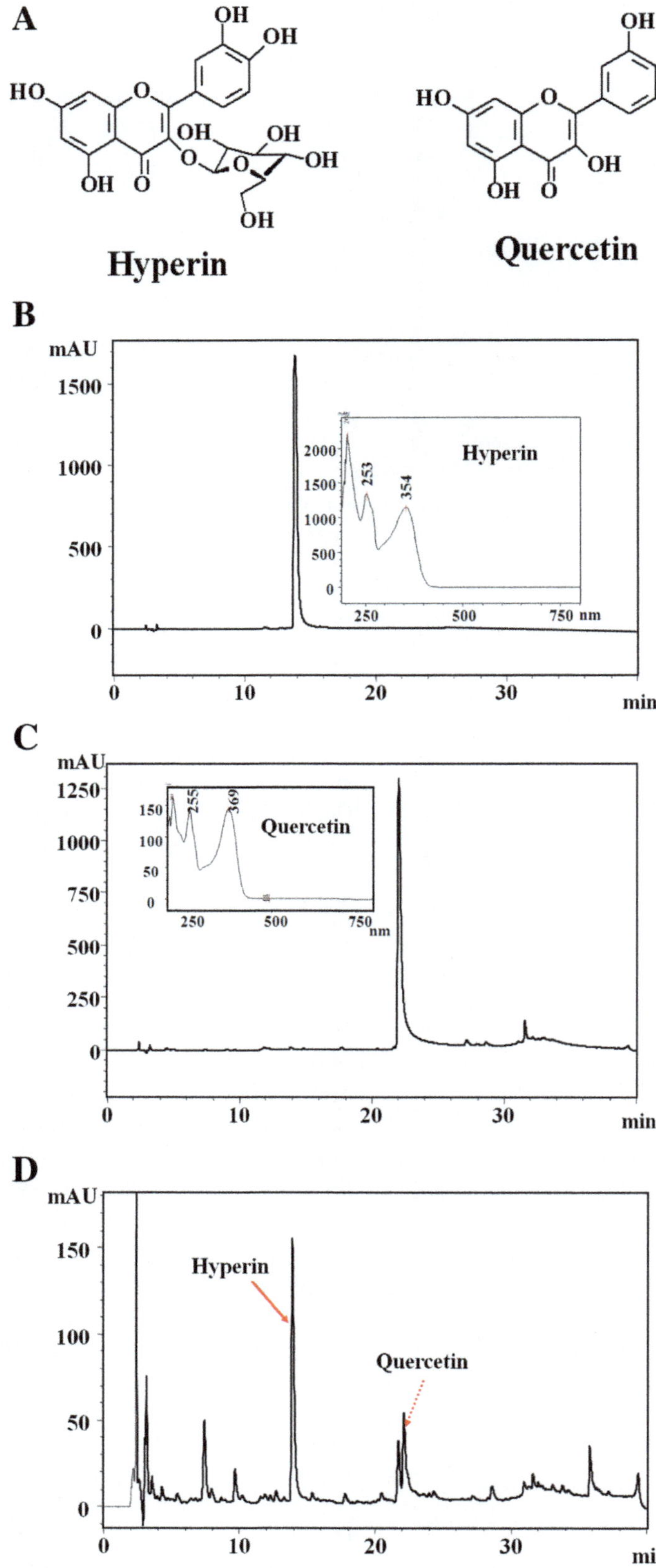

Figure 3 Hyperin and quercetin obtained from the established cell suspension cultures. Chemical structures of hyperin and quercetin **(A)** and chromatographic profile of hyperin (B), quercetin (C) and the suspension cell extracts **(D)**. The two insets are the UV spectrums of standard hyperin **(B)** and quercetin **(C)**.

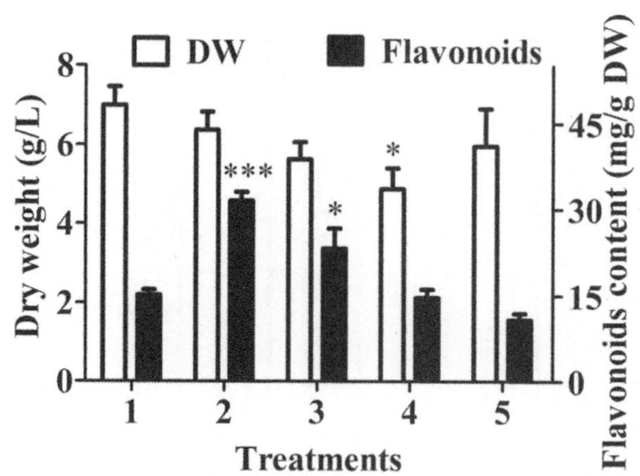

Figure 4 The effects of four abiotic elicitors on *H. perforatum* cell growth and flavonoid accumulation. 1 Control; 2 MeJA; 3 SA; 4 NH4VO$_3$; 5 NiSO$_4$ (*$P < 0.05$, **$P < 0.01$, and ***$P < 0.001$ indicate statistical significance compared to the control).

SA induction, which were 2.1 and 1.5 times higher in comparison to control cultures (15.36 mg/g DW), respectively. Several studies on other plant cell cultures have shown that MeJA and SA elicitation can enhance the production of secondary metabolites. For instance, MeJA was recognized as an effective elicitor that could increase the production of paclitaxel in *Taxus candensis* and *T. cuspidate* [27], anthocyanin in *Tulipa gesneriana* [28], and gymnemic acid in *Gymnema sylvestre* [29]. In SA-elicited cell suspension culture of *H. perforatum*, the production of both hypericin and pseudohypericin has doubled as compared to control cell suspension cultures [18]. As shown in Figure 4, MeJA and SA both induced flavonoid accumulation in *H. perforatum* cell suspension cultures, but SA was less efficient than MeJA in promoting flavonoid accumulation under the tested feeding concentration.

Some literatures reported the enhancement of metabolite production by NH_4VO_3 and $NiSO_4$ elicitation [30]. However, the two elicitors revealed negative effects on biomass and flavonoid accumulation in *H. perforatum* cell cultures (Figure 4). It was probably due to the significant biological toxicity caused by the inorganic salts [30]. Some other elicitors such as fluoro- and hydroxyl-containing derivatives of methyl jasmonate, which were proven more potent than methyl jasmonate in suspension culture of *Taxus* [31,32], may also serve as possible options for the enhancement of flavonoid biosynthesis in *H. perforatum*. In addition, some new elicitors have been reported to have significantly promoted secondary metabolite biosynthesis by plant cell cultures [33]. However, elicitation treatments may have varied effects among different cell lines as mentioned above. Therefore, efforts devoted into the enhancement of flavonoid production by *H. perforatum* cell suspension culture are still required considering its further application. In the present investigation, the addition of MeJA (50 μM) resulted in maximum flavonoid production, which indicated the potential of this abiotic elicitor for the enhancement of flavonoid biosynthesis in cell suspension culture of *H. perforatum* and was selected for further study.

Effects of MeJA feeding time on cell growth and flavonoid production

The elicitor induction time is one of the key factors that affect the cell growth and product yield for plant cell suspension culture [33]. The effects of MeJA addition time on cell growth and flavonoid accumulation was evaluated by the addition of 50 μM MeJA into the cell cultures on lag phase (day 5 and day 10) and exponential phase (day 15 and day 20). Figure 5A,B,C showed the biomass and flavonoid production of *H. perforatum* cell cultures after MeJA addition. As shown in Figure 5A,B, C, the cell growth and production of flavonoids were suppressed deleteriously when MeJA was added on the lag phase. MeJA treatment on the exponential phase (day 15 and day 20) led to a slow decrease in DW (19.0% and 9.7%, respectively); however, the flavonoid content and production were significantly increased as compared to control cultures (Figure 5A,B,C). A similar phenomenon was also observed by Huang and Zhong [30], in which elicitation treatment on the log phase of *P. ginseng* reduced the DW and enhanced ginsenoside accumulation. Notably, MeJA treatment on day 15 resulted in the highest value of flavonoid content (38.26 mg/g DW) or production (229.79 mg/L). Therefore, MeJA treatment on day 15 was taken as the optimal addition time for flavonoid production.

Effects of MeJA dosage on cell growth and flavonoid production

A suitable elicitor concentration is important for the cell growth and product yield in plant cell suspension culture process [34]. To study the effects of MeJA addition concentrations, cell suspension cultures of *H. perforatum* were treated with different levels (50-200 μM) of MeJA after 15 days of cultivation. The DW and flavonoid content on day 20 were shown in Figure 6A. It is clear that all the tested concentrations of MeJA had an inhibitory impact on the cell growth; similar results were also obtained in the SA-induced cell suspension culture of *H. perforatum* [18]. Induction with 50 or 100 μM of MeJA decreased DW by 11.9%-23.1%, while higher concentration of MeJA (150 μM) severely decreased the DW from 7.0 ± 0.5 g/L (control cultures) to 5.0 ± 0.5 g/L. Addition of low dosage (<150 μM) of MeJA elicitor to the cultured cells of *H. perforatum* increased the flavonoid accumulation. As shown in Figure 6A, the flavonoid content was increased about 1.1-fold with a dosage of 50 μM and

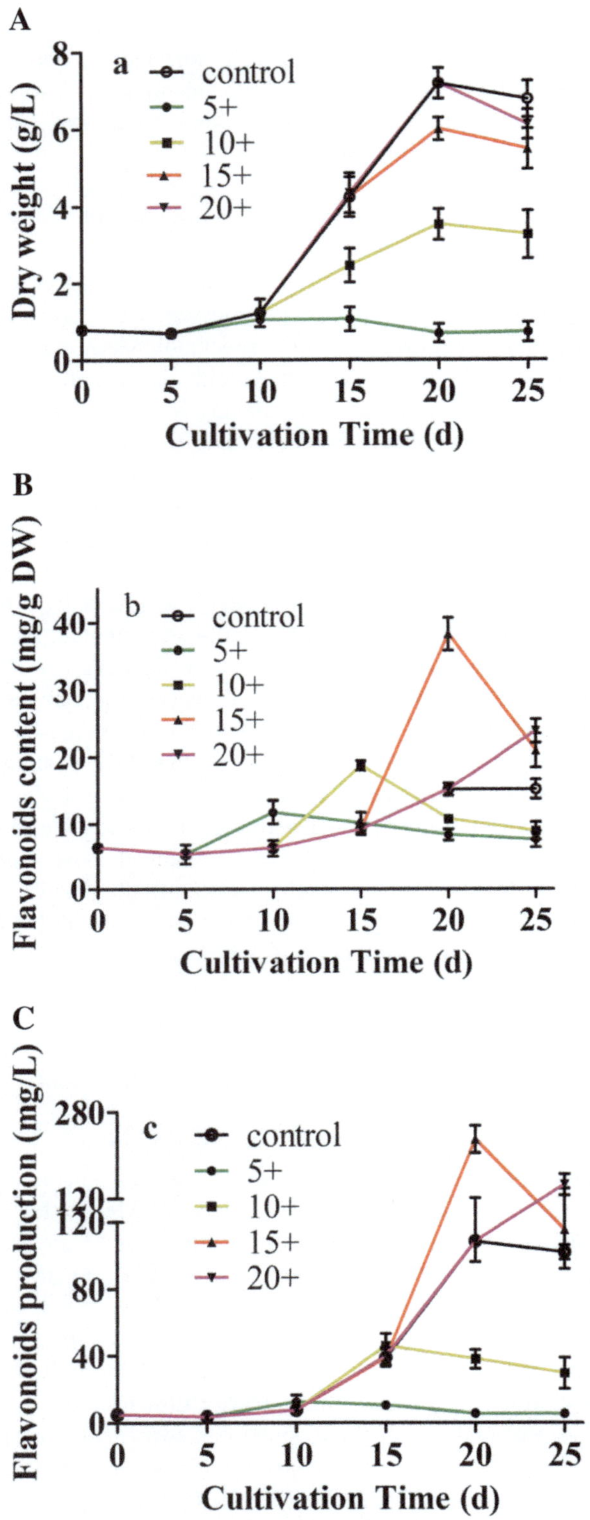

Figure 5 Effects of MeJA (50 μM) feeding time on cell growth (A), flavonoid content (B), and flavonoid production (C). 5+, 10+, 15+, and 20+ indicate MeJA feeding on day 5, 10, 15, and 20.

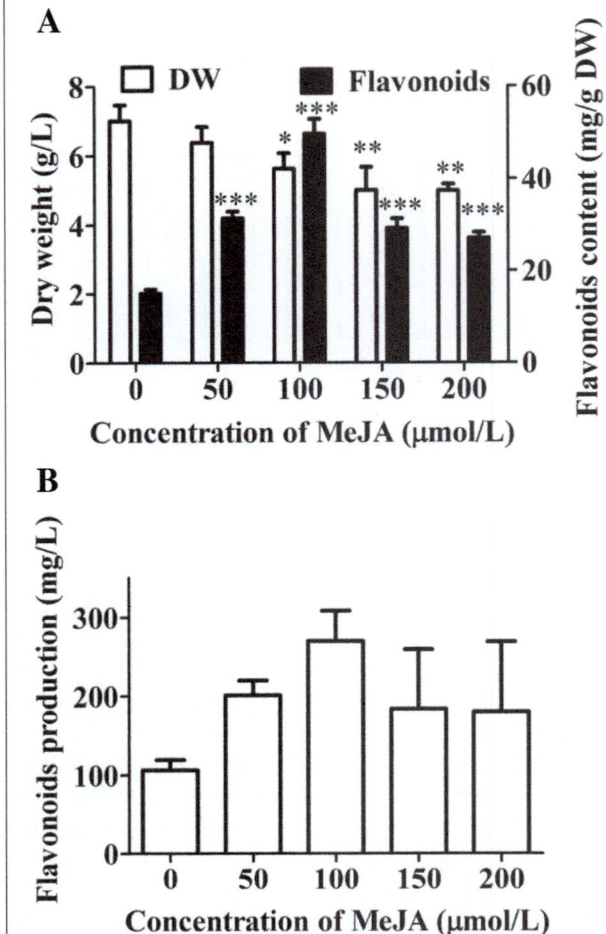

Figure 6 Effects of MeJA dosage on cell growth and flavonoid content (A) and flavonoid production (B). *$P < 0.05$, **$P < 0.01$, and *** $P < 0.001$ indicate statistical significance compared to the control.

about 2.3-fold with a dosage of 100 μM as compared to the control cultures. The flavonoid content reached maximum (52.8 mg/g DW) at the elicitation dose of 100 μM, which was relatively higher than the previously reported values in the adventitious root culture (i.e., 42.7 and 48.6 mg/g DW) [5,17]. However, further dosage increase (150-200 μM) decreased the flavonoid biosynthesis compared to moderate MeJA levels (50-100 μM). Sensitivity of suspension cell cultures to elicitor concentration differs with plant species [19,29]. The negative effect of MeJA at higher concentrations (200 μM) on cell growth and metabolite production was also reported in *Gymnema sylvestre* cell suspension cultures [29]. The best dose of the elicitor with maximum production (279.5 mg/L) of flavonoids was 100 μM as shown in Figure 6B, which was about 2.7-fold the value of control cultures. The above-mentioned results indicate that the optimal MeJA elicitation strategy was treatment on day 15 with a dosage of 100 μM.

Effects of MeJA addition on CAT and PAL activities

The enzyme activity changes were often related to the secondary metabolite accumulation [35]. For a better understanding of the elicitation effect of MeJA, the activities of two important enzymes (CAT and PAL) under MeJA treatment were determined. Reactive oxygen species (ROS) are toxic intermediates resulting from successive steps in the reduction of molecular O_2 in plant cells. The ROS exerts various effects on plant defense responses, including cell wall reinforcement, hypersensitive cell death, defensive gene activation, as well as defensive compound induction [36]. CAT is a well-known intracellular enzyme which protects against the ROS generated within cells [37]. As shown in Figure 7A, the enzyme activity increased with the cultivation time and reached maximum on day 19, thereafter decreased. MeJA down-regulated the CAT activity after its addition to the cell cultures on day 15 and kept at a relatively lower level as compared to the control (Figure 7A), which led to a higher accumulation of ROS and resulted in a promotion of secondary metabolites production. Similar results were observed on phytoalexin production in the suspension cell culture of *Glycine max*, which indicated that the ROS mediated elicitor-induced accumulation of secondary metabolites [38].

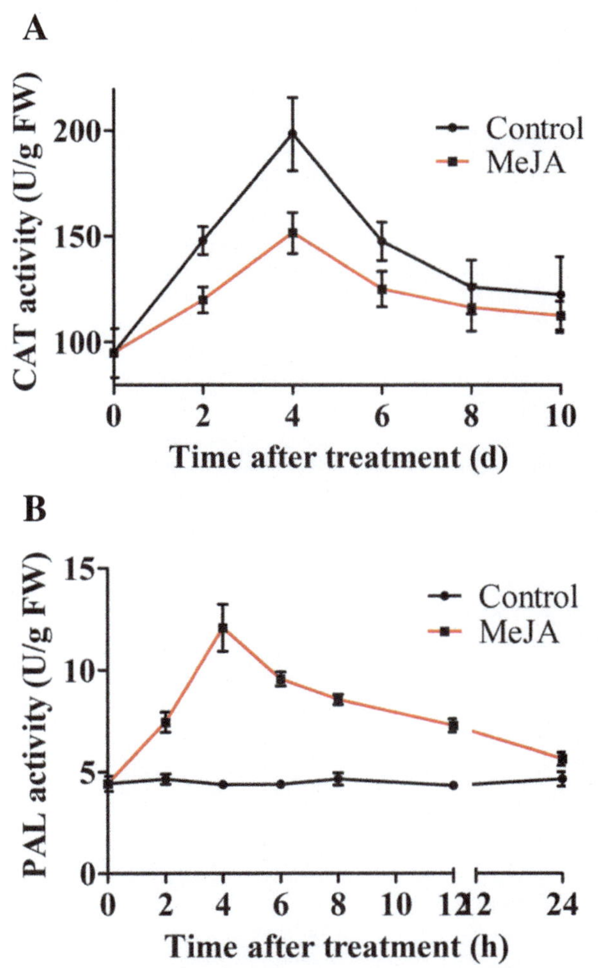

Figure 7 Dynamic changes of CAT (A) and PAL (B) activity with 100 μM of MeJA added on day 15 in cell suspension culture of *H. perforatum*.

PAL is the first enzyme of phenylpropanoid biosynthesis in plants and plays an important role in the biosynthesis of flavonoids, lignins, and many other compounds [36]. An increase in PAL activity after elicitor treatment often results in enhanced secondary metabolism in plant cells [19]. Figure 7B shows the changes of intracellular PAL activity in the suspension cultures after the treatment by MeJA (100 μM). As shown in Figure 7B, MeJA induced the increase of PAL activities and reached maximum (12.1 U/g FW) 4 days after treatment, thereafter decreased. However, cell cultures without MeJA treatment (control) revealed little change of PAL activities. In the presence of MeJA, PAL activity was much higher than that of control (Figure 7B), the elicitor-induced up-regulation of the PAL activity resulted in an improved flavonoid production in the suspension cultures (Figure 6B). Zhao et al. [19] have observed similar results in *Salvia miltiorrhiza* cell cultures, which indicated that both biotic and abiotic elicitation treatment resulted in the up-regulation of PAL activity and improved tanshinone accumulation as compared to the control cultures. These results demonstrated that MeJA is an efficient elicitor for the induction of flavonoid production in the suspension cell culture of *H. perforatum*.

Conclusions

Plant cell suspension culture is an efficient alternative method for the production of useful biochemicals. In this work, MeJA-elicited cell suspension cultures of *H. perforatum* enhanced the production of flavonoids through the down-regulation of the CAT activity and up-regulation of the PAL activity. The flavonoid production reached 280 mg/L after the optimal elicitation conditions, which was 2.7-fold of the control cultures. These results would be useful for the hyperproduction of flavonoids from *H. perforatum* and would be useful for the production of valuable compounds from other plant cell cultures.

Abbreviations

CAT: Catalase; DW: Dry cell weight; FW: Fresh cell weight; MeJA: Methyl jasmonate; PAL: Phenylalanine ammonia lyase; SA: Salicylic acid.

Competing interests

The authors declare that they have no competing interests.

Authors' contributions

J W was in charge of the experiments and paper writing. LY Y participated in the experiments and paper writing. J Q kindly offered experimental plant and guided the experiments. YH L directed the study as the tutor. All authors read and approved the final manuscript.

Acknowledgements

This work was supported by the National Special Fund for State Key Laboratory of Bioreactor Engineering (2060204) and the Fundamental Research Funds for the Central Universities (WF1113010).

Author details

[1]State Key Laboratory of Bioreactor Engineering, Department of Bioengineering, East China University of Science and Technology, 130 Meilong Rd., Shanghai 200237, China. [2]Shanghai Collaborative Innovation Center for Biomanufacturing Technology, 130 Meilong Road, Shanghai 200237, People's Republic of China. [3]School of Life Science and Technology, Tongji University, Shanghai 200092, China.

References

1. Wu SQ, Yu XK, Lian ML, Park SY, Piao XC (2014) Several factors affecting hypericin production of *Hypericum perforatum* during adventitious root culture in airlift bioreactors. Acta Physiol Plant 36:975–981
2. Zou Y, Lu Y, Wei D (2004) Antioxidant activity of a flavonoid-rich extract of *Hypericum perforatum* L. in vitro. J Agric Food Chem 52:5032–5039
3. Cui XH, Murthy HN, Wu CH, Paek KY (2010) Adventitious root suspension cultures of *Hypericum perforatum*: effect of nitrogen source on production of biomass and secondary metabolites. In Vitro Cell De-Pl 46:3728–3741
4. Saddiqe Z, Naeem I, Maimoona A (2010) A review of the antibacterial activity of *Hypericum perforatum* L. J Ethnopharmacol 131:511–521
5. Cui XH, Murthy HN, Jin YX, Yim YH, Kim JY, Paek KY (2011) Production of adventitious root biomass and secondary metabolites of *Hypericum perforatum* L. in a balloon type airlift reactor. Bioresour Technol 102:10072–10079
6. Bagdonaitė E, Mártonfi P, Repčák M, Labokas J (2012) Variation in concentrations of major bioactive compounds in *Hypericum perforatum* L. from Lithuania. Ind Crops Prod 35:302–308
7. Savio LEB, Astarita LV, Santarém ER (2011) Secondary metabolism in micropropagated *Hypericum perforatum* L. grown in non-aerated liquid medium. Plant Cell Tiss Org 108:465–472
8. Zou YP, Lu YH, Wei DZ (2010) Protective effects of a flavonoid-rich extract of *Hypericum perforatum* L. against hydrogen peroxide-induced apoptosis in PC12 cells. Phytother Res 24:6–10
9. Baque MA, Moh SH, Lee EJ, Zhong JJ, Paek KY (2012) Production of biomass and useful compounds from adventitious roots of high-value added medicinal plants using bioreactor. Biotechnol Adv 30:1255–1264
10. Wu J, Zhong JJ (1999) Production of ginseng and its bioactive components in plant cell culture: current technological and applied aspects. J Biotechnol 68:89–99
11. Priya B, Mantosh S, Aniruddha M (2014) Leaf extract mediated green synthesis of silver nanoparticles from widely available Indian plants: synthesis, characterization, antimicrobial property and toxicity analysis. Bioresour Bioprocess 1:3
12. Xu M, Yang B, Dong J, Lu D, Jin H, Sun L, Xu X (2011) Enhancing hypericin production of *Hypericum perforatum* cell suspension culture by ozone exposure. Biotechnol Prog 27:1101–1116
13. Cui XH, Chakrabarty D, Lee EJ, Paek KY (2010) Production of adventitious roots and secondary metabolites by *Hypericum perforatum* L. in a bioreactor. Bioresour Technol 101:4708–4716
14. Walker TS, Pal Bais H, Vivanco JM (2002) Jasmonic acid-induced hypericin production in cell suspension cultures of *Hypericum perforatum* L. (St. John's wort). Phytochemistry 60:289–293
15. Tocci N, D'Auria FD, Simonetti G, Panella S, Palamara AT, Pasqua G (2012) A three-step culture system to increase the xanthone production and antifungal activity of *Hypericum perforatum* subsp. *angustifolium* in vitro roots. Plant Physiol Biochem 57:54–58
16. Tocci N, Simonetti G, D'Auria FD, Panella S, Palamara AT, Valletta A, Pasqua G (2011) Root cultures of *Hypericum perforatum* subsp. *angustifolium* elicited with chitosan and production of xanthone-rich extracts with antifungal activity. Appl Microbiol Biotechnol 91:977–987
17. Cui XH, Murthy HN, Paek KY (2014) Pilot-scale culture of *Hypericum Perforatum* L. adventitious roots in airlift bioreactors for the production of bioactive compounds. Appl Biochem Biotechnol 14:1123–1128
18. Gadzovska S, Maury S, Delaunay A, Spasenoski M, Hagège D, Courtois D, Joseph C (2012) The influence of salicylic acid elicitation of shoots, callus, and cell suspension cultures on production of naphtodianthrones and phenylpropanoids in *Hypericum perforatum* L. Plant Cell Tiss Org 113:25–39
19. Zhao JL, Zhou LG, Wu JY (2010) Effects of biotic and abiotic elicitors on cell growth and tanshinone accumulation in *Salvia miltiorrhiza* cell cultures. Appl Microbiol Biotechnol 87:137–144
20. Gueven A, Knorr D (2011) Isoflavonoid production by soy plant callus suspension culture. J Food Eng 103:237–243
21. Lin L, Wu J (2002) Enhancement of shikonin production in single- and two-phase suspension cultures of *Lithospermum erythrorhizon* cells using low-energy ultrasound. Biotechnol Bioeng 5:78–81
22. Murashige T, Skoog F (1962) A revised medium for rapid growth and bio assays with tobacco tissue cultures. Physiol Plant 15:473–485
23. Georgiev M, Abrashev R, Krumova E, Demirevska K, Ilieva M, Angelova M (2009) Rosmarinic acid and antioxidant enzyme activities in *Lavandula vera* MM cell suspension culture: a comparative study. Appl Biochem Biotechnol 159:415–423
24. Lowry OH, Rosebrough NJ, Farr AL, Randall RJ (1951) Protein measurement with the Folin phenol reagent. J Biol Chem 193:265–271
25. Dong HD, Zhong JJ (2001) Significant improvement of taxane production in suspension cultures of *Taxus chinensis* by combining elicitation with sucrose feed. Biochem Eng J 8:145–162
26. Cai Z, Riedel H, Saw NMMT, Mewis I, Reineke K, Knorr D, Smetanska I (2011) Effects of elicitors and high hydrostatic pressure on secondary metabolism of *Vitis vinifera* suspension culture. Process Biochem 46:1411–1422
27. Ketchum RE, Tandon M, Gibson DM, Begley T, Shuler M (1999) Isolation of labeled 9-dihydrobaccatin III and related taxoids from cell cultures of *Taxus canadensis* elicited with methyl jasmonate. J Nat Prod 62:1395–1406
28. Saniewski M, Miszczak A, Kawa-Miszczak L, Wegrzynowicz-Lesiak E, Miyamoto K, Ueda J (1998) Effects of methyl jasmonate on anthocyanin accumulation, ethylene production, and CO_2 evolution in uncooled and cooled tulip bulbs. J Plant Growth Regul 17:33–42
29. Veerashree V, Anuradha C, Kumar V (2012) Elicitor-enhanced production of gymnemic acid in cell suspension cultures of *Gymnema sylvestre* R. Br. Plant Cell Tiss Org 108:27–33
30. Huang C, Zhong JJ (2013) Elicitation of ginsenoside biosynthesis in cell cultures of *Panax ginseng* by vanadate. Process Biochem 48:1227–1245
31. Qian ZG, Zhao ZJ, Xu YF, Qian XH, Zhong JJ (2005) A novel synthetic fluoro-containing jasmonate derivative acts as a chemical inducing signal for plant secondary metabolism. Appl Microbiol Biotechnol 68:98–103
32. Qian ZG, Zhao ZJ, Xu YF, Qian XH, Zhong JJ (2004) Novel chemically synthesized hydroxyl-containing jasmonates as powerful inducing signals for plant secondary metabolism. Biotechnol Bioeng 86:809–816
33. Huang C, Zhong JJ (2013) Enhancement of ginsenoside biosynthesis in cell cultures of Panax ginseng by N,N'-dicyclohexylcarbodiimide elicitation. J Biotechnol 165:30–36
34. Vasconsuelo A, Boland R (2007) Molecular aspects of the early stages of elicitation of secondary metabolites in plants. Plant Sci 172:861–877
35. Xu C, Zhao B, Ou Y, Wang X, Yuan X, Wang Y (2006) Elicitor-enhanced syringin production in suspension cultures of *Saussurea medusa*. World J Microb Biotechnol 23:965–979
36. Zhao J, Davis LC, Verpoorte R (2005) Elicitor signal transduction leading to production of plant secondary metabolites. Biotechnol Adv 23:283–294
37. Kim YH, Kim Y, Cho E, Kwak S, Kwon S, Bae J, Huh GH (2004) Alterations in intracellular and extracellular activities of antioxidant enzymes during suspension culture of sweetpotato. Phytochemistry 65:2471–2484
38. Guo ZJ, Lamb C, Dixon RA (1998) Potentiation of the oxidative burst and isoflavonoid phytoalexin accumulation by serine protease inhibitors. Plant Physiol 118:1487–1495

6

Synthesis of silver nanoparticles using seed exudates of *Sinapis arvensis* as a novel bioresource, and evaluation of their antifungal activity

Mehrdad Khatami[1,2*], Shahram Pourseyedi[1], Mansour Khatami[2], Hadi Hamidi[1], Mehrnaz Zaeifi[3] and Lida Soltani[3]

Abstract

Background: In general, silver nanoparticles (AgNPs) are particles of si ver with a size less than 100 nm. In recent years, synthesis of nanoparticles using plant extract has gained much interest in nanobiotechnology. In this concern, this study investigates green synthesis of AgNPs from silver nitrate using *Sinapis arvensis* as a novel bioresource of cost-effective nonhazardous reducing and stabilizing compounds. A stock solution of silver nitrate (0.1 M) was prepared. Different concentrations of silver nitrate (1, 2.5, 4, and 5 mM) were prepared from the above solution, then added to 5 mL of *S. arvensis* seed exudates. The mixtures were kept in 25°C. The synthesis of AgNPs was confirmed by the change in mixtures color from light yellow to brown. The antifungal activity of synthesized AgNPs was investigated *in vitro*.

Results: The resulting AgNPs were characterized by UV-vis spectroscopy, X-ray diffraction (XRD), transmission electron microscopy (TEM), and Fourier transform infrared spectroscopy (FTIR). Formation of the AgNPs was confirmed by the change in mixture color from light yellow to brown and maximum absorption at 412 nm due to surface plasmon resonance of AgNPs. The role of different functional groups in the formation of AgNPs was shown by FTIR. X-ray diffraction was shown that the AgNPs formed in our experiments were in the form of nanocrystal, and TEM analysis showed spherical particles with an average size of 14 nm. Our measurements indicated that *S. arvensis* seed exudates can mediate facile and eco-friendly biosynthesis of colloidal-spherical AgNPs with a size range of 1 to 35 nm. The synthesized AgNPs showed significance antifungal activity against *Neofusicoccum parvum* cultures.

Conclusion: The AgNPs were synthesized using a biological source. This synthesis method is nontoxic, eco-friendly, and a low-cost technology for the large-scale production. The AgNPs can be used as a new generation of antifungal agents.

Keywords: Synthesis; AgNPs; Spherical AgNPs; Antifungal

Background

Silver nanoparticles (AgNPs) are particles of silver which are in the range between 1 and 100 nm. Nanostructure materials indicate unique physicochemical and biological environmental properties, including optical, magnetic, electronic, catalytic activity, and biological properties [1], which have increased their applications in medicine, agriculture, environment, and industry [2]. AgNPs have high potential as commercial nanomaterials [3] and an effective antimicrobial agent.

Several techniques for synthesizing AgNPs have been proposed. Generally, AgNPs are prepared by different kinds of chemical and physical methods, but majority of these techniques are both expensive and environmentally hazardous [4]. Furthermore, the synthesized nanoparticles may be unstable and tend to agglomerate rapidly and become useless unless capping agents are applied for their stabilization [5]. Diverse chemical and physical methods have been used to prepare AgNPs with various sizes and shapes, such as UV irradiation [6,7], microware irradiation [8], chemical reduction [9],

* Correspondence: mehrdad7khatami@gmail.com
[1]Department of Biotechnology, Shahid Bahonar University of Kerman, End of 22-Bahman Blvd, 76169–133, Kerman 76189-18951, Iran
[2]Department of Enviroment, The Enviromental Researches Center, Havaniroz Street, Jomhori Blvd, Kerman 76189-18951, Iran
Full list of author information is available at the end of the article

electron irradiation [10], photochemical [11], and lithography methods [12]. However, most of these methods involve more than one step, high energy requirement, low material conversions, difficulty in purification, and hazardous chemicals [13]. The synthesis of nanoparticles by chemical methods may lead to the production of some toxic chemical compound that may have adverse effects on their applications [14].

The biological synthesis of nanoparticles can potentially eliminate these problems. Biological synthesis of nanoparticles is nontoxic, eco-friendly, and a low-cost technology for the large-scale production of well-characterized nanoparticles [14]. Therefore, there is a need to develop biological processes for nanoparticle synthesis. Recently, many live organisms such as bacteria, fungi, algae, and plants have been used for synthesis of nanoparticles [15-17]. The reduction of Ag^+ to Ag^0 took place by combinations of biomolecules such as proteins, polysaccharides, and flavonoids [18]. Green synthesis of AgNPs has become very important in the recent years. Green AgNPs have the potential for large scale applications in the formulation of dental resin composites, bone cement [19,20], water and air filters [21,22], clothing and textiles, medical devices and implants [23], cosmetics [4], and packaging [24]. Besides their antimicrobial properties, AgNPs and silver nanocomposites have other interesting characteristics which will further enable them to be used in catalysts, biosensors, conductive inks, and electronic devices [25,26]. They can be produced economically and in large industrial scale [14].

In this paper, we report biosynthesis of stable colloidal AgNPs using *Sinapis arvensis* seed exudates. This plant is an important medicinal crop in the southern regions of Iran. Antifungal activities of synthesized AgNPs were also investigated. Details of biosynthesis, physical characterizations, and antifungal activity of AgNPs are described.

Methods

Reagents

Silver nitrate and potato dextrose agar (PDA) medium were obtained from Merck, Darmstadt, Germany. Seeds of *S. arvensis* were obtained from Pakanbazr, Isfahan, Iran. Strain of the fungus *N. parvum* was prepared from the Department of Plant Protection, Shahid Bahonar University of Kerman, Iran.

Seed exudates preparation

The surface of *S. arvensis* seeds were disinfected using 30% sodium hypochlorite for 5 min and rinsed with sterile distilled water three times. In the next step, the seeds were placed in 70% alcohol for 2 min and then washed four times with sterile distilled water and then imbibed in deionized (DI) water. (1 g dry weight/10 mL DI water) After being incubated at 25°C for 48 h in the dark, seeds were removed from the soaking medium. The supernatant phase was collected and centrifuged at 4,500 rpm for 10 min to separate the liquid fraction from any large insoluble particles and filtered by Whatman filter paper no. 1 (Sigma Aldrich, St. Louis, MO, USA). During the experiment, pH was 4.5.

Biosynthesis of AgNPs

For this reason, a 50 mL stock solution containing 0.1 M silver nitrate was prepared. Different concentrations of silver nitrate (1, 2.5, 4, and 5 mM) were prepared from the above solution, then added to 5 mL of *S. arvensis* seed exudates and incubated at 25°C as described previously [27]. After treatment, the pale yellow color of reaction mixture was changed to brown indicating synthesis of AgNPs.

Characterization of AgNPs

UV-vis spectroscopy

The biosynthesis of AgNPs was monitored periodically using a UV-vis spectrophotometer (Scan Drop-type product, Analytik Jena, Germany) at different concentrations at room temperature. These measurements operated at a resolution of 1 nm and wavelength range between 300 and 650 nm [14].

X-ray diffraction

The formation and quality of compounds were investigated by X-ray diffraction (XRD) technique. For this purpose, AgNPs was centrifuged (at 13,000 rpm; 25°C) for 10 min, washed with DI water and re-centrifuged in three cycles. Then purified AgNPs were dried and subjected to XRD experiment. X-ray diffraction was performed by STOE Stadi P (STOE & Cie. GmbH, Darmstadt, Germany) ($\lambda = 1.54178$ Å). The scanning was done in the region of 2θ from 10° to 80° [17].

Transmission electron microscopy

Transmission electron microscopy (TEM) was performed by using of a Carl Zeiss (Jena, Germany, 80 kV) for determining the morphology and size of AgNPs.

Inductively coupled plasma spectrometry

Inductively coupled plasma spectrometry (ICP) Varian BV ES-700, Sydney, Australia, was used to determining the remaining concentration of silver ions after synthesizing AgNPs. Changing rate of metal ion to nanoparticles is calculated by the following equation:

$$Q = \left(\frac{C_0 - C_f}{C_0}\right) \times 100$$

where C_0 and C_f stands for initial and final concentration

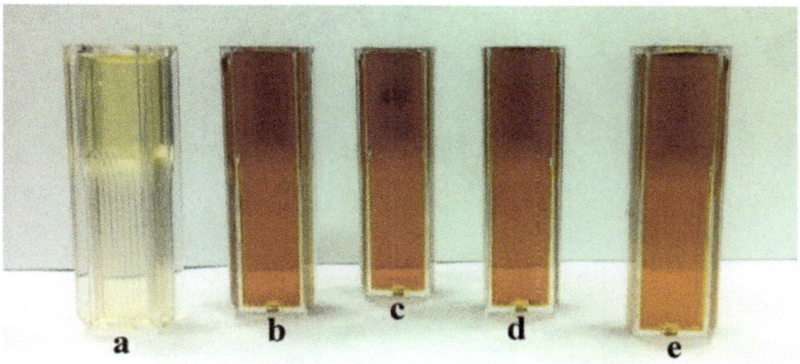

Figure 1 Reduction of Ag^+ to Ag^0. The color change of the *S. arvensis* exudates upon the formation of AgNPs from different concentrations of silver nitrate: **(a)** control, **(b)** 1 mM, **(c)** 2.5 mM, **(d)** 4 mM, and **(e)** 5 mM.

of silver ion (mg/L), respectively. Q is the conversion percentage of silver ion to AgNPs [28].

Antifungal activity of AgNPs

To determine the antifungal activity of AgNPs, mycelium growth inhibition test was used. Four concentrations of AgNPs (2.5, 5, 10, and 40 μg/mL) were prepared in PDA medium after autoclaving. 6-mm agar plugs of fresh culture of *N. parvum* prepared and transferred to centers of media containing different concentrations of synthesized AgNPs. Control plates contain no AgNPs. All plates incubated at 28°C. When in control, the fungus completely covered the entire surface of the medium, the mean radius of fungal growth in all plates measured and recorded [29]. All treatments were performed in triplicates.

The following formula was used to assess of the growth inhibition of mycelium:

$$\text{Mycelium growth inhibition }\% = \frac{R-r}{R} \times 100$$

R is the mean radius of control and r is the radius of samples treated with nanoparticle. Data processed with SAS statistical analysis using Duncan's test.

Results and discussion

Visual observation

Reduction of Ag^+ to Ag^0 was confirmed by color change of the reaction mixture from colorless to brown (Figure 1).

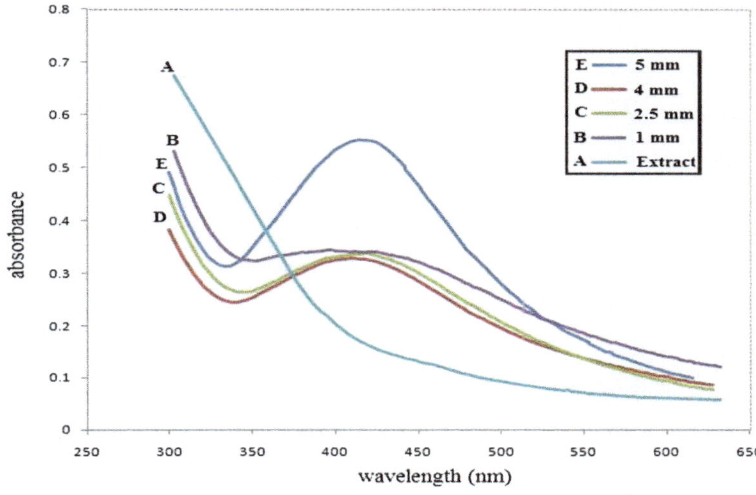

Figure 2 UV-vis absorption spectrum of AgNPs synthesized by treating 1, 2.5, 4, and 5 mM silver nanoparticles after 1 week.

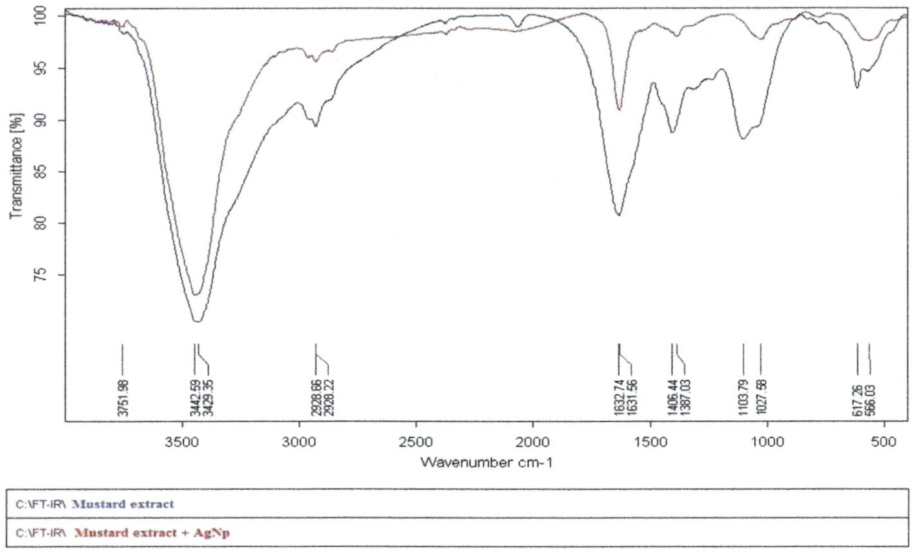

Figure 3 FTIR spectra of the *S. arvensis* exudates before and after the synthesis of AgNPs.

Ultraviolet visible scanning spectroscopy studies

It was observed that the maximum absorbance of reaction mixture occurs at 412 nm, indicating that AgNPs were produced. Figure 2 shows the UV-vis absorption spectra of synthesized AgNPs with different concentrations of silver nitrate.

FTIR analysis

Fourier transform infrared spectroscopy (FTIR) spectrum of biosynthesized AgNPs shows absorption peaks at 3,429, 2,928, 1,632, 1,406, 1,103, and 617 cm^{-1} (Figure 3). Strong absorption peak at 3,429 cm^{-1} is resulted from stretching of the N-H band of amino groups or is indicative of present O-H groups due to the presence of alcohols, phenols, carbohydrates, etc. The peak that appeared around 2,928 cm^{-1} is related to the stretching of the C-H bonds [30]. The peaks at 1,632 and 1,406 cm^{-1} are assigned for aliphatic amines. The absorption peak at 1,632 cm^{-1} is close to that reported for native proteins [31]. The peak at 1,406 cm^{-1} corresponds to C-C stretching vibrations for aromatic ring [32]. FTIR study indicates that probably the carboxyl (-C=O), hydroxyl (-OH), and amine (N-H) groups in seed exudates are mainly involved in the reduction of Ag^{+} ions to Ag^{0} nanoparticles.

XRD analysis

The formation of the nanocrystalline AgNPs was further confirmed by the XRD analysis as showed in Figure 4.

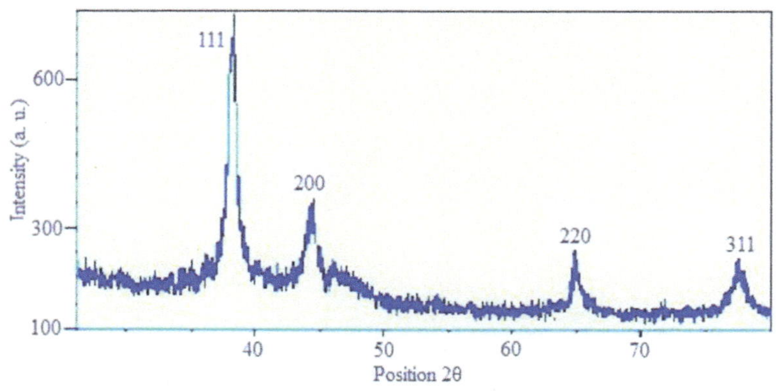

Figure 4 XRD pattern of AgNPs synthesized by treating silver nitrate with *S. arvensis* seed exudates.

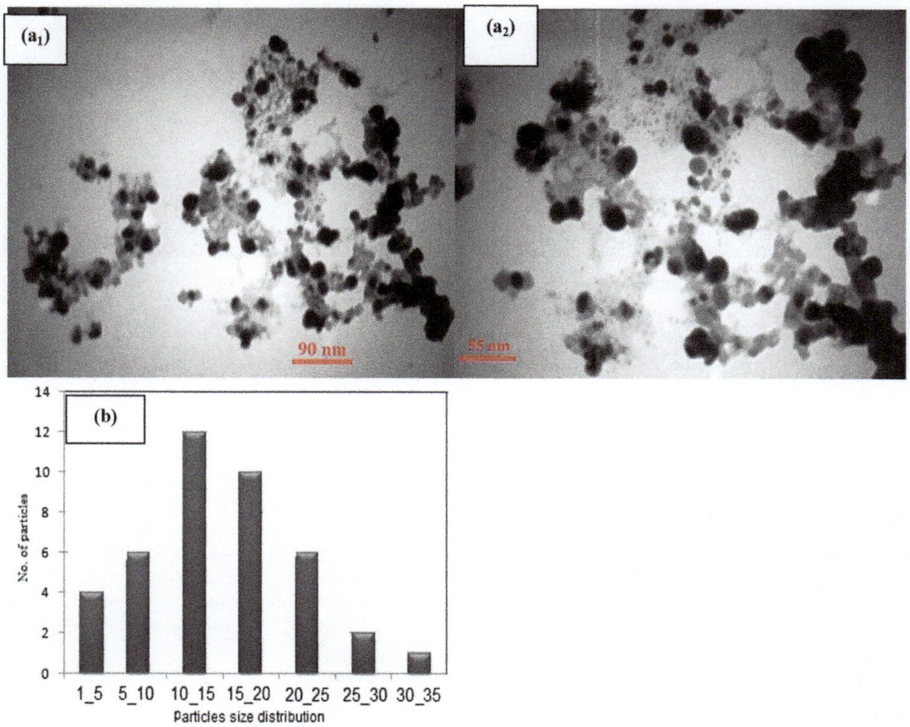

Figure 5 The TEM images of the prepared AgNPs. (a_1, a_2) TEM images of AgNPs that were synthesized by *S. arvensis* seed exudates in different scales. **(b)** Histogram of particle size distribution of the biosynthesized AgNPs.

Strong peaks were observed at 2θ values at 38.09°, 44.15°, 64.67°, and 77.54°, corresponding to (111), (200), (220), and (311) Bragg's reflection based on the face-centered-cubic (fcc) crystal structure of AgNPs [33]. The broadening of Bragg's peaks indicates the formation of AgNPs. The XRD pattern thus shows that the AgNPs formed by the reduction of Ag^+ ions by *S. arvensis* seed exudates are crystalline in nature.

Transmission electron microscopy analysis

TEM was used to determine the size and shape of nanoparticles. The TEM images of the prepared AgNPs at 55 and 90 nm scales are shown in the Figure 5a_1 and a_2. TEM images show that they have spherical shape. Particle size distribution histogram determined from TEM is shown in Figure 5b. AgNPs size is between 1 and 35 nm.

Table 1 ICP analysis

Q (%)	C_f (μg/mL)	C_0 (μg/mL)
0.955	23.85	535

Inductively coupled plasma spectrometry

ICP analysis revealed complete reduction of Ag ions within 50 days of the reaction, and more importantly, it showed that the conversion percentage of metal ion to metal nanoparticles is more than 95% (Table 1).

Antifungal activity of synthesized AgNPs

Inhibitory effects on fungal growth in a PDA medium containing 2.5, 5, 10, and 40 μg/mL concentrations of AgNPs were studied. The results showed a very significant effect of synthesized AgNPs on the mycelium growth of the fungus *N. parvum*. More than 83% mycelium growth inhibition of the fungus *N. parvum* was treated with a concentration of 40 μg/mL of AgNPs. The lowest level of growth inhibition was observed at a concentration of 2.5 μg/mL of AgNPs with the 15% mycelium growth inhibition. The growth inhibition for concentrations of 5 and 10 μg/mL, respectively, were 15% and 71%. Results clearly showed that with the increase in the concentration of AgNPs, the inhibitory effects on fungal mycelium growth increased (Figure 6).

Percentage of growth inhibition due to the effect of AgNPs was analyzed by statistical software SAS. Results confirmed a significant effect of AgNPs on fungal growth inhibition at 1% confidence interval.

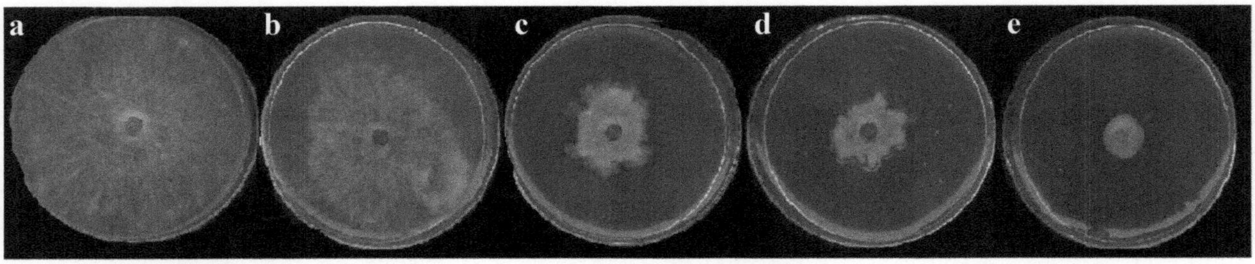

Figure 6 Inhibitory effects on fungal mycelium growth. Fungal growth on PDA medium containing control **(a)**. PDA medium containing **(b)** 2.5 μg/mL, **(c)** 5 μg/mL, **(d)** 10 μg/mL, and **(e)** 40 μg/mL of AgNPs.

Conclusions

Metallic nanoparticles are traditionally synthesized by wet chemical synthesis techniques where the chemicals used are quite often toxic and flammable. But in this study, the *S. arvensis* seed exudates were successfully used for the single-pot biosynthesis of spherical AgNPs in ambient conditions with the size range from 1 to 35 nm, as inferred from the TEM imaging. This was achieved without the use of external stabilizing or capping agents. We concluded that *S. arvensis* seed exudates are a bioreductant and capping agent as well as an easily available plant source playing an important role in the synthesis of highly stable AgNPs. X-ray diffraction pattern strongly indicated a high purity of biosynthesized AgNPs. This pristine method is facile, cost effective, clean and green, and therefore is applicable for a variety of purposes. Moreover, it is easy to scale up the production of AgNPs to industrial scale using this method.

This green chemistry approach towards the synthesis of AgNPs has many advantages such as ease with which the process can be scaled up, economic viability, etc. Application of such nanoparticles in medicine and other applications makes this method useful for the large scale synthesis of other inorganic nanomaterials. Toxicity studies of AgNPs on pathogen open a door for a new range of antifungal and antibacterial agents. So in the present study, we demonstrated that AgNPs have significant antifungal activity. It was determined that the growth inhibitory effect of AgNPs strongly depends on the concentration of AgNPs and with the increase concentration of AgNPs in the medium the inhibitory effect on fungal growth increased. So AgNPs can be used as excellent new antifungal agent.

Competing interests

The authors declare that they have no competing interests.

Authors' contributions

All of them have been also involved in the drafting and revision of the manuscript. All authors read and approved the final manuscript.

Acknowledgements

The authors are thankful to all members of the Department of Biotechnology Science, University of Kerman. We also acknowledge Mr. Abazari, Laboratory of Science, University of Kerman, for helping us. Also, the authors are thankful to the Iran Nanotechnology Council. Dr. Jian He Xu and Dr. Zolala J have made substantive intellectual contributions to this study, substantial contributions to the conception and design of it as well as to the acquisition, analysis, and interpretation of data.

Author details

[1]Department of Biotechnology, Shahid Bahonar University of Kerman, End of 22-Bahman Blvd, 76169–133, Kerman 76189-18951, Iran. [2]Department of Enviroment, The Enviromental Researches Center, Havaniroz Street, Jomhori Blvd, Kerman 76189-18951, Iran. [3]Department of Plant Protection, Shahid Bahonar University of Kerman, End of 22-Bahman Blvd, 76169–133, Kerman 76169-14111, Iran.

References

1. Phanjom P, Ahmed G (2015) Biosynthesis of silver nanoparticles by *Aspergillus oryzae* (MTCC No. 1846) and its characterizations. Nanoscience and Nanotechnology 5:14-21
2. Song JY, Kim BS (2009) Rapid biological synthesis of silver nanoparticles using plant leaf extract. Bioprocess Biosyst Eng 32:79–84
3. Chaloupka K, Malam Y, Seifalian AM (2010) Nanosilver as a new generation of nanoproduct in biomedical applications. Trends Biotechnol 28:580–588
4. Sintubin L, Verstraete W, Boon N (2012) Biologically produced nanosilver: current state and future perspectives. Bioeng 109:2422–2436
5. Kassaee M, Akhavan A, Sheikh N, Sodagar A (2008) Antibacterial effects of a new dental acrylic resin containing silver nanoparticles. J Appl Polym Sci 110:1699–1703
6. Alt V, Bechert T, Steinrücke P, Wagener M, Seidel P, Dingeldein E, Doman E, Schnettler R (2004) An in vitro assessment of the antibacterial properties and cytotoxicity of nanoparticulate silver bone cement. Biomaterials 25:4383–4391
7. Jain P, Pradeep T (2005) Potential of silver nanoparticle-coated polyurethane foam as an antibacterial water filter. Biotechnol Bioeng 90:59–63
8. Sharma VK, Yngard RA, Lin Y (2009) Silver nanoparticles: green synthesis and their antimicrobial activities. Adv Colloid Interfac 145:83–96
9. de Mel A, Chaloupka K, Malam Y, Darbyshire A, Cousins B, Seifalian AM (2012) A silver nanocomposite biomaterial for blood-contacting implants. J Biomed Mater Res A Part A 100:2348–2357
10. Kokura S, Handa O, Takagi T, Ishikawa T, Naito Y, Yoshikawa, T (2010) AgNPs as a safe preservative for use in cosmetics. Nanotechnol 6:6570–574
11. Azeredo H (2009) Nanocomposites for food packaging applications. Res Int 42:1240–1253
12. Tsuji M, Gomi S, Maeda Y, Matsunaga M, Hikino S, Uto K, Tsuji T, Kawazumi H (2012) Rapid transformation from spherical nanoparticles, nanorods, cubes, or bipyramids to triangular prisms of silver with PVP, citrate, and H_2O_2. Langmuir 28:8845–8861

13. Ghaffari-Moghaddam M, Hadi-Dabanlou R, Khajeh M, Rakhshanipour M, Shameli K (2014) Green synthesis of AgNPs using plant extracts. Korean J Chem Engineering 31:548–557
14. Banerjee P, Satapathy M, Mukhopahayay A, Das P (2014) Leaf extract mediated green synthesis of silver nanoparticles from widely available Indian plants: synthesis, characterization, antimicrobial property and toxicity analysis. Bioresources and Bioprocessing 1:3
15. Mohammadinejad R, Pourseyedi S, Baghizadeh A, Ranjbar S, Mansoori G (2013) Synthesis of silver nanoparticles using *Silybum marianum* seed extract. Int J Nanosci Nanotechnol 9:221–226
16. Swamy M.K, Sudipta K, Jayanta K, Balasubramanya S (2015) The green synthesis, characterization, and evaluation of the biological activities of silver nanoparticles synthesized from *Leptadenia reticulata* leaf extract. Appl. Nanosci 5:1–9
17. Lei B, Zhang X, Zhu M, Tan W (2014) Effect of fluid shear stress on catalytic activity of biopalladium nanoparticles produced by *Klebsiella Pneumoniae* ECU-15 on Cr(VI) reduction reaction. Bioresources and Bioprocessing 1:28
18. Speth TF, Varma RS (2011) Microwave-assisted green synthesis of silver nanostructures. Accou of Chem Rese Acc of Cheml Resea 44:469–478
19. Song K, Lee S, Park T, Lee B (2009) Preparation of colloidal silver nanoparticles by chemical reduction method. Kore J of Che Eng 26:153–155
20. Li K, Zhang FS (2010) A novel approach for preparing silver nanoparticles under electron beam irradiation. J of Nanopa Resea 12:1423–1428
21. Harada M, Kawasaki C, Saijo K, Demizu M, Kimura Y (2010) Photochemical synthesis of silver particles using water-in-ionic liquid microemulsions in high-pressure CO_2. J of Coll Interf Sci 343:537–545
22. Jensen MD, Malinsky CL, Haynes RPV (2000) Duyne, Nanosphere lithography: tunable localized surface plasmon resonance spectra of silver nanoparticles. J Phys Chem. 104 1059
23. Philip D (2011) *Mangifera indica* leaf-assisted biosynthesis of well-dispersed AgNPs. Spectrochimica Acta 78:327–331
24. Bankar A, Joshi B, Kumar AR, Zinjarde S (2009) Banana peel extract mediated novel route for synthesis of AgNPs. Colloid Surf A Physicochem Eng 368:58–63
25. Dwivedi AD, Gopal K (2010) Biosynthesis of silver and gold nanoparticles using *Chenopodium album* leaf extract. Colloid Surf A Physicochem Eng Aspect 369:27–33
26. Velmurugan P, Shim J, Kamala-Kannan S, Lee KJ, Oh BT, Balachandar V (2011) Crystallization of silver through reduction process using *Elaeis guineensis* biosolid extract. Biotechnol Prog 27:273–279
27. Singh K, Panghal M, Kadyan S, Chaudhary U, Parkash YJ (2014) Antibacterial activity of synthesized silver nanoparticles from *Tinospora cordifolia* against multi drug resistant strains of *pseudomonas aeruginosa* isolated from burn patients. J Nanomed Nanotechnol 5:2
28. Dubey M, Bhadauria S, Kushwah BS (2009) Green synthesis of nanosilver particles from extract of *Eucalyptus hybrida* (safeda) leaf. Dig J Nanomater Bios 4(34):537–543
29. Kim SW, Jung JH, Lamsal K, Kim YS, Min JS, Lee YS (2012) Antifungal effects of silver nanoparticles (AgNPs) against various plant pathogenic fungi. Koren j of Microbiol 40:53–58
30. Mulvaney P (1996) Surface plasmon spectroscopy of nanosized metal particles. Longmuir 12:788–800
31. Khatami M, Pourseyedi S (2015) *Phoenix dactylifera* (date palm) pit aqueous extract mediated novel route for synthesis high stable AgNPs with high antifungal and antibacterial activity. IET Nanobiotechnol 9:1–7
32. Yilmaz M, Turkdemir H, Kilic MA, Bayram E, Cicek A, Mete A, Ulug B (2011) Biosynthesis of AgNPs using leaves of *Stevia rebaudiana*. Mater Chem Phys 130:1195–1202
33. Kaviya S, Santhanalakshmi J, Viswanathan B, Muthumary J, Srinivasan K (2011) Biosynthesis of silver nanoparticles using *Citrus sinensis* peel extract and its antibacterial activity. Spectrochimica Acta 79:594–598

7

Biocatalytic synthesis of ethyl (*R*)-2-hydroxy-4-phenylbutyrate with a newly isolated *Rhodotorula mucilaginosa* CCZU-G5 in an aqueous/organic biphasic system

Liqun Wang[1*], Junjie Miao[1], Zhongqiang Wang[2], Lijuan Wang[1], Qing Qing[1] and Shang-Tian Yang[2*]

Abstract

Background: Optically active ethyl (*R*)-2-hydroxy-4-phenylbutyrate [(*R*)-HPBE] is an important chiral building block for the synthesis of angiotensin-converting enzyme (ACE) inhibitors. It is reported that microbial or enzymatic reduction of ethyl 2-oxo-4-phenyl-butyrate (OPBE) is an attractive way to produce optically active (*R*)-HPBE.

Results: The asymmetric reduction of OPBE to synthesize optically active (*R*)-HPBE with a newly isolated *Rhodotorula mucilaginosa* CCZU-G5 as catalyst was investigated in an aqueous/organic solvent biphasic system. *R. mucilaginosa* CCZU-G5 showed a good tolerance (the metabolic activity retention >80%) in the biphasic system composed of aqueous buffer and organic solvent with a log *P* value over 4.6. Isooctane was found to be the most suitable organic phase solvent. In the biphasic system, the volumetric phase ratio, OPBE concentration, cell concentration, reaction temperature, and buffer pH were optimized. Under the optimum conditions (volumetric phase ratio: 1/1, OPBE concentration: 100 mM, cell concentration: 0.075 g/mL, pH 7.5, 35°C), the final yield and the optical purity of (*R*)-HPBE reached 98.3% and >99.0% enantiomeric excess (*ee*), respectively, after 12 h of reaction.

Conclusions: All the results suggested that the OPBE-reducing enzymes in a newly isolated *R. mucilaginosa* cells possess highly stable and excellent stereoselectivity by establishing an aqueous/organic biphasic system.

Keywords: *Rhodotorula mucilaginosa* CCZU-G5; Biphasic system; Isooctane; Ethyl (*R*)-2-hydroxy-4-phenylbutyrate

Background

Optically active ethyl (*R*)-2-hydroxy-4-phenylbutyrate [(*R*)-HPBE] is an important chiral building block for the synthesis of angiotensin-converting enzyme (ACE) inhibitors such as benazepril, enalapril, and lisinopril [1]. In general, ACE inhibitors prevent the conversion of the precursor decapeptide angiotensin I to the powerful vasoconstrictor substance angiotensin II and have been demonstrated to be potent antihypertensive drugs [2]. In recent years, various chemical and biological approaches for (*R*)-HPBE preparation have been reported, mainly in two ways: kinetic resolution and synthesis. However, chemical synthesis usually involves multiple steps and stringent reaction conditions [3], and resolution methods are limited by theoretical maximum yield of only 50% [4].

Microbial or enzymatic reduction of ethyl 2-oxo-4-phenyl-butyrate (OPBE) is an attractive way to produce optically active (*R*)-HPBE, since OPBE can be easily synthesized and is relatively cheap. Several biocatalysts have been used in the synthesis of (*R*)-HPBE, including the hydrolysis and transesterification catalyzed by lipase [5] and the reduction of OPBE catalyzed by isolated dehydrogenase [6] and whole cells [7,8]. Since the reduction reaction requires stoichiometric amounts of nicotinamide cofactors, whole cells rather than isolated enzymes were used preferentially to avoid enzyme purification and cofactor addition [9].

In the past decade, however, only a few microorganisms have been reported as efficient biocatalysts in the reduction of OPBE to (*R*)-HPBE. Chadha et al. reported

* Correspondence: wangliqun567@gmail.com; yang.15@osu.edu
[1]School of Pharmaceutical Engineering and Life Sciences, Changzhou University, 1 Ge Hu Road, Jiangsu 213164, China
[2]William G. Lowrie Department of Chemical and Biomolecular Engineering, The Ohio State University, 140 W 19th Ave, Columbus, OH 43210, USA

that the enantioselective reduction of OPBE to (*R*)-HPBE could be achieved by using cell-free aqueous extracts of the callus of *Daucus carota* (wild carrot) with a high yield (90%) and enantiomeric excess (*ee*) (99%) [10]. However, their process required a high cell/substrate ratio of 100:1, a large amount of cells, and a long reaction time of 10 days. Dao et al. and Lacerda et al. reported the successful reduction of OPBE with *Pichia angusta* and *Saccharomyces cerevisiae*, respectively, to give (*R*)-HPBE with a moderate enantioselectivity (81% *ee*) [7,11]. Recently, Chen et al. described the successful preparation of (*R*)-HPBE with favorable *ee* (99%) and yield (92%) by using *Candida boidinii* CIOC21 [12]. However, the relatively low concentrations of the substrate (around 4.1 g/L) and product (around 3.8 g/L) in their process would restrict its application in large-scale production. In addition, Zhang et al. used *Candida krusei* SW2026 to produce (*R*)-HPBE from 20 g/L of OPBE with excellent *ee* (97.4%) and a moderate yield (82%) [8].

Recently, we have isolated a new yeast strain *Rhodotorula mucilaginosa* CCZU-G5 from vineyard soil samples and used it in preparing (*R*)-HPBE with high *ee* and yield. However, a severe substrate inhibition was observed when the tested OPBE concentration in an aqueous single-phase system was high due to the high hydrophobicity of the substrate and its toxicity to the cells, and the highest substrate concentration that the bacterium could transform was only 50 mM. The aqueous/organic solvent biphasic system is a good alternative to resolve the aforementioned problems occurred in the aqueous system. The organic solvent phase in the biphasic system acts as a substrate reservoir and prevents the cells in the aqueous phase from being damaged by high substrate concentration. This biphasic system has attracted great attention over the past few decades, and several successful examples have been reported [13-15].

In this study, resting cells of *R. mucilaginosa* CCZU-G5 were used as biocatalysts for asymmetric reduction of OPBE in an aqueous/organic solvent biphasic system. Various parameters such as substrate concentration, cell concentration, reaction temperature, and pH were investigated and optimized to improve the yield and optical purity (*ee*) of (*R*)-HPBE. Compared to the monophasic aqueous system, the asymmetric reduction of OPBE in the aqueous/organic solvent biphasic system gave excellent *ee* and a much higher yield due to reduced substrate and product inhibition. To our best knowledge, this is the first study using *R. mucilaginosa* cells for high-yield and high-purity production of (*R*)-HPBE.

Methods

Chemicals

(*R*)-HPBE was purchased from Sigma-Aldrich Chemical Co. (St. Louis, MO, USA) OPBE was supplied by Wujin Fine Chemical Factory Co., Ltd. (Changzhou, Jiangsu, China). All other reagents and solvents were commercially available, and were of analytical grade purity.

Microbial strain and cultivation conditions

The yeast *R. mucilaginosa* CCZU-G5 was isolated from vineyard soil samples and preserved in China General Microbiological Cultures Collection Center (CGMCC 6328). It was grown in the following medium: glucose 20 g/L, yeast extract 10 g/L, peptone 15 g/L, $(NH_4)_2SO_4$ 1 g/L, $MgSO_4{\cdot}7H_2O$ 1 g/L, pH 7.0. The strain was incubated aerobically at 30°C and 180 rpm in 500-mL Erlenmeyer flasks with 70 mL sterilized medium. After 72 h of growth, the cells were harvested by centrifugation (8,000×*g* for 10 min) at 4°C, washed twice with 0.85% (*w*/*v*) NaCl, and then stored at 4°C for further use.

Bioconversion in monophasic aqueous system

The reduction of OPBE in the aqueous system was conducted in a 50-mL Erlenmeyer flask capped with a septum. Two grams of wet cells was suspended in 20.0 mL phosphate-buffered saline (PBS) (0.1 M, pH 7.0) with 1 g glucose and 0.4 mmol OPBE. The cell suspensions were subsequently incubated in a rotary incubator at 35°C and 180 rpm. The mixture was centrifuged to remove the cells at different time intervals, and the supernatant was extracted three times with ethyl acetate and dried over anhydrous Na_2SO_4 for gas chromatography (GC) analysis.

Tolerance assay of *R. mucilaginosa* CCZU-G5 in biphasic systems

Four grams of harvested cells was suspended in 100 mL PBS (0.1 M, pH 7.0) to give a final concentration of 0.04 g/mL. Ten milliliters of the cell suspension was added to 10 mL of each of the different organic solvents in a 50-mL Erlenmeyer flask capped with a septum. The cell suspension (0.04 g/mL) in PBS without any organic solvent was used as a control. All cell suspensions were subsequently incubated in a rotary incubator at 30°C and 180 rpm for 24 h. The cell suspensions were then centrifuged (8,000×*g* for 10 min) at 4°C. Then the cells were transferred to 10 mL of 20 g/L glucose and incubated at 30°C and 180 rpm. After 4 h, the suspensions were centrifuged, and the supernatants were analyzed using a spectrophotometer to determine the concentration of glucose through DNS method to obtain the amount of consumed glucose [16]. The metabolic activity retention is defined as the ratio of the amount of glucose consumed by the cells pretreated in the biphasic system to that consumed by the cells pretreated in PBS.

Bioconversion in biphasic systems

Experiments examining the asymmetric reduction of OPBE in biphasic systems were performed in 50-mL

Erlenmeyer flasks. A certain amount of wet yeast cells were suspended in 0.1 M PBS and mixed with 10 mL organic solvent containing OPBE to form the aqueous/organic biphasic system (shown in Scheme 1). The flasks were incubated in a rotary shaker at 180 rpm and a specified temperature. Due to their high hydrophobicity, the substrate OPBE and the product (*R*)-HPBE were primarily partitioned in the organic phase, while the cells were mainly suspended in the aqueous phase. Unless otherwise specified, the concentration of the substrate and product were calculated based on the volume of the whole biphasic system. Six hundred-microliter aliquots of the sample were withdrawn from the organic phase at different time intervals: 100 μL aliquots were mixed with 100 μL dodecane as the internal standard for GC analysis, and 500 μL aliquots were used for high-performance liquid chromatography (HPLC) analysis.

Analytical methods

The concentrations of OPBE and HPBE were determined with a gas chromatography. (GC-950, Shanghai Haixin Chromatographic Instrument Co., Ltd., Shanghai, China) equipped with a flame ionization detector and a SE-30 capillary column (30 m, i.d. 0.5 mm). The (*R*)-HPBE and (*S*)-HPBE were analyzed using an Agilent 1260 HPLC system (Sta. Clara, CA, USA) equipped with a Chiralcel OD-H column (4.6 mm × 250 mm, 5 μm, Diacel, Hyogo, Japan) using *n*-hexane:isopropanol (95:5, *v*/*v*) as eluent at a flow rate of 1.0 mL/min. The detection was performed at 225 nm.

The *ee* value was calculated as follows:

$$ee = \frac{[R]-[S]}{[R]+[S]} \times 100\%$$

where [*R*] and [*S*] are the peak areas of (*R*)-HPBE and (*S*)-HPBE, respectively. The yield of (*R*)-HPBE was calculated from the final molar concentration of the product, C_p, and the initial molar concentration of the substrate, C_s as follows:

$$\text{Yield} = \frac{c_P}{c_s} \times 100\%$$

Results and discussion

Tolerance of *R. mucilaginosa* CCZU-G5 in different biphasic systems

The metabolic activity of cells could be affected by the organic solvent in a biphasic system [16,17]. It has been reported that solvents with a higher log *P* are more hydrophobic and thus more advantageous for enzymatic reactions [8]. In this work, 11 organic solvents with different log *P* values ranging from 0.68 to 5.6 were studied and compared for their toxicity to *R. mucilaginosa* CCZU-G5. As the toxicity indicator, cell viability or metabolic activity retention (*R* %) in the various biphasic systems was evaluated and the results were shown in Table 1. In general, there was a positive correlation between activity retention and the log *P* value of the organic solvent. Satisfactory metabolic activity retention over 80% was attained when the log *P* value of the solvent was over 4.6. The maximal metabolic activity retention reached 97.8% in the water/n-decane system. The metabolic activity retention was low when the log *P* value of the solvent was lower than 3.2. For example, it was just 14.7% in the water/ethyl acetate biphasic system because the cell membrane would be destroyed by the polar organic solvent with a low log *P* value [18].

Selection of organic solvents

The asymmetric reduction of OPBE to (*R*)-HPBE with resting cells of *R. mucilaginosa* CCZU-G5 in an aqueous medium has been optimized in our previous work (20 mM OPBE, yield 76.3%, *ee* 99.2%, productivity 6.8 g/L/day). The aqueous systems supported a desirable *ee* value, but the highest substrate concentration that could be transformed

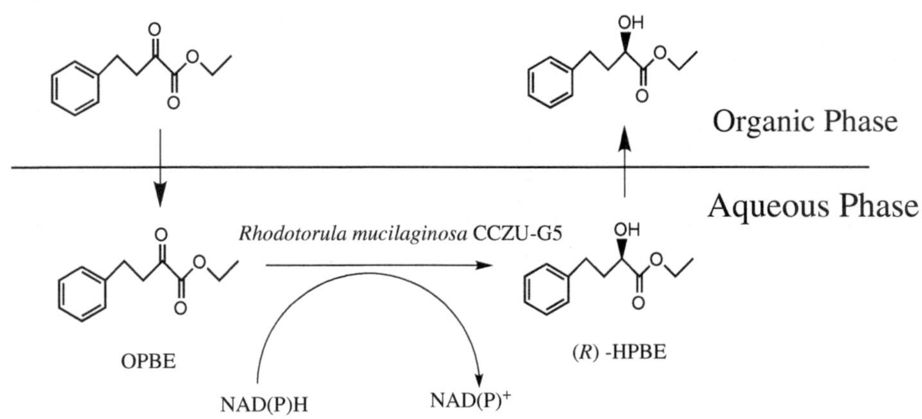

Scheme 1 Mechanism of *R. mucilaginosa* CCZU-G5 for asymmetric reduction of OPBE to (*R*)-HPBE biphasic system.

Table 1 Tolerance and asymmetric reduction of OPBE to (*R*)-HPBE with *R. mucilaginosa* CCZU-G5 in different organic/aqueous biphasic systems

Organic solvent	Log *P*	Metabolic activity retention (%)[a]	Yield (%)[b]	*ee* (%)[b]
EtOAC	0.68	14.5 ± 0.6	11.3 ± 0.7	16.5
Butyl acetate	1.7	17.1 ± 0.7	14.3 ± 1.3	18.4
Benzene	2.0	17.5 ± 0.6	36.1 ± 1.1	45.5
Toluene	2.5	22.3 ± 0.7	44.1 ± 1.0	49.4
Cyclohexane	3.2	50.4 ± 1.3	53.1 ± 1.5	68.9
n-Hexane	3.5	67.2 ± 0.9	69.3 ± 0.5	>99.0
n-Heptane	4.0	70.5 ± 0.8	89.3 ± 0.4	>99.0
n-Octane	4.5	78.8 ± 0.7	90.3 ± 0.4	>99.0
Isooctane	4.6	80.1 ± 0.6	98.1 ± 0.3	>99.0
n-Hydride	5.1	92.2 ± 0.4	63.1 ± 0.3	>99.0
n-Decane	5.6	98.0 ± 0.4	52.8 ± 0.9	>99.0

[a]Metabolic activity retention (*R*) of *R. mucilaginosa* CCZU-G5: 0.4 g wet cells suspended in 10 mL PBS buffer (0.1 M, pH 7.0) was mixed with 10 mL of different organic solvents incubated at 30°C and 180 rpm for 24 h. The cell suspension in PBS without any organic solvent was used as a control. Then the harvested cells were transferred to 10 mL of 20 g/L glucose and incubated at 30°C and 180 rpm for 4 h. [b]Reaction conditions: 2.0 g wet cells suspended in 10 mL PBS buffer (0.1 M, pH 7.0) containing 1.0 g glucose was mixed with 10 mL of an organic solvent containing 2 mmol OPBE and incubated at 30°C and 180 rpm for 24 h.

with 0.1 g/mL cell was merely 50 mM. To enhance the substrate concentration and overcome the substrate-tolerance obstacle [14,19], the reaction was explored in the aqueous/organic biphasic system. With the addition of organic solvents, the solubility of the substrate could be enhanced. Besides, the hydrophilic microbial cells (in aqueous phase) could be separated from the hydrophobic substrate and the product in the organic phase. The selection of the organic solvent as substrate and product carrier in an aqueous/organic solvent biphasic system is based mainly on its biocompatibility towards the biocatalyst and adequate solubility of both substrate and product [20]. Therefore, the influence of organic solvents on the catalytic activity and enantioselectivity was studied in 11 different biphasic systems with different log *P* values ranging from 0.68 to 5.6. In general, there was a positive correlation between the product yield and the log *P* value of organic solvent (Table 1), with the exceptions of *n*-hydride and *n*-decane, which have log *P* values of over 5 but gave a relatively low product yield. The maximal yield of 98.1% was achieved in the aqueous/isooctane biphasic system. The yield decreased significantly in the organic solvents with lower log *P* values such as butyl acetate and EtOAC, which gave a poor yield of 11.3%. The *ee* values of (*R*)-HPBE were higher than 99.0% when the log *P* values of the organic solvents were more than 3.2. Considering the yield and the *ee* value, the aqueous/isooctane biphasic system was selected for further study.

Effects of phase volume ratio

The volumetric phase ratio influences the phase interfacial area between two phases, which, in turn, affects biotransformation in the biphasic systems [21]. The effects of the volume ratio of the aqueous phase to the organic phase (V_{aq}/V_{org}) on the asymmetric reduction of OPBE were studied in the aqueous/isooctane biphasic system. A fixed volume (10 mL) of isooctane was mixed with different volumes (2 to 20 mL) of the aqueous phase containing 2 g cells to form biphasic aqueous/isooctane system. As seen in Figure 1, the product yield increased when the phase ratio (V_{aq}/V_{org}) went up from 0.2 to 0.8 due to the increased availability of substrate molecules in aqueous phase. With a phase ratio of 0.8 to 1.2, the maximum yield of (*R*)-HPBE (98.1%) was achieved. However, the yield decreased when the phase volume ratio was higher than 1.2 probably because of the increased sensitivity to the substrate toxicity when the cells were diluted. Although the yield was highly dependent on the phase volume ratio, excellent optical purity of the product (*ee* >99.0%) was observed in the aqueous/isooctane biphasic system, regardless of the phase ratio.

Effects of temperature

It is well known that the reaction temperature is an important factor affecting the catalytic characteristics such as the activity, enantioselectivity, and stability of a biocatalyst [22]. As shown in Figure 2, the product yield increased gradually when the temperature rose from 25°C to 35°C. However, at higher temperatures, the yield decreased significantly with increasing the temperature, indicating thermal deactivation of the cells at higher temperatures. The *ee* value was not significantly affected by temperature and remained above 99.0% at all temperatures studied.

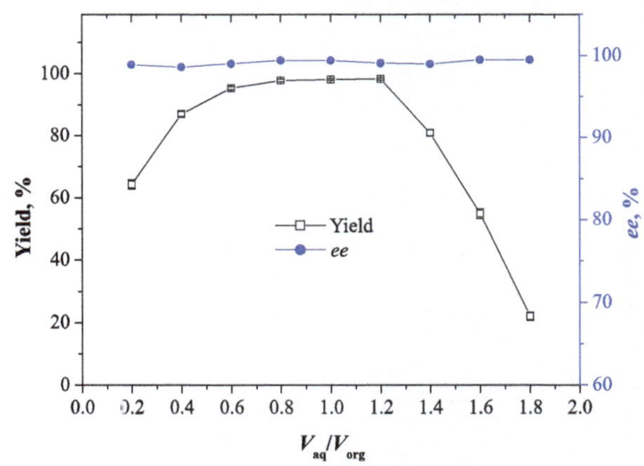

Figure 1 Effects of phase volume ratio on asymmetric reduction of OPBE to (*R*)-HPBE in aqueous/isooctane biphasic system. Reaction conditions: 2.0 g wet cells suspended in a certain volume of PBS (0.1 M, pH 7.0) containing 1.0 g glucose and mixed with 10 mL isooctane containing 2 mmol OPBE and incubated at 30°C and 180 rpm for 24 h.

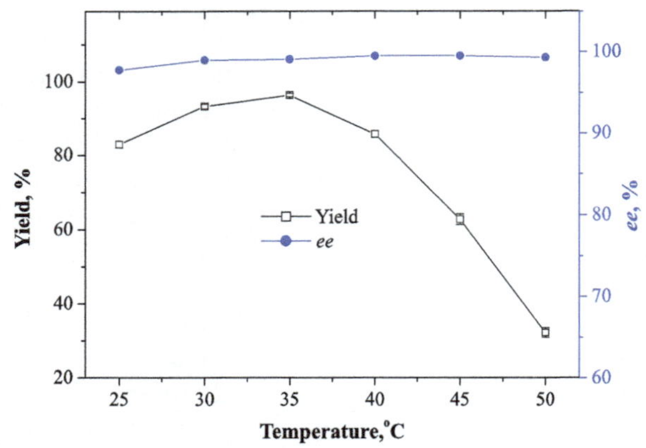

Figure 2 Effects of temperature on asymmetric reduction of OPBE to (*R*)-HPBE in aqueous/isooctane biphasic system. Reaction conditions: 2.0 g wet cells in 10 mL PBS (0.1 M, pH 7.0) containing 1.0 g glucose, mixed with 10 mL isooctane containing 2 mmol OPBE, and incubated at various temperatures and 180 rpm for 24 h.

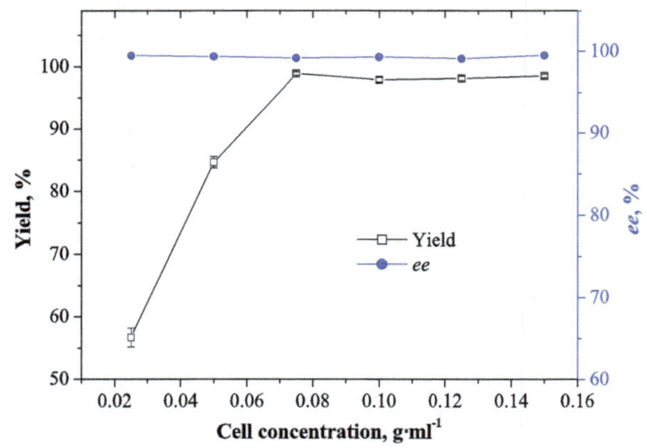

Figure 4 Effects of cell concentration on asymmetric reduction of OPBE to (*R*)-HPBE in aqueous/isooctane biphasic system. Reaction conditions: 2.0 g wet cells in 10 mL PBS (0.1 M, pH 7.5) containing 1.0 g glucose, mixed with 10 mL isooctane containing 2 mmol OPBE, and incubated at 35°C and 180 rpm for 24 h.

Effects of pH

The reaction pH plays a crucial role in bioreduction and influences not only enzymatic enantioselectivity and activity but also the regeneration of coenzymes [23-25]. In addition, variations in the pH may also alter the ionic state of substrates and the enzymes involved in the reactions [26]. Figure 3 shows the effects of aqueous medium pH (over the range of 5.5 to 8.0) on the asymmetric reduction of OPBE to (*R*)-HPBE. The highest product yield was obtained at pH 7.5. The product yield decreased dramatically at lower or higher pH values. It is clear that pH could exert a tremendous influence on the reaction rate and yield in the aqueous/isooctane system. However, the *ee* value was not significantly affected by the pH and remained above 99.0% for all pH values studied. Based on the results, pH 7.5 was the optimum for the reaction.

Effects of cell concentration

The concentration of biocatalyst is an important factor in enzymatic reactions because it affects both enantioselectivity and reaction rate [19]. In order to investigate the effects of cell concentration on OPBE reduction in aqueous/isooctane system, the *ee* and yield of the product were determined at different cell concentrations. As shown in Figure 4, the cell concentration in the aqueous phase had no effect on the *ee* value but had a marked

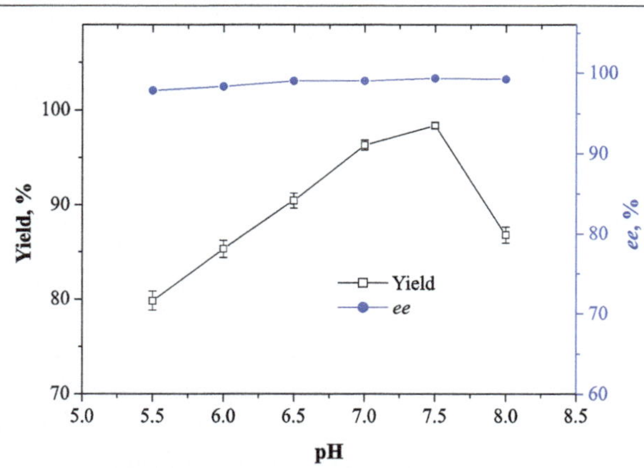

Figure 3 Effects of pH on the asymmetric reduction of OPBE to (*R*)-HPBE in aqueous/isooctane biphasic system. Reaction conditions: 2.0 g wet cells in 10 mL PBS containing 1.0 g glucose, mixed with 10 mL isooctane containing 2 mmol OPBE, and incubated at 35°C and 180 rpm for 24 h.

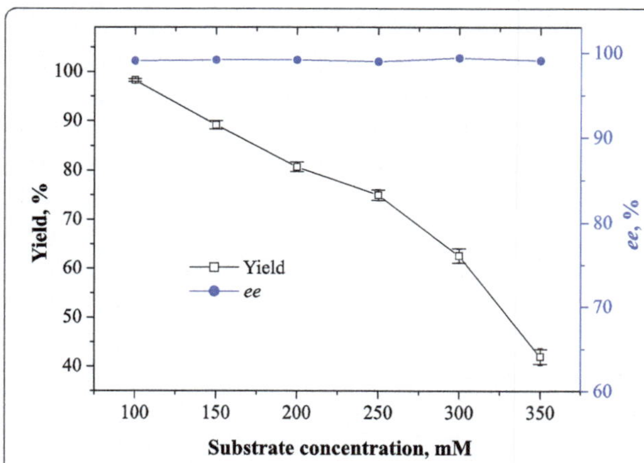

Figure 5 Effects of substrate concentration on asymmetric reduction of OPBE to (*R*)-HPBE in aqueous/isooctane biphasic system. Reaction conditions: 1.5 g wet cells and 1.0 g glucose in 10 mL PBS (0.1 M, pH 7.5), mixed with 10 mL isooctane containing different amounts of OPBE, 35°C, 180 rpm, 24 h. The OPBE concentration is based on the total liquid volume in the system.

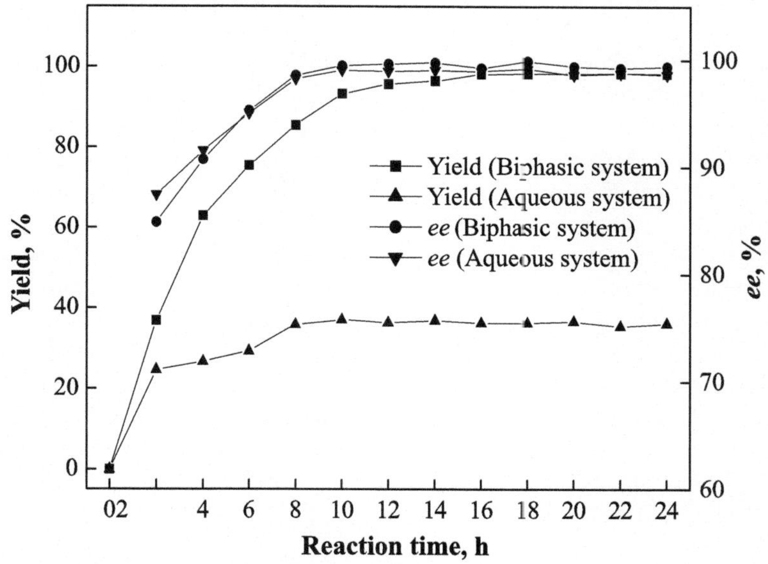

Figure 6 Time courses of the asymmetric reduction of OPBE to (*R*)-HPBE. Time courses of the asymmetric reduction of OPBE to (*R*)-HPBE in a monophasic aqueous system and in the aqueous/isooctane biphasic system. Reaction conditions: 1.5 g wet cells suspended in 10 mL PBS (0.1 M, pH 7.5) containing 1.0 g glucose mixed with 10 mL isooctane containing 2 mmol OPBE incubated at 35°C and 180 rpm. The monophasic aqueous system was 20 mL PBS (0.1 M, pH 7.0) containing 1.0 g glucose and 0.4 mmol OPBE.

effect on the reaction rate and thus yield, which increased to approximately 98% with increasing the cell concentration to 0.075 g/mL. Further increasing of the cell concentration had no effect on the yield.

Effects of substrate concentration

The substrate concentration in a reaction medium affects not only the reaction rate but also the enantioselectivity of the reaction [27]. Therefore, it is of great importance to investigate the effect of substrate concentration on the OPBE asymmetric reduction in the aqueous/isooctane biphasic system. Experiments with different initial substrate concentrations from 100 to 350 mM were conducted to investigate the effects of the substrate concentration on the asymmetric reduction of OPBE to (*R*)-HPBE under the optimal conditions obtained above. With the OPBE concentration increasing from 100 to 300 mM, the productivity increased from 10.2 to 20.3 g/L/day, although the yield reduced from 98% to 65% (Figure 5). When the substrate concentration was 350 mM, both yield and productivity decreased to 42% and 15.3 g/L/day, respectively, indicating that a high OPBE concentration (>300 mM) would significantly inhibit the catalytic activity of cells even in the biphasic system. However, no significant influence on the enantioselectivity was observed at high concentrations of both substrate and product, and the *ee* value of product was consistently above 99.0%.

Comparison of the aqueous/organic solvent biphasic system with the monophasic aqueous system

The reaction in the aqueous/isooctane biphasic system was compared with that in the monophasic aqueous

Table 2 Comparison between *R. mucilaginosa* CCZU-G5 and other reported microorganisms

Strain	Reaction media	OPBE conc	Reaction time (h)	*ee* (%)	Yield (conversion) (%)	References
R. mucilaginosa CCZU-G5	Aqueous phase	20 mM	8	99.2 (*R*)	76.3	This study
	Aqueous/isooctane	100 mM[a]	12	>99.0 (*R*)	98.3	This study
		200 mM[a]			>80	
Saccharomyces cerevisiae	Water/benzene	400 mM	48	87.5 (*R*)	(41.9)	Shi et al., 2009 [28]
Candida krusei SW2026	Water/dibutyl phthalate	100 mM	16	97.4 (*R*)	82	Zhang et al., 2009 [8]
	Aqueous phase	10 mM	14	99.7 (*R*)	95.1	
Candida boidinii CIOC21	Aqueous phase	20 mM	12	99.0 (*R*)	92	Chen et al., 2009 [12]
Pichia angusta	Aqueous phase	0.14% in volume	24	81 (*R*)	(100)	Lacerda et al., 2006 [7]

[a]Based on the total volume of the whole biphasic system.

system. As shown in Figure 6, the reaction in the monophasic aqueous system with 100 mM OPBE stopped after 8 h with a low yield (36.1%) and low productivity (11.4 g/L/day), which can be attributed to the substrate and product inhibition. In contrast, the reaction reached a high yield of 98.3% and a high productivity of 20.4 g/L/day after 12 h in the aqueous/isooctane biphasic system. Apparently, using isooctane as the substrate carrier eliminated the substrate inhibition and enhanced the reaction rate in the biphasic system. In addition, the *in situ* extraction of the product from the aqueous phase to the organic phase also decreased the product inhibition and allowed the reaction to continue until almost all substrate had been converted to the product, thus resulting in a high yield (98.3%) in the aqueous/isooctane biphasic system. Furthermore, the optical purity of the product has not changed (>99% *ee*) by introducing isooctane as the organic phase. It is noted that the reaction rate in the aqueous/isooctane biphasic system could be limited by mass transfer between the two phases [27], which can be increased by increasing the interfacial area between the two phases. Overall, the aqueous/isooctane biphasic system is advantageous for the asymmetric reduction of OPBE catalyzed by the microbial cells.

To date, only a few microorganisms have been described as efficient biocatalysts in the reduction of OPBE to (*R*)-HPBE. Table 2 compares the performances of the reported strains and *R. mucilaginosa* CCZU-G5 as biocatalysts in the preparation of (*R*)-HPBE from OPBE. Although excellent yield and *ee* value can be obtained with *C. krusei* SW2026 and *C. boidinii* CIOC21 in aqueous phase systems [8,12], the low substrate loading restricted their industrial applications. To increase the substrate concentration, Zhang et al. employed a water/dibutyl phthalate biphasic system [8]. Although the concentration of OPBE increased ninefold, the product yield decreased from 95.2% to 82%. To the best of our knowledge, only Shi et al. reduced OPBE at as high as 0.4 M with *S. cerevisiae* in water/benzene biphasic system [28]. However, the conversion of OPBE and *ee* of (*R*)-HPBE were only 41.9% and 87.5%, respectively. Moreover, the reaction time was very long. When the reduction of OPBE (100 mM) was carried out with *R. mucilaginosa* CCZU-G5 in aqueous/isooctane biphasic system, the shortest reaction time, excellent *ee* and product yield were achieved in the present study. Even if the concentration of OPBE was increased to 200 mM, the yield was still over 80%, and the *ee* was not affected. Compared with other strains listed in Table 2, *R. mucilaginosa* CCZU-G5 is a more competitive and promising biocatalyst for asymmetric reduction of OPBE to (*R*)-HPBE.

Conclusions

In this study, an aqueous/isooctane biphasic system was successfully established for asymmetric reduction of OPBE to (*R*)-HPBE with a newly isolated strain *R. mucilaginosa* CCZU-G5. Several factors such as volume ratio of the aqueous phase to the organic phase, reaction temperature, reaction pH, cell concentration, and substrate concentration significantly influenced the reaction rate and product yield. However, the optical purity of the product was not significantly affected and maintained at high levels of >99%. Under optimum reaction conditions (35°C, pH 7.5, 0.075 g/mL of cells, 100 mM OPBE, 1:1 of volume phase ratio), *R. mucilaginosa* CCZU-G5 exhibited excellent catalytic capability, giving product an excellent yield (98.3%) and *ee* (>99%).

Competing interests

The authors declare that they have no competing interests.

Authors' contributions

L-QW designed the study and drafted the manuscript. Z-QW revised the manuscript. QQ provided the experimental guidance. S-TY provided the experimental guidance and manuscript revision. J-JM and L-JW conducted the experiments and drafted the manuscript. All authors read and approved the final manuscript.

Acknowledgements

This work was supported by National Natural Science Foundation of China (No. 21102011) and the Science and Technology Supporting Project of Changzhou (No. CE20145041).

References

1. Herold P, Indolese AF, Studer M, Jalett HP, Siegrist U, Blaser HU (2000) New technical synthesis of ethyl (*R*)-hydroxy-4-phenylbutyrate of high enantiomeric purity. Tetrahedron 56:6497–6499
2. Iwasaki G, Kimura R, Numao N, Konda K (1989) A practical and diastereoselective synthesis of angiotensin converting enzyme inhibitors. Chem Pharm Bull 37:280–283
3. Zhang W (2009) Biocatalytic synthesis of ethyl (R)-2-hydroxy-4-phenylbutyrate. MS thesis. Jiangnan University, China
4. Liese A, Kraal U, Kierkelsc H, Schulze B (2002) Membrane reactor development for the kinetic resolution of ethyl 2-hydroxy-4-phenylbutyrate. Enzyme Microb Technol 30:673–681
5. Huang SH, Tsai SW (2004) Kinetic resolution of (R, S)-ethyl 2-hydroxyl-4-phenylbutyrate via lipase-catalyzed hydrolysis and transesterification in isooctane. J Mol Catal B Enzym 28:65–69
6. Kaluzna I, Andrew AA, Bonilla M, Martzen MR, Stewart JD (2002) Enantioselective reductions of ethyl 2-oxo-4-phenylbutyrate by *Saccharomyces cerevisiae* dehydrogenases. J Mol Catal B Enzym 17:101–105
7. Lacerda PSB, Ribeiro JB, Leite SGF, Ferrara MA, Coelho RB, Bon EPS, Lima ELS, Antunes OAC (2006) Microbial reduction of ethyl 2-oxo-4-phenylbutyrate. Searching for R-enantioselectivity. New access to the enalapril like ACE inhibitors. Tetrahedron Asymmetry 17:1186–1188
8. Zhang W, Ni Y, Sun ZH, Zheng P, Lin WQ, Zhu P, Ju NF (2009) Biocatalytic synthesis of ethyl (*R*)-2-hydroxy-4-phenylbutyrate with *Candida krusei* SW2026: a practical process for high enantiopurity and product titer. Process Biochem 44:1270–1275
9. Wang W, Zong MH, Lou WY (2009) Use of an ionic liquid to improve asymmetric reduction of 4-methoxyacetophenone catalyzed by immobilized *Rhodotorula* sp. AS2.2241 cells. J Mol Catal B Enzym 56:70–76
10. Chadha A, Manohar M, Soundararajan T, Lokeswari TS (1996) Asymmetric reduction of 2-oxo-4-phenylbutanoic acid ethyl ester by *Daucus carota* cell cultures. Tetrahedron Asymmetry 7:1571–1572

11. Dao DH, Kawai Y, Hida K, Hornes S, Nakamura K, Ohna A, Okamura M, Akasaka T (1998) Stereochemical control in microbial reduction. 30. Reduction of alkyl 2-oxo-4-phenylbutyrate as precursors of angiotensin converting enzyme (ACE) inhibitors. Bull Chem Soc Jpn 71:425–432
12. Chen YZ, Lin H, Xu XY, Xia XY, Wang LX (2009) Preparation the key intermediate of angiotensin-converting enzyme (ACE) inhibitors: high enantioselective production of ethyl (*R*)-2-hydroxy-4-phenylbutyrate with *Candida boidinii* CIOC21. Adv Synth Catal 350:426–430
13. Gong PF, Xu JH (2005) Bio-resolution of a chiral epoxide using whole cells of Bacillus megaterium ECU1001 in a biphasic system. Enz Microb Technol 36:252–257
14. He JY, Sun ZH, Ruan WQ, Xu Y (2006) Biocatalytic synthesis of ethyl (S)-4-chloro-3-hydroxy-butanoate in an aqueous-organic solvent biphasic system using *Aureobasidium pullulans* CGMCC 1244. Process Biochem 41:244–249
15. Kansal H, Banerjee UC (2009) Enhancing the biocatalytic potential of carbonyl reductase of Candida viswanathii using aqueous-organic solvent system. Bioresour Technol 100:1041–1047
16. Jiang Q, Yao SJ, Mei LH (2002) Tolerance of immobilized baker's yeast in organic solvents. Enz Microb Technol 30:721–725
17. Vermue M, Sikkema J, Verheul A, Bakker R, Tramper J (1993) Toxicity of homologous series of organic solvents for the gram-positive bacteria Arthrobacter and Norcardia sp. and the gram-negative bacteria *Acietobacter* and *Pseudomonas* sp. Biotechnol Bioeng 42:747–758
18. Gu LQ, Wei HF, Zhang XC, Xu G, Ma L (1998) Bioreduction of quinine derivatives by immobilized baker's yeast in hexane. Chin J Chem 16:45–50
19. Li AT, Zhang JD, Yu HL, Pan J, Xu JH (2011) Significantly improved asymmetric oxidation of sulfide with resting cells of *Rhodococcus* sp. in a biphasic system. Process Biochem 46:689–694
20. Pan J, Dang ND, Zheng GW, Cheng B, Ye Q, Xu JH (2014) Efficient production of l-menthol in a two-phase system with SDS using an immobilized Bacillus subtilis esterase. Bioresources and Bioprocessing 1:12
21. Fernandes P, Vidinha P, Ferreira T, Silvestre H, Cabral JMS, Prazeres DMF (2002) Use of free and immobilized *Pseudomonas putida* cells for the reduction of a thiophene derivative in organic media. J Mol Catal B Enzym 19–20:353–361
22. Philips RS (1996) Temperature modulation of the stereochemistry of enzymatic catalysis: prospects for exploitation. Trends Biotechnol 14:13–16
23. Hage A, Schoemaker HE, Field JA (2001) Optimization of stereoselective ketone reduction by the white-rot fungus *Merulius tremellosus* ono991. Appl Microbiol Biotechnol 57:79–84
24. Katz M, Sarvary I, Frejd T, Hahn-Hagerdal B, Gorwa-Grauslund MF (2002) An improved stereoselective reduction of a bicyclic diketone by Saccharomyces cerevisiae combining process optimization and strain engineering. Appl Microbiol Biotechnol 59:641–648
25. Patel RN, Robison RS, Szarka LJ, Kloss J, Thottathil JK, Mueller RH (1991) Stereospecific microbial reduction of 4,5-dihydro-4-(4-methoxyphenyl)-6-(trifluoromethyl-1H-1)-benzazepin-2-one. Enz Microb Technol 13:906–912
26. Kianmehr A, Pooraskari M, Mousavikoodehi B, Mostafavi SS (2014) Recombinant *D*-galactose dehydrogenase partitioning in aqueous two-phase systems: effect of pH and concentration of PEG and ammonium sulfate. Bioresources and Bioprocessing 1:6
27. Shimizu S, Kataoka M, Katoh M, Morikawa T, Miyoshi T, Yamada H (1990) Stereoselective reduction of ethyl 4-chloro-3-oxobutanoate by a microbial aldehyde reductase in an organic solvent-water diphasic system. Appl Environ Microbiol 56:2374–2377
28. Shi YG, Fang Y, Wu HP, Li F, Zuo XQ (2009) Improved production of ethyl-(*R*)-2-hydroxy-4-phenylbutyrate with pretreated *Saccharomyces cerevisiae* in water/organic solvent two-liquid phase systems. Biocatal Biotransfor 27:211–218

8

Mitochondrial fusion and fission are involved in stress tolerance of *Candida glabrata*

Shubo Li[1,2,3,4], Liming Liu[1,3,4*] and Jian Chen[1,3*]

Abstract

Background: Recently, cell tolerance toward environmental stresses has become the major problem in the development of industrial microbial fermentation. Acetoin is an important chemical that can be synthesized by microbes. Its toxicity was investigated using *Candida glabrata* as the model in this study.

Results: A series of physiological and biochemical experiments demonstrated that the organic solvent acetoin can inhibit cell growth by increasing intracellular reactive oxygen species (ROS) production and inducing damage to mitochondria and cell apoptosis. Integrating RT-PCR experiments, the genes *fzo*1 and *dnm*1 were overexpressed to regulate the balance between mitochondrial fusion and fission. Enhancement of mitochondrial fusion was shown to significantly increase cell tolerance toward acetoin stress by inhibiting ROS production and increasing the intracellular adenosine triphosphate (ATP) supply, which was also demonstrated by the addition of citrate.

Conclusions: Regulating mitochondrial fusion-fission may be an alternative strategy for rationally improving the growth performance of eukaryotes under high environmental stress conditions, and also expands our knowledge of the mechanisms of cell tolerance through the processes of energy-related metabolic pathways.

Keywords: Acetoin; ATP supply; *Candida glabrata*; Cell tolerance; Environmental stress; Mitochondrial fusion-fission

Background

With the development of metabolic engineering, microorganisms have been engineered to produce many important bulk chemicals, such as ethanol, acetone, and butanol. However, most of these are toxic to cells as they can damage cell membranes and walls and interfere with essential physiological processes [1]. Thus, improving cell tolerance has become a major challenge for microbial fermentation. To this end, genomic tools (transcriptomics, proteomics, and metabolomics) have been applied to investigate microbial responses to various solvents, and a number of strategies have been exploited to avoid solvent toxicity. Recently, a visual summary of cellular responses directed toward overcoming solvent stress and enhancing survivability was compiled [2-4], as follows: (i) metabolic detoxification, converting toxic compounds into less harmful chemicals; (ii) expression of heat shock proteins (HSP), assisting in protein folding and preventing aggregation; (iii) use of proton motive force and associated energy production; (iv) molecular efflux pumps, exporting solvents to the extracellular space; and (v) changes in cell membrane composition and biophysics, combating the fluidizing effects of solvents. But, interestingly, most of these involve energy-dependent processes, indicating that energy and energetic processes play crucial roles in protecting cells against environmental stress. For example, enhancing glucose transport and catabolism favors increased energy production and compensates for the energy expended in relieving stress in prokaryotes and eukaryotes [5-7]. Therefore, engineering the energy-generation machinery could be an alternative approach to improving cell tolerance toward environmental stress.

Mitochondria, as the power plants of the cell, can supply most of its energy through oxidative phosphorylation, and play an important role in maintaining cellular functions, such as the citric acid cycle and cell apoptosis [8,9]. As highly dynamic organelles, mitochondria maintain a balance between frequent cycles of fusion and fission, which tightly regulate mitochondrial number, morphology, and functions under ever-changing physiological conditions [10-12]. The

* Correspondence: mingll@jiangnan.edu.cn; jchen@jiangnan.edu.cn
[1]State Key Laboratory of Food Science and Technology, Jiangnan University, Wuxi 214122, China
[3]The Key Laboratory of Industrial Biotechnology, Ministry of Education, Jiangnan University, Wuxi 214122, China
Full list of author information is available at the end of the article

mitochondrial fusion process can be divided into three events: docking, fusion of the outer membrane, and fusion of the inner membrane. An increase in fusion events can result in mitochondrial inner connected networks, which are protective against apoptotic sensitivity and promote survival under conditions of stress [13]. In yeast, the mechanisms of mitochondrial fusion and fission have been identified and characterized. The fusion apparatus requires the activity of the large mitochondrial GTPase mitofusin (MFN), consisting of three core components, Fzo1, Ugo1, and Mgm1, of which Fzo1 is the key component regulating lipid bilayer fusion of mitochondrial outer membranes [11]. However, for mitochondrial fission, the core apparatus needs another large GTPase, a dynamin-related protein that includes three components, Dnm1, Fis1, and Mdv1, of which Dnm1 is required for mitochondrial fission and apoptosis [12,14]. Therefore, mitochondria have the ability to coordinate a balance between fusion and fission to handle cellular and environmental stressors [15,16].

Candida glabrata, a haploid multivitamin auxotrophic yeast, is an important strain used for industrial production of pyruvate [17] and α-ketoglutaric acid [18]. Additionally, *C. glabrata* was also used to produce malate [19] and acetoin [20], which can serve as a high value-added platform for the food, pharmaceutical, and chemical industries as a member of the C4-dicarboxylic acid family. However, high concentrations of acetoin were found to inhibit cell growth [20], and increasing the tolerance of cells may be favorable to improving the fermentative performance of a strain. To this aim, *C. glabrata* was used to study the effects of acetoin stress on cellular physiological characteristics (such as reactive oxygen species (ROS) and cell viability), and the balance between mitochondrial fusion and fission was also regulated to improve cellular physiological characteristics, thus enhancing cell robustness. These results demonstrated that enhancing mitochondrial fusion to increase the adenosine triphosphate (ATP) supply may be a novel approach to improving cell tolerance toward environmental stress.

Methods

Strains and plasmids

Escherichia coli JM109 was purchased from Invitrogen (Carlsbad, CA, USA) and used for plasmid construction. *C. glabrata* Δ*ura*3 was derived from *C. glabrata* CCTCCM 202019 (multivitamin auxotroph) and used as the receptor for gene overexpression. More information about the plasmids and strains is given in Table 1.

Plasmid construction and transformation

The primers used in this work are listed in Additional file 1: Table S1, and standard cloning and bacterial transformations were performed according to Sambrook and Russell [21]. The genes *fzo*1 and *dnm*1 were amplified by polymerase chain reaction (PCR) from genomic DNA of *S. cerevisiae* (CEN.PK2-1C), and then inserted into the desired plasmid multi-cloning sites. In all cases, PCR was performed using TaKaRa *Pyrobest* DNA Polymerase (Takara Bio Inc, Shiga, Japan). All genes were sequenced to ensure correct identify of the insert prior to transformations. Yeast strains were transformed using the lithium acetate method [22].

Culture medium and conditions

During construction, strains were grown in complex (YPD) medium consisting of 10 g/L yeast extract, 20 g/L peptone, and 20 g/L glucose. All engineered strains were screened on synthetic complete (SC) medium consisting of 20 g/L glucose, 7 g/L urea, 5 g/L KH_2PO_4, 0.8 g/L $MgSO_4 \cdot 7H_2O$, 3 g/L sodium acetate, and 15 g/L agar, at pH 6.0. They were fermented in a medium (medium A) consisting of 100 g/L glucose, 3 g/L urea, 7 g/L KH_2PO_4, 0.8 g/L $MgSO_4 \cdot 7H_2O$, and 5 g/L sodium acetate in shake-flask culture (200 rpm, 30°C) using $CaCO_3$ as the buffering agent. The inoculum size and the vitamin solution (0.04 mg/L thiamine-HCl, 0.16 mg/L biotin, 0.4 mg/L pyridoxine-HCl, and 8 mg/L nicotinic acid) were added to all media to constitute 15% *v*/*v* and 1% *v*/*v*, respectively. When necessary, different acetoin concentrations were

Table 1 Plasmids and strains used in this study

Strains/plasmids	Relevant characteristics	Reference
Plasmids		
pYX212	2 μm, Amp^R, *URA3*, P_{TPI}	Lab collection
pYX212-*fzo*1	Harboring the gene of *fzo*1 from *Saccharomyces cerevisiae* CEN.PK2-1C	This study
pYX212-*dnm*1	Harboring the gene of *dnm*1 from *S. cerevisiae* CEN.PK2-1C	This study
Strains		
C. glabrata CCTCC M202019	Multivitamin (thiamine, biotin, nicotinic acid and pyridoxine) auxotroph	[21]
C. glabrata Δ*ura*3	The mutant derived from *C. glabrata* CCTCC M202019	[19]
C-pYX	*C. glabrata* Δ*ura3* (pYX212)	This study
C-Fzo1	*C. glabrata* Δ*ura3* (pYX212- *fzo*1)	This study
C-Dnm1	*C. glabrata* Δ*ura3* (pYX212- *dnm*1)	This study

added to the culture. Transformed *E. coli* JM109 cells were grown at 37°C in Luria-Bertani (LB) medium containing 100 μg/mL ampicillin.

Spotting assay for evaluation of acetoin tolerance

To evaluate microbial tolerance, a spotting assay was applied in the presence of acetoin [23]. First, the cells were incubated overnight in SC medium with shaking (200 rpm, 30°C), and then they were collected and re-suspended by centrifugation (10,000×*g*, 20 s) in sterilized water. Second, the suspensions were serially diluted to an OD_{660} of 1×10^{0}, 1×10^{-1}, 1×10^{-2}, 1×10^{-3}, 1×10^{-4}, 1×10^{-5}, 1×10^{-6}, 1×10^{-7}, and 1×10^{-8}, and then spotted (5 μL each) onto agar plates containing different acetoin concentrations. The plates were sealed with vinyl plastic tape to prevent evaporation of acetoin and incubated at 30°C.

Analytical methods

The optical absorbance at 660 nm (A_{660}) was converted to dry cell weight (DCW) according to the following formula [17]: A_{660} DCW = 1: 0.23 (g/L).

Cell survivability

When the cells were incubated to $OD_{660} = 1.0$, some of cultures were diluted in sterile deionized water and spread evenly onto SC plates with varied acetoin concentrations (0%, 10%, and 15%), and the colony forming units (cfu) were determined after 24 h of incubation. Cell survival was expressed as survivability (cfu %), and defined as below:

$$\text{Cell survivability} = \frac{\text{The number of colonies on acetoin-containing agar plate}}{\text{The number of colonies on non-acetoin-containing agar plate}} \times 100\%;$$

Physiological parameters

When the cells had been incubated to the exponential phase ($OD_{660} = 2.0$), different acetoin concentrations were added and continual cultivation proceeded for 24 h (200 rpm, 30°C). Cells were then collected and washed three times with sterilized water. Afterwards, the OD was adjusted to 1.0 before measuring the following physiological parameters (parameters were expressed as mean values with standard error of the mean of at least three independent experiments):

(1) Mitochondrial membrane potential ($\Delta\Psi_m$): $\Delta\Psi_m$ was measured using Rh123 (Sigma, Shanghai, China) as described previously [24]. Cells were collected and washed twice with phosphate buffer saline (PBS), and the pellets were incubated with 500 μM Rh123 in the dark for 10 min at room temperature. Then cells were washed with PBS three times, and the suspensions were measured using flow cytometry (FCM, BD Biosciences, Shanghai, China).
(2) ROS level: cells were collected and prepared using the ROS Assay Kit (Beyotime Institute of Biotechnology, Jiangsu, China), and ROS production was detected via flow cytometry.
(3) ATP production: cells were collected and prepared according to the protocol of the ATP Assay Kit (Beyotime Institute of Biotechnology, Jiangsu, China), and ATP levels were measured with a luminometer. Protein concentrations were determined using a BCA Protein Assay Kit (Beyotime Institute of Biotechnology, Jiangsu, China).
(4) Cell apoptosis: cell apoptosis was measured by flow cytometric analysis using the FACS Calibur (BD Biosciences, Shanghai, China). Fluorescence emission was measured through a 500/50-nm bandpass filter for Rh123-labeled cells, and through a 660/16-nm bandpass filter for PI-labeled cells. Propidium iodine (PI, Sigma, Shanghai, China) staining was used to monitor cell membrane integrity. Samples (500 μL) were incubated with 3 μL PI stock solution (1 mg/mL) for 5 min at room temperature in the dark, followed by microscopy and flow cytometry, and cell viability was calculated by measuring PI fluorescence on a log scale. Before measuring cell apoptosis and viability using staining with flow cytometry, Rh123/PI dual staining was quantified using samples from fresh cultures with or without acetoin treatment. A minimum of 10,000 events were analyzed per sample at a low flow rate. CELL QUEST software was used for data acquisition and analysis [25].

Transcriptional analysis

Total RNA isolation was carried out using an RNAprep pure Plant Kit, and reverse transcription (cDNA synthesis) was performed according to the protocol of the Prime-Script®RT reagent kit Perfect Real Time (Takara Bio Inc, Shiga, Japan). Quantitative real-time PCR (RT-PCR) was done using the β-ACTIN gene as the internal control, and the primers used in RT-PCR are given in Additional file 1: Table S1. Each sample was tested in triplicate in a 96-well plate (Bio-Rad Corp, Hercules, CA, USA).

Statistical analysis

All experiments were carried out in triplicate, and the results are expressed as mean ± standard deviation. SPSS 18 (SPSS Statistics 18.0, SPSS Institute, Inc., Chicago, IL, 2010, USA) was used for one-way analysis of variance and canonical correlation analysis (CCA), and significant differences ($P < 0.05$) among means were determined by the least significant difference test.

Results

Tolerance of *C. glabrata* toward acetoin stress

To determine the toxicity of the important chemical acetoin, a spotting assay on agar plates containing different acetoin concentrations was used. As shown in Figure 1a, *C. glabrata* grew on agar plates containing 6 g/L of acetoin, but was significantly inhibited by 15 g/L of acetoin. Furthermore, cell tolerance was also evaluated in shake-flask culture, and it was found that cell growth decreased by 7.5% (6 g/L), 19.8% (12 g/L), 44.4% (15 g/L), and 59.2% (18 g/L), respectively, compared to the control conditions (without addition of acetoin) (Figure 1b). Therefore, 6 g/L of acetoin had a slight effect on cell growth, which was consistent with the spotting assay results.

Effects of acetoin stress on cellular physiological characteristics

To investigate the toxic effects of acetoin on cells in depth, the parameters cell viability and intracellular ROS were measured under conditions of acetoin stress. Cell viability was reduced to 22% after treatment with 18 g/L acetoin (Figure 2a), while intracellular ROS production increased to 32.5 from 14.1 compared to the control conditions (without addition of acetoin) (Figure 2b). Furthermore, double staining with Rh123 and PI, which could discriminate intact, apoptotic, and necrotic cells, was applied to observe cell apoptosis by flow cytometry. As shown in Figure 3, Rh123/PI double fluorescent staining with flow cytometry generated a contour diagram with three distinct regions: R1, necrotic cells; R2, apoptotic cells; R3, intact cells. When the concentration of acetoin increased from 0 to 18 g/L, the percentage of necrotic cells (R1) increased slightly from 0.8% to 2.15%, but the percentage of apoptotic cells (R2) increased significantly to 96.48%. Correspondingly, the percentage of intact cells (R3) decreased markedly from 99.22% to 1.37%. Therefore, the levels of cell apoptosis were effectively increased during high levels of acetoin stress. These results demonstrate that acetoin can dramatically change the levels of cell viability and intracellular ROS, and induce cell apoptosis and death, and then significantly inhibit cell growth.

Roles of mitochondrial fusion and fission in tolerance toward acetoin stress

In yeast, mitochondria are the major source of ROS production, and the processes of mitochondrial fusion-fission are closely related to cell apoptosis [26]. Therefore, the gene expression levels of *fzo*1 and *dnm*1 were measured by RT-PCR under different acetoin conditions. As shown in Figure 4a, the expression levels of *fzo*1 and *dnm*1 were dramatically enhanced, increased by 3.4- and 1.8-fold, respectively, at 12 g/L of acetoin. To confirm these results, engineered C-Fzo1 and C-Dnm1 overexpressing *fzo*1 and *dnm*1, respectively, were constructed using the strain C-pYX (harboring the plasmid pY212) as the control. As shown in Figure 4b, no distinct differences were observed in cell growth between C-pYX, C-Fzo1, and C-Dnm1 in the absence of acetoin stress. But under high acetoin stress, the engineered strain C-pFzo1 exhibited superior tolerance compared to the strain C-pYX, as it grew well on an agar plate containing 18 g/L of acetoin, and even grew in the presence of 20 g/L acetoin. In contrast, the strain C-Dnm1 was significantly inhibited in the presence of 15 g/L acetoin, showing lower tolerance than strain C-pYX. Therefore, increasing mitochondrial fusion improved cell tolerance, whereas increasing mitochondrial fission had a negative effect on cell tolerance toward acetoin stress.

Effects of mitochondrial fusion-fission on physiological characteristics

To elucidate the mechanisms by which mitochondrial fusion-fission regulate cell tolerance, the effects of

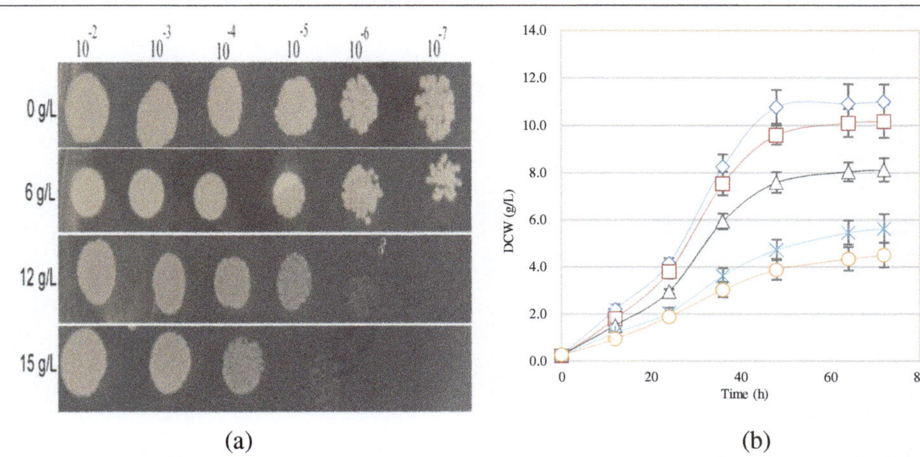

Figure 1 Cell tolerance toward acetoin stress. Application of a spotting assay **(a)** and time course of cell growth (DCW) in shake-flask culture with acetoin added at 0 h **(b)**. Error bars represent mean values and standard deviations of three independent experiments.

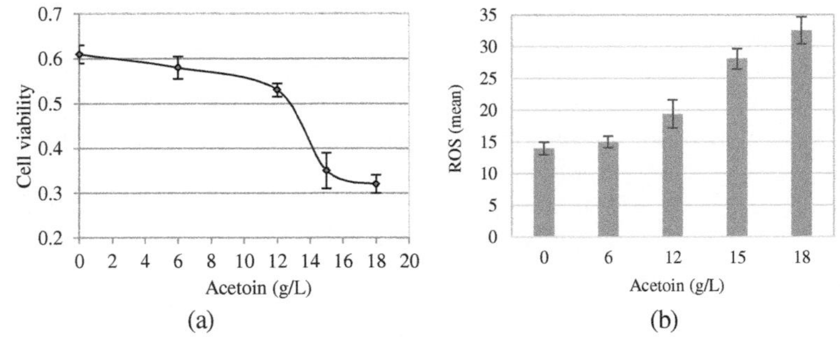

Figure 2 Effects of different acetoin concentrations on cell viability (a) and intracellular ROS production (b). Error bars represent mean values and standard deviations of three independent experiments.

mitochondrial fusion-fission on cellular physiological characteristics were further investigated. As shown in Figure 5, the intracellular environment significantly changed due to the expression of mitofusin in the engineered strains. Compared to the control conditions (without addition of acetoin), some different characteristics were observed in engineered strains cultured with 15 g/L of acetoin, as follows. (1) Strain C-Fzo1 effectively inhibited the production of intracellular ROS (increased by only 69.3%), which was far lower than that produced by strains C-pYX (increased by 144%) and C-Dnm1 (increased by 180%) (Figure 5a). (2) Strain C-Fzo1 significantly maintained cell survivability and ATP production, as these parameters decreased by 44.3% and 40.7%, respectively, to values that were higher than those of strains C-pYX (decreased by 64.4% and 69.7%, respectively) and C-Dnm1 (decreased by 75.7% and 76.1%, respectively) (Figure 5b,c). Furthermore, strain C-Fzo1 more effectively maintained the balance of mitochondrial membrane potential (ΔΨm, with a decrease of 18.6%) than strain C-Dnm1 (ΔΨm decreased by 48.2%), which may be due to lower ROS production and higher ATP formation (Figure 5d). These results demonstrate that the enhancement of mitochondrial fusion could improve the ATP supply and decrease ROS production to

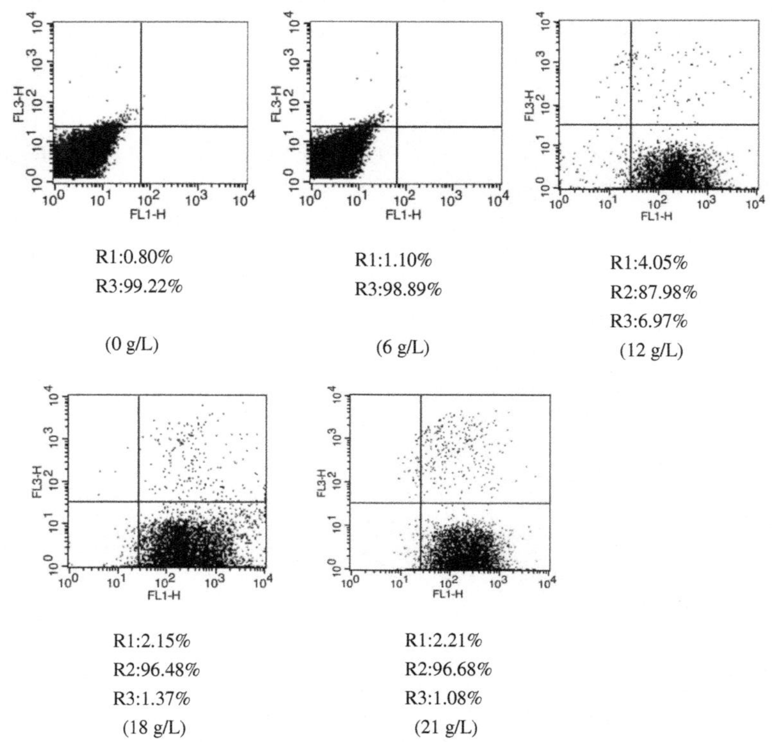

Figure 3 Contour diagram of the flow cytometry of *C. glabrata* cells under the conditions of different acetoin concentrations. FL1-H, Rh123 staining; FL3-H, PI staining. Region 1 (R1), percentage of necrotic cells; region 2 (R2), percentage of apoptotic cells; region 3 (R3), percentage of intact cells.

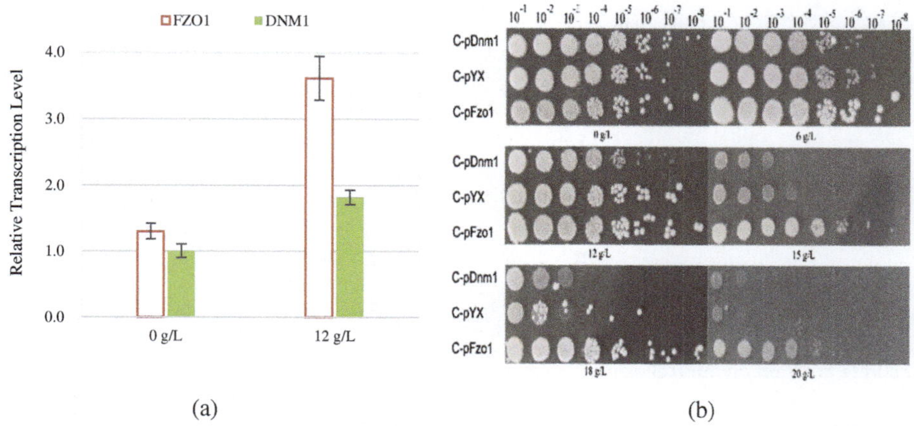

Figure 4 Roles of mitochondrial fusion-fission in cell tolerance toward acetoin stress. (a) The gene expression levels of *fzo1* and *dnm1*. **(b)** Spotting assay of the engineered strains.

maintain the balance of ΔΨm, and thus increase cell survivability and cell tolerance against acetoin stress.

Effect of citrate on *C. glabrata* adaptation to acetoin stress

According to the results of a previous study [27], 50 nmol/L of sodium citrate was added to determine its effects on cell growth and ATP production of the wild-type strain under different acetoin conditions. As shown in Figure 6a, compared to the control conditions (without addition of sodium citrate), sodium citrate effectively improved intracellular ATP supply, increasing ATP production by 18.8% (6 g/L of acetoin), 48.9% (12 g/L of acetoin), and 11.7% (15 g/L of acetoin). Correspondingly, cell tolerance of acetoin stress was also enhanced, so that cell growth increased by 18.2% (6 g/L of acetoin), 30.1% (12 g/L of acetoin), and 9.5% (15 g/L of acetoin) (Figure 6b). Therefore, improving the supply of cellular ATP could indeed increase cell tolerance toward acetoin stress.

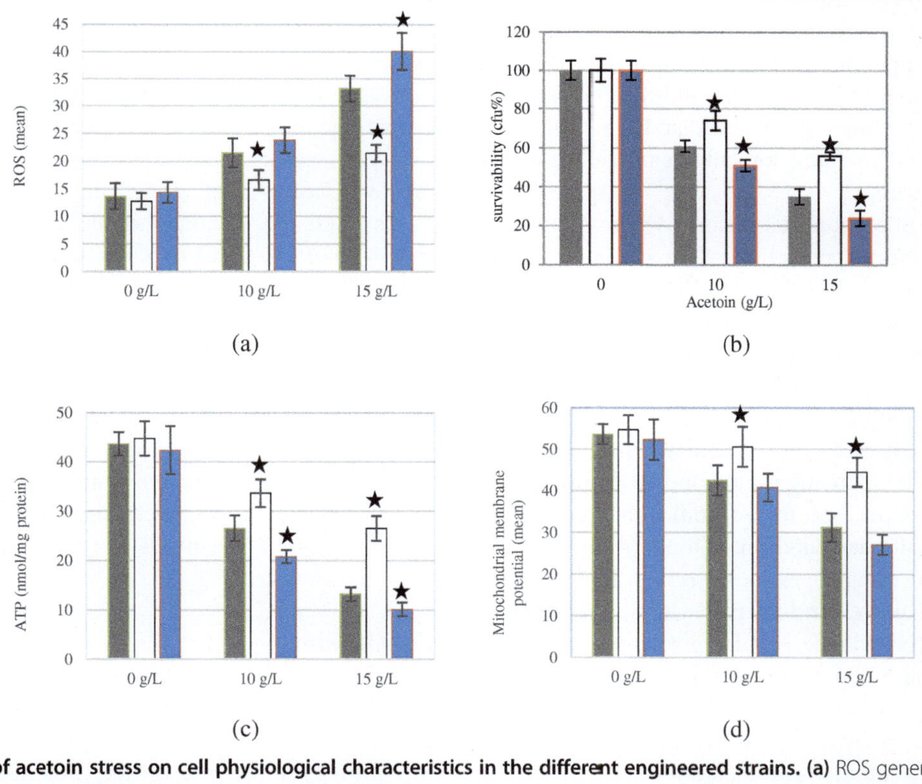

Figure 5 Effects of acetoin stress on cell physiological characteristics in the different engineered strains. (a) ROS generation. **(b)** Cell viability. **(c)** ATP production. **(d)** ΔΨm. Error bars represent mean values and standard deviations of three independent experiments (black square, C-pYX; white square, C-Fzo1; blue square, C-Dnm1) (*$P < 0.05$ vs. the group of strain C-pYX).

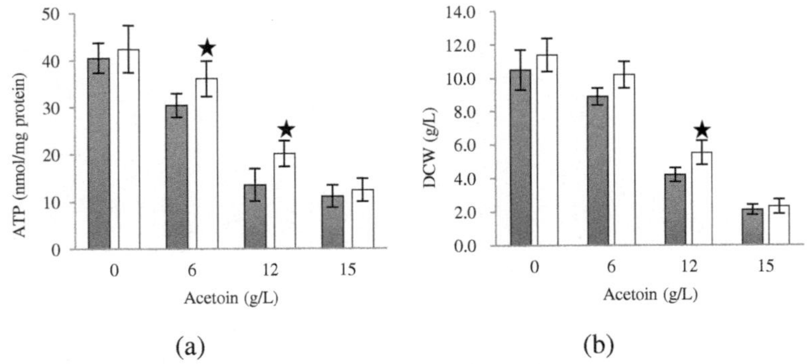

Figure 6 Effects of citrate addition on intracellular ATP production (a) and cell growth (b). (black square, without citrate addition; white square, with citrate addition). Error bars represent mean values and standard deviations of three independent experiments (*$P < 0.05$ vs. the group without citrate addition).

Discussion

In cells, mitochondria continually generate ROS as the byproduct of electron transport during oxidative phosphorylation. If ROS production increases it can damage mitochondrial components (such as proteins, lipids, and DNA) and then weaken metabolism and further increase ROS generation, causing a 'vicious downward spiral' of ROS generation and damage accumulation [28,29]. Additionally, intracellular ROS can participate in signaling effects to control the activity of protein kinases and phosphates and regulate gene expression, and thus induce cell apoptosis and cell death [30,31]. Here, the results indicated that acetoin stress could significantly increase intracellular ROS formation, and then decrease cell viability and induce cell apoptosis (Figures 2 and 3). Therefore, we hypothesized that reducing intracellular ROS production may be an alternative route for enhancing cell tolerance toward acetoin stress.

In eukaryotes, mitochondrial fusion-fission is important for maintaining mitochondrial numbers and morphology, and plays a critical role in sustaining functional mitochondria [32,33]. To date, the processes of mitochondrial fusion-fission have been used to investigate and explain cell death induced by different environmental stresses, such as ethanol [34], hyperosmotic [35], and acetic acid stress [36]. In such cases mitochondrial fission could facilitate apoptosis by inducing mitochondrial fragmentation and ROS production, but fusion could compensate the contents of partially damaged mitochondria and maintain energy output to mitigate various stresses. In this study, the processes of mitochondrial fusion-fission were regulated and used to enhance cellular properties to combat acetoin stress. As a result, the engineered strain C-Fzo1 overexpressing mitofusin *fzo1*, which regulates mitochondrial fusion and maintains mitochondrial morphology, could effectively inhibit intracellular ROS formation and increase cell viability, and thus enhance cell tolerance during high levels of acetoin stress.

But, interestingly, enhancement of mitochondrial fusion could also effectively improve ATP production and help maintain the mitochondrial membrane potential ($\Delta\Psi m$), which may be another reason for the high tolerance of strain C-Fzo1. Apoptosis was apparent through measurement of characteristic apoptotic indicators, such as decreased $\Delta\Psi m$, production of ROS, and caspase involvement, in which $\Delta\Psi m$ became transiently increased upon apoptotic stimulation in yeast cells, resulting in mitochondrial fragmentation, the release of apoptogenic factors including cytochrome c and apoptosis-inducing-factor (AIF) [37,38], and a permanent decrease in $\Delta\Psi m$ during subsequent apoptotic processes [39,40]. Therefore, maintaining the mitochondrial membrane potential may be favorable for enhancing the tolerance of cells toward environmental stress. Additionally, citrate was added and this demonstrated that improving the ATP supply had a positive effect on cell growth in the presence of acetoin stress. Actually, mitochondrial fission-fusion has been shown to alter energy requirements to regulate cell components, so that hyperfused mitochondria led to an increase in mitochondrial oxygen consumption and ATP formation that enhanced cell growth [41,42]. Therefore, the results reported here provide an alternative strategy for rationally improving the growth performance of eukaryotes, especially those that are organic solvent producers, under high organic solvent conditions, and also expand our knowledge of the mechanism of organic solvent tolerance through the processes of energy-related metabolic pathways.

Conclusions

In this study, the processes of mitochondrial fusion-fission were regulated and used to inhibit intracellular ROS production, raise the intracellular ATP supply, and maintain the mitochondrial membrane potential, and therefore increase cell tolerance toward acetoin stress. However, mitochondrial fission-fusion can also regulate many other physiological functions, such as Ca^{2+} homeostasis [43], metabolite transportation, and signaling exchange [32], to maintain the metabolic stability of mitochondria. Hence, the roles of mitochondrial fission-fusion in influencing the intracellular environment and physiological characteristics require further comprehensive investigation, exploiting some novel strategies to improve the performance of industrial microorganisms in a future study.

Additional file

Additional file 1: Table S1. Primers used for RT-PCR and PCR amplification in this study.

Competing interests

The authors declare that they have no competing interests.

Authors' contributions

SBL and LML conceived the study. SBL made contributions to the design of the experiments, the acquisition of data, the analysis and interpretation of data and contributed to the manuscript writing. LML and JC conceived and organized the study and helped to draft the manuscript, and have revised the manuscript. All authors read and approved the final manuscript.

Acknowledgements

This research was financially supported by the Major State Basic Research Development Program of China (973 Program, No. 2013CB733602), the Program for Young Talents in China, the Provincial Outstanding Youth Foundation of Jiangsu Province (BK2012002), the Priority Academic Program Development of Jiangsu Higher Education Institutions, the National Natural Science Foundation of China (31270079) and the Fund from Wuxi City (CLE01N1111).

Author details

[1]State Key Laboratory of Food Science and Technology, Jiangnan University, Wuxi 214122, China. [2]College of Light Industry and Food Engineering, Guangxi University, Nanning 530004, China. [3]The Key Laboratory of Industrial Biotechnology, Ministry of Education, Jiangnan University, Wuxi 214122, China. [4]Laboratory of Food Microbial-Manufacturing Engineering, Jiangnan University, Wuxi 214122, China.

References

1. Dunlop MJ (2011) Engineering microbes for tolerance to next-generation biofuels. Biotechnology for Biofuels 4:32
2. Ramos JL, Duque E, Gallegos MT, Godoy P, Ramos-Gonzalez MI, Rojas A, et al. (2002) Mechanisms of solvent tolerance in gram-negative bacteria. Annual Review of Microbiology 56:743–768
3. Ma MG, Liu ZL (2010) Mechanisms of ethanol tolerance in *Saccharomyces cerevisiae*. Appl Microbiol Biot 87(3):829–845
4. Nicolaou SA, Gaida SM, Papoutsakis ET (2010) A comparative view of metabolite and substrate stress and tolerance in microbial bioprocessing: from biofuels and chemicals, to biocatalysis and bioremediation. Metabolic Engineering 12(4):307–331
5. Segura A, Godoy P, van Dillewijn P, Hurtado A, Arroyo N, Santacruz S, et al. (2005) Proteomic analysis reveals the participation of energy- and stress-related proteins in the response of *Pseudomonas putida* DOT-T1E to toluene. Journal of Bacteriology 187(17):5937–5945
6. Piper PW (1995) The heat-shock and ethanol stress responses of yeast exhibit extensive similarity and functional overlap. FEMS Microbiology Letters 134(2–3):121–127
7. Alsaker KV, Paredes C, Papoutsakis ET (2010) Metabolite stress and tolerance in the production of biofuels and chemicals: gene-expression-based systems analysis of butanol, butyrate, and acetate stresses in the anaerobe *Clostridium acetobutylicum*. Biotechnology and Bioengineering 105(6):1131–1147
8. Green DR, Reed JC (1998) Mitochondria and apoptosis. Science 281(5381):1309–1312
9. Chan DC (2006) Mitochondria: dynamic organelles in disease, aging, and development. Cell 125(7):1241–1252
10. Detmer SA, Chan DC (2007) Functions and dysfunctions of mitochondrial dynamics. Nature Reviews. Molecular Cell Biology 8(11):870–879
11. Westermann B (2008) Molecular machinery of mitochondrial fusion and fission. Journal of Biological Chemistry 283(20):13501–13505
12. Westermann B (2010) Mitochondrial dynamics in model organisms: what yeasts, worms and flies have taught us about fusion and fission of mitochondria. Seminars in Cell & Developmental Biology 21(6):542–549
13. Chen HC, McCaffery JM, Chan DC (2007) Mitochondrial fusion protects against neurodegeneration in the cerebellum. Cell 130(3):548–562
14. Rapaport D, Brunner M, Neupert W, Westermann B (1998) Fzo1p is a mitochondrial outer membrane protein essential for the biogenesis of functional mitochondria in *Saccharomyces cerevisiae*. Journal of Biological Chemistry 273(32):20150–20155
15. Youle RJ, Karbowski M (2005) Mitochondrial fission in apoptosis. Nature Reviews. Molecular Cell Biology 6(8):657–663
16. Santel A, Frank S (2008) Shaping mitochondria: the complex posttranslational regulation of the mitochondrial fission protein DRP1. IUBMB Life 60(7):448–455
17. Liu LM, Li Y, Li HZ, Chen J (2004) Manipulating the pyruvate dehydrogenase bypass of a multi-vitamin auxotrophic yeast *Torulopsis glabrata* enhanced pyruvate production. Letters in Applied Microbiology 39(2):199–206
18. Xu N, Liu LM, Zou W, Liu J, Hua Q, Chen J (2013) Reconstruction and analysis of the genome-scale metabolic network of Candida glabrata. Molecular BioSystems 9(2):205–216
19. Chen XL, Xu GQ, Xu N, Zou W, Zhu P, Liu LM, et al. (2013) Metabolic engineering of *Torulopsis glabrata* for malate production. Metabolic Engineering 19:10–16
20. Li SB, Xu N, Liu LM, Chen J (2014) Engineering of carboligase activity reaction in *Candida glabrata* for acetoin production. Metabolic Engineering 22:32–39
21. Maniatis T, Fritsch EF, Sambrook J (1982) Molecular cloning: a laboratory manual. vol. 545. Cold Spring Harbor Laboratory , Cold Spring Harbor, NY
22. Zhou JW, Dong ZY, Liu LM, Du GC, Chen J (2009) A reusable method for construction of non-marker large fragment deletion yeast auxotroph strains: a practice in *Torulopsis glabrata*. Journal of Microbiological Methods 76(1):70–74
23. Nishida N, Ozato N, Matsui K, Kuroda K, Ueda M (2013) ABC transporters and cell wall proteins involved in organic solvent tolerance in *Saccharomyces cerevisiae*. Journal of Biotechnology 165(2):145–152
24. Ludovico P, Sansonetty F, Corte-Real M (2001) Assessment of mitochondrial membrane potential in yeast cell populations by flow cytometry. Microbiology-Sgm 147(12):3335–3343
25. Sharon LM, Thomas GC (1999) Inhibition of caspase activity delays apoptosis in a transfected NS/0 myeloma cell line. Biotechnology and Bioengineering 67(2):165–176
26. Breitenbach M, Laun P, Gimona M (2005) The actin cytoskeleton, RAS-cAMP signaling and mitochondrial ROS in yeast apoptosis. Trends in Cell Biology 15(12):637–639
27. Zhou J, Liu L, Chen J (2011) Improved ATP supply enhances acid tolerance of *Candida glabrata* during pyruvic acid production. Journal of Applied Microbiology 110(1):44–53
28. Andreyev AY, Kushnareva YE, Starkov AA (2005) Mitochondrial metabolism of reactive oxygen species. Biochemistry Biokhimiia 70(2):200–214
29. Sedensky MM, Morgan PG (2006) Mitochondrial respiration and reactive oxygen species in *C. elegans*. Experimental Gerontology 41(10):957–967
30. Allen RG, Tresini M (2000) Oxidative stress and gene regulation. Free Radical Bio Med 28(3):463–499
31. Ischiropoulos H, Beckman JS (2003) Oxidative stress and nitration in neurodegeneration: cause, effect, or association? Journal of Clinical Investigation 111(2):163–169
32. Otera H, Ishihara N, Mihara K (2013) New insights into the function and regulation of mitochondrial fission. Bba-Mol Cell Res 1833(5):1256–1268
33. Youle RJ, van der Bliek AM (2012) Mitochondrial fission, fusion, and stress. Science 337(6098):1062–1065
34. Kitagaki H, Araki Y, Funato K, Shimoi H (2007) Ethanol-induced death in yeast exhibits features of apoptosis mediated by mitochondrial fission pathway. FEBS Letters 581(16):2935–2942
35. Criollo A, Galluzzi L, Maiuri MC, Tasdemir E, Lavandero S, Kroemer G (2007) Mitochondrial control of cell death induced by hyperosmotic stress. Apoptosis 12(1):3–18

36. Ludovico P, Rodrigues F, Almeida A, Silva MT, Barrientos A, Corte-Real M (2002) Cytochrome c release and mitochondria involvement in programmed cell death induced by acetic acid in Saccharomyces cerevisiae. Molecular Biology of the Cell 13(8):2598–2606
37. Kroemer G, Reed JC (2000) Mitochondrial control of cell death. Nature Medicine 6(5):513–519
38. Creagh EM, Martin SJ (2001) Caspases: cellular demolition experts. Biochemical Society Transactions 29(6):696–702
39. Pozniakovsky AI, Knorre DA, Markova OV, Hyman AA, Skulachev VP, Severin FF (2005) Role of mitochondria in the pheromone- and amiodarone-induced programmed death of yeast. Journal of Cell Biology 168(2):257–269
40. Yu XX, Wang HJ, Liu LM (2014) Two non-exclusive strategies employed to protect *Torulopsis glabrata* against hyperosmotic stress. Appl Microbiol Biot 98(7):3099–3110
41. Mitra K, Wunder C, Roysam B, Lin G, Lippincott-Schwartz J (2009) A hyperfused mitochondrial state achieved at G(1)-S regulates cyclin E buildup and entry into S phase. Proceedings of the National Academy of Sciences of the United States of America 106(29):11960–11965
42. Mitra K (2013) Mitochondrial fission-fusion as an emerging key regulator of cell proliferation and differentiation. BioEssays 35(11):955–964
43. Jezek P, Plecita-Hlavata L (2009) Mitochondrial reticulum network dynamics in relation to oxidative stress, redox regulation, and hypoxia. International Journal of Biochemistry & Cell Biology 41(10):1790–1804

9

Bioreductive preparation of ACE inhibitors precursor (*R*)-2-hydroxy-4-phenylbutanoate esters: Recent advances and future perspectives

Guo-Chao Xu and Ye Ni*

Abstract

Optically active (*R*)-2-hydroxy-4-phenylbutanoate esters ((*R*)-HPBE) are key precursors for the production of angiotension-converting enzyme (ACE) inhibitors, which are important prescriptive drugs for preventing the formation of angiotensin II and lowering the blood pressure. The biocatalytic asymmetric reduction of ethyl 2-oxo-4-phenylbutanoate (OPBE) to (*R*)-HPBE with carbonyl reductases has several advantageous attributes, including high enantioselectivity, mild reaction condition, high catalytic efficiency, and environmental benignity. An increasing number of OPBE reductases have been discovered owing to the drastic achievements in genomics, screening and evolution technologies, and process engineering. The potential of (*R*)-HPBE production process has also been intensively evaluated. This review covers recent progress on the bioreductive preparation of (*R*)-HPBE, especially on various screening approaches for the identification of OPBE reductases and their characterization.

Keywords: ACE inhibitors; Asymmetric reduction; Reductases; (*R*)-HPBE

Introduction

Optically active secondary alcohols, especially hydroxyl acids or hydroxyl acid esters, are important compounds for introducing chiral elements to pharmaceuticals, agricultural pesticides, and other fine chemicals [1-4]. (*R*)-2-hydroxy-4-phenylbutanoate esters (OPBE), one class of chiral alcohols, are important precursors for the production of serials of angiotensin-converting enzymes (ACE) inhibitors, generally named as pril drugs (benazepril, cilazapril, quinapril, and ramipril), possessing (*S*)-homophenylalanine moiety as pharmacophore [5-7]. ACE (Kininase II, peptidyldipeptide hydrolase, EC 3.4.15.1) could catalyze the production of vasoconstrictor angiotensin II and inactivation of vasodilator bradykinin. Whereas inhibitors of ACE could prevent the formation of angiotensin II, hence decrease the blood pressure [8,9]. Along with the economic development, the incidence of hypertension is higher and younger which may induce stroke, heart failure, heart attack, and kidney failure. The annual revenues of prescriptive antihypertensive drugs increased every year, with $45 billion in total at the end of 2011 [10]. The ACE inhibitors drugs accounted for 10% among all the antihypertensive drugs. Hence, the preparation of their key chiral precursor (*R*)-HPBE is of special and increasingly interests.

Numbers of methods have been developed for the preparation of (*R*)-HPBE, including chemical, enzymatic, and chemoenzymatic approaches. According to the starting compounds, they can be divided into (1) enantioselective resolution of *rac*-HPBE, (2) asymmetric ene reduction, (3) asymmetric reduction of ene and carbonyl groups, (4) asymmetric reduction of OPBE, (5) hydrogenolysis, (6) enantioselective hydrolysis of *rac*-hydroxynitrile, etc. (Figure 1) [11-16]. The enantioselective resolution of *rac*-HPBE and *rac*-hydroxynitrile might represent effective approaches, however, limited to maximal yield of 50% [12,17]. Asymmetric transfer hydrogenation (ATH) of ketones plays an important role in asymmetric synthesis of chiral secondary alcohols [18-22]. Enantioselective hydrogenation of ethyl 2-oxo-4-phenylbutanoate is likely one of the most economic and efficient method. Great efforts have been made to investigate the potential of this approach over the past

* Correspondence: yni@jiangnan.edu.cn
The Key Laboratory of Industrial Biotechnology, Ministry of Education, School of Biotechnology, Jiangnan University, Wuxi 214122, China

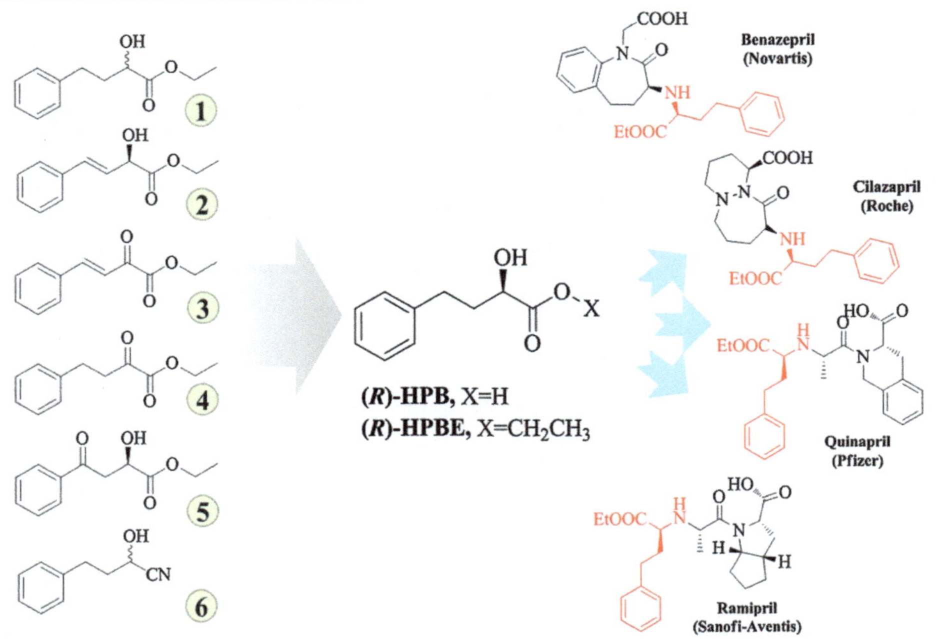

Figure 1 Various approaches for the synthesis of selected ACE inhibitors building blocks. (1) Enantioselective resolution of *rac*-HPBE, (2) asymmetric ene reduction, (3) asymmetric reduction of ene and carbonyl groups, (4) asymmetric reduction of OPBE, (5) hydrogenolysis, and (6) enantioselective hydrolysis of *rac*-hydroxynitrile.

few years. The highest enantioselectivity reported was 92% *ee* (*R*) using 10,11-dihydrocinchonidine (DHCd) as the chiral modified, and 94% *ee* (*R*) with the hydroxyl group in the DHCd modifier methylated. However, both cases were achieved using acetic acid as solvent, and under rigorous and commercially impractical conditions of high pressure (100 bar) and high temperature (400°C) hydrogen pretreatment of the Pt/Al_2O_3 catalyst [22]. After optimization, this process could be carried out at mild reaction condition (17°C and 5.8 bar) resulting enantioselectivity of merely 91% [23]. Therefore, seeking effective, highly enantioselective, and practical chemical approaches to (*R*)-HPBE is still of significance.

Tremendous endeavors have been committed to the preparation of (*R*)-HPBE employing biocatalytic asymmetric reduction systems, which is a reliable, scalable, and straightforward route to chiral alcohols. Compared with classic chemical processes, the asymmetric reduction of OPBE into (*R*)-HPBE, or the reduction of 2-oxo-4-phneylbutyrate (OPB) to (*R*)-2-hydroxy-4-phenylbutyrate [(*R*)-HPB] employing reductases and dehydrogenases (Figure 2) has several advantageous attributes, including high selectivity, high efficiency, mild reaction condition, and environmental friendless [24-27]. The complicated and chiral physiological structures endue the enzymes with relatively high chemo- and stereoselectivity and specificity. Various enzyme discovery and protein engineering tools have been proposed and proved, which would greatly shorten the developing period and increase the availability of robust enzymes. Several literatures on the biocatalytic asymmetric reduction for pharmaceuticals, protein engineering strategies for robust enzymes in chemical synthesis have been reviewed [28-32]. Also, various OPBE reductases have been successfully identified from natural samples or genome databases. The application potential and scalability of the biocatalytic processes have been systematically investigated. This mini-review was mainly focused on the discovery and properties of OPBE reductases. Future prospect of the application of bioreductive in preparation of optically pure (*R*)-HPBE synthesis was also discussed.

Screening of microorganisms for bioreductive preparation of (*R*)-HPBE

Due to the importance of (*R*)-HPBE as key chiral building blocks, searching for robust microorganisms has been extensively conducted as shown in Table 1. Some microorganisms, mainly yeasts, have been isolated from natural environment. Oda et al. screened 55 type culture yeasts from IFO and 499 isolated yeasts, among which two yeasts, *Rhodotorula minuta* IFO 0920 and *Candida holmii* KPY 12402, were identified with relatively high enantioselectivity. At 4.1 g L^{-1} OPBE loading, the *ee* and isolation yield could reach 90% and 58%, respectively, using *C. holmii* KPY 12402 [33,34]. *Kluyveromyces marxianus*, *Pichia pastoris*, *Pichia anomala*, and *Pichia angusta* have also been tested in the asymmetric reduction of OPBE into (*R*)-HPBE, and *P. angusta* displayed

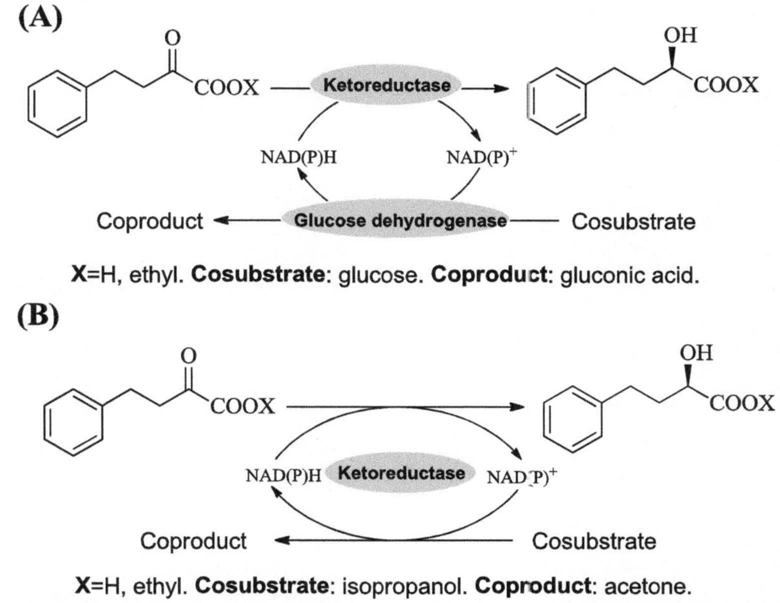

Figure 2 Asymmetric preparation of (*R*)-HPB derivatives employing oxidoreductases coupled with cofactor regeneration. (A) Enzyme-coupled system. **(B)** Substrate-coupled system.

the best performance of 100% conversion and 81% *ee* [35]. Besides, Chadha et al. reported the preparation of (*R*)-HPBE from OPBE by using *Daucus carota* with excellent chemical yield (90%) and *ee* (99%) [36]. Chen et al. investigated the ability of 40 yeasts kept in CIOC (Chinese Institute of Organic Chemistry, Chinese Academy of Sciences) in the reduction of OPBE. Most of the candidate yeasts shown good (*S*)-preference (>99%), while *Candida boidinii* CIOC21 was proved to be (*R*)-preference. As much as 2.0 g L^{-1} OPBE could be asymmetrically reduced into (*R*)-HPBE with 99.7% *ee* and 95.1% yield [37]. Target reaction oriented enrichment and screening of microorganisms have also been carried out in our laboratory, and one yeast strain *Candida krusei* SW2026 displayed desirable *ee* (99.7%) and yield (95.1%) in aqueous phase at 2.5 g L^{-1} of OPBE [17]. The highest record of OPBE loading in microbial system was achieved with *Bacillus pumilus* Phe-C3 by He and coworkers. At 6.2 g L^{-1} of OPBE and 12 g cdw L^{-1} catalyst, 95.3% OPBE could be reduced with 97.1% *ee* [38].

Discovery and characterization of OPBE reductases

Although many wild-type microbial cells could catalyze the asymmetric reduction of OPBE into (*R*)-HPBE, their industrial applications are limited by a number of disadvantageous. The expression level and enzymatic properties of OPBE reductases are often undesirable in parent strains [28]. Additionally, there might exist several enzymes with variable or opposite stereospecificity that could result HPBE with compromised chirality [39]. And the fermentation conditions of wild-type strains are usually strict and time-consuming. Studies on the discovery and properties of isolated OPBE reductases are therefore of special interests to prompt the biocatalytic preparation of chiral (*R*)-HPBE.

Table 1 Asymmetric reduction of OPBE into (*R*)-HPBE employing microorganisms

Entry	Strains	OPBE loading [g L^{-1}]	*ee*[%]	Yield[%]	Reference
1	*Candida krusei* SW2026	2.5	99.7	95.1	[17]
2	*Bacillus pumilus* Phe-C3	6.2	97.1	91	[37]
3	*Candida boidinii* CIOC21	2	99	92	[36]
4	*Baker's yeast*	1	92	90	[38]
5	*Candida holmi*	4.1	90	58	[33]
6	*Pichia angusta*	1.4	81	99	[34]
7	*Pichia pastoris*	1	95	99	[7]

Enzyme discovery strategies

Alongside the revolution of biotechnologies, a variety of tools are available for the quick screening and identification of biocatalysts. Several comprehensive reviews discussed the discovery and engineering tools developed to obtain naturally evolved or tailor-made enzymes with enhanced properties for chemical synthesis [29,30,40]. In the case of microbial strains isolated from natural environment, enzymes could be discovered through protein purification, insertional mutation, chromosome walking, shotgun library, etc. With regard to vast amount of genome information, genome hunting and genome database mining are two direct and time-saving strategies.

Protein purification

For a long time, purification of target enzymes from wild-type strains has been the dominant approach for enzyme discovery [41]. Through several purification steps, such as ammonium sulfate precipitation, ion exchange chromatography, hydrophobic chromatography, affinity chromatography and gel chromatography, and pure or partially purified enzymes could be harvested. After characterization with protein NMR of terminal peptide and MS, the encoding gene sequence of target enzymes could be acquired. Various carbonyl reductases of different origins (microorganisms and plant tissues) have been discovered through microorganism isolation and protein purification. A robust carbonyl reductase was purified from Adzuki bean, with high enantioselectivity (>99% *ee*) and tolerance to α-chloroacetophenone (0.2 M) [42].

Insertional mutation

This strategy utilizes a proper transposon to insert randomly into the genome sequence and disturb the normal properties or phenotypes of the strains. Through screening for mutants exhibiting expected changes and sequencing, the target genes could be identified. Various novel enzymes and pathways have been discovered employing this method, such as two novel genes *pao* and *sap* from *Pseudomonas* sp. Strain HZN6, coding for 6-hydroxy-L-nicotine oxidase and $NADP^+$-dependent 3-succinoylsemialdehyde-pyridine dehydrogenase in the catabolism of nicotine [43].

Chromosome walking

Based on conserved motif of certain enzymes, chromosome walking provided a new method to identify functional enzymes [44]. A functional P450 monooxygenase from *Rhodococcus* sp. ECU0066 was identified by chromosome walking by several rounds of PCR, with capability in the enantioselective oxidation of sulfides [45].

Shotgun library

Compared with other strategies, shotgun library was much more common and easier to operate in discover functional enzymes. Through construction the subclones of the genome in proper fragment length, a shotgun library could be developed. Many enzymes were identified using this approach. For example, Zhang and coworkers successfully cloned a novel arylamidase encoding gene from *Paracoccus* sp. strain FLN-7 which could hydrolyze amide pesticides [46].

Genome hunting

In the case of available genome data, genome hunting could be adopted to identify the target enzymes by cloning all of the potential genes from this single genome. Through genome hunting, Ni and coworkers heterogeneously expressed 13 reductases genes from *Bacillus* sp. ECU0013. Among them, the FabG (a β-ketoacyl-ACP reductase) exhibited high efficiency in the asymmetric reduction of 620 g L^{-1} OPBE to (*S*)-HPBE, with no addition of external expensive cofactor and 99% *ee*. Although the product configuration is not the desired one, it could be converted to (*R*)-HPBE through one-step configuration inversion [47].

Genome database mining

Another effective and promising strategy for quick discovery of suitable enzymes is genome database mining for homologous protein sequences of targeted enzymes. Owing to the revolution of sequencing technology, there are 58,117 sequencing projects (including 6,649 complete projects, 23,552 permanent drafts, 26,572 incomplete projects, and 1,404 targeted projects) in genome online database as of September 2014, providing massive data on protein-coding genes [48]. The selection of reported enzymes as probe and the choice of candidate proteins in database with moderate identity are the two key factors for the successful identification of new enzymes. The sequence of several known ethyl 4-chloro-3-oxobutanate (COBE) reductases was regarded as probes to search for novel enzymes in GenBank, resulting in an excellent reductase from *Streptomyces coelicolor* out of ten candidates for the synthesis of optically pure ethyl (*S*)-4-chloro-3-hydroxybutanoate, a chiral building blocks for HMG-CoA reductase inhibitors, with a stunning high productivity of 609 g L^{-1} day^{-1} [49].

Selected OPB(E) redutases

As shown in Table 2, a number of OPBE reductases have been discovered using above mentioned strategies. Most of the OPBE reductases are NADPH-dependent. A phenylacetaldehyde reductase (PAR) was purified from *Rhodococcus* ST-10 [50]. After amino-acid sequence analysis, the recombinant PAR was expressed in *Escherichia coli* [51]. One PAR mutant (HAR-1) could asymmetrically reduce 195 g L^{-1} OPBE to (*R*)-HPBE with 99% *ee* and 89% isolation yield [52]. The shotgun library of

Table 2 Enzymatic properties of oxidoreductases for producing (*R*)-HPB derivatives

Enzyme	Source strain	Accession no	Discovery strategy	Length [aa]	Cofactor	Family[a]	Specific activity [U mg^{-1}]	K_m [mM]	V_{max} [U mg^{-1}]	Opt. temp. [°C][b]	Sub. loading [g L^{-1}]	Time [h]	*ee* [%]	TTN[c]	STY [g L^{-1} d^{-1}][d]	Reference
Ypr1p	*S. cerevisae*	AHY75323.1	Genome hunting	312	NADPH	AKR	*nd* [e]	*nd*	*nd*	*nd*	2.0	*nd*	87	*nd*	*nd*	[54]
IolS	*B. subtilis*	CAB16014.1	Database mining	310	NADPH	AKR	5.1	2.61	4.18	30	330	12	>99	32039	660	[57]
CgKR2	*C. glabrata*	CAG61069.1	Database mining	311	NADPH	AKR	2.4	0.1	18.5	45	206	6	99	*nc* [f]	700	[59]
NcCR	*Neurospora crassa*	XM_954278	Genome hunting	331	NADPH	AKR	1.6	*nd*	*nd*	40	2.0	24	80	20	2	[56]
CgKR1	*C. glabrata*	CAG58832.1	Database mining	352	NADPH	MDR	23.1	*nd*	*nd*	*nd*	412	24	98.1	*nc* [f]	358	[58]
YOL151w	*S. cerevisiae*	CAA99172.1	Genome hunting	342	NADPH	MDR	*nd*	*nd*	*nd*	*nd*	1	48	99	10	0.5	[55]
PpADH	*Paracoccus pantotrophus* DSM 11072	EU427522	Shotgun library	262	NADH	SDR	0.0043	*nd*	*nd*	*nd*	5	24	99	24.3	5	[53]
PAR	*Rhodococcus* sp. ST-10	AB190261	Purification	348	NADH	MDR	*nd*	*nd*	*nd*	*nd*	195	24	99	947	174	[52]
YiaE	*E. coli* K12	BAE77742.1	Genome hunting	324	NADPH	HAD	18	7.2	24.6	45	20	24	99	*nc* [f]	20	[60]

[a]Enzyme families, AKR aldo-keto reductase, HAD 2-hydroxyacid dehydrogenase, SDR short chain dehydrogenase/reductase, MDR medium chain dehydrogenase/reductase. [b]Optimal temperature. [d]total turnover number of cofactor. [d]space-time yield. [e]nd, not determined. [f]not calculated because no cofactor was added.

Paracoccus pantotrophus DSM 11072 was developed to search for enzyme with OPBE reduction activity, resulting in the novel enzyme PpADH with 99% *ee* (*R*) at 5 g L^{-1} OPBE [53]. Through genome hunting, Yprlp and YOL151w were identified from *Saccharomyces cerevisiae* with 87% and 99% *ee* (*R*), respectively [54,55]. NcCR was cloned from *Neurospora crassa* with 1.6 U mg^{-1} specific activity and 80% *ee* (*R*) [56]. Employing genome database mining strategy, three OPBE reductases (IolS, CgKR1, and CgKR2) were discovered from genome database [57-59]. Both CgKR1 and CgKR2 were cloned from *Candida glabrata*, with specific activities of 23.1 and 2.4 U mg^{-1} and *ee* values of 98.7% and 99%, respectively. CgKR1 was stable enough to tolerate as high as 412 g L^{-1} OPBE, while CgKR2 could completely reduce 206 g L^{-1} OPBE into (*R*)-HPBE (>99% *ee*) with 700 g L^{-1} day^{-1} space-time yield. In our previous study, IolS could tolerate 330 g L^{-1} OPBE with 99% *ee* (*R*). The specific activity of IolS was 5.1 U mg^{-1} [57]. In addition, (*R*)-2-hydroxy-4-phenylbutyrate ((*R*)-HPB) is also a precursor for ACE inhibitors and could be synthesized through stereospecific reduction of 2-oxo-4-phenylbutyrate (OPB) as illustrated in Figure 1. One hydroxy acid dehydrogenase (YiaE) gene was cloned from *E. coli* K12, which could asymmetrically reduce 20 g L^{-1} OPB into its corresponding hydroxy acid, (*R*)-HPB, in >99% *ee* within 24 h. The specific activity was determined to be 20 U mg^{-1}, the optimum temperature was 45°C [60].

Table 2 Enzymatic properties of oxidoreductases for producing (*R*)-HPBE from OPBE.

Classification of OPB(E) reductases

According to the amino acid sequence and metal-ion dependency, the carbonyl reductases could be divided into short-chain dehydrogenase/reductase (SDR), medium-chain dehydrogenase/reductase (MDR), long-chain dehydrogenase/reductase (LDR), and aldo-keto reductase (AKR) [61-63]. Key motif search in online databases (SDR, http://www.sdr-enzymes.org/; MDR, http://www.bioinfo.ifm.liu.se/services/mdr/; AKR, http://www.med.upenn.edu/akr/) reveals that PpADH belongs to SDR family, CgKR1, YOL151w, and PAR belong to MDR family, while Yprlp, CgKR2, NcCR, and IolS are members of AKR. PAR is the only zinc-dependent carbonyl reductase. Tyr-Lys-Ser, Cys-His-Asp, and His-Tyr-Lys are common catalytic triads for SDR, MDR, and AKR, respectively [21,64,65]. Reductase YiaE from *E. coli* with reduction activity in the conversion of OPB into (*R*)-HPB was classified into 2-hydroxyacid dehydrogenase family (Table 2).

Regarding to crystal structure information of the OPBE reductases, protein crystal structure of IolS from *Bacillus subtilis* is the only one deposited in the PDB database (PDB accession no. 1PYF) [65]. In the structure of IolS, the most common fold, $(\beta/\alpha)_8$ barrel motif, seems to form a barrel conformation. The cofactor $(NAD(P)^+/NAD(P)H)$ is bound at the C-terminal of internal β-barrel, extending in an extended conformation from the center of the barrel with no intra-molecular hydrogen bonds. Besides to the conserved folding, the catalytic tyrosine at 58 sites acts as an acid in the reduction reaction. The acidity is enhanced through a hydrogen bond with the lysine of catalytic triad, which also in turn forms a salt bridge with the aspartate. In the apo form, the hydrogen bond between catalytic Tyr58 and Lys84 was 3.25 Å. When $NADP^+$ binds, the distance between the lysine ζ-amino group and the tyrosine hydroxyl is extended to 4.26 Å, which causes the disruption of this hydrogen bond due to a shift of Tyr58 towards the $NADP^+$. Two additional amino acids Asp51 and Lys84, which complete the catalytic triad by activating the phenolic proton on the tyrosine, also present in IolS. The interactions between both residues in the holo conformation are mediated by hydrogen bonds with a water molecule.

Since three other OPBE reductases (Yprlp, CgKR2, and NcCR) are also potential AKR members, the amino acid sequence alignment of them is shown in Figure 3. Four AKR members shared similar backbones, although the sequence identity is relatively low (<50%). The catalytic tyrosine and aspartate are quite conserved in all four OPBE reductases, while the lysines are located around the conserved site of IolS. Most of the AKR are NADP-dependent and share the similar loop at the C-terminal, with 2'-phosphate tightly bound via three salt bridges with Arg227, Arg282, and Lys214 and a hydrogen bond with the side-chain of Gln286.

Bioprocess engineering for improved efficiency

For a biocatalytic process, several parameters are critical to evaluate the efficiency, including space-time yield, total turnover number (TTN) of cofactor or reductase, productivity, etc. However, hydrophobic OPBE could cause the deactivation of enzymes and often requires longer reaction time to achieve a higher conversion rate [57]. Consequently, the efficiency is usually at low level. Various process engineering strategies have been developed to solve the problems and improve the efficiency in the preparation of chiral secondary alcohols employing reductases.

Cofactor regeneration

To achieve complete conversion of OPBE by carbonyl reductases, stoichiometric amount of reducing cofactors is usually required. The high cost of cofactors is one of the main hurdles for the wide application of bioreductive preparation. Introduction of proper cofactor cycling systems could reduce the addition of expensive cofactors and promote reaction towards the synthesis of product. Numerous cofactor regeneration systems have been

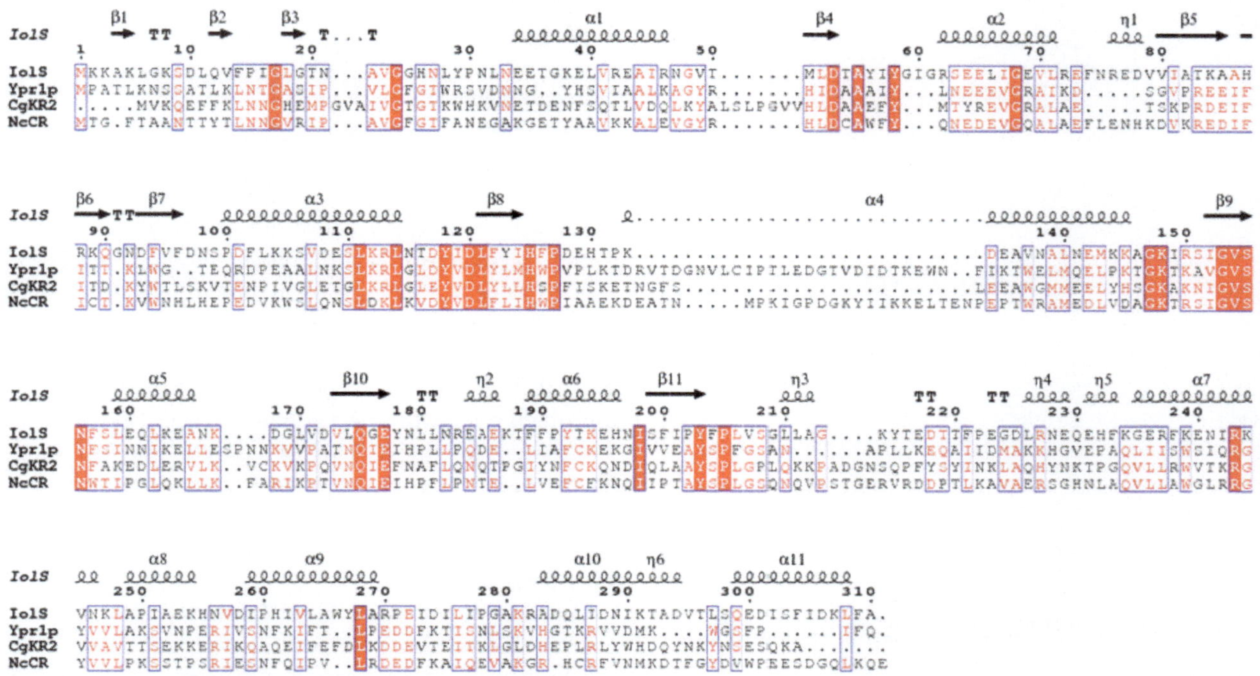

Figure 3 Amino acid sequence alignment of selected OPBE reductases belonging to AKR family using ESPript tool. *B. subtilis* (IolS) (CAB16014.1), *S. cerevisiae* (Ypr1p) (AHY75323.1), *C. glabrata* (CgKR2) (CAG61069.1), and *Neurospora crassa* (NcCR) (XM_954278). Secondary structure elements of IolS (PDB No. 1PYF) are indicated above the sequence with arrows for *β*-strands and cylinders for *α*-helices [66].

developed and systematical summarized in recent reviews [67-69]. In general, there are three approaches to regenerate cofactors, including substrate-coupled, enzyme-coupled, and reaction-internal regeneration accomplished by enzymatic, chemical, photochemical, or electrochemical reagents. Several principles need to be considered in order to select an appropriate regeneration system as following:

1. Catalysts as well as cosubstrates should be commercially available or easily producible.
2. Product formation should be thermodynamically as well as kinetically favored and has no interference with the enzymes or with the subsequent isolation of desired product.
3. Regeneration should be practical and inexpensive and stable over a long period of time.

In the view of all these factors, the feasibility of glucose dehydrogenase, formate dehydrogenase, and isopropanol-coupled ketoreductase technologies have been proven. Formate dehydrogenase, however, exhibited low recycling efficiency on $NADP^+$. Glucose dehydrogenase and isopropanol-coupled systems have been applied in the enantioselective synthesis of (*R*)-HPB derivatives. By using glucose dehydrogenases from *B. subtilis*, the TTN of $NADP^+$ could reach as much as 32,039 in the asymmetric reduction of 330 g L^{-1} OPBE employing IolS from *B. subtilis* under assistance of 0.05 mM $NADP^+$ (Figure 2A). PAR from *Rhodococcus* sp. ST-10 is a reductase with isopropanol oxidation activity. Itoh and coworkers adopted the substrate-coupled strategy in the preparation of (*R*)-HPBE with PAR (Figure 2B). Almost 195 g L^{-1} OPBE was completely reduced, and the TTN of NAD^+ was 947. Isopropanol could not only work as cosubstrate, but also as cosolvent to increase the solubility of OPBE.

Solvent engineering

Solvents constitute the microenvironment around the biocatalysts. By introducing organic solvents, ionic liquids, supercritical fluids, and glycerol into the biocatalytic reduction systems, not only the toxicity of substrate/product to hydrophilic biocatalysts could be alleviated, but also their spontaneously hydrolysis and volatilization could be relieved. Among them, organic/aqueous biphasic system was the most common used approach as illustrated in Figure 4A. The key of this system was to find a good suitable and biocompatibility solvents, which could balance the influence on the activity and enantioselectivity of enzymes and the partition behavior of substrate and product. *C. krusei* SW2026 was proved to display 99.7% *ee* ((*R*)-HPBE) and 95.1% yield at 2.5 g L^{-1} OPBE in aqueous system [17]. As substrate loading increased to 20 g L^{-1}, the *ee* and yield were reduced to 87.5% and 45.8%, respectively, which was

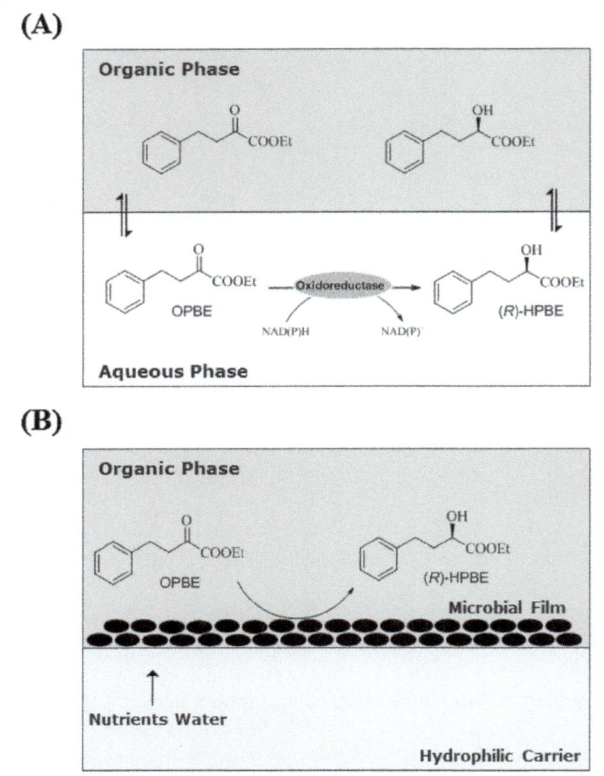

Figure 4 Scheme of bioreductive preparation of (*R*)-HPBE in organic solvent/aqueous biphasic system (A) and an interface bioreactor (B).

caused by the higher amount of toxic OPBE. Based on the solvent polarity, ten organic solvents with different Log*P* values ranging from 0.28 to 7.7 were selected. Dibutyl phthalate (Log*P* = 5.4) showed the highest biocompatibility and proper partition coefficient for OPBE and HPBE. Compared with aqueous system, less time was required to reach higher conversion, resulting in 95.0% *ee* and 92.5% isolation yield. This dibutyl phthalate/aqueous biphasic system could not only migrate the substrate toxicity effect from enzymes/cells, but also force the reaction equilibrium towards the reduction of OPBE by accumulating (*R*)-HPBE in the organic phase.

Substrate feeding

To realize a high substrate loading, substrate feeding at batches is an effective strategy to reduce the of substrate inhibition. A much higher catalytic efficiency could be achieved by substrate-feeding operation. In our studies using 1-L aqueous/octanol biphasic system, 30 g OPBE was fed once per hour for ten times (330 g in total). After 12 h of reaction, substrate conversion rate of >99% and space-time yield of 660 g L^{-1} day^{-1} were obtained at 330 g L^{-1} OPBE [57]. The substrate loading was 1.6 times higher than that of the highest record reported so for. Additionally, a S/C ratio of 31.7 was attained, ranking the highest catalyst yield in the asymmetric reduction of OPBE, which was 0.02–4.12 as reported [59].

Other processes

To alleviate the toxicity of hydrophobic OPBE/HPBE, interface bioreactors have been established. It is a biotransformation device between a hydrophilic carrier and a hydrophobic organic solvent as shown in Figure 4 [34]. In this system, the biocatalyst was immobilized on the carrier surface between the hydrophilic and hydrophobic solvents. Interestingly, the recycling of cofactors could proceed smoothly, and the product separation was relatively easier than that for aqueous systems.

The smart control of OPBE concentration in the reaction medium could also be achieved by adsorbing the substrate on hydrophobic resins (Figure 4B). D'Arrigo and coworkers proved that XAD™ 1180, a polystyrenic adsorbent selected among several Amberlite resins, was effective in controlling the OPBE amount presented in water phase. OPBE (1 g L^{-1}) could be reduced asymmetrically employing *Pichia pastoris* with 99% conversion and 95% *ee* (*R*) [7]. Continuous batch reactor (CBR) was established combining the biotransformation and OPBE/HPBE adsorption. In the CBR unit, the bioreduction of

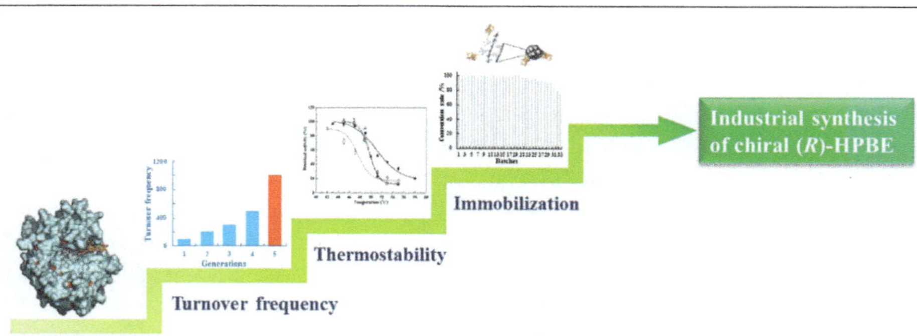

Figure 5 Further perspectives on the engineering of OPB(E) reductases for industrial synthesis of (*R*)-HPBE.

OPBE was carried out for 48 h at a substrate-feeding flow rate of 1 mL min^{-1}, resulting 95.1% *ee* (*R*) and conversion rate of 25%.

Prospects

Biocatalytic preparation of chiral alcohol is gaining momentum. Due to its high enantioselectivity, bioreductive preparation of chiral (*R*)-HPBE has been regarded as one of the most promising approaches. There are enormous difference between natural environment (i. e., cytosolic conditions) and industrial environment. Free enzymes might face hash environment, such as molar scale reactants and increased hydrophobicity instead of millimolar reagents and physiological aqueous environment. To become an efficient tool in large-scale synthesis of (*R*)-HPBE, OPBE reductases should display the following properties: (1) ≥100 g L^{-1} substrate loading, (2) ≤5 g L^{-1} biocatalyst loading (≥20 substrate to catalyst ratio), (3) ≥98% conversion rate, (4) ≤24 h reaction time, and (5) ≥99% *ee* [21,28]. Although several biocatalytic reduction systems could reach high substrate loading, enantioselectivity, and conversion, the biocatalyst loadings are still at relatively high and the S/C ratio is relatively low, which would hinder the product separation procedure.

Aiming at the industrial synthesis of (*R*)-HPBE, the biocatalytic procedure should be further engineered (Figure 5). In PDB database, extremely limited information on the crystal structure of OPBE reductases is available. The protein structure could broaden the knowledge of biocatalytic mechanism and provides guidance for further protein engineering. The catalytic efficiency (turnover frequency, k_{cat}) also remained to be improved to decrease the catalyst loading. As a robust industrial biocatalyst, thermostability and operational stability should also be high enough to tolerate rigorous reaction conditions. As shown in Table 2, most of the OPBE reductases are mesophilic enzymes, with optimum temperature of less than 45°C. Since the oxidoreduction reaction is cofactor-dependent, the cofactor engineering of bioreduction systems would fasten the reaction and decrease the addition of expensive cofactors. Importantly, reuse of the biocatalysts through proper immobilization strategies could also reduce the cost and simplify the product separation from reaction mixture.

Conclusion

Biocatalytic stereoselective reduction for (R)-2-hydroxy-4-phenylbutanate esters is gaining momentum. Various OPBE reductaes have been identified and some of them have indicated promising potential for practical manufacture of (R)-HPBE. The discovery strategies for OPBE reductases and bioprocessing engineering strategies have been reviewed. Further development of bioreductive preparation of (R)-HPBE will require the continued mining and designing industrially useful enzymes and elaboration of process engineering.

Competing interests

The authors declare that they have no competing interests.

Authors' contributions

GCX wrote this manuscript. YN contributed general advice and edited the manuscript. Both authors read and approved the final manuscript.

Acknowledgements

We are grateful to the Natural Science Foundation of China (21276112), National Basic Research and Development Program of China (2011CB710800), and New Century Excellent Talents in University (NCET-11-0658) for the financial support of this research.

References

1. Patel RN (2006) Biocatalysis: Synthesis of chiral intermediates for drugs. Curr Opin Drug Disc Dev 9:741–764
2. Tao JH, Xu JH (2009) Biocatalysis in development of green pharmaceutical processes. Curr Opin Chem Biol 13:43–50
3. Magano J, Dunetz JR (2012) Large-scale carbonyl reductions in the pharmaceutical industry. Org Process Res Dev 16:1156–1184
4. Reetz MT (2013) Biocatalysis in organic chemistry and biotechnology: Past, Present, and Future. J Am Chem Soc 135:12480–12496
5. Iwasaki G, Kimura R, Numao N, Kondo K (1989) A practical and diastereoselective synthesis of angiotensin converting enzyme inhibitors. Chem Pharm Bull 37:280–283
6. Bradshaw CW, Wong CH, Hummel W, Kula MR (1991) Enzyme-catalyzed asymmetric synthesis of (*S*)-amino-4-phenylbutanoic acid and (*R*)-2-hydroxy-4-phenylbutanoic acid. Bioorg Chem 19:29–39
7. D'Arrigo P, Pedrocchi-Fantoni G, Servi S (2010) Chemo-enzymatic synthesis of ethyl (R)-2-hydroxy-4-phenylbutyrate. Tetrahedron Asymmetry 21:914–918
8. Parmley WW (1998) Evolution of angiotensin-converting enzyme inhibition in hypertension, heart failure, and vascular protection. Am J Med 105:27S–31S
9. Maruyama S, Nakagomi K, Tomizuka N, Suzuki H (1985) Angiotensin I-converting enzyme inhibitor derived from an enzymatic hydrolysate of casein. II. Isolation and bradykinin-potentiating activity on the ileum of rats. Agric Bio Chem 49:1405–1409
10. Crandall MA (2007) The world cardiovascular drug market, 2nd edition, Kalorama Information, SKU:KLI1397872.
11. Baskar B, Pandian NG, Priya K, Chadha A (2004) Asymmetric reduction of alkyl 2-oxo-4-arylbutanoates and –but-3-enoates by *Candida parapsilosis* ATCC 7330: assignment of the absolute configuration of ethyl 2-hydeoxy-4-(*p*-methyl phenyl)but-3-enoate by ^{1}H NMR. Tetrahedron Asymmetry 15:3961–3966
12. Huang SH, Tsai SW (2004) Kinetic resolution of (R, S)-ethyl 2-hydeoxyl-4-phenylbutyrate via lipase-catalyzed hydrolysis and transesterification in isooctane. J Mol Catal B: Enzyme 28:65–69
13. Zhu LF, Meng QH, Fan WZ, Xie XM, Zhang ZG (2010) Direct asymmetric hydrogenation of 2-oxo-4-arylbut-3-enoic acids. J Org Chem 75:6027–6030
14. Aldea R, Alper H (1998) Hydrogenation of the carbonyl group in α-kektoesters and α-ketoamides catalyzed by ruthenium clay. J Org Chem 63:9425–9426
15. Rodrigues JAR, Milagre HMS, Milagre CDF, Moran PJS (2005) A highly enantioselective chemoenzymatic synthesis of syn-3-amino-2-hydroxy esters: Key intermediates of taxol side chain and phenylnorstatine. Tetrahedron Asymmetry 16:3099–3106
16. Meng QH, Zhu LF, Zhang ZG (2008) Highly enantioselective sequential hydrogenation of ethyl 2-oxo-4-arylbut-3-enoate to ethyl 2-hydroxy-4-arylbutyrate. J Org Chem 73:7209–7212
17. Zhang W, Ni Y, Sun ZH, Zheng P, Lin WQ, Zhu P, Ju NF (2009) Biocatalytic synthesis of ethyl (*R*)-2-hydroxy-4-phenylbutyrate with *Candida krusei* SW2-26: a practical process for high enantiopurity and product titer. Process Biochem 44:1270–1275
18. Noyori R (2003) Asymmetric catalysis: Science and opportunities. Adv Synth Catal 345:15–32

19. Noyori R, Yamakawa M, Hashiguchi S (2011) Metal-ligand bifuctional catalysis: a nonclassical mechanism for asymmetric hydrogen transfer between alcohols and carbonyl compounds. J Org Chem 66:7931–7944
20. Hall M, Bommarius AS (2011) Enantioenriched compounds via enzyme-catalyzed redox reactions. Chem Rev 111:4088–4110
21. Hollmann F, Arends IWCE, Holtmann D (2011) Enzymatic reductions for the chemist. Green Chem 13:2285–2314
22. Blaser HU, Jalett HP, Wiehl J (1991) Enantioselective hydrogenation of alpha-ketoesters with chinchona-modified platinum catalysts—Effect of acidic and basic solvents and additives. J Mol Catal 68:215–222
23. LeBlond C, Wang J, Liu J, Andrews AT, Sun YK (1999) Highly enantioselective heterogeneously catalyzed hydrogenation of *α*-ketoesters under mild conditions. J Am Chem Soc 121:4920–4921
24. Schoemaker HE, Mink D, Wubbolts MG (2003) Dispelling the myths—Biocatalysis in industrial synthesis. Science 299:1694–1697
25. Rollin JA, Tamb TK, Zhang PYH (2013) New biotechnology paradigm: Cell-free biosystems for biomanufacturing. Green Chem 15:1708–1719
26. Schmid A, Dordick JS, Hauer B, Kiener A, Wubbolts M, Witholt B (2001) Industrial biocatalysis today and tomorrow. Nature 409:258–268
27. Bai YL, Yang ST (2005) Biotransformation of (*R*)-2-hydeoxy-4-phenylbutyric acid be D-lactate dehydrogenase and *Candida boidinii* cells containing formate dehydrogenase coimmbolized in a fibrous bed bioreactor. Biotechnol Bioeng 92:137–146
28. Luetz S, Giver L, Lalonde J (2008) Engineered enzymes for chemical production. Biotechnol Bioeng 101:647–653
29. Behrens GA, Hummel A, Padhi SK, Schaetzle S, Bornscheuer UT (2011) Discovery and protein engineering of biocatalysts for organic synthesis. Adv Synth Catal 353:2191–2215
30. Bommarius AS, Blum JK, Abrahamson MJ (2011) Status of protein engineering for biocatalysts: how to design an industrially useful biocatalyst. Curr Opin Chem Biol 15:194–200
31. Bornscheuer UT, Huisman GW, Kazlauskas RJ, Lutz S, Moore JC, Robins K (2012) Engineering the third wave of biocatalysis. Nature 485:185–194
32. Davids T, Schmidt M, Boettcher D, Bornscheuer UT (2013) Strategies for the discovery and engineering of enzymes for biocatalysis. Curr Opin Chem Biol 17:215–220
33. Oda S, Ohta H (1992) Microbial transformation on interface between hydrophilic carriers and hydrophobic organic solvents. Biosci Biotechnol Biochem 56:2041–2045
34. Oda S, Inada Y, Kobayashi A, Ohta H (1998) Production of ethyl (*R*)-2-hydroxy-4-phenylbutanoate via reduction of ethyl 2-oxo-4-phenylbutanoate in an interface bioreactor. Biosci Biotechnol Biochem 62:1762–1767
35. de Lacerda PSB, Ribeiro JB, Leite SGF, Ferrara MA, Coelho RB, Bon EPS, da Silva Lima EL, Antunes OAC (2006) Microbial reduction of ethyl 2-oxo-4-phenylbutyrate. Searching for R-enantioselectivity. New access to the enalapril like ACE inhibitors. Tetrahedron Asymmetry 17:1186–1188
36. Chadha A, Manohar M, Soundararajan T, Lokeswari TS (1996) Asymmetric reduction of 2-oxo-4-phenylbutanoic acid ethyl ester by *Daucus carota* cell cultures. Tetrahedron Asymmetry 7:1571–1572
37. Chen YZ, Lin H, Xu XY, Xia SW, Wang LX (2008) Preparation the key intermediate of angiotensin-converting enzyme (ACE) inhibitors: High enantioselective production of ethyl (*R*)-2-hydeoxy-4-phenylbutyrate with *Candida boidinii* CIOC21. Adv Synth Catal 350:426–430
38. He CM, Chang DL, Zhang J (2008) Asymmetric reduction of substituted *α*- and *β*-ketoesters by *Bacillus pumilus* Phe-C3. Tetrahedron Asymmetry 19:1347–1351
39. Dao DH, Okamura M, Akasaka T, Kawai Y, Hida K, Ohno A (1998) Stereochemical control in microbial reduction. Part 31: Reduction of alkyl 2-oxo-4-arylbutyrates by baker's yeast under selected reaction conditions. Tetrahedron Asymmetry 9:2725–2737
40. Ni Y, Xu JH (2012) Biocatalytic ketone reduction: a green and efficient access to enantiopure alcohols. Biotechnol Adv 30:1279–1288
41. Nakamura K, Yamanaka R, Matsuda T, Harada T (2003) Recent developments in asymmetric reduction of ketones with biocatalysts. Tetrahedron Asymmetry 14:2659–2681
42. Xie Y, Xu JH, Lu WY, Lin GQ (2009) Adzuki bean: a new resource of biocatalyst for asymmetric reduction of aromatic ketones with high stereoselectivity and substrate tolerance. Bioresour Technol 100:2463–2468
43. Rommens JM, Iannuzzi MC, Kerem B, Drumm ML, Melmer G, Dean M, Rozmahel R, Cole JL, Kennedy D, Hidaka N (1989) Identification of cystic fibrosis gene: Chromosome walking and jumping. Science 245:1059–1065
44. Qiu JG, Yun M, Wen YZ, Chen LS, Wu LF, Liu WP (2012) Functional identification of two novel genes from *Pseudomonas* sp. strain HZN6 involved in the catabolism of nicotine. Appl Environ Microbiol 78:2154–2160
45. Zhang JD, Li AT, Yang Y, Xu JH (2010) Sequence analysis and heterologous expression of a new cytochrome P450 monooxygenase from *Rhodococcus* sp for asymmetric sulfoxidation. Appl Microbiol Biotechnol 85:615–624
46. Zhang J, Yin JG, Hang BJ, Cai S, He J, Zhou SG, Li SP (2012) Cloning of a novel arylamidase gene from *Paracoccus* sp. strain FLN-7 that hydrolyzes amide pesticides. Appl Environ Microbiol 78:4848–4855
47. Ni Y, Li CX, Zhang J, Shen ND, Bornscheuer UT, Xu JH (2011) Efficient reduction of ethyl 2-oxo-4-phenylbutyrate at 620 g/L by a bacterial reductase with broad substrate spectrum. Adv Synth Catal 353:1213–1217
48. Reddy TBK, Thomas AD, Stamatic D, Bertsch J, Isbandi M, Jansson J, Mallajosyula J, Pagani I, Lobos EA, Kyrpides NC (2014) The genomes online database (GOLD) v.5: a metadata management system based on a four level (meta)genome project classification. Nucl Acids Res doi: 10.1093/nar/gku950.
49. Wang LJ, Li CX, Ni Y, Zhang J, Liu X, Xu JH (2011) Highly efficient synthesis of chiral alcohols with a novel NADH-dependent reductase from *Streptomyces coelicolor*. Bioresour Technol 102:7023–7028
50. Itoh N, Morihama R, Wang JC, Okada K, Mizuguchi N (1997) Purification and characterization of phenylacetaldeyde reductase from a styrene-assimilating *Corynebacterium* strain, ST-10. Appl Environ Microbiol 63:3783–3788
51. Makino Y, Dairi T, Itoh N (2007) Engineering the phenylacetaldehyde reductase mutant for improved substrate conversion in the presence of concentrated 2-propanol. Appl Microbiol Biotechnol 77:833–843
52. Itoh N, Isotani K, Nakamura M, Inoue K, Isogai Y, Makino Y (2012) Efficient synthesis of optically pure alcohols by asymmetric hydrogen-transfer biocatalysis: application of engineered enzymes in a 2-propanol-water medium. Appl Microbiol Biotechnol 93:1075–1085
53. Lavandera I, Kern A, Schaffenberger M, Gross J, Glieder A, Glieder A, de Wildeman S, Kroutil W (2008) An exceptionally DMSO-tolerant alcohol dehydrogenase for the stereoselective reduction of ketones. ChemSusChem 1:431–436
54. Kaluzna IA, Andrew AA, Bonilla M, Martzen MR, Stewart JD (2002) Enantioselective reductions of ethyl 2-oxo-4-phenylbutyrate by *Saccharomyces cerevisiae* dehydrogenases. J Mol Catal B: Enzymatic 17:101–105
55. Kaluzna IA, Matsuda T, Sewell AK, Stewart JD (2004) Systematic investigation of *Saccharomyces cerevisiae* enzymes catalyzing carbonyl reductions. J Am Chem Soc 126:12827–12832
56. Richter N, Hummel W (2011) Biochemical characterization of a NADPH-dependent carbonyl reductase from *Neurospora crassa* reducing *α*- and *β*-keto esters. Enzyme Microbial Technol 48:472–479
57. Ni Y, Su YN, Li HD, Zhou JY, Sun ZH (2013) Scalable biocatalytic synthesis of optically pure ethyl (*R*)-2-hydroxy-4-phenylbutyrate using a recombinant *E. coli* with high catalyst yield. J Biotechnol 168:493–498
58. Ma HM, Yang LL, Ni Y, Zhang J, Li CX, Zheng GW, Yang HY, Xu JH (2012) Stereospecific reduction of methyl *o*-chlorobenzoylformate at 300 $g \cdot L^{-1}$ without additional cofactor using a carbonyl reductase mined from *Candida glabrata*. Adv Synth Catal 354:1764–1772
59. Shen ND, Ni Y, Ma HM, Wang LJ, Li CX, Zheng GW, Zhang J, Xu JH (2012) Efficient synthesis of a chiral precursor for angiotensin-converting enzyme (ACE) inhibitors in high space-time yield by a new reductase without external cofactors. Org Lett 14:1982–1985
60. Yun HD, Choi HL, Fadnavis NW, Kim BG (2005) Stereosepecific synthesis of (*R*)-2-hydroxy carboxylic acids using recombinant *E. coli* BL21 overexpressing YiaE from *Escherichia coli* K12 and glucose dehydrogenase from *Bacillus subtilis*. Biotechnol Prog 21:366–371
61. Persson B, Kallberg Y, Bray JE, Bruford E, Dellaporta SL, Favia AD, Duarte RG, Jörnvall H, Kavanagh KL, Kedishvili N, Kisiela M, Maser E, Mindnich R, Orchard S, Penning TM, Thornton JM, Adamski J, Oppermann U (2009) The SDR (short-chain dehydrogenase/reductase and related enzymes) nomenclature initiative. Chemico-Biol Interact 178:94–98
62. Hedlund J, Jörnvall H, Persson B (2010) Subdivision of the MDR superfamily of medium chain dehydrogenases/reductases through iterative hidden Markov model refinement. BMC Bioinformatics 11:534–549
63. Hyndman D, Bauman DR, Heredia VV, Penning M (2003) The aldo-keto reductase superfamily homepage. Chemico-Biol Interact 143–144:621–631
64. Hwang CC, Chang YH, Hsu CN, Hsu HH, Li CW, Pon HI (2005) Mechanistic roles of Ser-114, Tyr-155, and Lys-159 in 3-hydroxysteroid dehydrogenase/carbonyl reductase from *Comamonas testosteroni*. J Biol Chem 280:3522–3528

65. Ehrensberger AH, Wilson DK (2004) Structural and catalytic diversity in the two family 11 aldo-keto reductases. J Mol Biol 337:661–673
66. Robert X, Gouet P (2014) Deciphering key features in protein structures with the new ENDscript server. Nucl Acid Res 42:W320–W324
67. Moore JC, Pollard DJ, Kosjek B, Devine PN (2007) Advances in the enzymatic reduction of ketones. Acc Chem Res 40:1412–1419
68. Zhao HM, van der Donk WA (2003) Regeneration of cofactors for use in biocatalysis. Curr Opin Biotechnol 14:583–589
69. Huisman GW, Liang J, Krebber A (2010) Practical chiral alcohol manufacture using ketoreductases. Curr Opin Chem Biol 14:122–129

An immunochromatographic test strip for rapid detection of fish pathogen *Edwardsiella tarda*

Hesong Liu[1], Yue Wang[1], Jingfan Xiao[1*], Qiyao Wang[1], Qin Liu[1,2] and Yuanxing Zhang[1,2]

Abstract

Background: *Edwardsiella tarda*, the etiologic agent of edwardsiellosis, is a devastating fish pathogen prevailing in worldwide aquaculture industries and accounting for severe economic losses. There is a raising concern about *E. tarda* being a significant zoonotic pathogen, and it is urgent to develop a rapid detection of this pathogen. This is the first study to develop a test strip for rapid detection of *E. tarda* in turbot.

Results: Mouse monoclonal antibodies (MAbs) and rabbit polyclonal antibody (PAb) against *E. tarda* were generated from immunization of mice and rabbits with a virulent isolate of *E. tarda* EIB202. Two MAbs specific to isolates of *E. tarda* were obtained, and one of them (25C1) was selected to conjugate with colloidal gold as the detector antibody. Rabbit PAb was used as the capture antibody. It was found the strip had no cross-reactivity with non-*E. tarda* bacterial microbes and the limit of detection (LOD) was 1×10^5 colony-forming units (CFU)/ml. The detection could be visually observed by the naked eye within 5 min. This test strip was verified with a similar detection limit and much less analysis time compared with a dot blot immunoassay (1×10^5 CFU/ml for LOD and 120 min for reaction time). When the samples were mixed with turbot tissue homogenates, strong immunoreactivity was observed over 10^5 CFU/ml, which suggested that the turbot tissue homogenates did not affect the detection of the strip. Pre-enrichment with homogenized turbot tissue for 12 h could increase the detection limit of the *E. tarda* present in the sample up to 1 to 10 CFU/ml. In practice, in detecting 20 turbot ascite samples infected by *E. tarda*, the immunochromatographic test strip showed a high accuracy (100% positive).

Conclusions: The immunochromatographic test strip offers great promise for a rapid, simple, and economical method of *E. tarda* on-site detection, and with different antibodies, it might be used to detect other aquatic pathogens.

Keywords: Detection; Dot blotting; *Edwardsiella tarda*; Immunochromatographic test strip; Pathogen diagnosis

Background

Edwardsiella tarda, the etiologic agent of edwardsiellosis, is a devastating fish pathogen prevailing in worldwide aquaculture industries and accounting for severe economic losses. *E. tarda* can cause systemic hemorrhagic septicemia and emphysematous putrefactive disease in more than 20 species of freshwater and marine fishes, including turbot, tilapia, eel, flounder, and catfish. There is a raising concern about *E. tarda* being a significant fish pathogen, and it is urgent to develop a rapid detection of this pathogen. The traditional diagnostic test available for identification of *E. tarda* in diseased fish is bacterial culture and identification using biochemical tests [1]. During the past decade, a number of immunoassay methods, including enzyme-linked immunosorbent assay (ELISA) and dot blot immunoassay, have been used to detect *E. tarda* in infected organs like the liver, spleen, and kidney of the diseased fish [2,3]. Polymerase chain reaction (PCR) and modified PCR methods such as loop-mediated isothermal amplification (LAMP) were also used in the diagnoses of *E. tarda* [1,4,5], and the LAMP method was proved to be successful for *E. tarda* detection with a limit of detection (LOD) of 10 colony-forming units (CFU) [6]. Although the above methods are effective for *E. tarda* diagnosis, they are time-consuming (2 h to 7 days) and depend on expensive instruments and highly skilled personnel. Colloidal gold-based immunochromatographic assay (ICA) is a user-friendly format with few simple procedures and has been widely applied as a

* Correspondence: jfxiao@ecust.edu.cn
[1]State Key Laboratory of Bioreactor Engineering, East China University of Science and Technology, 130 Meilong Road, Shanghai 200237, China
Full list of author information is available at the end of the article

diagnostic tool especially for on-site identification of pathogens. The aim of our research was to develop a portable one-step method for the rapid detection of *E. tarda*. With a strip, the infecting pathogen of diseased fish could be determined on-site in several minutes without any instrument.

Methods

Strains, media, and chemicals

The strains used in this study are listed in Table 1. *E. tarda* strains were grown in tryptic soy broth (TSB) (Difco, Detroit, MI, USA) or tryptic soy agar (TSA) (Difco, Detroit, MI, USA) at 28°C, while *Escherichia coli* strains were cultured in Luria-Bertani (LB) broth at 37°C. *Edwardsiella ictaluri*, *Aeromonas hydrophila*, and *Chromobacterium violaceum* were cultured at 30°C for 20 h in brain-heart infusion (BHI) broth, LB medium with 1% NaCl, and LB medium with NaCl 2.5%, respectively. Then, the bacteria were collected by centrifugation at 10,000× *g* for 5 min at 4°C, followed by wash for three times with pre-chilled phosphate-buffered saline (PBS; 10 mmol/l, pH 7.4). The bacterial suspension in PBS was enumerated with a UV-visible spectrophotometer and inactivated with 1% (*v*/*v*) formalin for 1 h at 30°C. The formalin-killed bacteria were washed three times with pre-chilled PBS, resuspended, and stored at −80°C for further use.

Table 1 Strains used in this study

Bacteria	References or sources
Edwardsiella tarda	
EIB202	Xiao et al. [7]
WY37	Bai et al. [3]
ATCC15947	ATCC
ATCC23692	ATCC
DT	Guangdong, 2007
Edwardsiella ictaluri	
LH51	Liaoning, 2006
ATCC33202	ATCC
Vibrio anguillarum	
MVM425	Wu et al. [8]
VIB72	Wang et al. [9]
Vibrio parahaemolyticus	Ye et al. [10]
Vibrio alginolyticus EPGS	Rui et al. [11]
Vibrio harveyi VIB647	Pang et al. [12]
Chromobacterium violaceum CV026	Ravn et al. [13]
Aeromonas hydrophila	Wang et al. [14]
Escherichia coli	
CC118	Dennis and Zylstra [15]
DH5α	Winzer et al. [16]

The goat anti-mouse IgG was purchased from Tiangen Biotech (Beijing, China). Freund's complete adjuvant and Freund's incomplete adjuvant, $HAuCl_4{\cdot}3H_2O$, sodium citrate, and bovine serum albumin (BSA) were from Sigma (St. Louis, MO, USA). Nitrocellulose membranes were purchased from Millipore (Billerica, MA, USA). PVC sheets, absorbent paper, and glass paper were purchased from Shanghai Goldbio Tech (Shanghai, China). Other reagents were of analytical purity, and Milli-Q (Millipore purification system)-purified water was used for all experiments.

Preparation of outer membrane proteins

The outer membrane proteins (OMPs) of *E. tarda* EIB202 were extracted as described by Darwish et al. [17] with minor modifications. Briefly, *E. tarda* EIB202, isolated from diseased turbot, was cultured in LB medium containing 1% glucose at 30°C for 20 h. The bacteria were collected by centrifugation at 10,000× *g* for 5 min at 4°C and then washed three times with pre-chilled PBS. The pelleted bacteria were suspended in 10 ml of HEPES buffer (10 mmol/l, pH 7.4) with 2 mmol/l phenylmethanesulfonyl fluoride and sonicated at 350 W for 120 times (4 s once with 6-s interval). Intact cells and large debris were removed by centrifugation at 10,000× *g* for 15 min at 4°C. Supernatants were pelleted and centrifuged at 100,000× *g* for 1 h at 4°C. The precipitate was resuspended with 5 ml HEPES buffer, and the same volume of HEPES buffer containing 2% sodium lauroyl sarcosine was added to solubilize inner membrane proteins. Thirty minutes later, the pellets were ultra-centrifuged at 100,000× *g* for 1 h at 4°C. The purified OMPs were suspended in 1 ml PBS and aliquoted in small portions and stored at −80°C.

Production of antibodies against *E. tarda*

Monoclonal antibody (MAb) specific to *E. tarda* was produced according to previously described methods [18]. *E. tarda* was cultured in TSB at 28°C for 16 h, and the bacteria were then collected and resuspended in sterile water for deoxycholate hydrogen sulfide lactose (DHL) agar plate counting. It was inactivated at 37°C for 48 h in 1% formaldehyde and then used as the immunogen after washing three times with PBS to remove the formaldehyde. Five BALB/c mice were immunized by intraperitoneal injection of the 200 μl suspension containing 10^8 inactivated bacteria per ml or OMPs mixed with an equal volume of Freund's adjuvant (complete Freund's adjuvant for the first immunization and incomplete for the second and third immunization) for three times at 2-week intervals. One week after the third immunization, serum antibody titers of immunized mice were determined by indirect ELISA on inactivated *E. tarda* or OMP-coated plates. Mice producing the highest ELISA titer were given an intravenous boost (OMP, 50 μg) without adjuvant. One week after the intravenous boost, two BALB/c mice were sacrificed for the fusion with SP2/0-Ag14 myeloma cells. The dispersed spleen lymphocytes were prepared and fused with myeloma cells (SP2/0-Ag-14) to produce hybridomas. For the screen of MAbs against the somatic antigen of *E. tarda*, indirect ELISA was processed using supernatants of hybridomas with formalin-killed whole cells of *E. tarda* coating on plates. Two of the strongly reacting MAbs (designated as 25C1 and 15H10) were used for further characterizations and applications in immunoassays. Positive hybridoma cells selected by indirect ELISA and cloned by limiting dilution were cultured and injected intraperitoneally into the BALB/c mice with 5×10^6 cells each mouse for ascite production. After 7 to 14 days, ascite was collected. MAb in ascite fluid was precipitated with 50% ammonium sulfate, centrifuged, and dialyzed against PBS before being purified using a protein G agarose column.

New Zealand white rabbits were subcutaneously and intramuscularly injected with a total of 20 ml suspension containing 10^8 inactivated bacteria per ml mixed with an equal volume of Freund's adjuvant (complete Freund's adjuvant for the first immunization and incomplete for the second and third immunization) for four times at 3-week intervals, and then the rabbit antisera were collected by ear bleeding. The sensitivity and specificity of the antisera were determined by dot blotting and Western blotting. High-titer and highly specific antisera were pooled and purified using a protein A agarose column.

Reactivity and characterization of MAbs

Indirect ELISA tests were performed to assess the reactivity of MAbs to the formalin whole cells of *E. tarda* isolates. ELISA plates were coated with 50 μl per well of

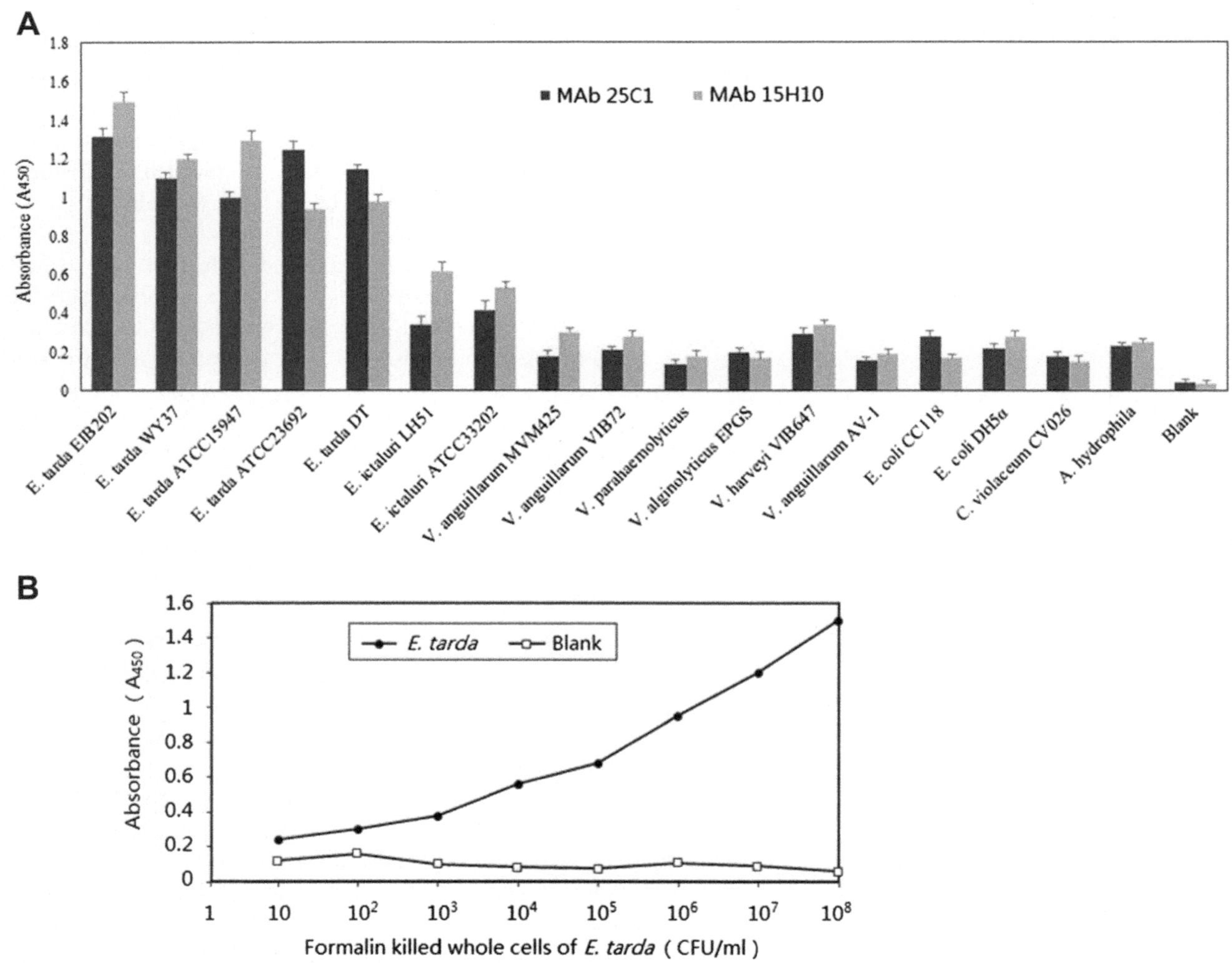

Figure 1 Reactivity and sensitivity of MAbs to *E. tarda* and test bacterial strains in indirect ELISA. **(A)** A fixed quantity of formalin-killed whole cells of *E. tarda* (10^8 CFU/ml) were coated to the plate, and 1:200 dilution of 25C1 and 15H10 MAbs was used as the primary antibody. **(B)** The ELISA plate was coated with tenfold diluted solutions (10^8 to 10 CFU/ml) of formalin-killed whole cells of *E. tarda* EIB202, and 1:200 dilution of 25C1 MAb was used as the primary antibody.

formalin-killed whole bacterial cells (10^8 CFU/ml) in carbonate-bicarbonate buffer (pH 9.6) and incubated overnight at 4°C. All tests were performed in duplicate. The indirect ELISA tests were performed as Kumar et al. described [18]. The isotyping of MAbs was carried out with a mouse monoclonal antibody isotyping kit (Sigma, St. Louis, MO, USA). Indirect ELISA was also used to test the sensitivity of the MAbs to detect formalin-killed whole cells of *E. tarda*. Briefly, ELISA plates were coated and dried overnight with 50 μl of tenfold diluted solutions of whole bacterial cells (10^8 to 10 CFU/ml). The subsequent steps were similar as mentioned above.

Preparation of test strip

The MAb specific to *E. tarda* was conjugated to colloidal gold particles (diameter = 40 nm) as the method described by Frens [19] and dispersed onto a glass fiber membrane as the conjugate pad. The rabbit polyclonal antibody (2 mg/ml) and goat anti-mouse IgG (1 mg/ml) were micro-sprayed with BioDot XYZ3050 (BioDot, Irvine, CA, USA) onto a nitrocellulose membrane separately at the position that would become the test line (T) at the upper position and control line (C) around the bottom of the membrane as the analytical membrane. A glass fiber membrane and absorbent paper were used as the sample pad and absorbent paper, respectively. The sample pad, conjugate pad, analytical membrane, and absorbent pad were assembled with 2 mm overlapping between each component on a plastic adhesive backing. The assembled master card was then cut into strips of 5 mm wide with a numerical controlled slitter (Shanghai Goldbio Tech, Shanghai, China) and stored in a desiccated plastic bag.

Specificity and sensitivity of test strip

In order to determine the cross-reactivity of the test strip, five strains of *E. tarda*, two strains of *E. ictaluri*, and ten other bacterial strains (Table 1) were diluted in the same volume of test to a concentration of 10^6 CFU/ml with PBS and applied to the test strips.

The sensitivity of the strip was determined qualitatively. The serially diluted *E. tarda* of several concentrations (1×10^6, 5×10^5, 1×10^5, and 5×10^4 CFU/ml) in sample buffer was used in sensitivity detection. The strip sensitivity results were compared with the results from dot blotting. All of the detections were performed in triplicate.

Dot blotting

The bacterial suspension used to determine the specificity of the test strips and *E. tarda* EIB202 of serial dilutions were spotted onto a nitrocellulose membrane (2 μl/spot), dried at 37°C, and quenched in 5% Blotto (5% nonfat dry milk, 0.05% Tween-20 in PBS) for 2 h. After being incubated with MAb at 37°C for 1 h, the membrane was extensively washed three times with PBST (0.05% Tween-20 in PBS) and incubated in goat anti-mouse IgG antibody horseradish peroxidase conjugate (GAM-HRP) at 37°C for 1 h. Then, the membrane was washed three times and incubated in TMB membrane substrate for 5 min. After revealing dark gray spots, the reaction was terminated by incubation of the membrane in water.

Detection of *E. tarda* in artificially contaminated fish tissue

Spleen tissues from fresh turbot were dissected and homogenized in TSB supplemented with 2% NaCl at a ratio of 1:9 (*w*/*v*). The turbot homogenate (10 ml) was spiked with *E. tarda* EIB202 to various original concentrations from 1 to 10^7 CFU/ml by tenfold serial dilutions. One milliliter from each tube was collected at 0, 6, 12, and 24 h and immediately heat killed at 60°C for 30 min and stored at −20°C before testing with the test strip. For the strip test, 400 μl from each sample was mixed with PBS at a ratio of 1:1, and 100 μl of the mixture was applied to the test strip.

On-site detection with test strip

Twenty ascite samples of diseased turbot from an aquaculture farm (Shandong Province, China) was prepared and applied directly to the test strip. Simultaneously, 10 μl ascite sample was spread on a DHL plate. The colonies with a black center (indicating H_2S production) were counted and identified to be *E. tarda* with the commercial API 20E kit (bioMérieux, Marcy-l'Étoile, France) and 16S rRNA sequencing.

Results and discussion

Preparation and characteristics of antibodies against *E. tarda*

In order to obtain MAb, BALB/c mice were immunized with the formalin-killed bacteria and purified OMPs of *E. tarda*, respectively. Compared with the antibody titer of anti-serum against OMP, the antibody titer of anti-serum against the formalin-killed bacteria was lower even after the fifth immunization and not suitable for hybridization. It might be attributed to the characteristics of *E. tarda* as an intracellular pathogen. MAbs against the OMP of *E. tarda* have been reported in detection of *E. tarda* from different geographic locations [18]. Therefore, OMP was used as the immunogen in this study. Out of the 1,536 wells of culture supernatants from two fusions, two hybridomas generated a strongly positive signal in indirect ELISA. Stable hybridoma cell lines

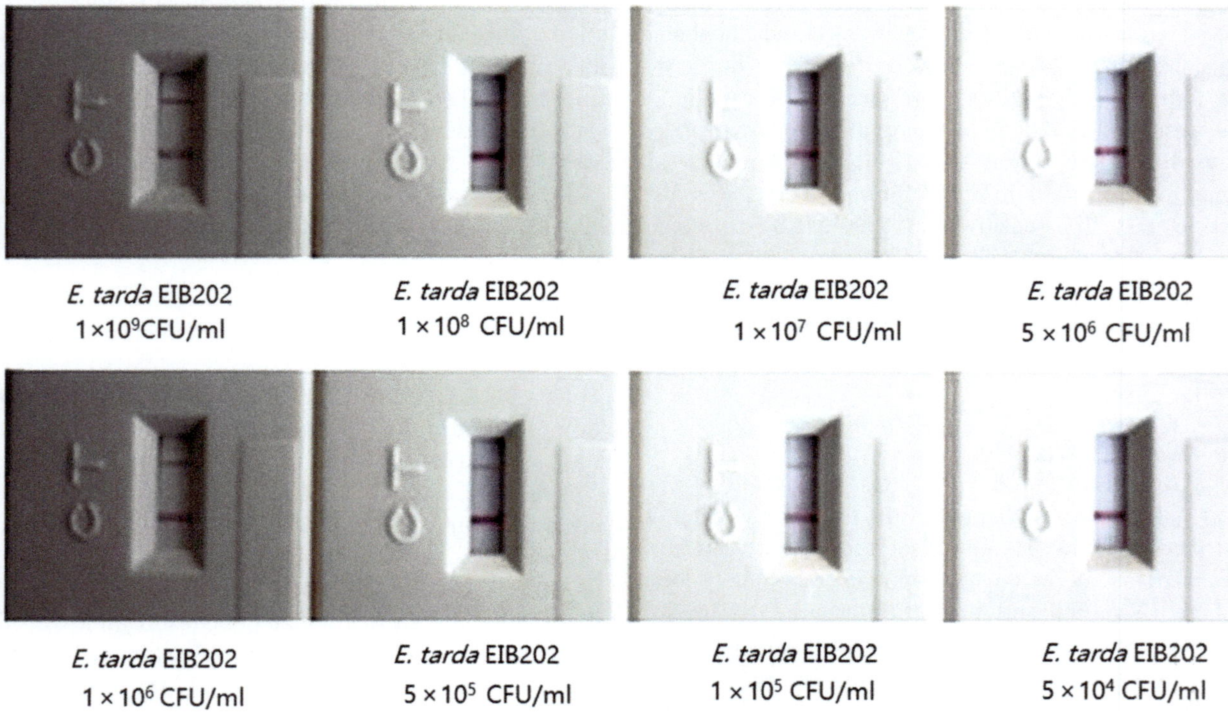

Figure 2 Sensitivity evaluation of the test strip prepared with monoclonal antibody 25C1 against *E. tarda* The standard bacterial suspension of *E. tarda* EIB202 was enumerated and serially diluted to 1×10^9, 1×10^8, 1×10^7, 1×10^6, 5×10^6, 5×10^5, 1×10^5, and 5×10^4 CFU/ml and directly applied to the test strips. The detection limit of the test strip was determined by the diminishing of red color at the test line.

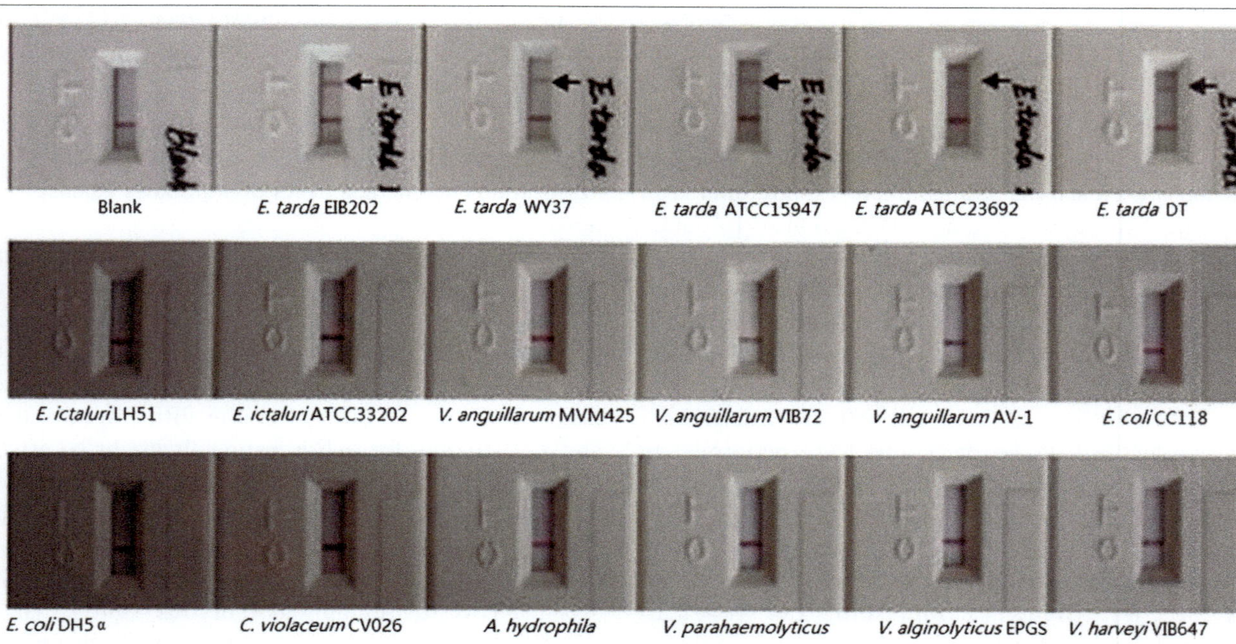

Figure 3 Specificity evaluation of the test strip prepared with monoclonal antibody 25C1 against *E. tarda*. Bacteria were diluted to 1×10^6 CFU/ml with PBS and applied to the sample pad of the test strip.

25C1 and 15H10 were obtained following three rounds of limiting dilution and the cloning procedure.

The isotyping results classified the two anti-*E. tarda* MAbs as subclass IgG2α. The reactivity of MAbs with different strains of *E. tarda* and with other bacteria was investigated using indirect ELISA. The two MAbs, 25C1 and 15H10, reacted strongly with formalin-killed whole bacterial cells from all isolates of *E. tarda* but not with the whole bacterial cells of *E. coli*, *E. ictaluri*, *Vibrio anguillarum*, *Vibrio parahaemolyticus*, *Vibrio alginolyticus*, *Vibrio harveyi*, *Chromobacterium violaceum*, and *Aeromonas hydrophila* (Figure 1A). The results showed that MAb 15H10 reacted with *E. ictaluri* more than MAb 25C1 (Figure 1A). The sensitivity of MAb 25C1 to detect whole cells of *E. tarda* was up to a level of 1×10^4 CFU/ml (Figure 1B). According to the results above, the cell line 25C1 produced an IgG2α subtype of MAb with high specificity to *E. tarda* which was selected to prepare MAb used in subsequent experiments. Rabbit anti-*E. tarda* EIB202 anti-sera were purified through a protein A column which demonstrated specificity to all of the tested *E. tarda* isolates and could be used as the capture antibody in test strip production.

Preparation of test strip

The test strip was assembled in a sandwich format. MAb 25C1 was used for conjugation with colloidal gold and served as the detection antibody. The rabbit anti-*E. tarda* polyclonal antibody (PAb) was used as the capture antibody. Transmission electron microscope (TEM) images revealed that the optimal diameter of colloidal gold particles ranged from 36 to 42 nm, with a mean diameter of 40 nm, to produce a distinguishable red color. The optimal concentrations of antibodies used to form the positive line and test line were determined by signal visualization. A clear bright-red display emerged with 6.0 μg/ml of gold-labeled MAb 25C1 and 1.0 μg/ml of rabbit anti-*E. tarda* PAb (data not shown).

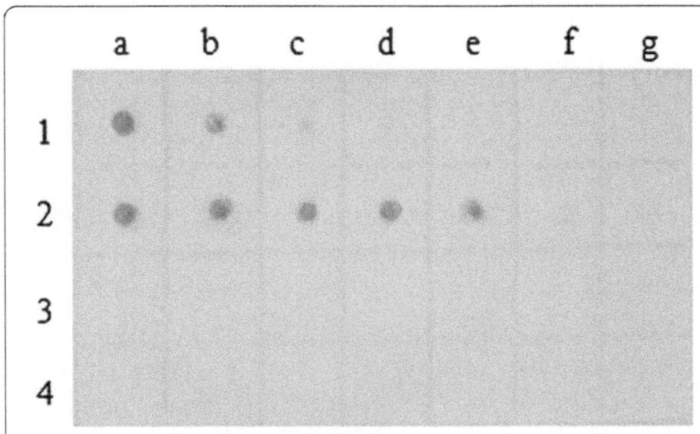

Figure 4 Dot blotting detection with monoclonal antibody 25C1 against *E. tarda*. Bacterial strains used for the cross-reactivity test at a concentration of 1×10^6 CFU/ml and *E. tarda* EIB202 of serial dilutions were spotted onto the nitrocellulose membranes (2 μl/spot) and processed dot blotting using MAb 25C1. Row 1: *E. tarda* EIB202 (a) 1×10^6 CFU/ml, (b) 5×10^5 CFU/ml, (c) 1×10^5 CFU/ml, (d) 5×10^4 CFU/ml, (e) 1×10^4 CFU/ml, (f) 5×10^3 CFU/ml, (g) 1×10^3 CFU/ml. Row 2: (a) *E. tarda* EIB202, (b) *E. tarda* WY37, (c) *E. tarda* ATCC15947, (d) *E. tarda* ATCC23692, (e) *E. tarda* DT, (f) *E. ictaluri* LH-51, (g) *E. ictaluri* 33202. Row 3: (a) *V. anguillarum* MVM425, (b) *V. anguillarum* VIB72, (c) *V. anguillarum* AV-1, (d) *E. coli* CC118, (e) *E. coli* DH5α, (f) *C. violaceum* CV026, (g) *A. hydrophila*. Row 4: (a) *V. parahaemolyticus*, (b) *V. alginolyticus* EPGS, (c) *V. harveyi* VIB647, (d) blank.

Specificity and sensitivity of test strip

According to the recently described method, the detections of all the *E. tarda* strains used in this study produced positive results when 100 μl bacterial suspension at a concentration of 1×10^6 CFU/ml was added on the strip. There were not any cross-reactions of the strip with *E. ictaluri*, *V. anguillarum*, *E. coli*, *C. violaceum*, *A. hydrophila*, *V. parahaemolyticus*, *V. alginolyticus*, and *V. harveyi* (Figure 2). The test strip showed a high specificity to *E. tarda* and supposed to detect different isolates of *E. tarda*.

To determine the sensitivity of the ICA test strip, different concentrations of bacterial suspension of *E. tarda* EIB202 were applied to the ICA strip and tested in triplicate. The sensitivity of the ICA test strip showed a positive correlation with bacterial concentration when the bacterial concentration was between 5×10^4 and 1×10^9 CFU/ml. The color intensity of the test line increased with *E. tarda* concentration. When the suspension with a concentration of 1×10^5 CFU/ml was added on the strip, the intensity of the test line was significantly weaker (within 5 min) than that of the control line (Figure 3), and the test line at 5×10^4 CFU/ml bacteria was visible just at about 60 min. Therefore, 1×10^5 CFU/ml of *E. tarda* could be considered as the low limit of the ICA test strip for visual detection (within 5 min).

Comparison to dot blot hybridization

As an easy technique, dot blot hybridization is currently used for pathogen detection. However, the dot blotting procedure is laboratory-dependent and time-consuming which needs long result turn-around time (approximately 6 h) [20,21]. In this study, the dot blotting using MAb 25C1 described earlier as a binding antibody was performed to determine the specificity and sensitivity of this method. The results showed that Mab 25C1 could recognize all isolates of *E. tarda* used in the study and had no cross-reaction with all the bacterial strains except for *E. tarda*, which displayed the perfect specificity of MAb 25C1. The LOD of dot blotting was 1×10^5 CFU/ml, which is similar to that of the ICA method (Figure 4). Compared with that of the dot blotting, the detection process of the ICA method was simple and time-saving (5 min vs 6 h). Although the dot blotting produced a similar LOD to that of the ICA strip, it was reported

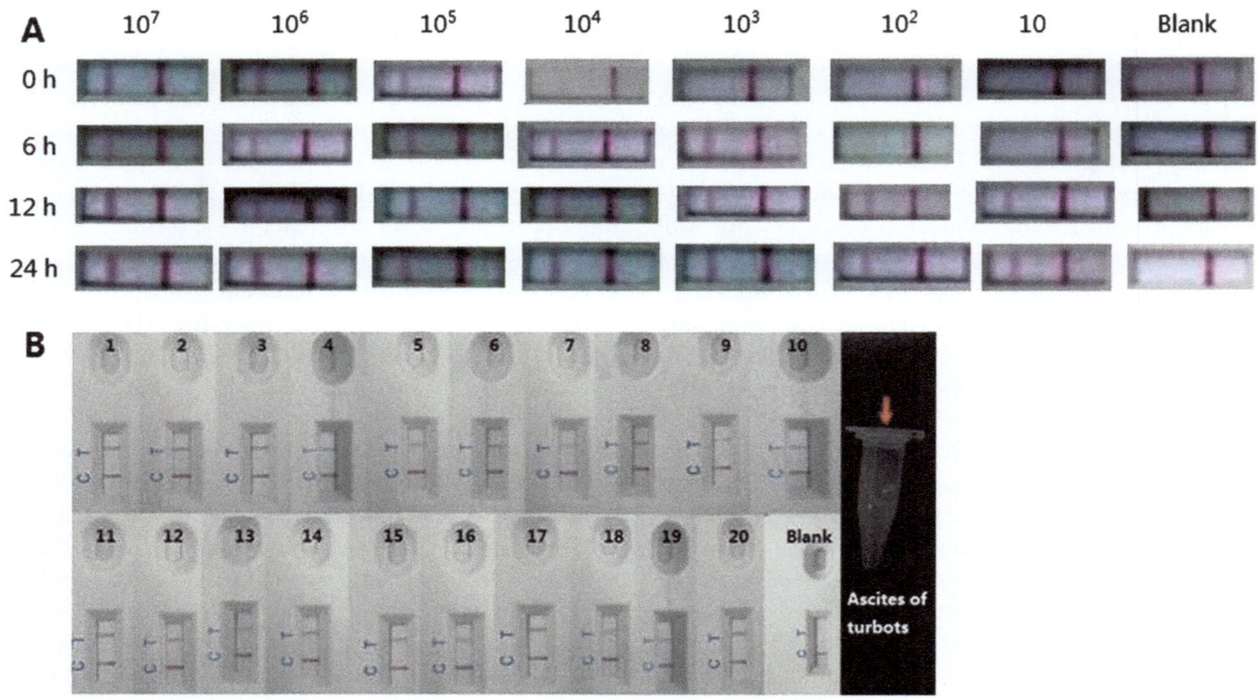

Figure 5 Strip test of *E. tarda* EIB202 with turbot tissue homogenates and on-site detection of turbot ascite. **(A)** *E. tarda* EIB202 at various concentrations (1 to 10^7 CFU/ml) were pre-enriched with turbot tissue homogenates for 0 to 24 h before being applied to the strip. PBS was used as the blank sample. **(B)** Ascites of diseased turbot collected from an aquaculture farm (Shandong Province, China) numbered 1 to 20 were prepared and applied directly to the test strip. The concentrations of *E. tarda* in ascites ranged from 5.1×10^5 to 7.8×10^7 CFU/ml by DHL agar counting. PBS was used as the blank sample.

that the limited capacity of the nitrocellulose membrane might reduce the detection sensitivity of this method because the high protein in the sample could compete with the bacterial antigen [22]. The test strip offered similar detection sensitivity to that of the dot blotting and would be a convenient tool for direct identification of fish pathogen *E. tarda*.

Detection of *E. tarda* in artificially contaminated fish tissue

The application of test strips to detect the targeted microorganisms in the fish tissue samples was evaluated. When mixed with homogenized turbot tissue, a strong immunoreactivity was observed in the samples inoculated with *E. tarda* EIB202 over 10^5 CFU/ml, which is similar to that of the sample in TSB. The results suggested that the existence of turbot tissue homogenates did not affect the detection of the strip. After pre-enrichment with the tissue of turbot for 6 h, a moderate immunoreactivity was observed in the turbot homogenate samples inoculated with *E. tarda* over 10^3 CFU/ml (Figure 5A). After pre-enrichment for 12 and 24 h, a moderate positive observation could be detected at the inoculate of 10′ CFU/ml in TSB. Therefore, pre-enrichment of the sample for 12 h could increase the detection limit of the *E. tarda* present in the sample up to 1 to 10 CFU/ml.

On-site detection of *E. tarda* infection with test strip

E. tarda is an important fish pathogen and has caused huge losses to the aquaculture industry throughout the world. Outbreaks of edwardsiellosis appear first in the form of infections affecting only a few fish, and these diseased fish are able to survive only for a few days. When early detection and proper measurement are not undertaken and the diseased fish not treated as soon as possible, the infection spreads to the surrounding areas to create an edwardsiellosis epidemic [6]. Since early detection is critical for controlling this disease, rapid and accurate methods for edwardsiellosis diagnosis would be very valuable. To evaluate the method developed in this work, the immunochromatographic test strip was practically utilized to detect *E. tarda* in an industrial turbot farm (Yantai, Shandong Province, China). Diseased turbot with typical symptoms of edwardsiellosis such as eye tumefaction, inflammation, hemorrhages, and ascites were sampled. Each of 20 ascite samples of diseased turbot was collected and applied directly to the test strip for *E. tarda* on-site detection, and PBS was used as blank control due to the absence of ascite in healthy turbot. Among the 20 ascite samples, 20 samples (100%) were positively detected, and the test lines were clearly indicated (Figure 5B). The concentrations of *E. tarda* in ascites ranged from 5.1×10^5 to 7.8×10^7 CFU/ml by DHL agar counting. The major colonies in the 20 ascite samples were further measured using API 20E and 16S rDNA sequencing to confirm that the diseased turbot were infected with *E. tarda* rather than other possible pathogens. Fortunately, the complete genome of *E. tarda* is available [23]. The results of 16S rRNA sequencing displayed that all the colonies isolated from the ascite samples belong to the same *E. tarda* strain with 99% identity to *E. tarda* FL6-60 [GenBank:CP002154.1] and EIB202 [GenBank:CP001135.1]. It was suggested that the diseased turbot from the aquaculture farm might be infected by the same *E. tarda* strain. In the conventional bacterial culture test, the procedure for isolation and identification of *E. tarda* from diseased fish samples usually requires 2 to 3 days, including bacterial enrichment, plating onto DHL agars, sequential subculture, and biochemical characteristic test, which is time-consuming and tedious. Therefore, the test strip is effective for the on-site detection of *E. tarda* in ascite samples.

Conclusions

The results in this work indicated that the detection limit, simplicity, and specificity of the test strips are sufficient for on-site use. In conclusion, the one-step immunodiagnostic test presented here could prove to be a useful tool for the inexpensive, rapid, and convenient on-site detection of *E. tarda* in clinical samples.

Competing interests

The authors declare that they have no competing interests.

Authors' contributions

HL carried out the preparation of the monoclonal antibody targeting *E. tarda* and development of the test strip. YW carried out the on-site detection of the test strip and helped in the manuscript preparation. JX, QW, QL, and YZ have made substantive intellectual contributions to this study and substantial contributions to the design of the study. All authors read and approved the final manuscript and helped design the whole study.

Acknowledgements

This work was supported by grants from the Ministry of Agriculture of China (Nos. CARS-50 and nyhyzx-201303047), Shanghai Rising-Star Program (No. 13QA1401000), and the Project of Marine Public Welfare Profession (No. 201405003-2).

Author details

[1]State Key Laboratory of Bioreactor Engineering, East China University of Science and Technology, 130 Meilong Road, Shanghai 200237, China. [2]Shanghai Collaborative Innovation Center for Biomanufacturing Technology (SCICBT), East China University of Science and Technology, 130 Meilong Road, Shanghai 200237, China.

References

1. Baird KD, Chikarmane HM, Smolowitz R, Uhlinger KR (2003) Detection of *Edwardsiella* infections in *Opsanus tau* by polymerase chain reaction. Biol Bull 205:235–236
2. Swain P, Mukherjee SC, Sahoo PK, Das BK, Pattnaik P, Murjani G, Nayak SK (1999) Dot enzyme linked immunosorbent assay (Dot-ELISA) for diagnosis of *Edwardsiella tarda* infection in fish. Asian Fish Science 14:89–93
3. Bai FF, Lan JX, Wang Y, Han Y, Zhang XH (2009) Indirect enzyme-linked immunosorbent assay (ELISA) for rapid detection of *Edwardsiella tarda*. J Fishery Sci of China 16:619–625
4. Chen JD, Lai SY (1998) PCR for direct detection of *Edwardsiella tarda* from infected fish and environmental water by application of the hemolysin gene. Zool Stud 37:169–176
5. Lan J, Zhang XH, Wang Y, Chen J, Han Y (2008) Isolation of an unusual strain of *Edwardsiella tarda* from turbot and establish a PCR detection technique with the *gyrB* gene. J Appl Microbiol 105:644–651
6. Savan R, Igarashi A, Matsuoka S (2004) Sensitive and rapid detection of edwardsiellosis in fish by a loop-mediated isothermal amplification method. Appl Environ Microbiol 70:621–624
7. Xiao J, Wang Q, Liu Q, Wang X, Liu H, Zhang Y (2009) Isolation and identification of fish pathogen *Edwardsiella tarda* from mariculture in China. Aquac Res 40:13–17
8. Wu H, Ma Y, Zhang Y, Zhang H (2004) Complete sequence of virulence plasmid pEIB1 from the marine fish pathogen *Vibrio anguillarum* strain MVM425 and location of its replication region. J Appl Microbiol 97:1021–1028
9. Wang Q, Yang M, Xiao J, Wu H, Wang X, Lv Y, Xu L, Zheng H, Wang S, Zhao G, Liu Q, Zhang Y (2009) Genome sequence of the versatile fish pathogen *Edwardsiella tarda* provides insights into its adaptation to broad host ranges and intracellular niches. PLoS One 4, e7646
10. Ye J, Ma Y, Liu Q, Zhao D, Wang Q, Zhang Y (2008) Regulation of *Vibrio alginolyticus* virulence by the LuxS quorum-sensing system. Journal of Fish Diseases 31:161–169
11. Rui H, Liu Q, Wang Q, Ma Y, Liu H, Shi C, Zhang Y (2009) Role of alkaline serine protease, asp, in *Vibrio alginolyticus* virulence and regulation of its expression by *luxO-luxR* regulatory system. J Microbiol Biotechnol 19:431–438
12. Pang L, Zhang X, Zhong Y, Chen J, Li Y, Austin B (2006) Identification of *Vibrio harveyi* using PCR amplification of the *toxR* gene. Lett Appl Microbiol 43:249–255
13. Ravn L, Christensen A, Molin S, Givskov M, Gram L (2001) Methods for detecting acylated homoserine lactones produced by gram-negative bacteria and their application in studies of AHL-production kinetics. J Microbio Methods 44:239–251
14. Wang Y, Xu Z, Jia A, Chen J, Mo Z, Zhang X (2009) Genetic diversity between two *Vibrio anguillarum* strains exhibiting different virulence by suppression subtractive hybridization. Acta Microbiologica Sinica 49:363–371
15. Dennis J, Zylstra G (1998) Plasposons: modular self-cloning minitransposon derivatives for rapid genetic analysis of gram-negative bacterial genomes. Appl Environ Microbiol 64:2710–2715
16. Winzer K, Sun Y, Green A, Delory M, Blackley D, Hardie K, Baldwin T, Tang C (2002) Role of *Neisseria meningitidis luxS* in cell-to-cell signaling and bacteremic infection. Infection and Immunology 70:2245–2248
17. Darwish AM, Newton JC, Plumb JA (2001) Effect of incubation temperature and salinity on expression of the outer membrane protein profile of *Edwardsiella tarda*. J of Aquaculture Animal Health 13:269–75
18. Kumar G, Rathore G, Sengupta U, Kapoor D, Lakra WS (2010) Production of monoclonal antibodies specific to major outer membrane protein of *Edwardsiella tarda*. Comp Immunol Microbiol Infect Dis 33:133–144
19. Frens G (1973) Controlled nucleation for the regulation of the particle size in monodisperse gold suspension. Nature, Physics Science 241:20–21
20. Chaivisuthangkura P, Senapin S, Wangman P, Longyant S, Sithigorngul P (2013) Simple and rapid detection of infectious myonecrosis virus using an immunochromatographic strip test. Arch Virol 158:1925–1930
21. Giagkas DC, Choli-Papadopoulou T, Pantazaki AA (2013) Development of an antibody for detection of rhamnolipids characterized as a major bacterial virulence factor. Antibodies 2:501–516
22. Parin C, Chalinan P, Siwaporn L, Paisarn S (2013) Evaluation of monoclonal antibody based immunochromatographic strip test for direct detection of *Vibrio cholerae* O1 contamination in seafood samples. J Microbiol Methods 2:304–311
23. Wang X, Wang Q, Xiao J, Liu Q, Wu H, Xu L, Zhang Y (2009) *Edwardsiella tarda* T6SS component *evpP* is regulated by *esrB* and iron, and plays essential roles in the invasion of fish. Fish and Shellfish Immunology 27:469–477

11

Ligninolytic enzymes: a biotechnological alternative for bioethanol production

Jersson Plácido* and Sergio Capareda

Abstract

Ligninolytic fungi and enzymes (i.e., laccase, manganese peroxidase, and lignin peroxidase) have been applied recently in the production of second-generation biofuels. This review contains the analysis of ligninolytic enzymes and their applications in second-generation biofuels. In here, each of the ligninolytic enzymes was described analyzing their structures, catalysis, and reaction mechanism. Additionally, delignification and detoxification, the two most important applications of ligninolytic enzymes, were reviewed and analyzed. The analysis includes an evaluation of the biochemical process, feedstocks, and the ethanol production. This review describes the current situation of the ligninolytic enzymes technology and its future applications in bioethanol industry.

Keywords: Ligninolytic fungi; Bioethanol; Laccases; Manganese peroxidase; Lignin peroxidase; Detoxification; Delignification

Review

Introduction

The United States regulation requires 36 billion gallons of biofuels, and 21 billion should be produced from lignocellulosic material or other new advanced fuels by 2020 [1]. Lignocellulose, the most abundant source of organic material in the world, is a polysaccharide combination of cellulose, hemicellulose, and lignin [2]. Bioethanol is one of the most essential biofuels produced from lignocellulosic material. In general, bioethanol production from lignocellulosic material includes three principal steps: 1) pretreatment, 2) saccharification, and 3) fermentation. The pretreatment process modifies the lignocellulose structure by removing the lignin and altering the cellulose and hemicellulose structure [3]. Saccharification is the enzymatic transformation of cellulose and hemicellulose into monosaccharides as glucose and xylose. Whereas, fermentation is the transformation of these monosaccharides into ethanol. The current pretreatment methodologies utilize energy-intensive processes (high pressures and temperatures) and harsh chemical compounds (NaOH, H_2SO_4). This combination generates undesirable compounds and process inefficiencies. To overcome these issues, biomass pretreatment has explored more sustainable technologies such ligninolytic enzymes.

In nature, different organisms degrade lignin; however, the most effective ones are the fungi belonging to the white rot fungi class [4]. These microorganisms possess an enzymatic system to effectively degrade lignin. This system contains three principal enzymes laccase, manganese peroxidase (MnP), and lignin peroxidase (LiP) [5]. These microorganisms have been used to delignify different substrates such paper, animal feed, and biofuels [3,6–8]. In biofuel production, the ligninolytic enzymes have two principal purposes, delignification and detoxification. Delignification methods apply ligninolytic fungus and their enzymes to reduce the lignin content in several feedstocks [3,5–8]. Detoxification utilizes the ligninolytic enzymes to reduce the toxic compounds present in the biomass hydrolysates after chemical or physicochemical pretreatments [9–14].

The aim of this review is to describe the relevant aspects of the ligninolytic enzymes laccase, MnP, and LiP and to analyze the present and future application of this technology in the production of biofuels.

Ligninolytic enzymes

Laccases

Laccases (E.C. 1.10.3.2) also known as benzenediol: oxygen oxidoreductase or p-diphenol oxidase belong to the

* Correspondence: plac324@tamu.edu
Department of Biological and Agricultural Engineering, Texas A&M University, 201 Scoates hall, 2117 TAMU, College Station, TX 77843, USA

oxidoreductase class. Yoshida, at the end of the nineteenth century, was the first to extract laccases from exudates of the Japanese tree *Rhus vernicifera.* Laccases are widely distributed in microorganisms, insects, and plants, showing a specific function in each of them. From this group, white rot fungi are the most studied laccases. These fungi use their enzymes to break the plant's lignocellulosic wall and obtain the host's nutrients. These fungal enzymes degrade the complex polyphenol structure that constitutes lignin, the principal recalcitrant component in the lignocellulosic wall. Besides their catalytic characteristics, these enzymes are attractive for biotechnological process, because they are extracellular and inducible, do not need a cofactor, and have low specificity. Laccase production is induced and increased by the addition of molecules as copper, dyes, or other recalcitrant compounds [15]. Laccase employs oxygen as an oxidizing agent and cofactor instead of other expensive cofactors such as pantothenic acid, thiamine, or biotin [16].

Laccases have low substrate specificity; this characteristic allows the degradation of several compounds with a phenolic structure [17]. Therefore, laccases have been employed in several areas such as bioremediation of aromatic recalcitrant compounds [18], treatment of effluents polluted with lignin [19], chemical synthesis [20,21], degradation of a wide number of textile dyes [22–25], and biomass pretreatment for biofuel production [26–28].

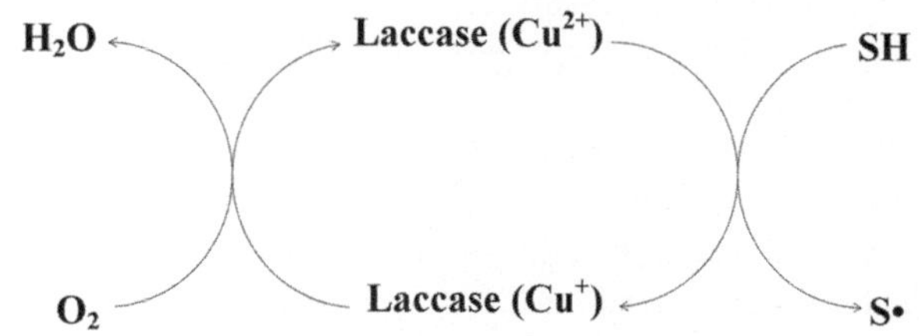

Figure 1 General diagram of laccase redox mechanism (SH reduced substrate, S• oxidized substrate).

Laccase structure and catalytic mechanism Laccases have a primary structure of approximately 500 amino acid residues organized in three consecutive domains, with a Greek key β barrel topology. The three domains are distributed in a first domain with 150 initial amino acids, a second domain between the 150 and 300 residue, and a third domain from the 300 to 500 amino acid. The structure is stabilized by two disulfide bridges localized between domains I and II and between domains I and III [16,29,30]. However, some laccases present three disulfide bridges. *Melanocarpus albomyces* has disulfide bridges inside domain I, another between domain I and domain III, and the last one between domain II and III [31].

Laccase belongs to multicopper oxidases (MCOs) and blue multicopper oxidases [21]. Laccases have four copper (Cu) molecules in their active site which participate in oxygen reduction and water production (Figure 1) [32]. The laccases' four copper atoms are disseminated in three types of cores or places: type 1 Cu (T1), type 2 Cu (T2), and type 3 Cu (T3). These cores are in two metallic active sites: the mononuclear location T1 and the trinuclear location T2/T3 [16,21,29,30]. Type 1 Cu (T1) or blue Cu is a paramagnetic copper with a strong absorption at 600 nm (blue coloration). This signal is generated by the covalent union between Cu-Cys in the mononuclear location [16,33]. In addition, T1 has the highest redox potential in the enzyme and participates in the enzymatic catalysis by oxidizing the substrate [16]. Laccase's oxidative power is affected by the amino acids surrounding T1 [16]. Similar to T1, T2 is a paramagnetic cooper; however, it lacks absorption in the visible region of the spectrum. This Cu exhibits a strong absorption under electron paramagnetic resonance (EPR); technique employed to identify this spot [16]. In addition, the T2/T3 core participates in the inhibition of the enzymatic activity by its interaction with anions such as fluoride or cyanide [34]. On the other hand, T3 Cu is a binuclear core formed by a Cu (II) dimer. This site presents absorption at 330 nm and a lack of EPR signal. Type 3 coppers are diamagnetic and have a Cu-Cu union which participates in the compound oxidation as an electron acceptor.

In laccases, the different types of coppers have variations associated with the coordination between them and the amino acids in the active site. T1 Cu exhibits a different triangular planar coordination similar to other MCOs. Laccases display two histidines (His) and one cysteine (Cys) as equatorial ligands. Meanwhile, the other MCOs have an additional axial ligand with a methionine. In contrast, laccases did not have this axial extra bond, they have a leucine (Leu) or a phenylalanine (Phe) substituting methionine [16,29–31]. On the other hand, T2 Cu exhibits coordination with two His and a water molecule; whereas, T3 Cu interacts with a water molecule and six His. In T3 Cu, the two Cu molecules share the water molecule and split the six His in two groups of three [16,29,30]. The two T3 Cu interact with the oxygen atom; however, the interaction between water and the two Cu at the same time only happens when the enzyme is in its oxidized form [16,29–31,33].

Laccase crystalline structures evidenced that T2 Cu (EPR active) could be any of the three Cu in the trinuclear site. In this site, two of the Cu reduce and become silent to EPR. Meanwhile, the other one remains active to EPR [35]. This interchange between T2 Cu and one of T3 coppers could be related to the reduction mechanism of the two oxygen molecules (cofactor), mechanism which is not clear for blue multicopper oxidases [35].

The enzymes' redox potential oscillates between 300 and 800 mV; these fluctuations depend on different factors such Cu ligands, distance among Cu, and the second sphere of amino acids in the active site. The last factor is associated with the degradation of recalcitrant compounds as lignin or industrial dyes [16].

Another structural characteristic in laccases is the presence of ligands: monosaccharides (mannose, N-acetyl glucosamine), ions like Ca^{2+}, and organic molecules like glycerol [16,29–31,35]. The presence of these ligands produce significant differences between laccases, an example of that is the glycosylations exhibited in the isoenzymes from *Trametes versicolor* [30].

Heme-peroxidases

After laccases, lignin peroxidase (LiP) and manganese peroxidase (MnP) are the most significant ligninolytic enzymes. These enzymes belong to the heme-proteins because they have the protoporphyrin IX as a prosthetic group. Similar to other heme-peroxidases such cytochrome c peroxidase and horseradish peroxidase, the catalytic cycle of LiP and MnP have three reactions [32,36–39].

Hydrogen peroxide causes the enzyme oxidation to produce the compound I and water:

Reduced peroxidase + H_2O_2 → Compound I + H_2O

The modified enzyme (compound I) catalyzes the production of a free radical (S•) and a second modified form of the enzyme (compound II) by an electron transfer from the substrate (SH: reduced substrate):

Compound I + SH → Compound II + S•

Finally, compound II reacts with a second substrate molecule to produce another free radical and water; meanwhile, the enzyme reduces to its original form.

Compound II + SH → Reduced peroxidase + S• + H_2O

Lignin peroxidase (LiP) LiPs (E.C. 1.11.1.14) were originally discovered in nitrogen- and carbon-limited cultures of *Phanerochaete chrysosporium* [32]. LiP possess high redox potential (700 to 1,400 mV), low optimum pH 3 to 4.5, and the ability to catalyze the degradation of a wide number of aromatic structures such veratryl alcohol (3,4-dimethoxybenzyl) and methoxybenzenes [32,36,38]. LiP oxidizes aromatic rings moderately activated by electron donating substitutes; in contrast, common peroxidases participate in the catalysis of aromatic substrates highly activated (ammine, hydroxyl, etc.). An explanation for this type of catalysis is the production of veratryl alcohol radicals. These radicals have higher redox potential than LiP's compounds I and II and can participate in the degradation of compounds with high redox potential [40].

LiPs are monomeric glycosylated enzymes of 40 kDa, with 343 amino acids residues, 370 water molecules, a heme group, four carbohydrates, and two calcium ions [39]. Their secondary structure is principally helicoidal. It contains eight major helixes, eight minor helixes, and two anti-parallel beta sheets. LiPs contain two domains at both sides of the heminic group. This group is inlaid in the protein but accessible to solvents via two small channels [39]. The heminic cavity includes 40 residues, and it bonds to the protein via hydrogen bridges. Additionally, the heminic iron (Fe) coordinates with a His and a molecule of water. This His is associated with the high redox potential of LiP. The enzyme's redox potential rises when the His has a reduced imidazol character [39]. In addition, a greater distance between the His and the heminic group increases the redox potential of the enzyme. This increment in the redox potential is a response to the electronic deficiency in the Fe of the porphyrin ring. In fact, this distance causes most differences among enzymes with similar porphyrin cores [39,41].

Another characteristic related with LiP's high redox potential is the invariant presence of a tryptophan residue (Trp171) in the enzymes' surface. Trp171 seems to facilitate electronic transference to the enzyme from substrates that cannot access into the heminic oxidative group [41]. Additionally, Trp171 participates with the catalysis of veratryl alcohol, a metabolite produced by some ligninolytic fungi. Veratryl alcohol participates in the oxidation of different aromatic molecules. Some researchers conceptualized that this alcohol protects the enzyme from the action of H_2O_2 and participate as a redox mediator between the enzyme and substrates which cannot get inside the heminic center [38,39].

Manganese peroxidase (MnP) Kuwahara in 1984 found the first MnP (E.C. 1.11.1.13) in batch cultures of *Phanerochaete chrysosporium* [32]. They are glycoproteins with a molecular weight between 38 and 62.5 kDa [42], approximately 350 amino acid residues, and a 43% of identity with LiP sequence. MnP structure has two domains with the heminic group in the middle, ten major helixes, a minor helix, and five disulfide bridges. One of those bridges participates in the manganese (Mn) bonding site. This site is a characteristic that distinguishes MnP from other peroxidases [37].

The enzyme's catalytic cycle starts with the transference of two electrons from the heminic group to H_2O_2; it produces compound I and water. After that, compound I catalyzes the oxidation of one substrate molecule with the production of a free radical and compound II. Compound II oxidizes Mn^{2+} to produce Mn^{3+}, the cation responsible to oxidize aromatic compounds. It is important to keep in mind that compound II demands Mn^{2+} presence for its reaction; in contrast, compound I can oxidize Mn^{2+} or the substrate. After Mn^{3+} is stabilized by organic acids, it reacts non-specifically with organic molecules by removing an electron and a proton from the substrates [42]. Mn^{3+} is a small size compound with high redox potential, which diffuses easily in the lignified cell wall. Therefore, Mn^{3+} starts the attack inside the plant cell wall which facilitates the penetration and action of the other enzymes [38,41].

Sources of ligninolytic enzymes

The ligninolytic enzymes are a ubiquitous group of enzymes found in different types of organisms as plants, bacteria, insects, and fungi. In plants, laccases are the most documented ligninolytic enzyme; these are extracellular glycoproteins composed by a monomeric protein with a sugar component. Laccases have been found in different types of plants: the Japanese lacquer tree, mango, mung bean, peach, tobacco, zea mays, etc. [43]. In plants, laccases realize different kind of functions such as lignin synthesis [43], iron oxidation from Fe(II) to Fe(III) [44], and regeneration of injured tissue [43]. In plants, lignin synthesis is laccases' most significant function. In these organisms, they catalyze the monolignol dimerization, the basic molecules of lignin polymers and oligomers [43].

In bacteria, lignin breakdown is exhibited by three groups: actinomycetes, α-proteobacteria, and γ-proteobacteria [45]. These three groups of bacteria were described in insects degrading wood microflora. These types of bacteria showed delignification in both *in vitro* and *in vivo* analyses [45]. *Streptomyces viridosporus*, actinomycetes species, and *Thermobifida fusca* exhibited LiP activities [45–47]. In contrast, *Azospirillum lipoferum*, *Thermus thermophilus*, *Marinomonas mediterranea*, *Bacillus subtilis*, and *Streptomyces cyaneus* showed laccase activity [48–51]. Bacterial laccases are intracellular enzymes with a monomeric, multimeric, or homotrimeric structure without carbohydrate moiety [43]. In nature, bacteria use laccases in spore protection and pigmentation. This type of enzymes exhibited resistance to high pH and temperatures [43,45,52].

In insects, the delignification process is done by ligninolytic enzymes produced by the insects or by the insects' microflora. The ligninolytic enzymes have been identified in different insects and different parts of them. An example of that are *Nephotettix cincticeps* (salivary glands) [53], *Manduca sexta* (Malpighian tubules, midgut, fat body, and epidermis) [54], *Reticulitermes flavipes* (gut) [55], and *Tribolium castaneum* (cuticles) [56]. The principal functions of insect's ligninolytic enzymes are cuticle sclerotization and pigmentation, toxic compound oxidation, and polymerization reactions [43,53].

The fungal ligninolytic enzymes are the most well-known enzymes, and they occur in ascomycetes, basidiomycetes, and deuteromycetes [57]. Inside these groups are several fungal species; however, the most studied fungi are: *Trametes versicolor, Phanerochaete chrysosporium, Pleurotus ostreatus, Dichomitus squalens, Lentinula edodes, Irpex lacteus*, and *Cerrena maxima* [32,58]. The principal functions of fungal ligninolytic enzymes are lignin degradation, spore and fruiting body formation, and degradation of plant's toxic compounds [32,57].

Biofuel applications

In the last years, ligninolytic fungi and their enzymes have appeared as a new alternative for pretreatment processes. This technology has been useful principally in two processes, delignification and detoxification. In delignification, ligninolytic fungi or their enzymes have been used to reduce the lignin content in different types of feedstock [3,4,59]. Whereas, in detoxification, ligninolytic enzymes have been used in the removal of chemical compounds from sugar hydrolysates after traditional pretreatments.

Delignification

In delignification process, ligninolytic enzymes have been applied in four different methods: 1) fungal delignification, 2) enzymatic delignification, 3) laccase-mediator system (LMS), and 4) integrated fungal fermentation (IFF). Table 1 resumes the application of these four delignification technologies. Fungal delignification utilizes the complete ligninolytic microorganisms for biomass delignification. In those cases, the microorganisms grow with the target biomass in a submerged culture [8,60] or a solid state fermentation [3,7]. Both cultivations achieved good delignification percentages; however, these were not always related with high glucose yields. The work of Salvachúa et al. (2011) is an example of that; in that case, the fungi with higher glucose yield were different from the ones with the highest delignification percentage. However, the lack of relation between high glucose yield and high delignification percentage occurs in different pretreatments and not only in microbial delignification. In fact, the principal issue with fungal pretreatment is the large duration of the process compared with the other pretreatment technologies [5]. The length of time used by the microorganisms to obtain high delignification percentages is not less than 13 days and can be up to 40 or 50 days [4,8] However, this period will depend on the strain used. The effectiveness of microbial delignification has been improved by adding an alkali treatment previous the fungal pretreatment. The alkali pretreatment benefits were the reduction of the process duration and the increment in the glucose and ethanol yield [61,62].

Enzymatic delignification is the application of enzymatic extracts and purified or semi-purified ligninolytic enzymes to realize the lignin degradation. Enzymatic delignification employs commercial or native ligninolytic enzymes [63]. Laccase is the most utilized enzyme followed by MnP and LiP; nevertheless, mixtures of two or three ligninolytic enzymes have been used for some delignification pretreatments. In those cases, the synergetic relationship between

Table 1 Delignification methods using ligninolytic fungi and ligninolytic enzymes

Feedstock	Fungal strain	Type of process	Mediator	Delignification %	Glucose yield	Ethanol yield	Process duration	Ref.
Wood pulp	*Trametes villosa*	*Enzymatic* Laccase	HBT HPA	50% 39%	-	-	15 h	[81]
Wood pulp	*Aspergillus fumigatus*	*Enzymatic* Laccase	HBT	14%	-	-	2 h	[59]
Hardwood chips	*Trametes versicolor*	*Enzymatic* Laccase	VA	54.2%,	-	-	2 h	[82]
Sugarcane bagasse	*Ceriporiopsis subvermispora*	*Fungal* MnP, LiP, Laccase	-	20%	-	-	30 d	[64]
Eucalyptus globulus	*Myceliophthora thermophila*	*Enzymatic* Laccase	MS	>25%	-	-	12 h	[19]
Paper pulp	*Pycnoporus cinnabarinus*	*Enzymatic* Laccase	AC SA CA HBT	21% 25% −9% 40%	-	-	4-24 h	[69]
Corn stover Switchgrass	*Ceriporiopsis subvermispora*	*Fungal* Laccase	-	>25% >25%	56% 37%	-	18 d	[4]
Ricinus communis	-	*Enzymatic* Laccase	-	85.69%	288.83 mg/g	-	4 h	[83]
Eucalyptus globulus	*Pycnoporus cinnabarinus*	*Enzymatic* Laccase	HBT	27%	-	-	2-4 h	[84]
Wheat straw	*Isolate Euc,* *Irpex lacteus*	*Fungal* Laccase	-	35% 35%	70% 70%	-	15-46 d	[2]
Wheat straw	*Ceriporiopsis subvermispora* *Irpex lacteus*	*Fungal* MnP, LiP, Laccase	-	34% 39%	69% 66%	90%	21 d	[3]
Corn stover	*Ceriporiopsis subvermispora1*	*Fungal* MnP Laccase	-	39.2%	66.9%	-	42 d	[7]
Corn stover (cs) Wheat straw (ws)	*Pycnoporus sanguineus H275*	*Fungal* Laccase	- - VA	27% (cs) 25% (ws) 71% (cs) 72% (ws)	-	-	21 d	[8]
Elephant grass (eg) Eucalypt wood (ew)	*Trametes villosa*	LMS	HBT	36% (eg) 58% (ew)	70.7% (eg) 57.8 (ew)	16.2% (eg) 12.3% (ew)	24 h	[72]
Corn stover	*Trametes versicolor*	LMS	HBT	-	31.7%	-	48 h	[70]
Wheat straw	*Sclerotium sp.*	LMS	ABTS	-	82.29%	-	48 h	[71]
Cotton gin trash	*Trametes versicolor*	LMS	PrimaGreen®	15%	23%	31.6%	96 h	[27]
Cotton gin trash	*Trametes versicolor*	LMS	PrimaGreen®	27%	41%	64%	96 h	[28]

Table 1 Delignification methods using ligninolytic fungi and ligninolytic enzymes *(Continued)*

Sugarcane bagasse	*Pleurotus ostreatus IBL-02*	Enzymatic MnP, LiP, Laccase	-	33.6%	72.9%	16.3 g/L	48 h	[67]
Wheat straw	*T. versicolor* *I. lacteus*	*Fungal* MnP LiP Laccase	-	-	68.33% 56.26%	-	21 d	[85]
Sugarcane bagasse	*Phlebia sp MG-60*	IFF		30%	-	45%	4 weeks	[86]
Sugarcane bagasse	*Phanerochaete chrysosporium* *Lentinula edode* *Pleurotus ostreatus*	*Fungal* MnP	-	64%	-	-	12 weeks	[87]
Oil palm empty fruit bunches	*Pleurotus floridanus*	*Fungal*		5%	15.4%	27.9%	4 weeks	[88]
Corn stover	*Coprinus comatus*	IFF		66.5%	82%		5-7 d	[76]
Wheat straw	*Irpex lacteus*	*Fungal* MnP Laccase	-	-	84%	74%	14-21 d	[62]
Corn stover	*Phanerochete chrysosporium* *Coridus versicolor*	Enzymatic MnP Laccase	-	-	323 mg/g	-	24 h	[66]
Bamboo culms	*Punctularia sp.*	*Fungal*		-	10%	-	12 weeks	[89]
Poplar wood	*Trametes velutina*	*Fungal*	-		43.6%	22.2%	8 weeks	[90]
Sugarcane bagasse	*Pleurotus florida*	*Fungal* MnP Laccase	-	7.91%	303.33 mg/g	-	25 d	[91]
Corn stover	*Irpex lacteu*	*Fungal* MnP		25.48%	308.98 mg/g	144.03 mg/g	28 d	[92]
Sugarcane Bagasse	*Phlebia sp. MG-60*	IFF	-	-	-	44%	4 weeks	[75]
Hardwood kraft pulp, Waste newspaper	*Phlebia sp. MG-60*	IFF	-	-	-	71.8% 51.1%	68 h	[73]
Sugarcane bagasse	*Phlebia sp. MG-60*	IFF	-	-	-	65.7%	20 d	[61]
Oak wood	*Phlebia sp. MG-60*	IFF	-	-	-	43.9%	56 d	[74]

LMS: Laccase-Mediator System, IFF: Integrated fungal fermentation, HBT: 1-Hydroxybenzotriazole, HPA: N-Hydroxyphthalimide, MS: Methyl syringate, VA: Violuric acid, AC: Acetosyringone, SA: Syringaldehyde, CA: p-coumaric acid.

the ligninolytic enzymes improved biomass delignification [64,65]. In wheat straw, ligninolytic enzymes generated a reduction in the cellulose conversion during the saccharification process. To avoid cellulases' inhibition, ligninolytic enzymes were deactivated before the saccharification process [66]. Similar to fungal delignification, alkali pretreatments have been used in enzymatic delignification as an initial pretreatment to facilitate the delignification process [66]. Enzymatic processes achieved similar delignification percentages than microbial pretreatment. However, enzymatic processes employ less amount of time (between 24 and 96 h) to produce the same delignification achieved by fungal pretreatment [67]. Although this methodology reduces the delignification process duration, the current development of enzymatic delignification cannot compete with the conventional pretreatment technologies in terms of timespan and costs [5].

LMS utilize fungal laccases and redox mediators to achieve the delignification process. Redox mediators are chemical compounds, which act as electron carriers between the enzyme and the final substrate (Figure 2) [19]. Mediators' redox potential increases when the enzyme oxidizes them; after that, oxidized mediators react with the final substrate to recuperate the lost electrons. The increment in the redox potential allows the degradation of recalcitrant compounds such as lignin and different kinds of aromatic compounds that the enzyme by itself cannot break [19,68]. The mediator compounds most used are 1-hydroxybenzotriazole (HBT), 2,2′-azino-bis(3-ethylbenzothiazoline-6-sulphonic acid) (ABTS), and some natural mediators as syringaldehyde or vanillin [69]. The principal applications of laccase-mediator system are bleaching and delignification in the paper industry. However, this system has been used for delignification of some biofuel feedstocks. In cotton gin trash, the delignification was performed by a commercial LMS from Genencor (Rochester, NY, USA). In that case, LMS was used after an ultrasonication and hot water pretreatments [27]. LMS was improved by the addition of alkali-ultrasonication process before the hot water and LMS pretreatment. Alkali-ultrasonication duplicated the ethanol and cellulose yield compared with the previous method [28].

LMS pretreatment was utilized over corn stover by using HBT as mediator [70]. In there, a correlation between laccase loading and enzymatic digestibility was found. Additionally, they found the importance of corn stover previous silage to increase LMS effectiveness [70]. In wheat straw, LMS pretreatment was used after the steam explosion pretreatment and was performed before and simultaneously with the enzymatic saccharification. The results in both processes were similar. Thus, LMS can participate in a semi-consolidate process for delignification and saccharification [71]. The application of LMS in elephant grass and eucalypt wood is the first report of LMS achieving significant delignification without a previous pretreatment. This improvement was achieved by using four steps of LMS pretreatment and NaOH washing [72]. LMS is less used in biofuels than the paper industry. This lack of applications is related with the small development of specific mediators for feedstock delignification.

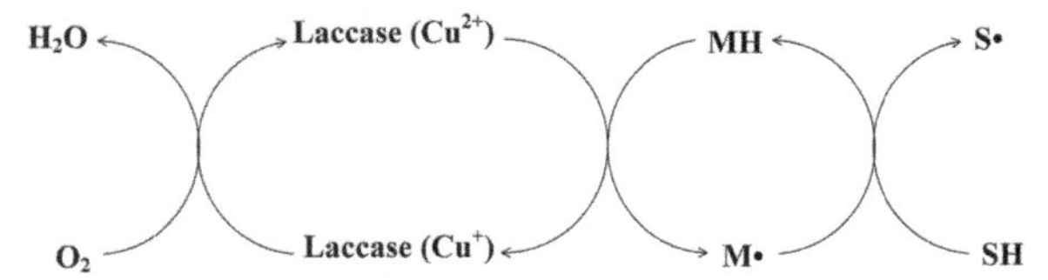

Figure 2 Laccase-mediator system's cycle.

Integrated fungal fermentation (IFF) is a consolidated process where a fungus or a group of fungi transform biomass into ethanol without the participation of other treatments or microorganisms [73]. *Phlebia* sp. MG-60 is a white rot fungus with the ability of selectively transforming lignin under solid state fermentation and produce ethanol from delignified biomasses under semi-aerobic submerged fermentation [74]. *Phlebia* sp. MG-60 was selected from a group of other 12 fungi. From this group, *Phlebia* sp. MG-60 was the only one to exhibit the uncommon characteristic of degrade lignin and produce ethanol from cellulose [73]. This fungal strain has been evaluated in the production of ethanol from different feedstocks such as hardwood kraft pulp, waste newspaper, sugarcane bagasse, and hard wood [73–75]. The addition of basal media, organic compounds, and minerals to the culture media increased the ethanol yield and the delignification percentage [75]. *Phlebia* sp. MG-60 produced an ethanol yield between 30% and 70% depending of the type of biomass utilized [61]. The process duration is between 6 and 29 days and is associated with the type of biomass and the presence of pretreatment. The addition of alkali pretreatment generated a reduction in the process timespan from 21 days to 240 h [61]. On the other hand, fungi co-culture is an alternative to generated and integrated process for ethanol production. In this case, ethanol is produced by mixing two types of fungi to develop delignification, saccharification, and fermentation. *Coprinus comatus* and *Trametes reesei* were cultivated to produce ethanol from corn stover. In this case, co-culture achieved greater delignification and cellulose conversion than monoculture [76].

Delignification using ligninolytic enzymes can be upgraded by improving the enzymatic catalysis using protein engineering. This area utilizes three types of approaches to modify ligninolytic enzymes: rational approaches, semi-rational approaches, and directed evolution [77]. Rational approaches are the molecular modification (site direct mutation) of the ligninolytic enzyme sequence using prior structural information as foundation for the specific sequence modification. Rational approaches have been used to improve the laccase ability to degrade non-phenolic substrates and to enhance the capacity to oxidize bulky phenolic molecules [78,79]. Semi-rational approaches employ saturation mutagenesis to modify hotspot residues in the enzymes. Saturation mutagenesis replaces the selected amino acids by all the codons that can generate the additional 19 amino acids [77]. This approach exhibited the production of enzymes with three- to eightfold higher catalytic efficiencies [80]. Directed molecular evolution is an approach that utilizes random mutation, gene recombination, and selection as fundaments. This molecular approach allows realizing the enzymatic design in absence of structural information. Direct molecular evolution has been used to improve catalytic activity and solvent tolerance [77].

The four delignification processes are compared in Table 2. In enzymatic and microbial process, the pros observed are: positive environmental impact, high delignification and detoxification, low sugar losses, and the possibility of develop a consolidate process. On the other hand, microbial and enzymatic have different cons. The enzymatic processes cons are: high costs, low commercial availability, and mediator necessity. Meanwhile, in microbial processes, the principal limitation is the process duration. Besides the limitations and assets, ligninolytic fungi and their enzymes represent an interesting option to the conventional chemical and physical processes used in biofuel production.

Detoxification

Some of the conventional biomass pretreatment technologies produce toxic compounds after treat biomass. These toxic compounds are classified in four groups: 1) furan derivatives, 2) pentose and hexose degradation, 3) weak acids, and 4) phenolic compounds from lignin [81]. These compounds affect the fermentative microorganisms and the cellulolytic enzymes, which generate an overall ethanol yield reduction [26].

The detoxification process employs chemical, physical, and biological strategies. Chemical and physical techniques utilize technologies such as filtration, anion exchange chromatography, and NaOH precipitation; these techniques are expensive, produce sugar losses, and do not remove all the inhibitors in the hydrolysates [12]. The biological detoxification (Table 3) process employs ligninolytic fungi or their enzymes to reduce the concentration of toxic compounds in feedstock hydrolysates [14]. The detoxification process is performed after or before the saccharification process and has a duration between 1 to 12 h [11,13,14,93]. Detoxification after or before saccharification is a controversial point. Jurado et al. [93] recommended enzymatic detoxification after cellulose hydrolysis because enzymatic detoxification leads to a reduction in the saccharification process [93]. In contrast, Moreno et al. [26] suggested enzymatic detoxification before saccharification, because it enhances ethanol yield, reduces microbes lag phase, and generates an increment in substrate loading. The authors considered these positive factors more important than the reduction in glucose recovery [26].

From the four types of toxic compounds produced in biomass hydrolysates, phenolic compounds are the most degraded type by ligninolytic enzymes [14,93]. Phenolic compounds inhibited fermentation at concentration of 1 ppm [98]. On the other hand, the other types of toxic compounds exhibit a variable behavior. These compound detoxification is affected by feedstock nature, enzyme characteristics, and pretreatment harshness [26]. Enzymatic detoxification has been applied in the reduction of toxicity after different pretreatments such as steam explosion, strong acids, organosolv, and hot liquid water. In addition, ligninolytic enzyme detoxification increase fermentation rates and ethanol yield [26]. As an example, in sugarcane bagasse, enzymatic detoxification generated an ethanol yield five times greater than ion exchange detoxification [14]. Lee et al. [96] improved the detoxification of rice straw hydrolysates by optimizing the temperature, enzyme concentration, and mediator compounds. The optimization increased the detoxification of phenolic compounds up to 92% by using HBT as mediator [96]. Ligninolytic enzyme detoxification is a future option for biofuel industries. However, this enzymatic option needs to improve their ability to degrade other types of toxic compounds from hydrolysates.

In recent years, enzymatic detoxification of furan derivatives has been evaluated. This type of detoxification is performed by aryl-alcohol oxidase (AAO) which is a less investigated group of ligninolytic enzymes. AAO have been assessed in the degradation of 5-hydroxymethylfurfural (HMF) and polyunsaturated alcohols [99–101]. Carro et al. [101] evaluated the transformation of HMF into 2,5-furandicarboxylic acid (FDCA) by AAO catalysis. AAO

Table 2 Delignification technique comparison

Delignification process	Delignification	Sugar losses	Process duration	Economic	Environmental impact
Enzymatic	High	Low	2 to 48 h	High costs	Positive
Fungal	High	Low	6 to 45 days	Low costs	Positive
LMS	High	Low	2 to 48 h	High costs	Depends of the mediator
IFF	Medium	Some	1 to 12 weeks	Low costs	Positive

Table 3 Feedstock hydrolysates detoxification

Feedstock	Microorganism	Enzyme	Pretreatment	Detoxification	Ethanol yield	Ref.
		Laccase		71%	0.55	
Wood	*Trametes versicolor*	LiP	Steam explosion, 205°C for 6 min.	53%	0.40	[9]
		Laccase+LiP		66%	0.47	
Spruce chips	*Trametes reesei*	Laccase	H_2SO_4 (0.5%) Steam explosion, 222°C, 23 bar 7 min.	80%	0.47	[10]
				80%	0.43	
Wheat straw	*Trametes versicolor*	Laccase	Hot liquid water at 180°C, 15psi, 30 min.	65%	-	[11]
Sugarcane bagasse	*Trametes versicolor*	Laccase	Steam explosion, 215°C during 10 min.	80%	0.43	[13]
Sugarcane bagasse	*Cyathus stercoreus*	Laccase	HCl (0.5–3.5% v/v),140°C for 30 min.	77.5%	0.374	[14]
Wheat straw	*Coriolopsis rigida*	Laccase	H_2SO_4 (1% w/w), 190°C, 12 bars, 10 min.	75%	-	[93]
Wheat straw	*Trametes villosa*	Laccase	Steam explosion, 220°C, 2.5 min.	88-92%	-	[26]
Rice straw	*Coltricia perennis*	Laccase	H_2SO_4 (0.5% w/w), 121°C, 15 min	37-76%	-	[94]
Wheat straw	*Trametes versicolor*	Laccase	Ethanol organosolv process, 220°C 120 min	20-90%	0.374g/g	[95]
Rice straw	*Yarrowia lipolytica*	Laccase	H_2SO_4 (0.5%, w/w) 5 h, steam explosion 121°C 30 min	52%		[96]
		LMS		92%		
Wheat straw	*Pycnoporus cinnabarinus*	Laccase	Steam explosion 200°C, 10 min	-	-	[97]

(*Pleurotus eryngii*) transformed efficiently HMF into 2,5-formylfurancarboxylic acid (FFCA); however, it was not capable of transforming directly from HMF into FDCA. To complete the catalysis of HMF, a fungal heme peroxygenase (*Agrocybe aegerita*) was added to the reaction. This enzyme transforms FFCA into FDCA using the H_2O_2 previously generated by AAO [101]. Ferreira et al. [100] determined the ability of AAO to reduce polyunsaturated alcohols. They evaluated 12 different compounds; the best degradations were observed in benzyl alcohol, m-fluorobenzyl alcohol, and p-fluorobenzyl alcohol. The enzyme has the ability to catalyze 'non-activated' alcohols which is an important factor for future detoxification procedures [100]. Other enzymatic process to transform furfurals have been studied; however, these processes utilize non-ligninolytic enzymes from different types of bacteria. In the future, furfural detoxification using ligninolytic enzymes need to be tested in biomass hydrolysates to have a more accurate vision of their real efficiency and applicability.

Conclusions

Ligninolytic fungi and their enzymes exhibited several biophysics and biochemical characteristics that have been applied in the delignification and detoxification of biofuel feedstocks. Ligninolytic processes (enzymatic and fungal) produce ethanol yield, glucose conversion, and delignification and detoxification percentages similar or superior than conventional detoxification and delignification methodologies. Nevertheless, ligninolytic processes have low cellulose modification, the presence of mediators, and long duration process (fungal processes). The transformation of ligninolytic enzymes as viable option for biofuel industry requires the improvements in the process duration, cost reduction, and more specialized mediators.

Competing interests

The authors of this manuscript do not have any financial competing interest to declare.

Authors' contributions

JP manuscript drafting and writing. SC was responsible of final manuscript approval. Both authors read and approved the final manuscript.

Acknowledgements

The Colombian government and the Fulbright organization are acknowledged for the financial support via the granting of FULBRIGHT-COLCIENCIAS for the Ph.D. studies of Jersson Plácido in the USA.

References

1. Sissine F, Energy Independence and Security Act of 2007: a summary of major provisions. 2007; 2011.
2. Dias AA, Freitas GS, Marques GSM, Sampaio A, Fraga IS, Rodrigues MAM, Evtuguin DV, Bezerra RMF (2010) Enzymatic saccharification of biologically pre-treated wheat straw with white-rot fungi. Bioresour Technol 101:6045–6050. doi:10.1016/j.biortech.2010.02.110
3. Salvachúa D, Prieto A, López-Abelairas M, Lu-Chau T, Martínez ÁT, Martínez MJ (2011) Fungal pretreatment: an alternative in second-generation ethanol from wheat straw. Bioresour Technol 102:7500–7506. doi:10.1016/j.biortech.2011.05.027
4. Wan C, Li Y (2011) Effectiveness of microbial pretreatment by Ceriporiopsis subvermispora on different biomass feedstocks. Bioresour Technol 102:7507–7512. doi:10.1016/j.biortech.2011.05.026
5. Nigam P, Gupta N, Anthwal A. Pre-treatment of agro-industrial residues. In: Nigam P and Pandey A. Biotechnology for Agro-Industrial Residues Utilisation, Springer Netherlands; 2009, p. 13–33.
6. Li L, Li X, Tang W, Zhao J, Qu Y (2008) Screening of a fungus capable of powerful and selective delignification on wheat straw. Lett Appl Microbiol 47:415–420. doi:10.1111/j.1472-765X.2008.02447.x

7. Wan C, Li Y (2010) Microbial delignification of corn stover by Ceriporiopsis subvermispora for improving cellulose digestibility. Enzyme Microb Technol 47:31–36. doi:10.1016/j.enzmictec.2010.04.001
8. Lu C, Wang H, Luo Y, Guo L (2010) An efficient system for pre-delignification of gramineous biofuel feedstock in vitro: application of a laccase from Pycnoporus sanguineus H275. Process Biochemistry 45:1141–1147. doi:10.1016/j.procbio.2010.04.010
9. Jönsson LJ, Palmqvist E, Nilvebrant N- and Hahn-Hägerdal B (1998) Detoxification of wood hydrolysates with laccase and peroxidase from the white-rot fungus *Trametes versicolor*. Appl Microbiol Biotechnol 49:691–697. doi:10.1007/s002530051233.
10. Larsson S, Reimann A, Nilvebrant N, Jönsson L (1999) Comparison of different methods for the detoxification of lignocellulose hydrolyzates of spruce. Appl Biochem Biotechnol 77:91–103. doi:10.1385/ABAB:77:1-3:91
11. Kolb M, Sieber V, Amann M, Faulstich M, Schieder D (2012) Removal of monomer delignification products by laccase from Trametes versicolor. Bioresour Technol 104:298–304. doi:10.1016/j.biortech.2011.11.080
12. Palmqvist E and Hahn-Hägerdal B (2000) Fermentation of lignocellulosic hydrolysates. I: inhibition and detoxification. Bioresour Technol 74:17–24. doi:10.1016/S0960-8524(99)00160-1.
13. Martín C, Galbe M, Wahlbom CF, Hahn-Hägerdal B and Jönsson LJ (2002) Ethanol production from enzymatic hydrolysates of sugarcane bagasse using recombinant xylose-utilising Saccharomyces cerevisiae. Enzyme Microb Technol 31:274–282. doi:10.1016/S0141-0229(02)00112-6.
14. Chandel AK, Kapoor RK, Singh A, Kuhad RC (2007) Detoxification of sugarcane bagasse hydrolysate improves ethanol production by Candida shehatae NCIM 3501. Bioresour Technol 98:1947–1950. doi:10.1016/j.biortech.2006.07.047
15. Minussi RC, Pastore GM, Durán N (2007) Laccase induction in fungi and laccase/N-OH mediator systems applied in paper mill effluent. Bioresour Technol 98:158–164
16. Matera I, Gullotto A, Tilli S, Ferraroni M, Scozzafava A, Briganti F (2008) Crystal structure of the blue multicopper oxidase from the white-rot fungus Trametes trogii complexed with p-toluate. Inorg Chim Acta 361:4129–4137. doi:10.1016/j.ica.2008.03.091
17. Zouari-Mechichi H, Mechichi T, Dhouib A, Sayadi S, Martínez AT, Martínez MJ (2006) Laccase purification and characterization from Trametes trogii isolated in Tunisia: decolorization of textile dyes by the purified enzyme. Enzyme Microb Technol 39:141–148. doi:10.1016/j.enzmictec.2005.11.027
18. Corvini P, Schäffer A, Schlosser D (2006) Microbial degradation of nonylphenol and other alkylphenols - our evolving view. Appl Microbiol Biotechnol 72:223–243. doi:10.1007/s00253-006-0476-5
19. Babot ED, Rico A, Rencoret J, Kalum L, Lund H, Romero J, del Río JC, Martínez ÁT, Gutiérrez A (2011) Towards industrially-feasible delignification and pitch removal by treating paper pulp with Myceliophthora thermophila laccase and a phenolic mediator. Bioresour Technol 102:6717–6722. doi:10.1016/j.biortech.2011.03.100
20. Sergio R (2006) Laccases: blue enzymes for green chemistry. Trends Biotechnol 24:219–226
21. Couto S, Toca-Herrera J (2006) Lacasses in the textile industry. Biotechnol Mol Biol Rev 1:115-120
22. Kokol V, Doliška A, Eichlerová I, Baldrian P, Nerud F (2007) Decolorization of textile dyes by whole cultures of Ischnoderma resinosum and by purified laccase and Mn-peroxidase. Enzyme Microb Technol 40:1673–1677. doi:10.1016/j.enzmictec.2006.08.015
23. Michniewicz A, Ledakowicz S, Ullrich R, Hofrichter M (2008) Kinetics of the enzymatic decolorization of textile dyes by laccase from Cerrena unicolor. Dyes and Pigments 77:295–302. doi:10.1016/j.dyepig.2007.05.015
24. Shanmugam S, Palvannan T, Sathish Kumar T, Michael A (2005) Biological decolourization of textile and paper effluents by *Pleurotus florida* and *Agaricus bisporus* (white-rot Basidiomycetes). World Journal of Microbiology and Biotechnology 21:1149–1151. doi:10.1007/s11274-005-0298-0
25. Chanagá Vera X, Plácido Escobar J, Marín Montoya M, Pérez Y, del Socorro M (2012) Native fungi with industrial dye degrading potential in the Aburrá Valley, Colombia. Revista Facultad Nacional de Agronomía, Medellín 65:6811–6821
26. Moreno AD, Ibarra D, Fernández JL, Ballesteros M (2012) Different laccase detoxification strategies for ethanol production from lignocellulosic biomass by the thermotolerant yeast Kluyveromyces marxianus CECT 10875. Bioresour Technol 106:101–109. doi:10.1016/j.biortech.2011.11.108
27. Plácido J, Imam T, Capareda S (2013) Evaluation of ligninolytic enzymes, ultrasonication and liquid hot water as pretreatments for bioethanol production from cotton gin trash. Bioresour Technol 139:203–208. doi:10.1016/j.biortech.2013.04.012
28. Plácido J, Capareda S (2014) Analysis of alkali ultrasonication pretreatment in bioethanol production from cotton gin trash using FT-IR spectroscopy and principal component analysis. Bioresources and Bioprocessing 1:1–9
29. Ferraroni M, Myasoedova N, Schmatchenko V, Leontievsky A, Golovleva L, Scozzafava A, Briganti F (2007) Crystal structure of a blue laccase from Lentinus tigrinus: evidences for intermediates in the molecular oxygen reductive splitting by multicopper oxidases. BMC Structural Biology 7:60
30. Bertrand T, Jolivalt C, Briozzo P, Caminade E, Joly N, Madzak C, Mougin C (2002) Crystal structure of a four-copper laccase complexed with an Arylamine: insights into substrate recognition and correlation with kinetics. Biochemistry (N Y) 41:7325–7333. doi:10.1021/bi0201318
31. Hakulinen N, Kiiskinen L, Kruus K, Saloheimo M, Paananen A, Koivula A, Rouvinen J (2002) Crystal structure of a laccase from Melanocarpus albomyces with an intact trinuclear copper site. Nat Struct Mol Biol 9:601–605
32. Dias A, Sampaio A, Bezerra R. Environmental applications of fungal and plant systems: decolourisation of textile wastewater and related dyestuffs. In: Singh S and Tripathi R. Environmental Bioremediation Technologies, Springer Berlin Heidelberg; 2007, p. 445–463.
33. Holwerda RA, Wherland S, Gray HB (1976) Electron transfer reactions of copper proteins. Annu Rev Biophys Bioeng 5:363–396. doi:10.1146/annurev.bb.05.060176.002051
34. Alcalde M. Laccases: biological functions, molecular structure and industrial applications. In: Polaina J and MacCabe AP. Industrial Enzymes, Springer Netherlands; 2007, p. 461–476.
35. Garavaglia S, Teresa Cambria M, Miglio M, Ragusa S, Iacobazzi V, Palmieri F, D'Ambrosio C, Scaloni A, Rizzi M (2004) The structure of Rigidoporus lignosus laccase containing a full complement of copper ions, reveals an asymmetrical arrangement for the T3 copper pair. J Mol Biol 342:1519–1531. doi:10.1016/j.jmb.2004.07.100
36. Piontek K, Glumoff T, Winterhalter K (1993) Low pH crystal structure of glycosylated lignin peroxidase from Phanerochaete chrysosporium at 2.5 Å resolution. FEBS Lett 315:119–124. doi:10.1016/0014-5793(93)81146-Q
37. Sundaramoorthy M, Kishi K, Gold MH, Poulos TL (1994) The crystal structure of manganese peroxidase from Phanerochaete chrysosporium at 2.06-A resolution. Journal of Biological Chemistry 269:32759–32767
38. Angel TM (2002) Molecular biology and structure-function of lignin-degrading heme peroxidases. Enzyme Microb Technol 30:425–444. doi:10.1016/S0141-0229(01)00521-X
39. Choinowski T, Blodig W, Winterhalter KH, Piontek K (1999) The crystal structure of lignin peroxidase at 1.70 Å resolution reveals a hydroxy group on the Cβ of tryptophan 171: a novel radical site formed during the redox cycle. J Mol Biol 286:809–827. doi:10.1006/jmbi.1998.2507
40. Khindaria A, Yamazaki I, Aust SD (1996) Stabilization of the veratryl alcohol cation radical by lignin peroxidase. Biochemistry (N Y) 35:6418–6424. doi:10.1021/bi9601666
41. Hammel KE, Cullen D (2008) Role of fungal peroxidases in biological ligninolysis. Curr Opin Plant Biol 11:349–355. doi:10.1016/j.pbi.2008.02.003
42. Martin H, (2002) Review: lignin conversion by manganese peroxidase (MnP). Enzyme Microb Technol 30:454–466. doi:10.1016/S0141-0229(01)00528-2.
43. Polak J, Jarosz-Wilkolazka A (2012) Fungal laccases as green catalysts for dye synthesis. Process Biochemistry 47:1295–1307. doi:10.1016/j.procbio.2012.05.006
44. Hoopes JT, Dean JFD (2004) Ferroxidase activity in a laccase-like multicopper oxidase from Liriodendron tulipifera. Plant Physiology and Biochemistry 42:27–33. doi:10.1016/j.plaphy.2003.10.011
45. Bugg TD, Ahmad M, Hardiman EM, Singh R (2011) The emerging role for bacteria in lignin degradation and bio-product formation. Curr Opin Biotechnol 22:394–400. doi:10.1016/j.copbio.2010.10.009
46. Ramachandra M, Crawford DL, Hertel G (1988) Characterization of an extracellular lignin peroxidase of the lignocellulolytic actinomycete Streptomyces viridosporus. Appl Environ Microbiol 54:3057–3063
47. Adav SS, Ng CS, Arulmani M, Sze SK (2010) Quantitative iTRAQ secretome analysis of cellulolytic thermobifida fusca. Journal of Proteome Research 9:3016–3024
48. Gupta N, Farinas ET (2010) Directed evolution of CotA laccase for increased substrate specificity using Bacillus subtilis spores. Protein Engineering, Design and Selection 23:679–682

49. [49] Faure D, Bouillant M- and Bally R (1995) Comparative study of substrates and inhibitors of Azospirillum lipoferum and Pyricularia oryzae laccases. Appl Environ Microbiol 61:1144–1146
50. Miyazaki K (2005) A hyperthermophilic laccase from Thermus thermophilus HB27. Extremophiles 9:415–425
51. Arias ME, Arenas M, Rodríguez J, Soliveri J, Ball AS, Hernández M (2003) Kraft pulp biobleaching and mediated oxidation of a nonphenolic substrate by laccase from Streptomyces cyaneus CECT 3335. Appl Environ Microbiol 69:1953–1958
52. Dwivedi UN, Singh P, Pandey VP, Kumar A (2011) Structure-function relationship among bacterial, fungal and plant laccases. J Molec Catal B 68:117–128. doi:10.1016/j.molcatb.2010.11.002
53. Hattori M, Tsuchihara K, Noda H, Konishi H, Tamura Y, Shinoda T, Nakamura M, Hasegawa T (2010) Molecular characterization and expression of laccase genes in the salivary glands of the green rice leafhopper, Nephotettix cincticeps (Hemiptera: Cicadellidae). Insect Biochem Mol Biol 40:331–338. doi:10.1016/j.ibmb.2010.02.009
54. Dittmer NT, Suderman RJ, Jiang H, Zhu Y, Gorman MJ, Kramer KJ, Kanost MR (2004) Characterization of cDNAs encoding putative laccase-like multicopper oxidases and developmental expression in the tobacco hornworm, Manduca sexta, and the malaria mosquito, Anopheles gambiae. Insect Biochem Mol Biol 34:29–41
55. Coy MR, Salem TZ, Denton JS, Kovaleva ES, Liu Z, Barber DS, Campbell JH, Davis DC, Buchman GW, Boucias DG, Scharf ME (2010) Phenol-oxidizing laccases from the termite gut. Insect Biochem Mol Biol 40:723–732. doi:10.1016/j.ibmb.2010.07.004
56. Arakane Y, Muthukrishnan S, Beeman RW, Kanost MR, Kramer KJ (2005) Laccase 2 is the phenoloxidase gene required for beetle cuticle tanning. Proc Natl Acad Sci U S A 102:11337–11342
57. dos Santos AB, Cervantes FJ, van Lier JB (2007) Review paper on current technologies for decolourisation of textile wastewaters: perspectives for anaerobic biotechnology. Bioresour Technol 98:2369–2385. doi:10.1016/j.biortech.2006.11.013
58. Martínez ÁT, Ruiz-Dueñas FJ, Martínez MJ, del Río JC, Gutiérrez A (2009) Enzymatic delignification of plant cell wall: from nature to mill. Curr Opin Biotechnol 20:348–357. doi:10.1016/j.copbio.2009.05.002
59. Vivekanand V, Dwivedi P, Sharma A, Sabharwal N, Singh R (2008) Enhanced delignification of mixed wood pulp by *Aspergillus fumigatus* laccase mediator system. World Journal of Microbiology and Biotechnology 24:2799–2804. doi:10.1007/s11274-008-9809-0
60. Martín-Sampedro R, Eugenio ME, Carbajo JM, Villar JC (2011) Combination of steam explosion and laccase-mediator treatments prior to Eucalyptus globulus kraft pulping. Bioresour Technol 102:7183–7189. doi:10.1016/j.biortech.2011.04.053
61. Khuong LD, Kondo R, De Leon R, Kim Anh T, Shimizu K and Kamei I (2014) Bioethanol production from alkaline-pretreated sugarcane bagasse by consolidated bioprocessing using Phlebia sp. MG-60. Int Biodeterior Biodegrad 88:62–68. doi:http://dx.doi.org/10.1016/j.ibiod.2013.12.008.
62. López-Abelairas M, Pallín MÁ, Salvachúa D, Lú-Chau T, Martínez M, Lema J (2013) Optimisation of the biological pretreatment of wheat straw with white-rot fungi for ethanol production. Bioprocess and biosystems engineering 36:1251–1260
63. Mattinen M, Maijala P, Nousiainen P, Smeds A, Kontro J, Sipilä J, Tamminen T, Willför S, Viikari L (2011) Oxidation of lignans and lignin model compounds by laccase in aqueous solvent systems. J Molec Catal B 72:122–129. doi:10.1016/j.molcatb.2011.05.009
64. Costa SM, Goncalves AR, Esposito E. *Ceriporiopsis Subvermispota* used in delignification of sugarcane bagasse prior to soda/anthraquinone pulping. In: Davison BH, Evans BR, Finkelstein M and McMillan JD. Twenty-Sixth Symposium on Biotechnology for Fuels and Chemicals, Humana Press; 2005, p. 695–706.
65. Archibald FS, Bourbonnais R, Jurasek L, Paice MG and Reid ID (1997) Kraft pulp bleaching and delignification by Trametes versicolor. J Biotechnol 53:215–236. doi:10.1016/S0168-1656(97)01675-1.
66. Wang F, Xie H, Chen W, Wang E, Du F, Song A (2013) Biological pretreatment of corn stover with ligninolytic enzyme for high efficient enzymatic hydrolysis. Bioresour Technol 144:572–578
67. Asgher M, Ahmad Z and Iqbal HMN (2013) Alkali and enzymatic delignification of sugarcane bagasse to expose cellulose polymers for saccharification and bio-ethanol production. Industrial Crops and Products 44:488–495. doi:http://dx.doi.org/10.1016/j.indcrop.2012.10.005.
68. Fillat U, Blanca Roncero M (2009) Effect of process parameters in laccase-mediator system delignification of flax pulp: Part I. Pulp properties Chem Eng J 152:322–329. doi:10.1016/j.cej.2009.05.036
69. Camarero S, García O, Vidal T, Colom J, del Río JC, Gutiérrez A, Gras JM, Monje R, Martínez MJ and Martínez ÁT (2004) Efficient bleaching of non-wood high-quality paper pulp using laccase-mediator system. Enzyme Microb Technol 35:113–120. doi:10.1016/j.enzmictec.2003.10.019.
70. Chen Q, Marshall MN, Geib SM, Tien M, Richard TL (2012) Effects of laccase on lignin depolymerization and enzymatic hydrolysis of ensiled corn stover. Bioresour Technol 117:186–192. doi:10.1016/j.biortech.2012.04.085
71. Qiu W, Chen H (2012) Enhanced the enzymatic hydrolysis efficiency of wheat straw after combined steam explosion and laccase pretreatment. Bioresour Technol 118:8–12. doi:10.1016/j.biortech.2012.05.033
72. Gutiérrez A, Rencoret J, Cadena EM, Rico A, Barth D, del Río JC and Martínez ÁT (2012) Demonstration of laccase-based removal of lignin from wood and non-wood plant feedstocks. Bioresour Technol 119:114–122. doi:http://dx.doi.org/10.1016/j.biortech.2012.05.112.
73. Kamei I, Hirota Y, Mori T, Hirai H, Meguro S and Kondo R (2012) Direct ethanol production from cellulosic materials by the hypersaline-tolerant white-rot fungus Phlebia sp. MG-60. Bioresour Technol 112:137–142. doi:http://dx.doi.org/10.1016/j.biortech.2012.02.109.
74. Kamei I, Hirota Y and Meguro S (2012) Integrated delignification and simultaneous saccharification and fermentation of hard wood by a white-rot fungus, Phlebia sp. MG-60. Bioresour Technol 126:137–141. doi:http://dx.doi.org/10.1016/j.biortech.2012.09.007.
75. Khuong LD, Kondo R, Leon RD, Anh TK, Meguro S, Shimizu K and Kamei I (2014) Effect of chemical factors on integrated fungal fermentation of sugarcane bagasse for ethanol production by a white-rot fungus, Phlebia sp. MG-60. Bioresour Technol 167:33–40. doi:http://dx.doi.org/10.1016/j.biortech.2014.05.064.
76. Ma K and Ruan Z (2015) Production of a lignocellulolytic enzyme system for simultaneous bio-delignification and saccharification of corn stover employing co-culture of fungi. Bioresour Technol 175:586–593. doi:http://dx.doi.org/10.1016/j.biortech.2014.10.161.
77. Mate DM and Alcalde M (2015) Laccase engineering: from rational design to directed evolution. Biotechnol Adv 33:25–40. doi:http://dx.doi.org.scopeesprx.elsevier.com/10.1016/j.biotechadv.2014.12.007.
78. Madzak C, Mimmi MC, Caminade E, Brault A, Baumberger S, Briozzo P, Mougin C, Jolivalt C (2006) Shifting the optimal pH of activity for a laccase from the fungus Trametes versicolor by structure-based mutagenesis. Protein Engineering, Design and Selection 19:77–84
79. Galli C, Gentili P, Jolivalt C, Madzak C, Vadalà R (2011) How is the reactivity of laccase affected by single-point mutations? Engineering laccase for improved activity towards sterically demanding substrates. Appl Microbiol Biotechnol 91:123–131
80. Andberg M, Hakulinen N, Auer S, Saloheimo M, Koivula A, Rouvinen J, Kruus K (2009) Essential role of the C-terminus in Melanocarpus albomyces laccase for enzyme production, catalytic properties and structure. FEBS Journal 276:6285–6300
81. Annunziatini C, Baiocco P, Gerini MF, Lanzalunga O, Sjögren B (2005) Aryl substituted N-hydroxyphthalimides as mediators in the laccase-catalysed oxidation of lignin model compounds and delignification of wood pulp. J Molec Catal B 32:89–96. doi:10.1016/j.molcatb.2004.11.002
82. Oudia A, Queiroz J, Simões R (2008) The influence of operating parameters on the biodelignification of *Eucalyptus globulus* kraft pulps in a laccase-violuric acid system. Appl Biochem Biotechnol 149:23–32. doi:10.1007/s12010-008-8138-y
83. Mukhopadhyay M, Kuila A, Tuli DK, Banerjee R (2011) Enzymatic depolymerization of Ricinus communis, a potential lignocellulosic for improved saccharification. Biomass Bioenergy 35:3584–3591. doi:10.1016/j.biombioe.2011.05.013
84. Ibarra D, Romero J, Martínez MJ, Martínez AT, Camarero S (2006) Exploring the enzymatic parameters for optimal delignification of eucalypt pulp by laccase-mediator. Enzyme Microb Technol 39:1319–1327. doi:10.1016/j.enzmictec.2006.03.019
85. Pinto PA, Dias AA, Fraga I, Marques G, Rodrigues MAM, Colaço J, Sampaio A, Bezerra RMF (2012) Influence of ligninolytic enzymes on straw saccharification during fungal pretreatment. Bioresour Technol 111:261–267. doi:10.1016/j.biortech.2012.02.068
86. Kondo R, De Leon R, Anh TK, Meguro S, Shimizu K, Kamei I (2014) Effect of chemical factors on integrated fungal fermentation of sugarcane bagasse

for ethanol production by a white-rot fungus, Phlebia sp. MG-60. Bioresour Technol 167:33–40
87. Dong XQ, Yang JS, Zhu N, Wang ET, Yuan HL (2013) Sugarcane bagasse degradation and characterization of three white-rot fungi. Bioresour Technol 131:443–451
88. Ishola MM, Isroi and Taherzadeh MJ (2014) Effect of fungal and phosphoric acid pretreatment on ethanol production from oil palm empty fruit bunches (OPEFB). Bioresour Technol 165:9–12. doi:http://dx.doi.org/10.1016/j.biortech.2014.02.053.
89. Suhara H, Kodama S, Kamei I, Maekawa N, Meguro S (2012) Screening of selective lignin-degrading basidiomycetes and biological pretreatment for enzymatic hydrolysis of bamboo culms. Int Biodeterior Biodegrad 75:176–180
90. Wang W, Yuan T, Cui B, Dai Y (2012) Pretreatment of Populus tomentosa with Trametes velutina supplemented with inorganic salts enhances enzymatic hydrolysis for ethanol production. Biotechnol Lett 34:2241–2246
91. Deswal D, Gupta R, Nandal P and Kuhad RC (2014) Fungal pretreatment improves amenability of lignocellulosic material for its saccharification to sugars. Carbohydr Polym 99:264–269. doi:http://dx.doi.org/10.1016/j.carbpol.2013.08.045.
92. Song L, Ma F, Zeng Y, Zhang X and Yu H (2013) The promoting effects of manganese on biological pretreatment with Irpex lacteus and enzymatic hydrolysis of corn stover. Bioresour Technol 135:89–92. doi:http://dx.doi.org/10.1016/j.biortech.2012.09.004.
93. Jurado M, Prieto A, Martínez-Alcalá Á, Martínez ÁT, Martínez MJ (2009) Laccase detoxification of steam-exploded wheat straw for second generation bioethanol. Bioresour Technol 100:6378–6384. doi:10.1016/j.biortech.2009.07.049
94. Kalyani D, Dhiman SS, Kim H, Jeya M, Kim I and Lee J (2012) Characterization of a novel laccase from the isolated Coltricia perennis and its application to detoxification of biomass. Process Biochemistry 47:671–678. doi:http://dx.doi.org/10.1016/j.procbio.2012.01.013.
95. Ludwig D, Amann M, Hirth T, Rupp S and Zibek S (2013) Development and optimization of single and combined detoxification processes to improve the fermentability of lignocellulose hydrolyzates. Bioresour Technol 133:455–461. doi:http://dx.doi.org/10.1016/j.biortech.2013.01.053.
96. Lee K, Kalyani D, Tiwari MK, Kim T, Dhiman SS, Lee J and Kim I (2012) Enhanced enzymatic hydrolysis of rice straw by removal of phenolic compounds using a novel laccase from yeast Yarrowia lipolytica. Bioresour Technol 123:636–645. doi:http://dx.doi.org/10.1016/j.biortech.2012.07.066.
97. Oliva-Taravilla A, Moreno AD, Demuez M, Ibarra D, Tomás-Pejó E, González-Fernández C and Ballesteros M (2015) Unraveling the effects of laccase treatment on enzymatic hydrolysis of steam-exploded wheat straw. Bioresour Technol 175:209–215. doi:http://dx.doi.org/10.1016/j.biortech.2014.10.086.
98. Delgenes JP, Moletta R and Navarro JM (1996) Effects of lignocellulose degradation products on ethanol fermentations of glucose and xylose by Saccharomyces cerevisiae, Zymomonas mobilis, Pichia stipitis, and Candida shehatae. Enzyme Microb Technol 19:220–225. doi:10.1016/0141-0229(95)00237-5.
99. Dijkman WP, Groothuis DE, Fraaije MW (2014) Enzyme-catalyzed oxidation of 5-hydroxymethylfurfural to furan-2,5-dicarboxylic acid. Angewandte Chemie International Edition 53:6515–6518. doi:10.1002/anie.201402904
100. Ferreira P, Medina M, Guillén F, Martinez M, Van Berkel W, Martinez A (2005) Spectral and catalytic properties of aryl-alcohol oxidase, a fungal flavoenzyme acting on polyunsaturated alcohols. Biochem J 389:731–738
101. Carro J, Ferreira P, Rodríguez L, Prieto A, Serrano A, Balcells B, Ardá A, Jiménez-Barbero J, Gutiérrez A, Ullrich R, Hofrichter M, Martínez AT (2015) 5-hydroxymethylfurfural conversion by fungal aryl-alcohol oxidase and unspecific peroxygenase. FEBS Journal:n/a-n/a. doi:10.1111/febs.13177

12

Surface physicochemical properties of selected single and mixed cultures of microalgae and cyanobacteria and their relationship with sedimentation kinetics

Ana L Gonçalves, Carla Ferreira, Joana A Loureiro, José CM Pires and Manuel Simões*

Abstract

Background: Microalgae are photosynthetic microorganisms presenting a diversity of biotechnological applications. However, microalgal cultivation systems are not energetically and economically feasible. Possible strategies that can be applied to improve the feasibility of microalgal production include biofouling control in photobioreactors, the use of attached growth systems and bioflocculation. These processes are ruled by surface physicochemical properties. Accordingly, the surface physicochemical properties of *Chlorella vulgaris*, *Pseudokirchneriella subcapitata*, *Synechocystis salina* and *Microcystis aeruginosa* were determined through contact angle and zeta potential measurements. Additionally, mixed cultures of the selected microorganisms were performed. Sedimentation kinetics of the studied cultures was also evaluated to understand how surface physicochemical properties influence microalgal recovery.

Results: All studied microorganisms, except *S. salina*, presented a hydrophilic surface. The co-culture of *S. salina* with the other studied microorganisms resulted in a more hydrophobic algal suspension. Regarding zeta potential determinations, all studied suspensions presented a negatively charged surface (approximately -40.8 ± 4.4 mV). Sedimentation experiments have shown that all microalgal suspensions presented low microalgal recovery efficiencies. However, a negative linear relationship between microalgal removal percentage and free energy of hydrophobic interaction was obtained.

Conclusions: The evidence of a relationship between microalgal removal percentage and free energy of hydrophobic interaction demonstrates the importance of surface physicochemical properties on microalgal settling. However, the low recovery efficiencies achieved, as well as the high net zeta potential values determined, indicate that another factor to consider in microalgal settling is the ionic strength of the culture medium, which play an important role in suspensions' stability.

Keywords: Cyanobacteria; Microalgae; Mixed cultures; Sedimentation kinetics; Surface physicochemical properties

Background

Microalgal culturing has been the focus of several research studies worldwide, due to the wide variety of biotechnological applications described for these photosynthetic microorganisms [1,2]. When growing autotrophically, microalgae convert CO_2 (from atmosphere or flue gas emissions) into organic carbon compounds, thus reducing the CO_2 accumulation in the atmosphere [3-6]. Additionally, microalgae assimilate other compounds, such as nitrogen and phosphorus, frequently found in wastewaters, meaning that these microorganisms may play an important role in wastewater treatment processes [7-10]. Finally, microalgal biomass has several applications [1,11-13]: (i) human food and animal feed; (ii) production of cosmetics, drugs and functional food; and (iii) biofuels. However, microalgal cultivation still presents high process costs, which are mainly due to the low biomass productivities and the associated harvesting costs, accounting for 20% to 30% of biomass production costs [14]. Moreover, it requires large amounts of water and nutrients, which is the reason to be considered a process with high environmental impact [15].

To improve biomass productivities in microalgal photobioreactors (PBRs), new strategies should be adopted to

* Correspondence: mvs@fe.up.pt
LEPABE, Departamento de Engenharia Química, Faculdade de Engenharia, Universidade do Porto, Rua Dr. Roberto Frias, 4200-465 Porto, Portugal

avoid biofouling (the attachment of cells to the walls of PBRs). This phenomenon is responsible for lower biomass productivities, since the amount of light (the energetic source required for photosynthesis) that effectively penetrates into microalgal suspension is significantly reduced [16-18]. On the other hand, several authors have reported the growth of microalgae in biofilms. These immobilized growth systems facilitate further harvesting processes [7,16,19,20]. For both promotion and control of microalgal attachment, knowledge about surface physicochemical properties is essential [16,21,22]. Knowing the surface characteristics may also be very helpful in metabolite extraction. For example, the use of solvents with polarities similar to those of target metabolites may facilitate the contact between both, increasing extraction efficiency [23].

Regarding the harvesting process, currently applied methods include chemical coagulation/flocculation, gravity sedimentation, flotation, filtration, centrifugation and electrical-based processes [14,24,25]. Selection of an appropriate harvesting method depends on the end product, namely its value and properties [14]: it is important to consider the acceptable level of moisture, salt concentrations, cell damage and strain features, such as their density and size. Additionally, selection of an effective harvesting procedure must take into account that microalgal biomass must be further processed. Therefore, these procedures must not be toxic or contaminate microalgal biomass. It is also desirable that the selected harvesting method allows the recycling of the culture medium [25]. In this sense, bioflocculation appears as a viable alternative for the commonly used procedures. This method consists in the addition of other microorganisms or microbial metabolites to microalgal cultures, trying to stimulate microbial aggregation and flocs formation. The use of bioflocculation as harvesting method is more energetically efficient and reduces the impact of the addition of chemical flocculants to microalgal biomass, which tend to be an expensive and toxic alternative [26]. The ability of microorganisms to form flocs strongly depends on their surface physicochemical properties.

Information about the surface physicochemical properties of microalgae and about their interaction mechanisms may be very helpful in the development of energetically and economically feasible biomass production systems and harvesting procedures, as well as in the development of new strategies to prevent biofouling. This study provides information about the surface physicochemical properties of the microalgae *Chlorella vulgaris* and *Pseudokirchneriella subcapitata* and the cyanobacteria *Synechocystis salina* and *Microcystis aeruginosa* determined experimentally, according to the method proposed by van Oss et al. [27-29]. Selection of these microorganisms was based on the following factors [30-33]: (i) these microalgae and cyanobacteria can be easily grown in laboratory cultures and (ii) several authors have reported the use of these microorganisms in a wide variety of biotechnological applications, such as CO_2 capture, wastewater treatment, biofuel production and synthesis of bioactive compounds. Additionally, algal suspensions resulting from the co-culture of different associations between the selected microorganisms were evaluated, to determine if these associations have influence on surface physicochemical properties and on their ability to form aggregates. Sedimentation kinetics of the studied cultures was also determined, to assess the influence of surface physicochemical properties on the recovery of microalgal biomass.

Methods

Microorganisms and culture medium

The microalgae *C. vulgaris* CCAP 211/11B and *P. subcapitata* CCAP 278/4 were obtained from the Culture Collection of Algae and Protozoa (UK), while the cyanobacteria *S. salina* LEGE 06079 and *M. aeruginosa* LEGE 91344 were obtained from the Laboratory of Ecotoxicology, Genomic and Evolution - CIIMAR (Centre of Marine and Environmental Research of the University of Porto, Portugal). Stock solutions of these microorganisms were prepared in OECD (Organisation for Economic Co-operation and Development) test medium [34], with the following composition (per litre): 250 mg $NaNO_3$, 12 mg $MgCl_2{\cdot}6H_2O$, 18 mg $CaCl_2{\cdot}2H_2O$, 15 mg $MgSO_4{\cdot}7H_2O$, 45 mg KH_2PO_4, 0.08 mg $FeCl_3{\cdot}6H_2O$, 0.1 mg $Na_2EDTA{\cdot}2H_2O$, 0.185 mg H_3BO_3, 0.415 mg $MnCl_2{\cdot}4H_2O$, 3 μg $ZnCl_2$, 1.5 μg $CoCl_2{\cdot}6H_2O$, 0.01 μg $CuCl_2{\cdot}2H_2O$, 7 μg $Na_2MoO_4{\cdot}2H_2O$ and 50 mg $NaHCO_3$. The culture medium was sterilized by autoclaving at 121°C for 15 min. The cells were incubated in 500-mL flasks at room temperature (24.0 ± 1.0°C), under continuous fluorescent light with an irradiance of 120 μE m^{-2} s^{-1} at the surface of the flasks. Agitation was obtained by bubbling atmospheric air (filtered through 0.22-μm cellulose acetate membranes, Orange Scientific, Belgium) in the bottom of the flasks.

Microalgal culturing in single and mixed cultures

Previously described stock solutions were used to prepare microalgal suspensions for further characterization of surface physicochemical properties and sedimentation kinetics determination. In these experiments, single cultures, as well as different associations between the selected microorganisms, were used. Batch experiments were performed in 500-mL flasks (VWR, Carnaxide, Portugal) with a working volume of 450 mL. Cells were cultivated for 7 days with an initial cell concentration of approximately 1.0×10^6 cells mL^{-1} (in mixed cultures, testing all possible associations of two microorganisms, the initial cell concentration was approximately 2.0×10^6 cells mL^{-1}). Temperature, light and agitation conditions were the same as for the stock solutions preparation. All the experiments were performed in duplicates under aseptic conditions.

Surface physicochemical properties

Zeta potential measurements

In the last day of each culture, microalgal suspensions were harvested, washed twice and resuspended in MilliQ water (Millipore, Billerica, MA, USA) to obtain a final concentration of about 5.0×10^6 cells mL^{-1}. Zeta potential was measured using a ZetaSizer Nano ZS (Malvern Instruments, Worcestershire, UK). All the determinations were performed in a clear disposable zeta cell at approximately 25°C. Mean values of each studied suspension were obtained by at least triplicate measurements of two independent experiments.

Surface contact angle measurements

In the last day of each culture, microalgal suspensions were harvested, washed twice and resuspended in saline solution (0.85% *w*/*v* NaCl) to obtain a final concentration of about 5.0×10^6 cells mL^{-1}. Algal lawns were prepared by filtering the previously washed suspensions using 0.45-μm nitrocellulose membrane filters (Advantec MFS, Inc., Tokyo, Japan) until complete clogging of the membranes. Contact angle measurements were performed using the sessile drop method, as described by Busscher et al. [35]. The measurements were carried out at room temperature using water, formamide and α-bromonaphthalene (Sigma-Aldrich, Sintra, Portugal) as the reference liquids. Determination of contact angles was performed automatically using a model OCA 15 Plus (DataPhysics, Filderstadt, Germany) video-based optical contact angle measuring instrument, allowing image acquisition and data analysis. Contact angle measurements (at least 12 determinations for each liquid and for each algal suspension) were performed for two independent experiments.

Surface parameters and hydrophobicity determinations

After contact angle measurements, the values of surface hydrophobicity of the studied algal suspensions were determined using the approach of van Oss [36], which allows the assessment of the absolute degree of hydrophobicity of any surface in comparison with their interaction with water. In this approach, the degree of hydrophobicity of a given surface (*s*) is expressed as the free energy of hydrophobic interaction between two entities of that surface when immersed in water (*w*): ΔG_{sws}^{TOT}, in mJ m^{-2}. When $\Delta G_{sws}^{TOT} < 0$, the interaction between the two entities is stronger than the interaction of each entity with water and the material is considered hydrophobic. Alternatively, if $\Delta G_{sws}^{TOT} > 0$, the material is hydrophilic. ΔG_{sws}^{TOT} can be calculated through the surface tension components of the interacting entities, according to Equation 1 [27-29]:

$$\Delta G_{sws}^{TOT} = -2\left(\sqrt{\gamma_s^{LW}}-\sqrt{\gamma_w^{LW}}\right)^2 + 4\left(\sqrt{\gamma_s^+\gamma_w^-} + \sqrt{\gamma_s^-\gamma_w^+}-\sqrt{\gamma_s^+\gamma_s^-}-\sqrt{\gamma_w^+\gamma_w^-}\right) \quad (1)$$

where γ^{LW} accounts for the Lifshitz-van der Waals (LW) component of the surface free energy and γ^+ and γ^- are the electron acceptor and electron donor parameters, respectively, of the Lewis acid-base component (γ^{AB}), being $\gamma^{AB} = 2\sqrt{\gamma^+\gamma^-}$. LW forces, usually attractive, result from instantaneous asymmetrical distribution of electrons in molecules (the higher the value of LW component, the more apolar is the surface and, therefore, the lower would be its affinity for polar liquids) [37]. Electron donor and acceptor parameters give information about the molecules present in the studied surface: higher γ^+ indicates the presence of positively charged molecules and higher γ^- indicates the presence of negatively charged molecules [36]. Acid-base (AB) forces result from electron transfer interactions between polar components of the involved surfaces. These interactions can be attractive (hydrophobic attraction) or repulsive (hydrophilic repulsion), depending on the free energy of hydrophobic interaction [37,38].

The surface tension components of a surface (*s*) were obtained by measuring the contact angles of three pure liquids (*l*), water and formamide (both polar) and α-bromonaphthalene (apolar), followed by the simultaneous resolution of three equations of the form Equation 2:

$$(1 + \cos\theta)\gamma_l^{TOT} = 2\left(\sqrt{\gamma_s^{LW}\gamma_l^{LW}} + \sqrt{\gamma_s^+\gamma_l^-} + \sqrt{\gamma_s^-\gamma_l^+}\right) \quad (2)$$

where θ is the contact angle and $\gamma^{TOT} = \gamma^{LW} + \gamma^{AB}$. The surface tension of liquid components was obtained from the literature [39]. The contact angle of the apolar liquid, α-bromonaphthalene, was used to quantify the apolar energy component γ_s^{LW}, since γ_l^- and γ_l^+ for this probe liquid are equal to zero. On the other hand, contact angles measured with the other probe liquids were used to determine the other surface parameters, γ_s^+ and γ_s^-.

Sedimentation kinetics

Microalgal recovery

In the last day of each culture, microalgal suspensions were also used to evaluate sedimentation kinetics. 1-mL samples were allowed to settle in polystyrene cuvettes at room temperature. Microalgal recovery within the settling period (8 h) was determined by measuring the turbidity (optical density (OD)) of the samples at 750 nm using a V-1200 spectrophotometer (VWR, Carnaxide, Portugal), according to the method proposed by Salim et al. [40]. Turbidity of the samples was measured at the same height in the cuvette. Microalgal removal percentage or microalgal recovery was calculated according to Equation 3 [40]:

$$\%R = \frac{OD_{750}(t_0)-OD_{750}(t)}{OD_{750}(t_0)} \cdot 100 \quad (3)$$

where $OD_{750}(t_0)$ is the turbidity of the sample determined in the beginning of microalgal settling and $OD_{750}(t)$ is the turbidity of the sample determined at time *t*.

Modelling microalgal sedimentation

Kinetic constants associated to the recovery of microalgal biomass were determined by fitting the experimental data, corresponding to the time-course evolution of microalgal removal percentages, to the Gompertz model, expressed by Equation 4 [41]:

$$y = a \cdot \exp(-\exp(b - ct)) \tag{4}$$

where y is the output value or microalgal removal percentage (%), a is the upper asymptote (%), b ($b > 0$) sets the displacement along the x-axis and corresponds to the lag time observed in the beginning of settling experiments (h) and c ($c > 0$) sets the tangent at the inflection point and corresponds to the sedimentation rate (h^{-1}). The kinetic parameters, a, b and c, were determined by minimizing the sum of squared residuals using the Solver supplement of Microsoft Excel 2013. The quality of the model fit was evaluated through analysis of the coefficient of determination (R^2).

Statistical analysis

Contact angles, surface tension parameters, zeta potential values and microalgal removal percentages were analysed using paired-samples t-test from the statistical software SPSS 22.0 (SPSS Inc., Chicago, IL, USA). Statistical tests were carried out at a significance level of 0.05.

Results and discussion

Surface physicochemical properties

The surface physicochemical properties determined using the approach of van Oss et al. [27-29] for the studied suspensions are shown in Table 1. Data is divided into single and mixed cultures, and the results are presented in the decreasing order of free energy of hydrophobic interaction (from the most hydrophilic to the most hydrophobic suspensions). Except for the microalga *C. vulgaris* and for the suspensions *P. subcapitata* + *M. aeruginosa* and *P. subcapitata* + *S. salina*, the contact angles measured with the polar probe liquids (water and formamide) were in agreement with the values of free energy of hydrophobic interaction, $\Delta G_{sws}^{\mathrm{TOT}}$. Higher contact angles were measured for microorganisms presenting more hydrophobic surfaces (lower $\Delta G_{sws}^{\mathrm{TOT}}$ values). On the other hand, contact angles measured with the probe liquid α-bromonaphthalene for the different algal suspensions were not statistically different ($p > 0.05$).

Looking at electron donor and acceptor parameters, the values determined for the studied microorganisms ranged from 22.6 ± 3.2 to 57.4 ± 7.6 mJ m^{-2} and from 0.0 ± 0.0 to 0.5 ± 0.4 mJ m^{-2}, respectively. For the studied consortia, γ_s^- and γ_s^+ values ranged from 7.6 ± 2.2 to 53.0 ± 8.9 mJ m^{-2} and from 0.0 ± 0.0 to 0.9 ± 0.3 mJ m^{-2}, respectively. These components were not statistically different from those obtained for single cultures ($p = 0.61$ and $p = 0.99$, respectively). Higher values determined for γ_s^- indicate that the studied suspensions are electron donors [42] and may be attributed to the presence of excessive molecules of oxygen in the surface of the microorganisms and to the neutralization of γ_s^+ sites by the dominant γ_s^- sites through intermolecular and intramolecular interactions [36]. On the other hand, γ_s^+ values close to zero are in accordance with the van Oss et al. [27-29] model, which predicts almost the nonexistence of electron acceptor parameters [43]. According to Volpe and Siboni [43], the assumption of a monopolar surface is unrealistic. However, the authors concluded that the van Oss et al. [27-29] approach can be applied when the aim is to compare the ability of different microorganisms to interact with other microorganisms or surfaces. When the aim is to perform molecular interpretations, these parameters (determined through the van Oss et al. [27-29] method) should not be used. Additionally, plotting γ_s^- as a function of $\Delta G_{sws}^{\mathrm{TOT}}$ resulted in a coefficient of determination (R^2) of 0.94, indicating that the van Oss et al. [27-29] approach can be correctly applied to determine the surface physicochemical properties of the selected algal suspensions. As for the contact angle measurements with the polar probe liquids, γ_s^- values correlated well with the free energy of hydrophobic interaction, since lower values of the electron donor parameter were observed for the microorganisms presenting lower $\Delta G_{sws}^{\mathrm{TOT}}$.

Regarding the Lifshitz-van der Waals component of the surface free energy, γ_s^{LW}, the values determined for the studied species were not statistically different ($p > 0.05$), ranging from 35.0 ± 1.5 to 35.9 ± 0.7 mJ m^{-2}. Similar results determined for this component indicate that cell wall composition of the studied microorganisms may not have significant differences. In the study performed by Ozkan and Berberoglu [44], differences in the LW component determined for green algae and diatoms were attributed to differences in the chemical composition of their cell walls. Several authors have reported that green algae, such as *C. vulgaris* and *P. subcapitata*, present cellulose-based cell walls, whereas diatoms present silica-based ones [45-48]. γ_s^{LW} values determined in this study for microalgae were very similar to those determined by Ozkan and Berberoglu [44] for green algae (ranging from 20.7 to 37.8 mJ m^{-2}). Additionally, γ_s^{LW} values determined for the studied cyanobacteria were also similar to those obtained for the cyanobacteria *Synechocystis* sp. and *Anabaena variabilis*

Table 1 Contact angles and surface physicochemical properties determined for each of the studied suspensions

Microalgal suspensions	Contact angles (°)			Surface tension parameters and free energy of hydrophobic interaction (mJ m^{-2})					Zeta potential (mV)
	θ_W	θ_B	θ_F	γ_s^{LW}	γ_s^{AB}	γ_s^-	γ_s^+	ΔG_{sws}^{TOT}	
Single cultures									
C. vulgaris	46.5 ± 2.8 a	37.1 ± 1.5 a	61.0 ± 0.3 a	35.0 ± 1.5 a	0.0 ± 0.0 a	48.7 ± 6.9 a	0.1 ± 0.1 a	52.5 ± 14.3 a	−35.4 ± 0.4 a
M. aeruginosa	44.8 ± 5.5 a	37.1 ± 1.6 a	36.4 ± 8.4 b	35.9 ± 0.7 a	10.7 ± 2.8 b	57.4 ± 7.6 a	0.5 ± 0.4 a	43.4 ± 15.5 a	−40.0 ± 2.4 a
P. subcapitata	48.7 ± 1.1 a	37.3 ± 2.2 a	45.8 ± 3.6 b	35.8 ± 1.0 a	5.4 ± 1.5 b	35.9 ± 1.7 b	0.2 ± 0.2 a	13.8 ± 3.8 b	−48.1 ± 0.9 b
S. salina	51.9 ± 8.0 a	37.3 ± 1.0 a	73.6 ± 1.1 a	35.8 ± 0.4 a	0.0 ± 0.0 a	22.6 ± 3.2 b	0.0 ± 0.0 a	−10.2 ± 6.6 b	−43.1 ± 2.4 b
Mixed cultures									
C. vulgaris + *M. aeruginosa*	41.9 ± 2.8 a	34.2 ± 4.4 a	51.6 ± 3.6 c	37.0 ± 1.8 a	0.5 ± 0.3 a	53.0 ± 8.9 a	0.0 ± 0.0 a	40.6 ± 11.8 a	−35.4 ± 1.0 a
C. vulgaris + *S. salina*	76.8 ± 4.2 b	35.3 ± 2.1 a	57.0 ± 3.1 c	36.6 ± 0.8 a	0.0 ± 0.0 a	52.2 ± 1.5 a	0.0 ± 0.0 a	40.1 ± 2.5 a	−45.1 ± 1.4 b
P. subcapitata + *M. aeruginosa*	41.9 ± 2.8 a	37.3 ± 0.9 a	35.5 ± 1.5 d	35.8 ± 0.4 a	11.7 ± 0.8 b	37.0 ± 4.8 b	0.9 ± 0.3 a	13.4 ± 7.2 b	−42.7 ± 1.4 b
C. vulgaris + *P. subcapitata*	57.0 ± 0.1 a	39.6 ± 0.1 a	56.0 ± 0.2 c	34.0 ± 0.1 a	0.0 ± 0.0 a	32.7 ± 0.3 b	0.0 ± 0.0 a	10.5 ± 0.5 b	−42.7 ± 1.5 b
P. subcapitata + *S. salina*	54.2 ± 1.9 a	37.6 ± 0.3 a	50.7 ± 1.1 c	35.6 ± 0.1 a	2.8 ± 0.5 c	32.0 ± 1.9 b	0.1 ± 0.0 a	8.3 ± 3.3 b	−40.5 ± 0.5 b
S. salina + *M. aeruginosa*	93.5 ± 3.3 c	35.0 ± 3.2 a	84.0 ± 1.3 a	36.7 ± 1.3 a	0.0 ± 0.0 a	7.6 ± 2.2 c	0.0 ± 0.0 a	−50.4 ± 8.6 c	−34.9 ± 1.0 a

Values are presented as the mean ± standard deviation of two independent experiments. Different letters within the same column represent statistically different values ($p < 0.05$). θ_W, contact angle with water; θ_B, contact angle with α-bromonaphthalene; θ_F, contact angle with formamide; γ_s^{LW}, Lifshitz-van der Waals component of the surface free energy; γ_s^{AB}, Lewis acid-base component of the surface free energy; γ_s^-, electron donor component; γ_s^+, electron acceptor component; ΔG_{sws}^{TOT}, free energy of hydrophobic interaction.

(28.3 and 37.0 mJ m^{-2}, respectively). γ_s^{LW} values determined for the different combinations of microalgae and cyanobacteria ranged from 34.0 ± 0.1 to 37.0 ± 1.8 mJ m^{-2}. These values were not statistically different ($p = 0.24$) from those determined for each individual strain, and in addition, no statistical differences ($p > 0.05$) were observed between the different studied consortia. As previously stated, similar results determined for this component indicate that cell wall composition of the studied microorganisms, even when grown in mixed cultures, may not have significant differences.

Values of γ_s^{AB} determined for the studied microalgae ranged from 0.0 ± 0.0 to 5.4 ± 1.5 mJ m^{-2}, whereas the same value determined for the cyanobacteria ranged from 0.0 ± 0.0 to 10.7 ± 2.8 mJ m^{-2}. The values obtained for microalgae were similar to those obtained by Ozkan and Berberoglu [44] for green algae (ranging from 0.0 to 5.1 mJ m^{-2}). Regarding the values obtained by the authors for cyanobacteria, γ_s^{AB} determined in this study for the cyanobacterium *S. salina* was equal to the one reported for *Synechocystis* sp. A γ_s^{AB} of 5.4 ± 1.5 mJ m^{-2} determined for the cyanobacterium *M. aeruginosa*, which is statistically higher than the one determined for *S. salina* ($p = 0.01$) may be related to the hydrophilic character of this cyanobacterium ($\Delta G_{sws}^{TOT} > 0$). The γ_s^{AB} values determined for the microalgae and for the cyanobacterium *M. aeruginosa* are a measure of the hydrophilic repulsion, since these microorganisms presented a hydrophilic surface. On the other hand, γ_s^{AB} determined for *S. salina* is a measure of hydrophobic attraction, due to the negative value of ΔG_{sws}^{TOT} determined for this cyanobacterium. The magnitude of AB interactions increases with increasing surface hydrophobicity (decreasing γ_s^+ and γ_s^-), and species having hydrophilic surface properties experience repulsive acid-base interactions [49]. γ_s^{AB} values determined for the studied consortia were not statistically different ($p = 0.64$) from those determined for microorganisms grown in single cultures, ranging from 0.0 ± 0.0 to 11.7 ± 0.8 mJ m^{-2}. Since positive ΔG_{sws}^{TOT} values were determined for all studied consortia, except for the one composed by both cyanobacteria, γ_s^{AB} values are a measure of hydrophilic repulsion. In the mixed culture of both cyanobacteria, γ_s^{AB} indicates the degree of hydrophobic attraction.

The values determined for the free energy of hydrophobic interaction have demonstrated that the microalgae and the cyanobacterium *M. aeruginosa* presented hydrophilic surfaces, while the surface of *S. salina* was considered hydrophobic. According to the determined values of ΔG_{sws}^{TOT}, the microorganisms can be listed by increased degree of hydrophobicity, as follows: *C. vulgaris*, *M. aeruginosa*, *P. subcapitata*, and *S. salina*. However, no statistical differences ($p = 0.60$) were observed in ΔG_{sws}^{TOT} values determined for *C. vulgaris* and *M. aeruginosa* (52.5 ± 14.3 and 43.4 ± 15.5 mJ m^{-2}, respectively). Negative values of ΔG_{sws}^{TOT} determined for *S. salina* suggest the potential of this microorganism to form flocs and, therefore, improve harvesting processes and also its ability to grow as sessile cells, which may have an application in biofilm formation for wastewater treatment and metabolite production. In mixed cultures, these values ranged from −50.4 ± 8.6 to

40.6 ± 11.8 mJ m^{-2}. Three distinct groups can be defined among the studied consortia: (i) the consortia composed of *C. vulgaris* + *M. aeruginosa* and of *C. vulgaris* + *S. salina* constitute the most hydrophilic group, with ΔG_{sws}^{TOT} values of about 40 mJ m^{-2}; (ii) the consortia of *P. subcapitata* + *M. aeruginosa*, *C. vulgaris* + *P. subcapitata* and *P. subcapitata* + *S. salina* constitute the group presenting a slightly hydrophilic surface (ΔG_{sws}^{TOT} values close to 10 mJ m^{-2}); and (iii) the more hydrophobic group, represented by the consortium of *S. salina* + *M. aeruginosa* (ΔG_{sws}^{TOT} of −50 mJ m^{-2}). These results have shown that the co-culture of a more hydrophobic microorganism with another one presenting a more hydrophilic surface results in a decrease in free energy of hydrophobic interaction values and, hence, in an increase in the degree of hydrophobicity. For example, the co-culture of the hydrophobic cyanobacterium *S. salina* with *C. vulgaris*, *M aeruginosa* and *P. subcapitata* resulted in a reduction of ΔG_{sws}^{TOT} values from 52.5 ± 14.3 to 40.1 ± 2.5, 43.4 ± 15.5 to −50.4 ± 8.6 and 13.8 ± 3.8 to 8.3 ± 3.3 mJ m^{-2}, respectively. The huge decrease of ΔG_{sws}^{TOT} observed for the co-culture of both cyanobacteria may be related to the high contact angles measured with polar probe liquids (θ_W and θ_F of 93.5° ± 3.3° and 84.0° ± 1.3°, respectively), which are considered unusual [44]. The co-culture of the most hydrophilic microorganisms (*C. vulgaris* + *M. aeruginosa*) resulted in a hydrophilic surface, with ΔG_{sws}^{TOT} of 40.6 ± 11.8 mJ m^{-2}.

Regarding zeta potential values determined for the studied microorganisms, these values ranged from −35.4 ± 0.4 to −48.1 ± 0.9 mV. Zeta potential measurements give information about the charge of cell surfaces: negative or positive values of zeta potential depend on the functional groups present on cell surfaces and also on the pH of the culture medium [44]. Functional groups commonly found on cell surfaces include hydroxyl (-OH), carboxyl (-COOH) and amine (-NH_2) [44,50,51]. Since these groups are ionisable, when cells are exposed to low pH values, functional groups are protonated and net surface charge becomes positive, and on the other hand, when cells are exposed to high pH values, functional groups are deprotonated and the resulting net surface charge is negative. At the point of zero charge (PZC), corresponding to an intermediate pH, some groups are protonated while others are deprotonated and the surface charge is neutralized [52,53]. Negative values observed for the studied microalgae and cyanobacteria were expected, since the pH of the culture medium measured when the samples were collected was high (9.64 ± 0.65) and PZC reported for algae was approximately pH 3 [52], indicating that functional groups on the microorganisms' surface were deprotonated. Additionally, net zeta potential values give information about suspensions' stability. When absolute value of zeta potential is high, repulsive forces prevail over van der Waals forces, and hence, particles/cells are stable in the dispersed form. On the other hand, for low net zeta potential values, van der Waals forces (usually attractive) are higher than repulsive ones and the stability of the suspension is affected, resulting in the formation of aggregates and further settling [54,55]. Net zeta potential values determined for the studied microalgae and cyanobacteria are substantial, and therefore, it is expected that the microorganisms keep stable in suspension. These results are in accordance with the non-evidence of flocs formation in the studied cultures, especially in that of the hydrophobic cyanobacterium *S. salina*. Therefore, zeta potential measurements are very important for a better understanding of the interactions among microalgae and cyanobacteria. Although microorganisms may present hydrophobic surfaces, their ability to form aggregates or to attach to surfaces may be strongly affected by the ionic strength of the culture medium. As for single cultures, zeta potential values determined for the studied consortia were negative, indicating an overall surface charge negative. In fact, the pH of the culture medium in these cultures was about 9.50 ± 0.33, which is higher than the reported PZC for microalgae (pH 3). Therefore, functional groups on microorganisms' surface may be deprotonated, conferring a negative surface charge. Zeta potential values determined ranged from −34.9 ± 1.0 to −45.1 ± 1.4 mV. These values were not statically different (p = 0.66) from those determined for single cultures, which may be related to similar ionic strengths of the culture medium, since both single and mixed cultures were cultured in OECD test medium [34] with the same composition. Additionally, no statistical difference (p > 0.05) was observed in zeta potential values for the studied microalgal associations. Net zeta potential values determined were also high, determining the stability of the suspensions and, therefore, the non-evidence of aggregates formation. This information may be very useful in the design of new strategies to control biofilm formation and development and to improve microalgal harvesting.

Determining the surface physicochemical properties of different consortia has shown that the co-culture of microorganisms with different degrees of hydrophobicity or hydrophilicity can alter the physicochemical properties of the single-cultured microorganisms. These findings can be very important in several applications, such as promoting aggregates formation and, therefore, bioflocculation and biofilm formation and, on the other hand, prevent and control biofouling. However, it is very important to take into account the ionic strength of the culture medium, since this parameter influences surface charge and, hence, the higher or lesser extent of electrostatic repulsions.

Sedimentation kinetics

Microalgal removal percentages determined for single and mixed cultures within the 8-h settling period are

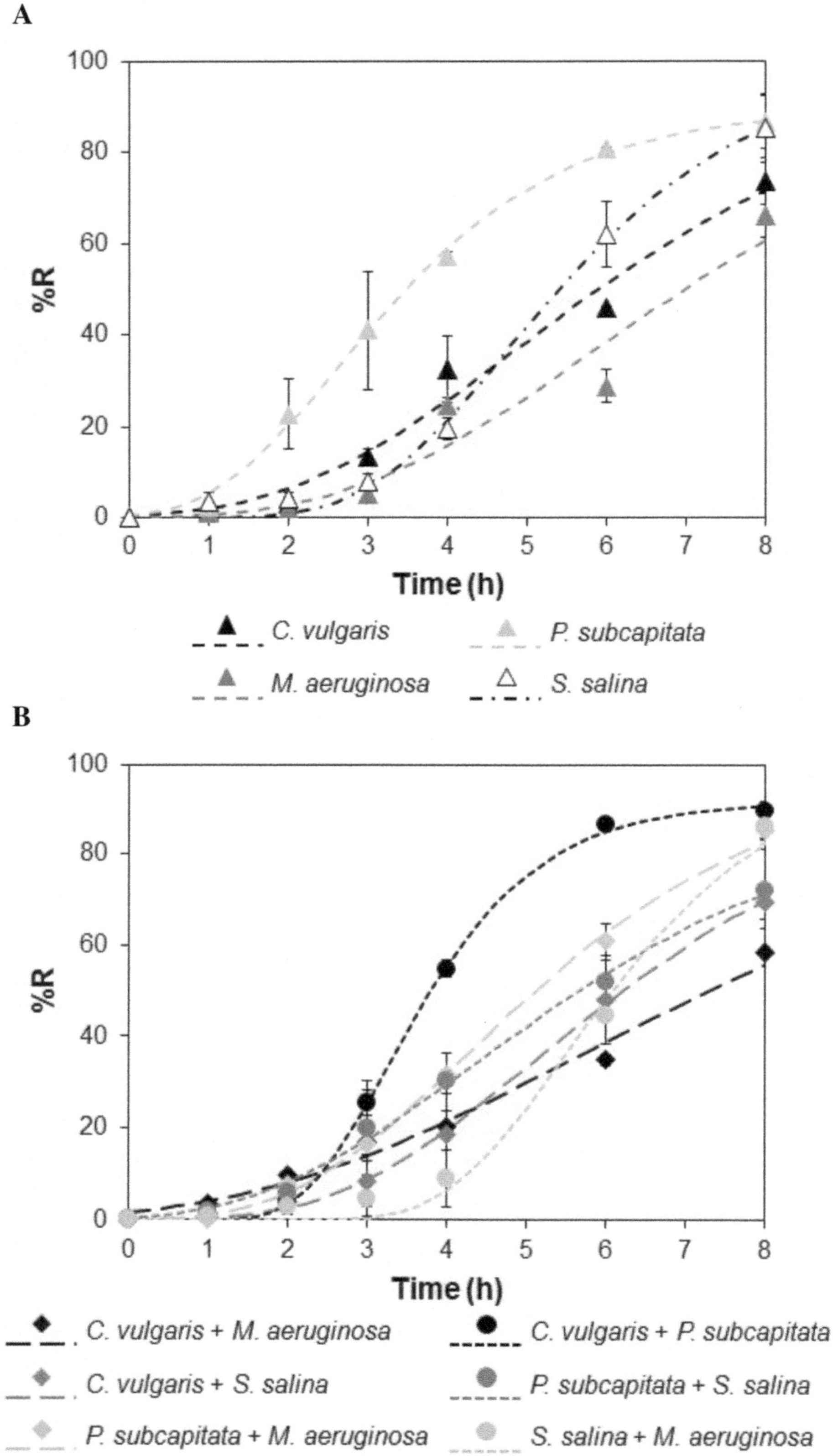

Figure 1 Microalgal removal percentages determined within the settling period for each of the studied suspensions. (**A**) Single cultures and (**B**) Mixed cultures. Error bars correspond to the standard deviation of two independent experiments. Dashed lines represent the model fit of the Gompertz model to the experimental data.

presented in Figure 1A and B, respectively. Analysis of these data indicates that sedimentation rates of the studied algal suspensions are very low. After 8 h of settling, maximum removal percentages achieved in single and mixed cultures were 86.6% ± 6.0% and 89.6% ± 2.2%, respectively. Additionally, as it is possible to see from Figure 1A and B, all microalgal suspensions presented a lag time, starting to sediment after, at least, 1 h of the beginning of the experiment. These results are in accordance with the zeta potential values determined for these suspensions. As it was referred in the 'Surface physicochemical properties' section, when absolute value of zeta potential is high, repulsive forces prevail over van der Waals forces, and hence, particles/cells are stable in the dispersed form. Similar removal percentages have been reported by Salim et al. [40] and Manheim and Nelson [56] when harvesting microalgae through bioflocculation. Higher removal percentages can be achieved using chemical coagulants/flocculants. In the study performed by Papazi et al. [57], chemical coagulants, such as aluminium and ferric sulphate, were used to harvest the microalga *Chlorella minutissima*, resulting in a removal percentage of 80% after 2 to 4 h of settling. In the study performed by Garzon-Sanabria et al. [58], after 1 h of settling, approximately 90% of the microalga *Nannochloris oculata* was recovered from the culture medium. These authors used aluminium chloride as coagulant.

Relationship between surface physicochemical properties and microalgal settling

Table 2 presents microalgal removal percentages determined after 8 h of settling and the kinetic parameters obtained through the model fit of the Gompertz model to the experimental data. Looking at R^2 values determined through the model fit (values close to 1) and to the model curves superimposed to the experimental data (in Figure 1A and B), it is possible to conclude that the Gompertz model can be correctly applied to describe microalgal recovery in the studied conditions. The sedimentation kinetic parameters determined through this model confirm the low settleability of the studied microalgal suspensions, which may be related to the high net values determined for zeta potential. According to these data, lag times, *b*, observed in the beginning of the sedimentation process were high, ranging from 1.0 to 3.0 h for single cultures and from 1.7 to 4.0 h for mixed cultures. Additionally, low sedimentation rates, *c*, were determined for both single and mixed cultures: 0.33 to 0.63 h^{-1} and 0.24 to 0.94 h^{-1}, respectively. Although microalgal recovery using these suspensions was not effective, the values determined through contact angle measurements can be correlated with sedimentation results. From Tables 1 and 2, it is possible to observe that, in general, an increase in the degree of hydrophobicity (lower ΔG_{sws}^{TOT} values) corresponds to an increase in microalgal removal percentages, % *R*, and in sedimentation rates, *c*. Higher removal percentages for the most hydrophobic surfaces were expected since, as it was stated in the 'Surface parameters and hydrophobicity determinations' section, when ΔG_{sws}^{TOT} decreases, the interaction between the two entities is stronger than the interaction of each entity with water. Therefore, particles in suspension tend to form aggregates, improving settling rates. In fact, plotting removal percentages determined

Table 2 Sedimentation kinetics determined for each of the studied suspensions

Microalgal suspensions	% R^a	Gompertz model parameters			
		a (%)	*b* (h)	*c* (h^{-1})	R^2
Single cultures					
C. vulgaris	73.6 ± 5.2 a	100.0	2.0	0.35	0.990
M. aeruginosa	66.1 ± 4.8 a	100.0	2.8	0.33	0.965
P. subcapitata	86.6 ± 6.0 b	89.7	1.0	0.63	0.999
S. salina	85.0 ± 7.5 b	100.0	3.0	0.56	0.997
Mixed cultures					
C. vulgaris + *M. aeruginosa*	58.6 ± 0.8 c	100.0	1.7	0.24	0.993
C. vulgaris + *S. salina*	69.6 ± 13.7 a	95.3	2.7	0.41	0.999
P. subcapitata + *M. aeruginosa*	85.2 ± 2.2 b	100.0	2.1	0.45	0.999
C. vulgaris + *P. subcapitata*	89.6 ± 2.2 b	91.8	2.2	0.94	1.000
P. subcapitata + *S. salina*	72.4 ± 8.7 a	82.1	1.7	0.38	0.998
S. salina + *M. aeruginosa*	86.1 ± 2.7 b	100.0	4.0	0.66	0.994

Microalgal removal percentage was determined after 8 h of settling, and recovery kinetic parameters were determined through the Gompertz model. [a]Values are presented as the mean ± standard deviation of two independent experiments. Different letters within the same column represent statistically different values ($p < 0.05$). % *R*, microalgal removal percentage determined after 8 h of settling; *a*, upper asymptote; *b*, displacement along the *x*-axis or lag time ($b > 0$); *c*, tangent at the inflection point or sedimentation rate ($c > 0$); R^2, coefficient of determination obtained through the model fit of Gompertz model to the experimental data.

after 8 h of settling as a function of the free energy of hydrophobic interaction suggests a negative relationship between these parameters. By excluding the values corresponding to the three most hydrophobic surfaces and to the most hydrophilic one, a linear relationship between microalgal removal percentage and free energy of hydrophobic interaction was obtained (R^2 = 0.91). These results show the importance of studying surface physicochemical properties to understand the flocculation process and improve harvesting procedures.

Conclusions

This study presented a broad characterization of surface physicochemical properties of microalgae and cyanobacteria, as well as how these properties vary when these microorganisms are co-cultured. Regarding the surface properties of single cultures, *S. salina* presented a hydrophobic surface, suggesting the ability of this microorganism to form aggregates. However, macroscopic observations have shown no evidence of aggregates formation, which was confirmed through the study of sedimentation kinetics. This might be due to the high net zeta potential values determined, which are responsible for the stability of the suspensions in their dispersed form. The co-culture of different microalgae and cyanobacteria has shown that when a more hydrophobic microorganism is co-cultured with another one presenting a more hydrophilic surface, free energy of hydrophobic interaction tends to decrease, resulting in a more hydrophobic surface. This study has also shown a negative linear correlation between microalgal removal percentages and free energy of hydrophobic interaction, reinforcing the importance of surface physicochemical properties on the sedimentation process. Due to the low recovery efficiencies achieved, another factor to consider in the prevention or promotion of cell attachment is the ionic strength of the culture medium, since this parameter strongly influences surface charge.

Competing interests

The authors declare that they have no competing interests.

Authors' contributions

ALG carried out the experimental assays and statistical analysis and drafted the document. CF helped in the experiments regarding contact angle measurements. JAL was involved in zeta potential determinations and respective analysis. JCMP participated in the design of the study and coordination. MS conceived the study and participated in its coordination. All authors read and approved the final manuscript.

Acknowledgements

This work was funded by FEDER funds through the Operational Programme for Competitiveness Factors - COMPETE, ON.2 - O Novo Norte - North Portugal Regional Operational Programme and National Funds through FCT - Foundation for Science and Technology under the projects: PEst-C/EQB/UI0511, NORTE-07-0124-FEDER-000025 - RL2_Environment&Health, Novomar - ref. 0687-Novomar-1-P, SFRH/BD/88799/2012, FEUP-ON2-25MC-EngªQuimicaBiologica and SFRH/BPD/66721/2009. The authors would like to thank CIIMAR (Centre of Marine and Environmental Research of the University of Porto) for providing the cyanobacteria *Synechocystis salina* LEGE 06079 and *Microcystis aeruginosa* LEGE 91344.

References

1. Brennan L, Owende P (2010) Biofuels from microalgae - a review of technologies for production, processing, and extractions of biofuels and co-products. Renew Sust Energ Rev 14:557–77
2. Demirbas A (2011) Biodiesel from oilgae, biofixation of carbon dioxide by microalgae: a solution to pollution problems. Appl Energy 88:3541–7
3. Bilanovic D, Andargatchew A, Kroeger T, Shelef G (2009) Freshwater and marine microalgae sequestering of CO_2 at different C and N concentrations - response surface methodology analysis. Energy Conv Manag 50:262–7
4. Gonçalves AL, Simões M, Pires JCM (2014) The effect of light supply on microalgal growth, CO_2 uptake and nutrient removal from wastewater. Energy Conv Manag 85:530–6
5. Ho S-H, Chen C-Y, Lee D-J, Chang J-S (2011) Perspectives on microalgal CO_2-emission mitigation systems - a review. Biotechnol Adv 29:189–98
6. Sayre R (2010) Microalgae: the potential for carbon capture. Bioscience 60:722–7
7. Boelee N, Temmink H, Janssen M, Buisman C, Wijffels R (2011) Nitrogen and phosphorus removal from municipal wastewater effluent using microalgal biofilms. Water Res 45:5925–33
8. Park JBK, Craggs RJ, Shilton AN (2011) Wastewater treatment high rate algal ponds for biofuel production. Bioresour Technol 102:35–42
9. Rawat I, Ranjith Kumar R, Mutanda T, Bux F (2011) Dual role of microalgae: phycoremediation of domestic wastewater and biomass production for sustainable biofuels production. Appl Energy 88:3411–24
10. Silva-Benavides A, Torzillo G (2012) Nitrogen and phosphorus removal through laboratory batch cultures of microalga *Chlorella vulgaris* and cyanobacterium *Planktothrix isothrix* grown as monoalgal and as co-cultures. J Appl Phycol 24:267–76
11. Hu Q (2004) Industrial production of microalgal cell-mass and secondary products - major industrial species. In: Richmond A (ed) Handbook of microalgal culture: biotechnology and applied phycology. Blackwell Science Ltd., Oxford, UK, pp 268–71
12. Singh S, Kate B, Banerjee U (2005) Bioactive compounds from cyanobacteria and microalgae: an overview. Crit Rev Biotechnol 25:73–95
13. Spolaore P, Joannis-Cassan C, Duran E, Isambert A (2006) Commercial applications of microalgae. J Biosci Bioeng 101:87–96
14. Molina Grima E, Belarbi EH, Acién Fernández FG, Robles Medina A, Chisti Y (2003) Recovery of microalgal biomass and metabolites: process options and economics. Biotechnol Adv 20:491–515
15. Lardon L, Hélias A, Sialve B, Steyer J-P, Bernard O (2009) Life-cycle assessment of biodiesel production from microalgae. Environ Sci Technol 43:6475–81
16. Ozkan A, Berberoglu H (2013) Adhesion of algal cells to surfaces. Biofouling 29:469–82
17. Posten C (2009) Design principles of photo-bioreactors for cultivation of microalgae. Eng Life Sci 9:165–77
18. Pulz O (2001) Photobioreactors: production systems for phototrophic microorganisms. Appl Microbiol Biotechnol 57:287–93
19. Irving TE, Allen DG (2011) Species and material considerations in the formation and development of microalgal biofilms. Appl Microbiol Biotechnol 92:283–94
20. Roeselers G, Van Loosdrecht M, Muyzer G (2008) Phototrophic biofilms and their potential applications. J Appl Phycol 20:227–35
21. Finlay JA, Bennett SM, Brewer LH, Sokolova A, Clay G, Gunari N et al (2010) Barnacle settlement and the adhesion of protein and diatom microfouling to xerogel films with varying surface energy and water wettability. Biofouling 26:657–66
22. Ista LK, Callow ME, Finlay JA, Coleman SE, Nolasco AC, Simons RH et al (2004) Effect of substratum surface chemistry and surface energy on attachment of marine bacteria and algal spores. Appl Environ Microbiol 70:4151–7
23. Mercer P, Armenta RE (2011) Developments in oil extraction from microalgae. Eur J Lipid Sci Technol 113:539–47
24. Christenson L, Sims R (2011) Production and harvesting of microalgae for wastewater treatment, biofuels, and bioproducts. Biotechnol Adv 29:686–702
25. Uduman N, Qi Y, Danquah MK, Forde GM, Hoadley A (2010) Dewatering of microalgal cultures: a major bottleneck to algae-based fuels. J Renew Sustain Energy 2:012701–15
26. Vandamme D, Foubert I, Muylaert K (2013) Flocculation as a low-cost method for harvesting microalgae for bulk biomass production. Trends Biotechnol 31:233–9
27. van Oss C, Chaudhury M, Good R (1987) Monopolar surfaces. Adv Colloid Interface Sci 28:35–64

28. van Oss C, Ju L, Chaudhury M, Good R (1989) Estimation of the polar parameters of the surface tension of liquids by contact angle measurements on gels. J Colloid Interface Sci 128:313–9
29. van Oss CJ, Good RJ, Chaudhury MK (1988) Additive and nonadditive surface tension components and the interpretation of contact angles. Langmuir 4:884–91
30. Chinnasamy S, Ramakrishnan B, Bhatnagar A, Das KC (2009) Biomass production potential of a wastewater alga *Chlorella vulgaris* ARC 1 under elevated levels of CO_2 and temperature. Int J Mol Sci 10:518–32
31. McLarnon-Riches CJ, Rolph CE, Greenway DL, Robinson PK (1998) Effects of environmental factors and metals on *Selenastrum capricornutum* lipids. Phytochemistry 49:1241–7
32. Philippis R, Vincenzini M (1998) Exocellular polysaccharides from cyanobacteria and their possible applications. FEMS Microbiol Rev 22:151–75
33. Wahlen BD, Willis RM, Seefeldt LC (2011) Biodiesel production by simultaneous extraction and conversion of total lipids from microalgae, cyanobacteria, and wild mixed-cultures. Bioresour Technol 102:2724–30
34. OECD (2011) Freshwater alga and cyanobacteria, growth inhibition test. In: Test Guideline 201: Organisation for Economic Co-operation and Development
35. Busscher H, Weerkamp A, van der Mei H, van Pelt A, de Jong H, Arends J (1984) Measurement of the surface free energy of bacterial cell surfaces and its relevance for adhesion. Appl Environ Microbiol 48:980–3
36. van Oss C (1995) Hydrophobicity of biosurfaces - origin, quantitative determination and interaction energies. Colloids Surf B Biointerfaces 5:91–110
37. Ozkan A, Berberoglu H (2013) Cell to substratum and cell to cell interactions of microalgae. Colloids Surf B Biointerfaces 112:302–9
38. van Oss CJ (2003) Long-range and short-range mechanisms of hydrophobic attraction and hydrophilic repulsion in specific and aspecific interactions. J Mol Recognit 16:177–90
39. Janczuk B, Chibowski E, Bruque J, Kerkeb M, Caballero FG (1993) On the consistency of surface free energy components as calculated from contact angles of different liquids: an application to the cholesterol surface. J Colloid Interface Sci 159:421–8
40. Salim S, Bosma R, Vermuë MH, Wijffels RH (2011) Harvesting of microalgae by bio-flocculation. J Appl Phycol 23:849–55
41. Gompertz B (1825) On the nature of the function expressive of the law of human mortality, and on a new mode of determining the value of life contingencies. Philos Trans Royal Soc London 115:513–83
42. Simões LC, Simões M, Oliveira R, Vieira MJ (2007) Potential of the adhesion of bacteria isolated from drinking water to materials. J Basic Microbiol 47:174–83
43. Volpe CD, Siboni S (1997) Some reflections on acid–base solid surface free energy theories. J Colloid Interface Sci 195:121–36
44. Ozkan A, Berberoglu H (2013) Physico-chemical surface properties of microalgae. Colloids Surf B Biointerfaces 112:287–93
45. Becker E (2007) Microalgae as a source of protein. Biotechnol Adv 25:207–10
46. Latała A, Nędzi M, Stepnowski P (2009) Toxicity of imidazolium and pyridinium based ionic liquids towards algae: *Chlorella vulgaris, Oocystis submarina* (green algae) and *Cyclotella meneghiniana, Skeletonema marinoi* (diatoms). Green Chem 11:580–8
47. Loos E, Meindl D (1982) Composition of the cell wall of *Chlorella fusca.* Planta 156:270–3
48. Northcote D, Goulding K, Horne R (1958) The chemical composition and structure of the cell wall of *Chlorella pyrenoidosa.* Biochem J 70:391
49. Bos R, Mei HC, Busscher HJ (1999) Physico-chemistry of initial microbial adhesive interactions - its mechanisms and methods for study. FEMS Microbiol Rev 23:179–230
50. Ferreira C, Pereira A, Pereira M, Melo L, Simões M (2011) Physiological changes induced by the quaternary ammonium compound benzyldimethyldodecylammonium chloride on *Pseudomonas fluorescens.* J Antimicrob Chemother 66:1036–43
51. González-Fernández C, Ballesteros M (2012) Microalgae autoflocculation: an alternative to high-energy consuming harvesting methods. J Appl Phycol 25:1–9
52. Hadjoudja S, Deluchat V, Baudu M (2010) Cell surface characterisation of *Microcystis aeruginosa* and *Chlorella vulgaris.* J Colloid Interface Sci 342:293–9
53. Stumm W, Morgan JJ (2012) Aquatic chemistry: chemical equilibria and rates in natural waters. John Wiley & Sons, New Jersey
54. de Schryver P, Crab R, Defoirdt T, Boon N, Verstraete W (2008) The basics of bio-flocs technology: the added value for aquaculture. Aquaculture 277:125–37
55. Zita A, Hermansson M (1994) Effects of ionic strength on bacterial adhesion and stability of flocs in a wastewater activated sludge system. Appl Environ Microbiol 60:3041–8
56. Manheim D, Nelson Y (2013) Settling and bioflocculation of two species of algae used in wastewater treatment and algae biomass production. Environ Prog Sustain Energy 32:946–54
57. Papazi A, Makridis P, Divanach P (2010) Harvesting *Chlorella minutissima* using cell coagulants. J Appl Phycol 22:349–55
58. Garzon-Sanabria AJ, Davis RT, Nikolov ZL (2012) Harvesting *Nannochloris oculata* by inorganic electrolyte flocculation: effect of initial cell density, ionic strength, coagulant dosage, and media pH. Bioresour Technol 118:418–24

13

UV mutagenesis treatment for improved production of endoglucanase and β-glucosidase from newly isolated thermotolerant actinomycetes, *Streptomyces griseoaurantiacus*

Adepu K. Kumar

Abstract

Background: Bioconversion of cellulosic biomass into fuel ethanol involves several steps, among which enzymatic breakdown of cellulose into fermentable sugars play a significant role. The key enzymes involved in cellulosic breakdown are mainly endoglucanases and β-glucosidases. Even though the biochemical and molecular characterization of number of endoglucanases and β-glucosidases was extensively studied, still there is a demand for novel microbial cellulases for industrial applications. Among the group of actinomycetes, *Streptomyces* spp. are well known as a cellulase producer. The advantage of using actinomycetes is being that production process could be easily scaled-up to commercial levels. However, recent research studies have shown that the production of cellulases from actinomycetes could also be significantly improved by employing different types of strain improvement methods, thus achieving high yields of extracellular proteins. Besides this, highly thermostable and broad pH range cellulases are required for bioethanol application.

Results: A lignocellulose degrading actinomycetes strain was newly isolated and identified as *Streptomyces griseoaurantiacus*. Strain improvement using UV mutagenesis developed two mutants (SG_{UV30} and SG_{UV5}) with 57.4 % and 12.8 % higher endoglucanase and β-glucosidase activities. The cellulases (endoglucanases and β-glucosidases) were found to be highly thermostable with no loss in enzyme activities at 80 °C for 60 min and nearly 80 % of initial activity was retained at 90 °C. Enzyme assays in presence of additives showed that $CoCl_2$, $CaCl_2$, and FeSO4 increased β-glucosidase activity but showed negative effect on endoglucanase activity. However, both the enzyme activities were significantly enhanced by addition of PEG 8000, sodium azide and $MnSO_4$.

Conclusions: Strain improvement of *S. griseoaurantiacus* was performed by UV mutagenesis where two mutant strains (SG_{UV30} and SG_{UV5}) were developed with improved endoglucanase and β-glucosidase activities. Cellulase production in submerged fermentation was carried out using a cheap lignocellulosic biomass residue, rice straw as a sole source carbon. The results clearly show that the mutant strains produced high-efficient cellulases that are stable at a broad pH range at very high temperatures. Besides, the mutants also showed high extracellular protein secretions, which could be promising in reducing the overall cellulase production costs at large scale.

Keywords: Cellulase; Endoglucanase; β-glucosidase; Actinomycetes; Lignocellulosic biomass

Correspondence: kiranbio@gmail.com
Bioconversion Technology Division, Sardar Patel Renewable Energy Research Institute, Near BVM Engineering College, Post Box No. 2, Vallabh Vidyanagar 388120, Gujarat, India

Background

Lignocellulosic biomass is available in plenty and is an inexpensive renewable bioresource for bioconversion to biofuels. However, this conversion is quite difficult owing to the complexity of the plant cell wall materials that are innately designed to resist microbial degradation. Among these, celluloses are one of the major structural components which could be converted into fuel ethanol with the help of microbial cellulose degrading enzymes, namely cellulases. This group of enzymes catalyze the hydrolysis of cellulose and other cellooligosaccharides into fermentable sugars. Primarily, but not limited to, the cellulase system consists of three major enzymes, *viz.* exoglucanase (avicelase), endoglucanase (CMCase) and β-glucosidase (cellobiase). In nature, a number of bacteria and fungi produce cellulases to hydrolyze the insoluble polysaccharides to soluble oligomers and subsequently to monomeric sugars [1]. The production and application potential of diverse microbial cellulases is largely explored using a variety of growth substrates, and voluminous research reports and review articles are published in this area [2–5]. Actinomycetes, especially *Streptomyces* spp, are known to produce cellulose-degrading enzymes and have attracted considerable interest among the researchers due to their potential applications in the recovery of fermentable sugars from the hydrolysis mixture. *Streptomyces* spp. are also capable of producing an array of different extracellular enzymes including cellulases, xylanases, and chitinases [6]. Synergistic action of these enzymes is vital for complete enzymatic hydrolysis of cellulose [7]. During the recent past the intense basic and applied research studies revealed the commercial significance and industrial applicability of novel potential cellulases to a greater extent [8-10]. The production economics of cellulosic ethanol from lignocellulosic residues is largely dependent on the enzyme cost especially in the bioconversion processes [11]. However, continuous efforts towards cost reduction have been directed for increasing the enzyme production levels, identification of hyperactive cellulose-degrading microbial strains, efficient ethanol fermentation techniques and enzyme recovery systems, etc. [12]. Besides these, use of different mutagenic agents for microbial strain improvement and fermentation processes was demonstrated [13], where simultaneous treatments with *N*-methyl-*N'*-nitro-*N*-nitrosoguanidine (NTG), ethidium bromide, and UV either alone or in combination were employed for higher endoglucanases [14].

The present investigation is on isolation of cellulase-producing actinomycete strains from agricultural and plant waste residues such as decaying woods and aged tree trunk burrows and to screen the microbe for robust utilization of different locally available lignocellulosic biomass residues. We also report on increasing cellulase production levels by applying traditional UV mutagenesis treatment and biochemical characterization of the crude cellulase enzymes (endoglucanase and β-glucosidase) from two mutant strains.

Methods

Substrate, chemicals, and media

Rice straw was obtained from local agricultural fields and was used as carbon substrate for cellulase production. Sodium dodecyl sulphate-polyacrylamide gel electrophoresis (SDS-PAGE) molecular weight markers were purchased from Bio-Rad Laboratories, Hercules, CA. Bovine serum albumin (BSA), *p*-nitrophenyl-β-D-glucopyranoside (PNPG), carboxymethyl cellulose (CMC), cellobiose, and 4-methylumbelliferyl-β-D-glucopyranoside (MUG) were purchased from Sigma (St. Louis, MO, USA). All other reagents were obtained from commercial sources and were of analytical grade.

Microbial isolation and screening

Cellulase producing microbes were isolated from diverse environmental places such as soil and water samples, decaying carton box scrapings, degrading wood pieces, and salt pans near the coastal marine areas of state of Andhra Pradesh, India. The samples (1 g each) were crushed with mortar and pestle and were then suspended in 10 ml of sterile distilled water. A 10^3 dilution of the samples was made and spread plated on IGA medium agar plates comprising 0.5 % peptone, 0.5 % yeast extract, 0.5 % malt extract, 0.5 % glycerol, 3 % 10× mineral salt medium, and 1.5 % agar. The composition of mineral salt medium is given in Additional file 1: Table S2. All plates were then incubated at 40 °C for at least 2–3 weeks. Further, the microbes are sub-cultured onto fresh agar plates in order to ensure the purity of the selected colonies. All the cultures were periodically maintained on sterile actinomycete isolation agar plates (HiMedia Laboratories, Mumbai, India) and stored at 4 °C for further use.

Cellulose-degrading microbes were screened using actinomycete isolation agar plates containing 0.5 % CMC and 10 mM MUG, separately. The plates were then incubated for 48 h at 40 °C. For endoglucanase production, the plates were stained with 0.05 % congo red dye, and for β-glucosidase production, the plates were directly viewed under UV light. A potential cellulase-producing microbe for the current investigation was selected based on the maximum cellulolytic index and intensity of fluorescence. The cellulolytic index value of the actinomycetes was calculated using the microbial colony size and the size of zone of clearance.

Molecular identification

Among the potential cellulase-producing microorganisms, an actinomycete strain denoted by SG14 was selected in the present study. The genomic DNA was isolated using genomic DNA isolation kit (Bangalore Genei, India). Consensus primers (forward primer: 5′C GCGGCCTATCAGCTTGTTG′3 and reverse primer: 5′ CCGTACTCCCCAGGCGGGG′3) were used for PCR amplification of 16S rRNA. The PCR product was then purified and subjected to DNA sequencing. The 16 SrRNA was submitted to NCBI database with GenBank ID KR094972. The nucleotide sequence was analyzed using Blastn, and multiple sequence analysis was performed using clustalX software using the top 20 consensus sequences obtained from the NCBI databank (>98 % sequence identity). A phylogenetic tree was prepared using neighbor-joining method with 1000 bootstrap replicates. Mega 5.0 software was used for tree analysis [15].

Media optimization and growth on different lignocellulosic residues

For cellulase production in submerged fermentation, seed inoculum was prepared in actinomycete broth medium (HiMedia Laboratories, Mumbai, India). Preliminary cellulase productions were tested using 1 % (*w*/*v*) CMC as carbon source in different growth media compositions viz. minimal salt media, Mandels mineral media [16], Czapek media [17], and Vogel's mineral media [18]. The detailed composition of growth media were given in Additional file 1: Tables S2 to S5. Production of cellulase was tested on different ligocellulosic biomass residues. The carbon sources tested were rice straw (RS), wheat bran (WB), rice husk (RH), wheat husk (WH), coco pith (CP), saw dust (SD), Walseth cellulose (WC), cardboard waste (CW), and newspaper waste (NW), respectively. Except WC, other carbon sources were desized to 0.5–1 cm using a hammer mill machine and subjected to extensive washing with distilled water and finally dried at 60 °C overnight until the moisture content was <5 %, while the WC substrate was prepared using commercial cellulose powder (Sigma Aldrich, USA) and pre-treated with 83 % (*w*/*v*) phosphoric acid for 25 h at 40 °C. The precipitate obtained from the Buchner funnel filtration method was then collected and washed twice with 70 % (*v*/*v*) ethanol and neutralized to pH 7.0. Finally, the precipitate was dehydrated by acetone rinsing and dried under vacuum. For submerged fermentation, 100 ml of minimal media was placed in a 500-ml Erlenmeyer flask supplemented with 1 % (*w*/*v*) rice straw. The growth media were inoculated with 4 % spore suspension and incubated at 40 °C at 160 rpm for 5 days. The culture broth was filtered through Whatman filter paper (9 μm size), and the filtrate was then centrifuged at 10,000 rpm for 30 min. The clear supernatant was collected and treated as the crude cellulase enzyme for biochemical analysis.

Determination of cellulase activities

Endoglucanase activity was assayed by incubating 0.5 ml of crude enzyme with 0.5 ml of 2 % CMC in citrate buffer (50 mM, pH 5.0) for 30 min at 60 °C. The reducing sugars generated were determined by standard dinitrosalicylic acid (DNSA) method [19]. β-Glucosidase assay was carried out by mixing 0.2 ml of culture supernatant with 0.2 ml of 0.01 M PNPG and 1.6 ml of citrate buffer (50 mM, pH 5.0). The reaction mixture was then incubated for 30 min at 60 °C. Reaction was stopped by the addition of 4 ml NaOH-glycine buffer (0.2 M, pH 10.6) and measured by colorimetric method [20]. The amount of enzyme required for liberating either 1 μg of glucose or 1 μg *p*-nitrophenol per minute under the standard assay conditions was considered as one unit of endoglucanase or β-glucosidase activity unit, respectively. All the assay experiments were carried out in triplicates and mean values are given.

Zymogram analysis using non-denaturing SDS-PAGE

Non-denaturing SDS-PAGE was performed for identification of endoglucanase and β-glucosidase proteins. Endoglucanase protein bands were detected using the congo red dye staining method of an overlaying gel containing 2 % CMC, while β-glucosidase active protein bands were detected in the gel by incubating with 10 mM MUG for 30 min at 45 °C, and then the gel was viewed directly under UV light. Briefly, a 10 % separating gel was prepared, and 30 μg of crude cellulase enzyme protein from native and UV-mutated samples was loaded. After electrophoresis, the gel was equilibrated with sodium citrate buffer (50 mM, pH 5.0) for 30 min and was overlayered on a CMC-containing polyacrylamide gel. This cassette was then enclosed in a plastic wrap and incubated at 60 °C for 2 h. Post incubation, the CMC-containing gel was peeled off and was stained with 0.2 % congo red solution for 1 h. To visualize the zone of clearance corresponding to the endoglucanase activity, destaining was performed with 1 M NaCl. The developed gels were visualized and digitally imaged. Protein concentration was determined using BSA as standard following the method described by Bradford [21].

Effect of UV treatment on cellulase production

Spore suspension from an overnight grown culture was prepared by serial dilution method. A 0.5-ml culture sample with a 10^6 dilution (approximately 10^5–10^6 spores ml^{-1}) was spread plated onto a 0.5 % CMC agar plate under sterile conditions. Each petri plate was then exposed to a UV tube (6 W, 240 nm) with a constant 10-cm distance for a treatment period of 15 s to 60 min. The UV-treated petri plates were taken out at regular intervals (15, 30, and 45 ; 1,5,15,30, and 60 min) and incubated at 40 °C for 48 h. The survivor microbial colonies after the UV treatment were then selected for cellulase production by a similar zone of clearance method either using 0.05 % congo red dye as mentioned earlier. The selected colonies were further screened by analyzing the endoglucanase and β-glucosidase activities using rice straw as sole source of carbon in submerged fermentation.

Determination of optimum pH, temperature, and stability

The enzyme reactions were carried out separately in buffer solutions at varying pH (pH 2.0, pH 5.0, pH 7.0, pH 9.0, and pH 11.0) and temperature (40 to 100 °C with 10 °C increments). The endoglucanase and β-glucosidase activities were measured using the assay methods as mentioned in the above section. For pH stability studies, the enzyme samples were incubated in different test buffer solutions with varying pH (pH 2.0, pH 5.0, pH 7.0, pH 9.0, and pH 11.0) for 30 to 60 min at 60 °C, and then the residual enzyme activities were measured in sodium citrate buffer (50 mM, pH 5.0). For temperature stability studies, the enzyme samples were incubated at different test temperatures (60, 80, 90, and 100 °C) in sodium citrate buffer (50 mM, pH 5.0) for different time periods, and residual activities were measured.

Effect of metal ions and chemical reagents

The enzyme samples were mixed with different metal ions, chemical reagents, and solvents, and then the effect of these additives on the enzyme activity was analyzed. The test compounds used were (10 mM each) $MgSO_4$, $MgCl_2$, $CoCl_2$, $CaCl_2$, $CuSO_4$, $MnSO_4$, $FeSO_4$, 0.02 % sodium azide, 0.8 % Tween 20, and 20 % PEG 8000. Control experiments were also analyzed voiding either the chemical agent or the enzyme to nullify the background absorbance on the activity measurements.

Results and discussion

Microbial identification

The present investigation is on a cellulase-producing actinomycete strain, *Streptomyces griseoaurantiacus*, which could tolerate high UV radiation for a maximum period of 60 min. The extracellular cellulase production and corresponding enzyme activities were found to be increased significantly after UV treatment. A total of 78 cellulase-producing microbes (bacteria and actinomycetes) were isolated from various natural habitats. Among these isolates, an actinomycete with maximum zone of clearance ratio on the CMC agar plate with a cellulolytic index value of 34 mm was selected (Additional file 1: Figure S1). Molecular identification and phylogenetic analysis revealed that the microbial strain belongs to *Streptomyces* spp. with a close relation to *S. griseoaurantiacus* NBRC 15440 (Fig. 1). Use of purified cellulose or its products like CMC as a growth carbon source is uneconomical for large-scale production of cellulases. Hence, different locally available cheap lignocellulosic residues were tested for microbial cellulase production (Fig. 2). Among the test carbon substrates, rice straw-grown culture extracts showed maximum endoglucanase activity (2.84 ± 0.2 U ml^{-1}).

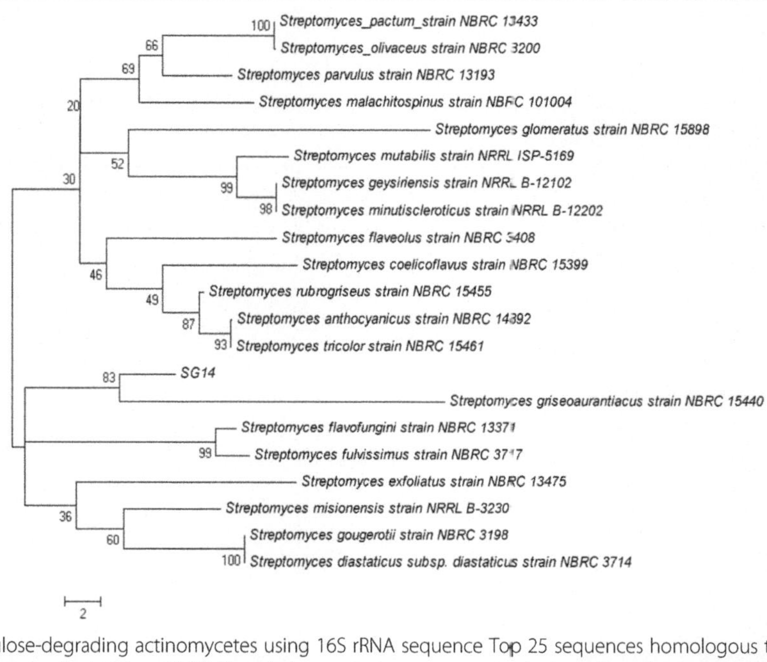

Fig. 1 Identification of cellulose-degrading actinomycetes using 16S rRNA sequence Top 25 sequences homologous to microbial 16S ribosomal partial sequence database were collected from NCBI. The phylogenetic tree was prepared using NJ method with 1000 bootstrap replicates in MEGA 5.2 software [15]. The microbe used in the current study was designated as SG14

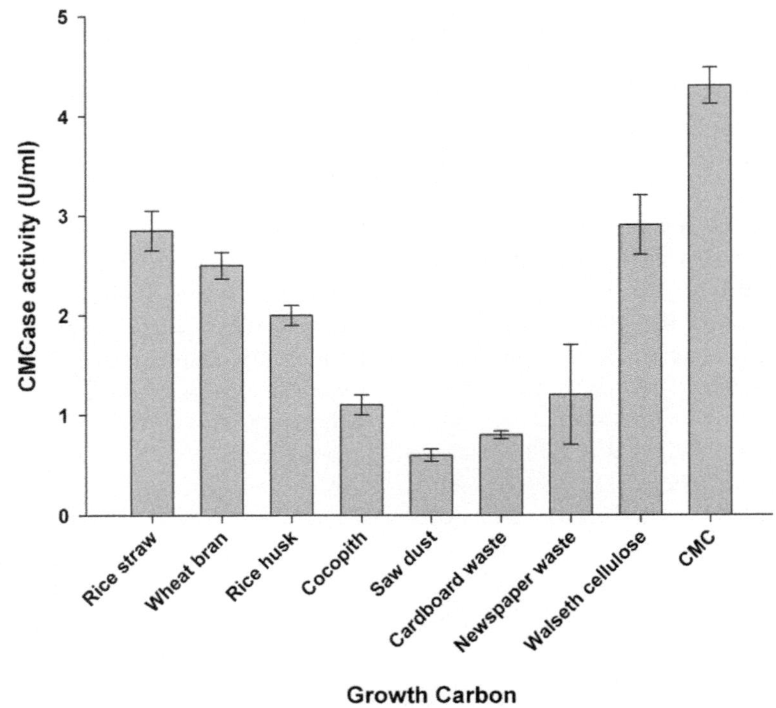

Fig. 2 Production of cellulases in different cellulosic waste materials. The final concentration of the growth substrate used was 1 % (*w/v*)

Microbial growth on rice straw

Growth optimization studies showed that 1 % (*w*/*v*) rice straw and 4 % (*v*/*v*) seed inoculum produced maximum endoglucanase (Additional file 1: Figure S2). Thereafter, increase in either rice straw or inoculum concentration had no significant effect on the endoglucanase activity. Similarly, maximum endoglucanase production was observed when grown in minimal growth medium (2.7 ± 0.2 U ml^{-1}), followed by Vogel's mineral medium (2.2 ± 0.1 U ml^{-1}) (Table 1). Jaradat *et al* also reported the influence of growth conditions and medium composition on the cellulase production from *Streptomyces* sps [22]. In general, the enzyme production depends on the amount of nitrogen supplementation and its accessibility during microbial growth. Unlike other complex growth mediums, minimal media were economical and comparatively simple in their composition (Additional file 1: Table S2). Since no significant difference was observed in cellulase activities, further experiments were performed using rice straw as the sole source of carbon and minimal medium for microbial growth.

UV mutagenesis and zymogam analysis

Mutational analysis was performed using traditional UV treatment of the actinomycete for increasing cellulase production. Based on the endoglucanase and β-glucosidase enzyme activities the prominent UV mutant strains were selected for further biochemical studies. However, initial screening of the mutants was established on the index ratio of the zone of clearance. The zone of clearance is indicated by the visible growth of the microbial colony to the degraded CMC agar plate as described in the "Methods" section. The isolated UV mutants were then tested subsequently in submerged fermentation for a minimum of 20 sub-culturing experiments. The variation in the production of endoglucanase and β-glucosidase and reproducibility of the results were evaluated. Similar results were obtained with all the tested sub-cultured experiments, and no significant alteration in cellulase production was observed. Maximum cellulase production from the mutant cells was obtained within 94 h as compared to the longer incubation period (120 h) with the native strain. One of the most intriguing features of *S. gresioaurantiacus* is its high resistance to the UV radiation for a maximum treatment period of 60 min. After 48 h growth incubation, >40 % of microbial

Table 1 Cellulase production in different growth media

Growth media[a]	Enzyme activity (U ml^{-1})	
	CMCase	β-glucosidase
Minimal media	2.7 ± 0.2	13.4 ± 0.5
Mandels mineral media	1.8 ± 0.4	6.0 ± 0.8
Czapek media	1.9 ± 0.2	10.5 ± 0.3
Vogel's mineral media	2.2 ± 0.1	11.2 ± 0.6

[a]All media contain 1 % CMC as substrate

cells was found to survive and form colonies after UV treatment. These data clearly indicated that the mutants were not only able to survive harsh conditions but also stable after several generations without any back mutation. Very few microbes in nature could survive such harsh treatments, and to date, no reports are available on such high-UV-tolerant actinomycetes. In most of the microbial mutagenesis studies, UV treatment was generally followed by a chemical treatment to avoid back mutation [14]. However, our studies showed that these mutagens were stable without any drastic changes in the cellulase production profiles and corresponding enzyme activities for several subsequent generations indicating that the developed mutants were highly stable.

Among the tested UV mutants, maximum endoglucanase and β-glucosidase activities were observed in 30-min and 5-min treated cells (Table 2). Hereafter, the mutant strains for endoglucanase and β-glucosidase production are referred as SG_{UV30} and SG_{UV5}, respectively. The endoglucanase activity in native, SG_{UV30}, and SG_{UV5} strains were found to be 2.84 ± 0.5 U ml^{-1}, 4.47 ± 0.3 U ml^{-1}, and 3.88 ± 0.4 U ml^{-1}, respectively, whereas the β-glucosidase activity in native, SG_{UV30}, and SG_{UV5} strains were 15.6 ± 0.7 U ml^{-1}, 10.4 ± 0.4 U ml^{-1}, and 17.6 ± 0.5 U ml^{-1}, respectively (Table 2). From these results, it is clearly evident that the endoglucanase activity in SG_{UV30} and β-glucosidase activity in SG_{UV5} were nearly 57.4 and 12.8 % higher compared to the native strain. These results were in close agreement with the recent findings from a UV-mutated *Aspergillus niger* strain, where a twofold increase in endoglucanase activity was reported [23]. Our results suggest that modification of *S. griseoaurantiacus* for cellulase production by UV mutagenesis has a profound effect on overall enzyme production. Similarly, there are few reports on high UV treatment of fungi for enhancing cellulase production. In *Trichoderma viride* ZY-1, the UV mutant produced 98.1 % higher FPase than the native strain [24], while in *Aspergillus nidulans*, the UV mutant showed 3.96 times more endoglucanase activity than the wild type [25].

The enzyme activity results from the native and UV mutants were also supported by the zymogram analysis data. Non-denaturing SDS-PAGE analysis with equal protein loadings of culture filtrates on a CMC gel from the UV-treated cells showed maximum zone of clearance with the SG_{UV30} strain (Fig. 3). These results further provided evidence for enhanced cellulase production by UV treatment. As shown in Fig. 2, the endoglucanase activity in the native strain and SG_{UV30} strain from the CMC growth substrate was 4.3 ± 0.2 U ml^{-1}, which was found to be nearly similar (4.47 ± 0.3 U ml^{-1}) when grown on rice straw. For cost-effective production of enzymes at a commercial scale, use of such cheap lignocellulosic residues is preferred over synthesized carbon substrates.

Effect of pH, temperature and additives on cellulase activity

Although ~36.6 % higher endoglucanase activity was observed in SG_{UV5} cells, the β-glucosidase activity was not found to be increased. This variation in activity levels could be mostly due to the UV-dependent effect on the cellulase-encoding genes, since endoglucanase and β-glucosidase are two different proteins and are expressed by separate genes. It clearly indicated that UV mutagenesis is a random mutation process. As the mutation occurs at the gene level and mostly involved in thymine dimerization, the translation process of some of the genes might have inhibited by the UV treatment. Further detailed molecular investigation and proteomic studies are required to understand these expressed endoglucanase and β-glucosidase proteins. Moreover, UV treatment showed significant effect on overall increase in the extracellular protein production.

Table 2 Effect of UV treatment on cellulase activities and protein production

UV treatment time (min)	Enzyme activity (U ml^{-1})		Protein ($mg\ ml^{-1}$)
	CMCase	β-glucosidase	
Control	2.84	15.6	0.081
0.15	3.20	12.0	0.062
0.30	2.84	11.2	0.074
0.45	2.80	4.8	0.064
1.0	2.32	16.0	0.079
5.0	3.88	17.6	0.096
15.0	3.76	12.0	0.097
30.0	4.40	10.4	0.109
45.0	3.32	12.8	0.115
60.0	3.08	11.6	0.086

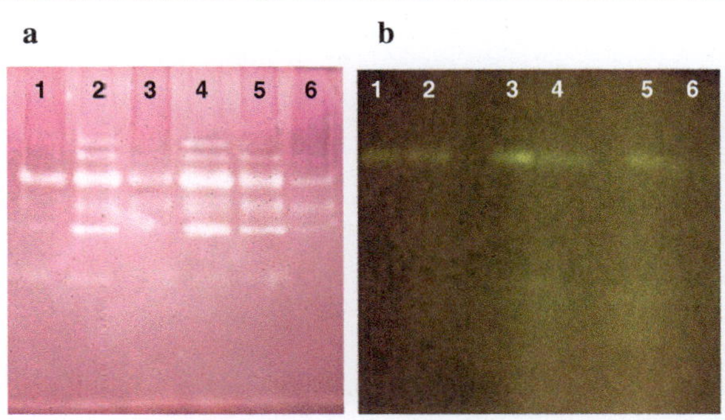

Fig. 3 Zymogram analyses of the culture filtrate after UV treatment. Cells were treated with UV light for different time periods. **a** CMC gel with congo red dye staining was used for endoglucanase protein detection, and **b** MUG gel with UV fluorescence was used for β-glucosidase protein detection. *Lane 1*: 0 min, *lane 2*: 5 min, *lane 3*: 15 min, *lane 4*: 30 min, *lane 5*: 45 min, and *lane 6*: 60 min, respectively

The protein concentration in the culture supernatant was increased from 0.081 to 0.115 mg ml^{-1} after UV treatment (Table 2). Besides this, the reason for low protein production with decreased enzyme activities at higher UV treatment periods may be due to the longer exposures of the microbe to harsh UV radiation which resulted in undesirable genetic modifications inside the cells [26].

The optimum temperature for endoglucanase and β-glucosidase activity was found to be 60 ± 5 °C and 70 ± 3 °C, respectively (Fig. 4a). These results were in close agreement with other published literatures on actinomycetes. Kluepfel [27] reported that *Streptomyces lividans* grown on xylan showed an optimum endoglucanase activity at 60 °C. However, endoglucanase from *Streptomyces* T3-1 had an optimum temperature of 50 °C [28], and *Streptomyces reticuli* endoglucanase showed an optimum temperature of 55 °C [29]. At 70 °C, more than 90 % of the endoglucanase was retained, while for β-glucosidase, a similar residual activity was observed at slightly lower temperatures (50–60 °C). These data clearly indicated that the enzymes work together effectively at their optimum temperature without compromising the activities of the enzyme complex. Cellulases from thermophilic microorganisms are generally reported to be highly thermostable in nature. Our studies also comply with this nature of

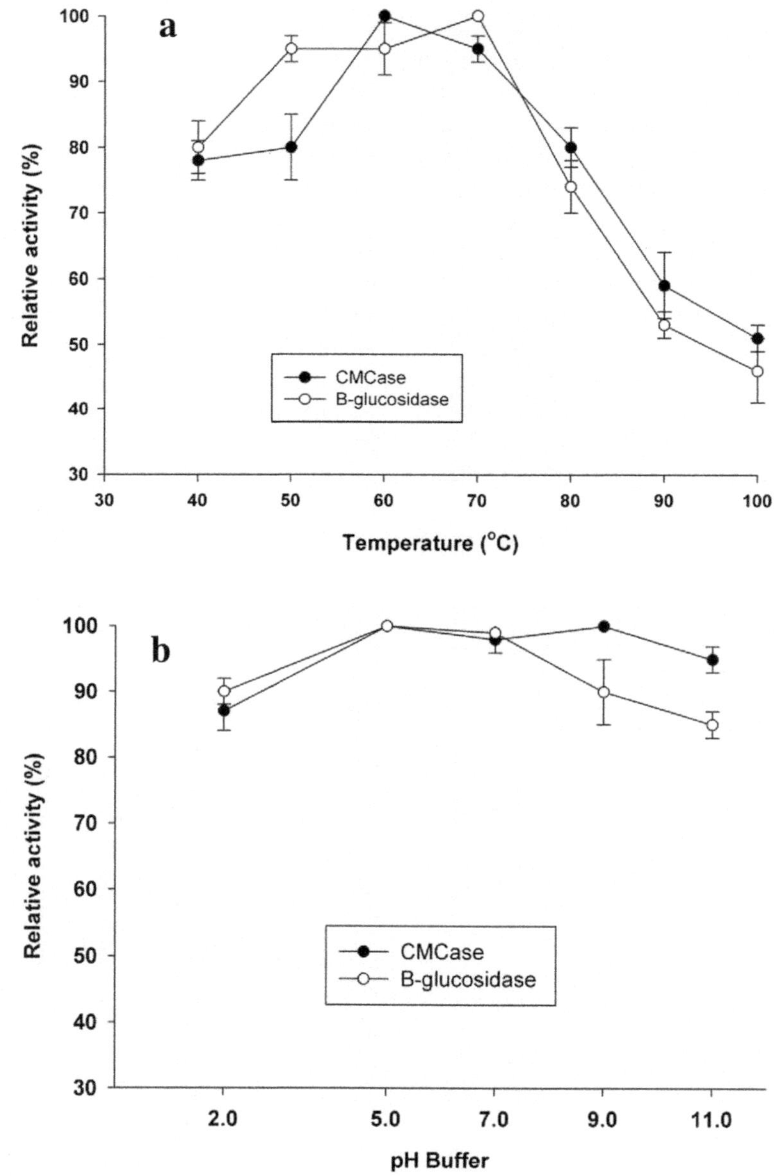

Fig. 4 Effect of temperature (**a**) and pH (**b**) on endoglucanase and β-glucosidase activities

thermostability. Above 60 °C incubation, the endoglucanase and β-glucosidase enzyme activities from the native strain were found to decrease drastically. In comparison to these, the proteins secreted from SG_{UV30} and SG_{UV5} were found to be highly stable and no loss in enzyme activities was observed after incubation at 80 °C for 60 min, whereas at 90 °C, 80 % initial activity was retained for endoglucanase and β-glucosidase. However, at 100 °C, 50 % endoglucanase and 70 % β-glucosidase residual activities were observed after 30 and 60 min incubation (Fig. 5a, b). Although both the enzymes were highly thermostable, the data indicated that β-glucosidase had better stability than endoglucanase.

The stability of the protein in different pH buffer systems was analyzed, and it was found that >70 % endoglucanase and >80 % β-glucosidase residual activities were retained after 60 min incubation period in pH 2.0 to pH 11.0 (Fig. 5c, d), except where no significant loss of endoglucanase and β-glucosidase activity was observed at pH 5.0 and pH 7.0, respectively. The data indicated that these proteins were highly stable in broad pH solutions. In contrary, the endoglucanase and β-glucosidase proteins from the native strain were not stable either above or below pH 5.0. The reason behind this drastic improvisation in pH stability from the mutant strains is not clearly known.

In general, based on the enzyme activity at a specific pH, the cellulase proteins are categorized into three groups, namely acidic, basic, and neutral cellulases. However, it is highly intriguing to ascertain that both the crude endoglucanase and β-glucosidase have no distinct maximum activity at a specific pH under the test conditions (Fig. 4b). The enzyme activities were found to be nearly similar in all test buffers with varying pH. The data is strongly evident from the production of multiple endoglucanase and β-glucosidase enzymes in the crude culture filtrate (Fig. 3a). Although there were multiple β-glucosidase proteins present in the culture filtrate, only a single prominent β-glucosidase protein with maximum enzyme activity was clearly visible, while other proteins were not digitally captured due to their low activity (Fig. 3b). However, multiple low molecular weight proteins with β-glucosidase activity were observed with the naked eye. In nature, microorganisms capable of degrading lignocellulosic materials contain multiple cellulase-producing genes in the genome and

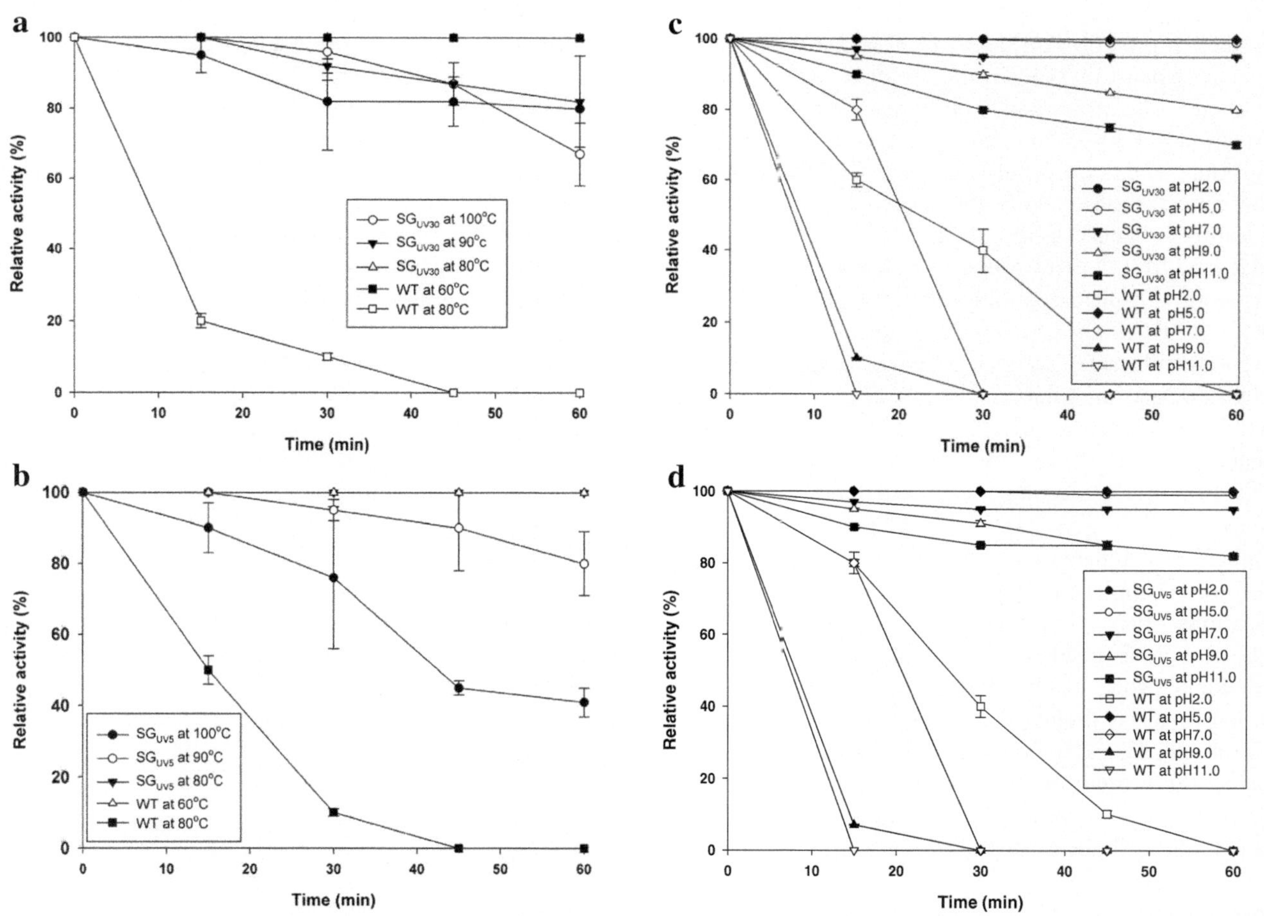

Fig. 5 Thermal and pH stability studies of endoglucanase and β-glucosidase. Temperature stability analysis of **a** endoglucanase and **b** β-glucosidase; pH stability analysis of **c** endoglucanase and **d** β-glucosidase

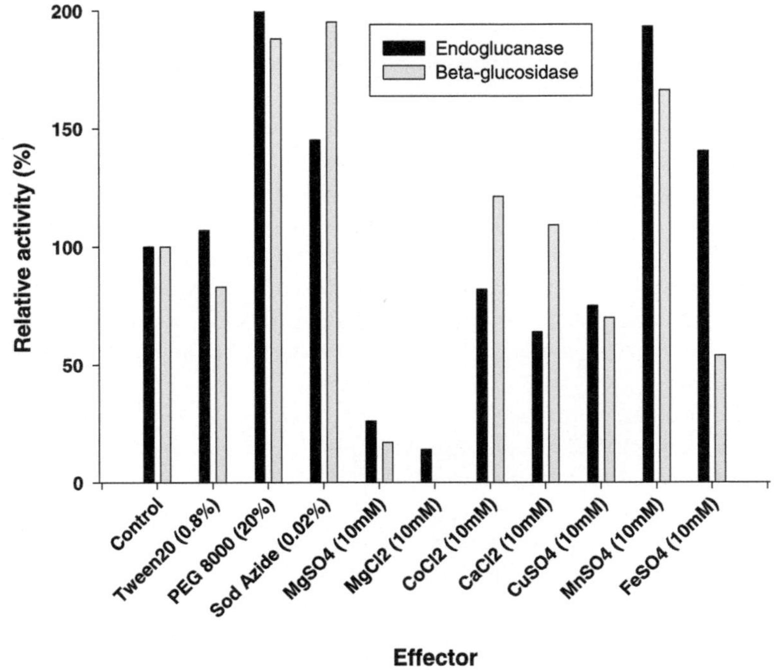

Fig. 6 Effect of metal ions and chemical agents on endoglucanase and β-glucosidase activities

secrete the proteins into the extracellular environment for efficient synergistic bioconversion of complex lignocellulosic materials into simple substrates for their metabolic growth [30]. Further, the effect of different divalent metals and ionic agents on cellulase activity was also examined. The endoglucanase and β-glucosidase activities were significantly enhanced in the presence of PEG 8000, sodium azide, and $MnSO_4$. However, $CoCl_2$, $CaCl_2$, and $FeSO_4$ decreased endoglucanase and increased β-glucosidase activities (Fig. 6), while $CuSO_4$ alone decreased the enzyme activities. Tween 80 had a marginal effect on the cellulase activities. A strong inhibitory effect was observed for both the enzymes when incubated with $MgSO_4$ and $MgCl_2$, indicating that Mg^{+2} ions may bind to the critical amino acids near the active site region and inhibit the enzyme catalysis.

Since the cellulases are stable in a broad pH range, this enzyme cocktail could be used in diverse industrial applications where a specific acidic, basic, or neutral cellulase protein is needed to be supplemented during an individual process step. Moreover, single cellulase enzyme cocktail is preferred over separate cellulase protein with a specific pH condition, thus reducing the overall cost of the protein. In addition, the crude cellulase protein cocktail also showed broad pH stability with high potential in cellulosic ethanol production.

Conclusions

S. griseoaurantiacus was mutated by UV treatment, and two mutant strains (SG_{UV30} and SG_{UV5}) were developed with improved endoglucanase and β-glucosidase production and activity. Rice straw was used as a cheap lignocellulosic residue for microbial growth for cellulase production in submerged fermentation. Both the native and mutant strains were able to utilize rice straw very efficiently. The UV mutants showed 57.4 % and 12.8 % higher endoglucanase and β-glucosidase activities compared to the wild-type enzymes. There was no loss in endoglucanase and β-glucosidase activities at 80 °C were found to be highly thermostable with no loss in enzyme activities at 80 °C for 60 min and nearly 80 % of initial activity was retained at 90 °C. Studies on purification of the endoglucanases and β-glucosidases from these mutant strains and their synergistic action on enzymatic saccharification will provide the industrial applicability in cellulosic ethanol production.

Additional file

Additional file 1: Supporting information. A file showing two supplementary figures and five supplementary tables.

Competing interests
The authors declare that they have no competing interests.

Author's contributions
AKK designed and performed the research experiments. AKK wrote the paper and approved the final manuscript for publication.

Author's information
AKK is a senior scientist at the Bioconversion Technology Division, Sardar Patel Renewable Energy Research Institute, near BVM Engineering College, Vallabh Vidyanagar, Gujarat, India.

Acknowledgements
The author is thankful to the Director, SPRERI for providing the necessary facilities to carry out this research work. The financial support from the start-up research fund (SPRERI/AKK/RP-2) of SPRERI, Gujarat is highly acknowledged.

References

1. Rathan RK, Ambili M (2011) Cellulase enzyme production by *Streptomyces* sp using fruit waste as substrate. Aust J Basic Appl Sci 5:1114–1118
2. Horn SJ, Vaaje-Kolstad G, Westereng B, Eijsink VG (2012) Novel enzymes for the degradation of cellulose. Biotechnol Biofuel 5:45
3. Miettinen-Oinonen A, Londesborough J, Joutsjoki V, Lantto R, Vehmaanpera J (2004) Three cellulases from *Melanocarpus albomyces* for textile treatment at neutral pH. Enzym Microbial Technol 34:332–341
4. Sukumaran RK, Singhania RR, Pandey AK (2005) Microbial cellulases - production, application and challenges. J Sci Indus Res 64:832–844
5. Bhat MK (2000) Cellulases and related enzymes in biotechnology. Biotechnol Adv 18:355–383
6. Manivasagan P, Gnanan S, Sivakumar K, Thangaradjou T, Vijaylaxmi S, Balasubramanian T (2010) Isolation, identification and characterization of multiple enzyme producing actinobacteria from sediment samples of Kodiyakarai coast, the Bay of Bengal. Afr J Microbiol Res 4:1550–1559
7. Berg B, Laskowski R (2006) Litter decomposition: a guide to carbon and nutrient turnover, vol 38, CA, Adv. Ecol. Res. Elsevier Academic Press, San Diego
8. Bajpai P (1999) Applications of enzymes in the pulp and paper industry. Biotechnol Prog 15:147–157
9. Harman GE, Kubicek CP (1998) *Trichoderma* and *Gliocladium*: enzymes, biological control and commercial applications, vol 2. Taylor and Francis Ltd, London, pp 393
10. Uhlig H (1998) Industrial enzymes and their applications. In: John Wiley & Sons, Inc, New York, pp 435
11. Reith JH, den Uil H, van Veen H, de Laat WTAM, Niessen JJ, de Jong E, Elbersen HW, Weusthuis R, van Dijken JP, Raamsdonk L (2002) Co-production of bio-ethanol, electricity and heat from biomass residues. In: Proceedings of the 12th European conference on biomass for energy, industry and climate protection. 17–21 June 2002, Amsterdam, The Netherlands, 1118–1123.
12. Xu ZH, Bail YL, Xu X, Shi JS, Tao WI (2005) Production of alkali-tolerant cellulase-free xylanase by *Pseudomonas* sp. UN024 with wheat bran as the main substrate. World J Microbiol Biotechnol 21:575–581
13. Parekh S, Vinci VA, Strobel RJ (2000) Improvement of microbial strains and fermentation processes. Appl Microbiol Biotechnol 54:287–301
14. Chand P, Aruna A, Maqsood AM, Rao LV (2005) Novel mutation method for increased cellulase production. J Appl Microbiol 98:318–323
15. Tamura K, Peterson D, Peterson N, Stecher G, Nei M, Kumar S (2011) MEGA5: molecular evolutionary genetics analysis using maximum likelihood, evolutionary distance, and maximum parsimony methods. Mol Biol Evol 28:2731–2739
16. Mandels M, Hontz I, Nystrom J (1974) Enzymatic hydrolysis of waste cellulose. Biotechnol Bioeng 16:1471–1493
17. Czapek F (1902) Beitr Chem Physiol Pathol 1:540
18. Vogel HJ (1956) A convenient growth medium for *Neurospora* (medium N). Microbial Genetics Bulletin 13:42–43
19. Dashtban M, Maki M, Leung KT, Mao C, Qin W (2010) Cellulase activities in biomass conversion: measurement methods and comparison. Crit Rev Biotechnol 30:302–309
20. Kubicek CP (1982) Beta-glucosidase excretion by *Trichoderma pseudokoningii*: correlation with cell wall bound beta-1,3-glucanase activities. Arch Microbiol 132:349–354
21. Bradford MM (1976) A rapid and sensitive method for the quantitation of microgram quantities of protein utilizing the principle of protein-dye binding. Anal Biochem 72:248–254
22. Jaradat Z, Dawagreh A, Ababneh Q, Saadoun I (2008) Influence of culture conditions on cellulase production by *Streptomyces* sp. strain J2. Jordan J Biol Sci 1:141–146
23. Irfan M, Javed J, Syed Q (2011) UV mutagenesis of *Aspergillus niger* for enzyme production in submerged fermentation. Pak J Biochem Mol Biol 44:137–140
24. Yao R, Li M, Deng S, Hu H, Wang H, Li F (2012) Mutagenesis of *Trichoderma viride* by ultraviolet and plasma. Plasma Sci Tech 14:353–356
25. Lui j, Feng Y, Yu Y, Zhou X, He W (2011) Selection of fungus with high ability of cellulase activity production using UV mutagenesis. International Conference on Consumer Electronics, Communications and Networks (CECNet), 1643–1645
26. Ikehata H, Ono T (2011) The mechanisms of UV mutagenesis. J Radiat Res 52:115–125
27. Kluepfel D, Shareck F, Mondou F, Morosoli R (1986) Characterization of cellulase and xylanase activities of *Streptomyces lividans*. Appl Microbiol Biotechnol 24:230–234
28. Jang HD, Chen KS (2003) Production and characterization of thermostable cellulases from *Streptomyces* transformant T3-1. World J Microbiol Biotechnol 19:263–268
29. Schrempf H, Walter S (1995) The cellulolytic system of *Streptomyces reticuli*. Int J Macromolecules 15:353–355
30. Maki M, Leung KT, Qin W (2009) The prospects of cellulase-producing bacteria for the bioconversion of lignocellulosic biomass. Int J Biol Sci 5:500–516

14

An overview on biofuel and biochemical production by photosynthetic microorganisms with understanding of the metabolism and by metabolic engineering together with efficient cultivation and downstream processing

Dayanidhi Sarkar[1] and Kazuyuki Shimizu[2,3*]

Abstract

Biofuel and biochemical production by photosynthetic microorganisms such as cyanobacteria and algae is attractive to improve energy security and to reduce CO_2 emission, contributing to the environmental problems such as global warming. Although biofuel production by photosynthetic microorganisms is called as the third generation biofuels, and significant innovation is necessary for the feasibility in practice, these fuels are attractive due to renewable and potentially carbon neutral resources. Moreover, photosynthetic microorganisms are attractive since they can grow on non-arable land and utilize saline and wastewater streams. Highly versatile and genetically tractable photosynthetic microorganisms need to capture solar energy and convert atmospheric and waste CO_2 to high-energy chemical products. Understanding of the metabolism and the efficient metabolic engineering of the photosynthetic organisms together with cultivation and separation processes as well as increased CO_2 assimilation enables the enhancement of the feasibility of biofuel and biochemical production.

Keywords: Microalgae; Cyanobacteria; Biofuels; Metabolic engineering; CO_2 fixation; Metabolic regulation

Background

International Panel on Climate Change (IPCC) keeps warning the global society on global warming caused by green-house gases such as CO_2 based on the accumulating data and the reliable prediction model. IPCC asks world societies to make decisions to invest for the reduction of CO_2 emissions mostly caused by human activities. This may be also considered from the point of view of future cost caused by the severe climate change due to global warming. Namely, the global warming may cause serious local climate change as well as the rise in the sea level, which give severe damage to the societies worldwide. In fact, we have often experienced disastrous climate change year by year, and it seems to be more and more severe.

The global carbon cycle has been perturbed by emissions from the combustion of fossil fuels and by changes in land use and land intensity. These perturbations have led to cumulative anthrogenic CO_2 emissions of 570 ± 70 petagrams carbon since 1750 to 2012 [1]. Seventy percent of these emissions originated from the combustion of fossil fuels [1].

According to the data of International Energy Agency (OECD, 2011), total energy consumption in the world increased more than 78% over the last three decades. Major usage of fossil fuels causes serious environmental problems worldwide, and much attention has been focused on reducing their usage by alternative clean fuels. Namely, due to the global warming problem caused by the increased use of fossil fuels together with limited amount of fossil fuels and the fluctuating cost caused by unstable political disturbances, alternative renewable

* Correspondence: shimi@bio.kyutech.ac.jp
[2]Kyushu Institute of Technology, Iizuka, Fukuoka 820-8502, Japan
[3]Institute of Advanced Bioscience, Keio University, Tsuruoka, Yamagata 997-0017, Japan
Full list of author information is available at the end of the article

energy sources have recently been paid much attention [2]. In fact, at the present staggering rate of consumption, the world fossil oil reserves will be exhausted in less than 50 years [3]. Carbon neutral biofuels are needed to replace the petroleum oil which causes global warming caused by the emission of green house gases. Currently, the world consumes about 15 terawatts of energy per year, and only 7.8% of this is derived from renewable energy sources [4]. Moreover, in comparison with other forms of renewable energy such as wind, tidal, and solar energy, liquid biofuels allow solar energy to be stored and also to be used directly in existing engines and transport infrastructure [5].

Annually, about $5{,}500 \times 10^{21}$ J of solar energy reaches the Earth's atmosphere [6]. Photosynthetic organisms including higher plants, microalgae, and cyanobacteria play the crucial roles of capturing solar energy and storing it as chemical energy [7]. The amount of solar energy currently captured by arable crops is limited by arable land area (about 3.9% of the Earth's surface area), fresh water (about 1% of global water), nutrient supply, and solar energy-to-biomass conversion efficiency [8-10]. Terrestrial plants capture 121.7×10^9 metric tons of carbon from the atmosphere each year [11] using solar light and CO_2 as the energy and carbon sources. Photosynthesized carbon is then chemically converted to a variety of chemical compounds, and it is attractive to use photosynthetic organisms as green factories for producing carbohydrates, liquid fuels, and pharmaceutical drugs as well as food and feed, thus contributing to the balancing of the atmospheric carbon [12].

The advantages of using photosynthetic microorganisms include the photosynthetic efficiency, location on non-arable land (about 25% of the Earth's surface), and the use of saline and wastewater source [7], where less than 1% of the available solar energy flux is converted into chemical energy by photosynthesis [13], and much effort has been focused on the enhancement of photosynthetic carbon fixation.

The so-called first generation biofuels have been produced from corn starch and sugarcane. However, this causes the problem of the so-called 'food and energy issues' as the production scale increases. The second generation biofuels production from lignocellulosic biomass has thus been paid recent attention. However, it requires energy-intensive pretreatment for the degradation of lignocellulosic biomass [14]. The third generation biofuel production from photosynthetic organisms such as cyanobacteria and algae has been also attracted some attention, but the cell growth rate is quite low, and thus the productivity of the metabolites is significantly low [15].

Although the biofuel and biochemical production by photosynthetic organisms has a big hurdle to overcome, it is still highly attractive due to CO_2 fixation with sunlight (and water) from environmental protection point of view, and thus contributing to the global warming problem as well. A variety of host organisms such as bacteria, fungi, and microalgae may be considered for the production of biofuels and biochemicals from CO_2 with sunlight. Although photosynthetic organisms offer the ability to produce biofuels and biochemicals directly from CO_2 and sunlight, significant innovation is inevitable for the process development in relation to large-scale cultivation, harvesting, and product separation, since the production rate is significantly low.

The commonly used photosynthetic organisms for biofuel and biochemical production are algae and cyanobacteria [16,17]. Microalgae are photosynthetic eukaryotic organisms with size ranging from 1 to 100 μm, while cyanobacteria are prokaryotic organisms with size ranging from 1 to 10 μm. Cyanobacteria gave rise to the chloroplasts of eukaryotic algae and also land plants, and they share many features such as the ability to drive photosynthetic water photolysis and thereby contribute to the production of both atmospheric oxygen and reduced organic carbon [7].

Microalgae are unicellular photosynthetic microorganisms that can convert solar energy to chemical energy with efficiency of 10 to 50 times greater than terrestrial plants [18]. Algae have far higher cell growth rates than plants and, therefore, have much smaller footprints for land required for producing energy [19-21]. Many microalgae are rich in oil especially under nitrogen-starved condition, which can be converted to biodiesel using existing technology. The productivity of these photosynthetic microorganisms in converting CO_2 into carbon-rich lipids, only a step or two away from biodiesel, significantly exceeds that of agricultural oleaginous crops, without competing for arable land [22]. They require aquatic environments that may vary from freshwater to seawater. Not only do these organisms fix CO_2, but they also have the potential to be used for the production of inexpensive bulk chemicals, because the major inputs to the system (light and CO_2) are essentially free [23]. Microalgae cells contain approximately 50% of carbon, in which 1.8 kg of CO_2 is fixed by producing 1 kg of microalgae biomass [19].

Recent studies have reported that *Chlorella* sp., *Scenedesmus* sp., and *Botryococcus braunii* are among the microalgae strains that have shown promising result to bio-mitigate O_2 emission with typical CO_2 consumption rate of 200 to 1,300 mg/L/day [24-28]. Successful commercial utilization of microalgae has been established in low-volume, high-value derivatives such as nutritional supplements, antioxidants, cosmetics, natural dyes, and polyunsaturated fatty acids (PUFA) [29].

In the case of plants, it is frequently found that metabolite pools exist in more than one location or that the subcellular location of one or more reactions is uncertain

[30]. Entire sections of metabolic pathways like glycolysis are duplicated between organelles, particularly the plastid and cytosol, with both being potentially active and carrying flux [31]. The simplest way is to examine metabolites which are formed in only one of the compartments [32-36]. Another method involves the fraction of cellular material prior to metabolite analysis [37]. Unfortunately, even with the supplemental information provided by analyzing compartment-specific metabolites, it may still be difficult to statistically distinguish different configurations of the metabolic map [38].

In the present article, we focus on the typical photosynthetic microorganisms such as algae and cyanobacteria and attempted to make a review on the metabolic regulation, metabolic engineering, and process development with efficient operation for the production of biofuels and chemicals to understand the current status and expect future perspectives.

Candidate photosynthetic microorganisms for biofuel production

Perhaps the most comprehensive evaluation of algal species has been orchestrated by the US Department of Energy's Aquatic Species Program (ASP) to develop microalgae as a source of biodiesel [29]. Over 3,000 strains of microalgae have been isolated from ponds and seas. Cellular oil content varies with growth phases [39]. The cells of the chlorophyte microalga *Parietochloris incise* synthesize almost twice as many triacylglycerols (TAGs) in the stationary phase than in the exponential growth phase [40]. Although microalgae have a high level of biodiversity, only a few species can be subjected to genetic manipulation [41]. The algae with the best developed genetic toolbox are the unicellular green microalgae *Chlamydomonas reinhardtii* [42]. It is a well-established model organism for the study of various cellular processes such as photosynthesis, flagella, starch metabolism, and photobiological production of hydrogen [42]. Like many other algal species, *C. reinhardtii* can accumulate significant amount of oil when subjected to unfavorable environmental conditions [43-47]. *C. reinhardtii* has proven to be a useful model organism to study the improvement of biodiesel production by microalgae [48,49]. It is unicellular and stays as haploid during most of its life cycle [42], thus it is particularly useful in the context of a forward genetic approach, because the mutant phenotype can be observed during the first generation and does not need to reach a diploid homozygous stage [50]. The freshwater green microalgae *P incise* enhances not only its production of TAG under nitrogen starvation but also the production of arachidonic acid, a valuable nutraceutical [51]. Green algae including *Spirogyra* sp. and *Chlorococum* sp. have been shown to accumulate high levels of polysaccharides both in their complex cell walls and as starch [4]. This starch can be used for bioethanol production. Bioethanol production from algae shows significant potential due to their low percentage of lignin and hemicelluloses as compared to other lignocellulosic plants [52]. Microalgae and cyanobacteria are also able to directly produce biohydrogen through photofermentation in an anaerobic process involving oxidation of ferredoxin by the hydrogenase enzyme [53].

Many species of macroalgae are known to have high levels of carbohydrate, although in many cases these carbohydrates consist of galactose [54]. Recent research has shown that the red algae *Gelidium amansii* and the brown algae *Laminaria japonica* are both a potential biomass source for biohydrogen production through anaerobic fermentation [54,55]. Recently, microalgae have also been paid attention from the point of view of biogas production in the anaerobic fermentation [4].

Metabolism of photosynthetic microorganisms

Oxygenic photosynthesis is the process by which plants, algae, and cyanobacteria convert sunlight and CO_2 into chemical energy and biomass. The algal photosynthesis is at least able to convert approximately 5% to 7% of incident light energy to biomass, where a systems-based approach to understand the stresses and efficiencies associated with light energy harvesting, CO_2 fixation, and carbon partitioning is necessary to make headway toward improving photosynthetic yields [56].

The cell growth conditions are roughly classified as *autotrophic* condition for the case of using only CO_2 under light condition, *mixotrohphic* condition for the case of using both carbohydrate and CO_2 under light condition, and *heterotrphic* condition for the case of using carbohydrate under dark condition. Although autotrophic condition is preferred from the environmental protection point of view using only CO_2 as a carbon source, the cell growth rate is significantly low, and thus the productivity of the metabolic products is low.

The atmospheric CO_2 is fixed either by C_3 photosynthesis where the three carbon molecule such as 3-phosphogrycerate (3PG) is used as the product of ribulose 1,5-bisphosphate carboxylase (RubisCO) reaction or by C_4 photosynthesis where four carbon molecule such as oxaloacetate (OAA) is used as the product of phosphoenol pyruvate (PEP) carboxylase (Ppc) reaction followed by the decarboxylation at malic enzyme (Mez) from malate yielding pyruvate. The C_4 photosynthesis may be created by evolution from ancestral C_3 photosynthesis during a global decline in atmospheric CO_2 level [57]. The C_4 pathway will have higher efficiency than the C_3 pathway in CO_2 fixing with which they consume water and nitrogen [58].

Since CO_2 fixation is attractive from the environmental protection point of view, several strategies have been considered for the efficient carbon fixation for the cell synthesis [59], where carbon fixation can be enhanced by amplifying carboxysome expression [60] or heterologous expression of RubisCO gene *rbcLS* [61]. The CO_2 fixation can be also enhanced by a hybrid RubisCO, which contains both plant and microalgae subunits [62].

The green organisms such as plants and algae gain energy via aerobic respiration, and the metabolism changes depending on the oxygen availability. The green organisms are exposed to a variety of oxygen availability in the environment that may vary from fully aerobic state (normoxia) to oxygen deficiency (hypoxia) or the anaerobic condition (anoxia). In the context of recent climate change, excess rainfall and frequent flooding may cause the green cells subjected to hypoxia or anoxia condition [63]. Oxygen is the final acceptor of electrons in the mitochondrial oxidative phosphorylation to generate ATP, while under hypoxia and anoxia conditions, the small amount of ATP is generated through glycolysis, and NAD(P)H must be reoxidized by the fermentative pathways. Under oxygen-limiting condition, the glycolysis flux is accelerated by the so-called 'Pasteur effect', and plant metabolism uses pyruvate to direct towards ethanolic and lactic fermentations [64]. In almost all plant, a rapid activation of lactate dehydrogenase (LDH) has been observed under oxygen limitation [63]. Lactate production causes damage to the cell by lowering cytoplasmic pH, and thus lactate production is transient eventually replaced by ethanolic fermentation. The α-ketoglutaric acid (αKG) in the TCA cycle can be oxidized with the incorporation of NH_4^+ and NADH to form glutamate (Glu), which is then decarboxylated to ɤ-amino butyric acid (GABA) by glutamate decarboxylase (GDC), where some protons are utilized in GDC reaction and stabilizes the cytosolic pH [63].

Metabolism of algae

It is critical to properly understand the metabolism in response to culture environment, where the systems biology approaches including metabolite profiling [65] and integration of different levels of information such as metabolites, fluxes, transcript, and protein abundance [66] are useful.

Consider the metabolism of photosynthetic microorganisms [67]. The light energy is incorporated into the cell, where light quanta absorbed by pigments drive the photosynthetic electron transport, where NADPH instead of NADH is used to generate ATP at the respiratory chain. The primal pathway for CO_2 fixation is the Calvin-Benson-Bassham (CBB) cycle, where the first step is catalyzed by RubisCO (Figure 1). This enzyme is also an oxygenase, which can react with O_2 and lead to a different pathway called photorespiration. Algae have the photorespiration pathway, and photosynthesis is inhibited by high O_2 concentration. Photosynthesis reactions such as light reactions, CBB cycle, and starch synthesis occur in chloroplasts. Algae and plant cells have subcellular compartments such as chloroplast, mitochondria, and cytoplasm. After the export of GAP from the chloroplast to cytoplasm, the carbon flow is divided into the sugar synthesis pathway or the glycolytic pathway to form pyruvate. Sugars such as sucrose are the major storage products in the cytoplasm of plant cells. In plant cells, replenishment of carbon to maintain the operation of the TCA cycle is achieved by anaplerotic reactions involving CO_2 fixation by PEP carboxylase (Ppc). The pentose phosphate (PP) pathway operates in the cytoplasm, where CBB cycle is functioning in the chloroplast.

Of all the pigments, chlorophyll takes a major fraction. δ-aminolevulinic acid (δ-ALA) is the key chlorophyll precursor molecule. The classical succinate-glycine pathway is the condensation of glycine and succinyl-CoA catalyzed by δ-ALA synthetase. In addition, glutamate and αKG are incorporated into δ-ALA much more efficiently than are glycine and succinate in many green cells. Although most of the fatty acid synthesis occurs in the chloroplast, the source of acetyl-coenzyme. A (AcCoA) derives from its synthesis in the mitochondria. The fatty acid composition of the lipids of *Chlorella* cells varies considerably, particularly for the α-linolenic acid (C18:3) content [67].

Under autotrophic condition, significant ATP is formed from mitochondrial oxidative phosphorylation. The CBB cycle is the main ATP sink in the autotrophic culture. The ATP yield decreases in the following order: heterotroph > mixotroph > autotroph [67].

Metabolism of cyanobacteria

Cyanobacteria are commonly used as model systems for the metabolism of higher plants. Cyanobacteria possess certain promising properties such as (1) large amounts of lipids, commonly present in thylakoid membranes, (2) higher photosynthetic levels and the cell growth rates compared to algae and higher plants, (3) easy growth with basic nutritional requirements such as air ($CO_2 + N_2$), water, and mineral salts with light [68].

The central metabolic network in *Synechocystis* is shown in Figure 2, which includes those of the glycolysis, PP pathway, CBB cycle, part of TCA cycle, and the C1 metabolism. Cyanobacteria have an incomplete TCA cycle lacking αketoglutarate dehydrogenase (KGDH) [69,70]. The enzymes, isocitrate lyase (Icl) and malate synthase (MS), which form the glyoxylate pathway, function in cyanobacteria [71]. The Mez and PEP synthase (Pps) are responsible for the gluconeogenetic steps, where the PEP carboxy kinase (Pck) is absent in *Synechocystis*. The PP pathway operates for glucose catabolism mainly in the heterotrophic conditions, while the CBB cycle is active under mixotrophic and autotrophic conditions.

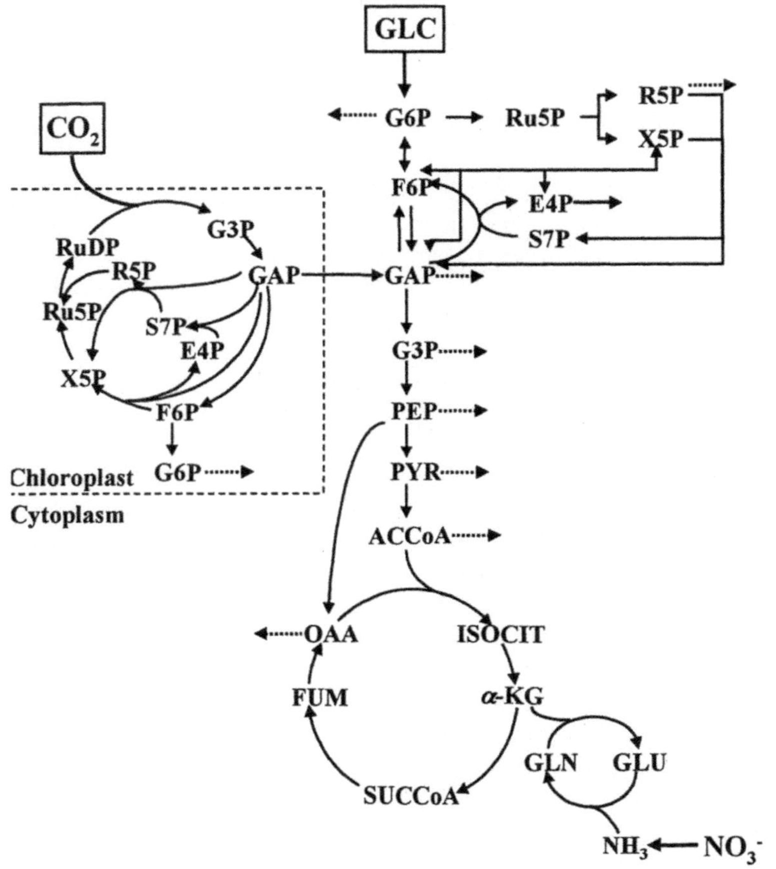

Figure 1 Metabolic pathways of photosynthetic microorganism (adapted from [67]).

The metabolic flux analysis of *Synechocystis* cultivated under heterotrophic and mixotrophic conditions has been made based on ^{13}C-metabolic flux analysis (^{13}C-MFA) [72,73], and the metabolic regulation analysis has also been made with integration of different levels of information [74]. In the heterotrophic cultivation of *Synechocystis*, more than 90% of glucose is channeled through the PP pathway (Figure 3) [72,73]. The high flux through the oxidative PP pathway yields a large amount of NADPH, as well as biosynthetic precursors such as ribose 5-phospate (R5P) and erythrose 4-phosphate (E4P). In the mixotrophic culture, CO_2 is fixed through the CBB cycle. The conventional ^{13}C-MFA is based on the steady state and thus limited to heterotrophic and mixotrophic conditions [72,73,75], while ^{13}C-MFA for autotrophic condition can be made by the isotopically nonstationary metabolic flux analysis (MFA) [76] with transient measurements of isotope incorporation following a step change from unlabeled to labeled CO_2 (Figure 3) [77].

Since cyanobacteria have negligible photorespiration and produce little or no glycolate during photosynthesis, it is unlikely that serine is synthesized, as in higher plants, from glycine by the glycolate pathway. Serine is synthesized directly from 3PG through a phosphorylated route in cyanobacteria [78].

Under both heterotrophic and mixotrophic conditions, the relative flux through Ppc is high. The reaction catalyzed by Ppc contributes to about 25% of the assimilated CO_2 under mixotrophic condition [72,73], indicating that Ppc is important for the fixation of CO_2 in cyanobacterial cells [79], where cyanobacterial cells fix significant amounts of carbon as C4 acids under light conditions. Considering that Mez in cyanobacteria is NADP-linked [80], it is more likely that Ppc and Mez serve as a device to fix a large amount of CO_2 as C4 acids and then release CO_2 and produce NADPH by the decarboxylation of malate. This Ppc and Mez pathways can effectively bypass the Pyk reaction, where its activity is repressed under light condition [77]. This is similar to the carbon metabolism in C4 plants, for which CO_2 and NADPH generated by Mez are utilized by the CBB cycle. In fact, although the major pathway of CO_2 fixation in the CBB cycle is similar to that in C3 plants, cyanobacteria have many of the physiological characteristics of C4 plants.

In the mixotrophic culture, copious amounts of reducing power are required in the CBB cycle to fix CO_2 to

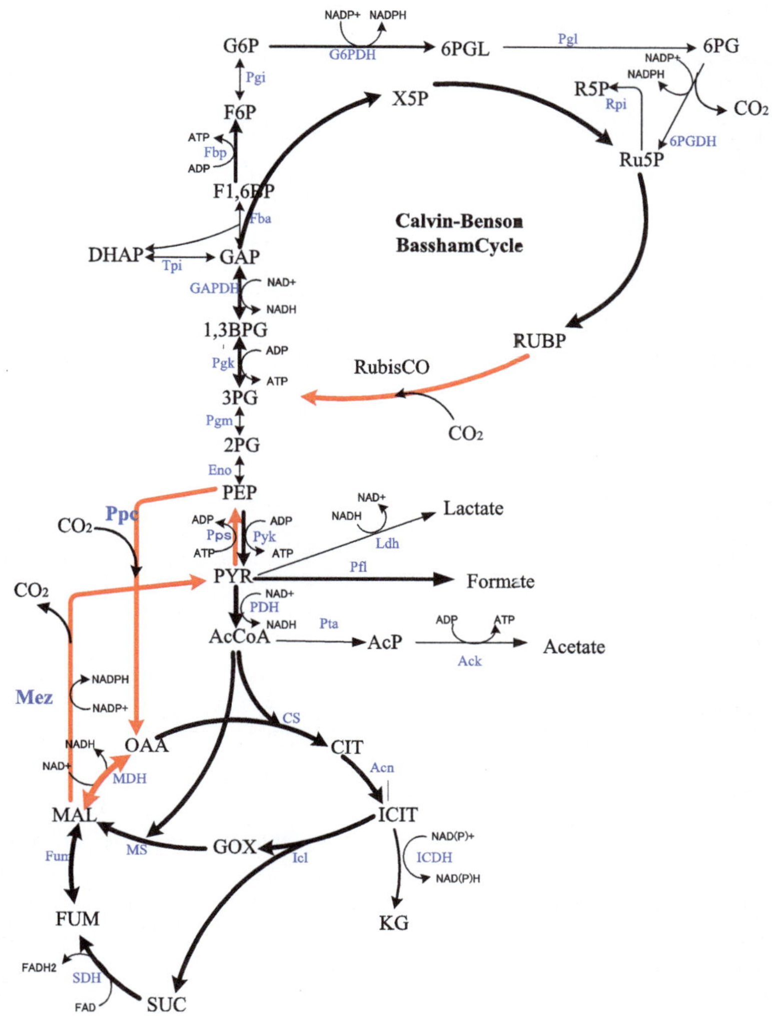

Figure 2 Metabolic pathways of cyanobacteria.

carbohydrates. Hence, there has to be a supply of NADPH in large amounts to fulfill the biosynthetic demands. The PP pathway in the heterotrophic culture and the photosynthetic electron transport in the mixotrophic culture accounts for a major fraction of NADPH production. Moreover, cyanobacteria utilize NADPH as an electron donor of the respiratory electron transport chain. Therefore, the excess NADPH is reoxidized during respiration to provide energy. The CBB cycle is the main ATP sink in the mixotrophic culture.

Catabolic regulation and carbon storage regulation under nitrogen limitation

Like many bacteria on earth, photosynthetic organisms are found in diverse ecological habitats, where the organisms are exposed to periods of severe nutrient starvation. In particular, cyanobacteria are found in a wide range of ecological habitats including oceans and lakes [81]. They also survive in deserts, polar regions, and hot springs, where the nutrient starvation is much more severe.

Under unfavorable growth condition such as nitrogen starvation, TAG is typically produced in microalgae, where its fraction ranges from 20% to 60% (weight/dry weight) [82]. The efficient production of TAG in microalgae requires a thorough understanding of lipid metabolism and TAG accumulation [83]. It is important to analyze different levels of information to uncover the molecular mechanism underlying the increased TAG accumulation for microalgae such as *C. reinhardtii* and its starchless and cell wall-deficient mutant strains [84].

Cyanobacteria have sophisticated mechanisms to cope with nitrogen limitation, where the primary step is the capture of nitrogen-containing compounds with high affinity, where nitrate, nitrite, and ammonium are the typical nitrogen sources with a preference for ammonium [85]. Some strains can fix dinitrogen gas and may use also urea, cyanate, and amino acids as additional nitrogen

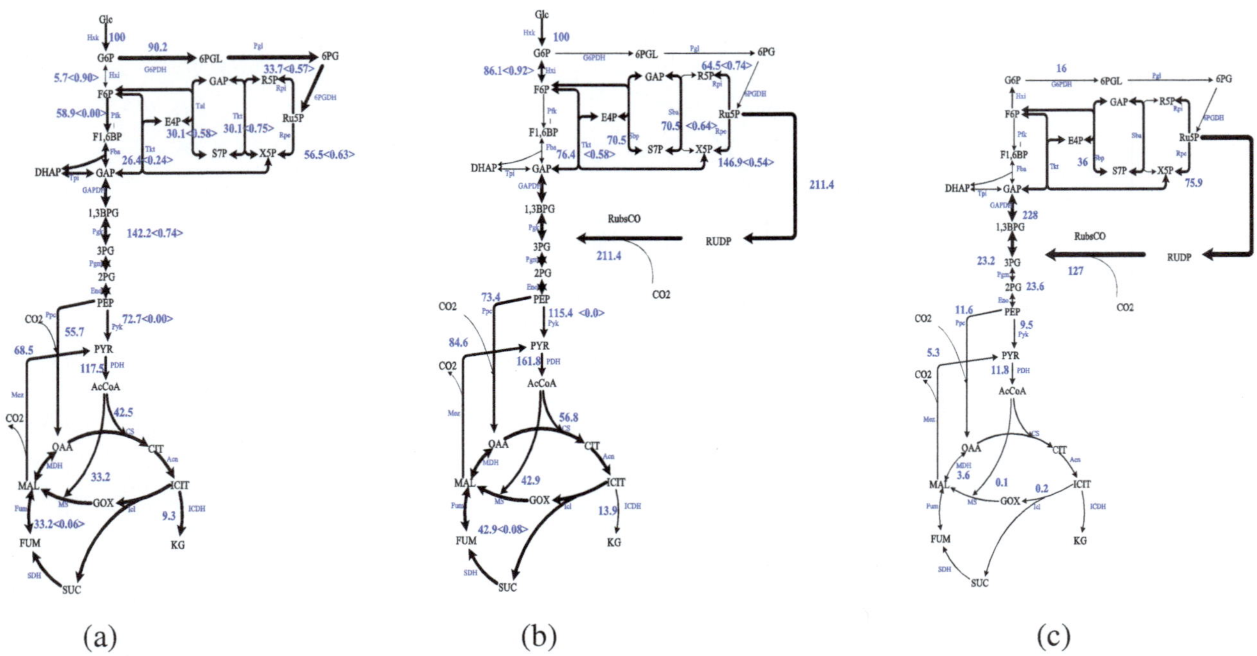

Figure 3 **^{13}C-Metabolic flux distribution of *Synecocystis* sp.** PCC6803 cultivated under heterotrophic **(a)**, mixotrophic **(b)**, and **(c)** autotrophic conditions. The flux values were obtained from [73] for (a) and (b) and from [77] for (c).

sources [85-87]. Nitrogen compounds are eventually converted to ammonium and assimilated for biosynthesis via the glutamin synthetase (GS)-glutamine oxoglutarate aminotransferase or glutamate synthase (GOGAT) cycle, where glutamate dehydrogenase (GDH) pathway does not function, probably due to low affinity to ammonium.

For survival under nitrogen starvation, cyanobacteria accumulate reserve materials in the form of inclusions and granurs, where the induction for their accumulation is made upon high light or CO_2, nutrient starvation as well as addition of arginine or chloramphenicol [88]. Cyanophycin (multi-*l*-arginyl-poly-[L-aspartic acid]) is a nitrogen reserve and is a non-ribosomally synthesized peptide consisting of equi-molar quantities of alginine (Alg) and aspatic acid (Asp), where cyanobacteria may consume internal storage compounds such as cyanophycin as nitrogen source upon nitrogen starvation [88,89].

After cyanophycin is exhausted, cells degrade the phycobilisomes that are large protein-rich right-harvesting antennae attached to the outside of the thylakoid membranes and support the light-dependent reactions of photosynthesis [90], where it is composed of rod and core proteins to provide nitrogen, which leads to a color change of cells from blue-green to yellow-green, known as bleaching [90]. Upon availability of nitrogen source again, cyanophycin is immediately synthesized [89].

Nutrient balance is important for the cell growth, since proteins, nucleic acids, carbohydrates, lipids, and pigments must be supplied in a suitable ratio for the balanced growth. In eukaryotic microalgae, autophagy is induced by nitrogen starvation to degrade cytoplasmic components including plastids in the large vacuoles [91]. In cyanobacteria, a unique Nb1A-dependent mechanism is induced to degrade certain phycobiliproteins, where the non-bleaching phenotype gene, *nblA* plays an important role for the degradation of phycobiliprotein [92,93]. The phycobilisome has a role in nitrogen storage as well as photosynthetic antenna [90]. Moreover, NblA1/A2-dependent protein turnover contributes to the maintenance of many amino acids (AAs) in NblA1/A2-dependently, while Lys pool markedly increased under sulfur starvation in cyanobacteria [94].

The internal C/N ratio is sensed by the PII protein, GlnB, in particular under N-limitation [95]. The global nitrogen regulator NtcA plays important roles for nitrogen regulation, where it senses αKG levels and regulates the genes involved in nitrogen assimilation. NtcA directly regulates the expression of *nrrA* gene which encodes a nitrogen-regulated response regulator of the OmpR family. NrrA is involved in induction of sugar catabolic genes as well under nitrogen starvation [96]. NrrA also regulates glycogen catabolism in *Anabaena* sp. by directly regulating expression of *glgP* gene encoding glycogen phosphlylase and *sigE* gene encoding a group 2 σ factor of RNA polymerase [97]. Nrr controls cyanophycin accumulation and glycogen catabolism in cyanobacteria [98], where glycogen is accumulated, whereas the expression of sugar catabolic genes is widely upregulated under nitrogen starvation [99].

Microalgae produce certain biomass compounds under nutrient limitation [100]. In cyanobacteria such as *Synechocystis* sp., polyhydoxy butyric acid (PHB), one of polyhydroxy alcanoate (PHA), is accumulated under nitrogen or phosphate starvation, where PHB is formed from AcCoA via β-keto-thiolase (PhaA), acetoacetyl-CoA reductase (PhaB), and PHA synthase (PhaC), where PHA synthase is activated by acetyl phosphate (AcP) [101]. Since acetoacetyl-CoA reductase requires NADPH, the pathway modification that causes excess NADPH yields higher PHB production [102].

Cyanobacteria have nine sigma factors such as SigA-I, where RNA polymerase sigma factor SigE plays important roles under nitrogen starvation [103]. SigE activates the expression of the genes associated with degradation of glycogen and catabolic genes of glycolysis and PP pathway [104]. Moreover, SigE activates PHB synthetic pathway gene expression [105], and thus the overexpression of *sigE* allows higher PHB production under nitrogen limitation [106]. Moreover, SigE also activates the expression of hydrogenase gene, and thus the overexpression of *sigE* also allows higher hydrogen production under anaerobic condition [107].

Cyanobacteria have two types of sunscreen pigments such as scytonemin and mycosporine-like amino acids, where these secondary metabolites play roles against environmental stresses such as UV radiation and desiccation [108].

Systems biology approach and modeling of the metabolism

Although algae and cyanobacteria have been paid recent attention for the potential to the sustainable biosynthesis, unknown and uncharacterized gene and protein functions hamper the progress toward the future era of algae industrial biotechnology. The systems biology approach plays a crucial role for function prediction based on the database with proper metabolic modeling [109].

Some attempts have been made for the modeling of photosynthetic organisms [110,111], while mechanistic model of photosynthesis in microalgae has also been developed [112,113]. The sequential statistical analysis based on experimental design coupled with least squares multiple regression has been made to analyze the dependence of respiratory and photosynthetic responses upon concomitant modulation of light intensity as well as acetate, CO_2, nitrate, and ammonia concentrations in the culture of *C. reinhardtii* [114].

MFA may be considered to gain insight into the metabolism, where the optimal light intensity can be identified for the biomass yield of *C. reinhardtii* by considering the cell maintenance and biomass formation [115]. A mixed integer linear programming method was used to find the optimal flux distributions of *C. reinhardtii* cultivated under photoautotrophic conditions in photobioreactors functioning in physical light limitation based on the constraint-based model, which includes thermodynamic and energetic constraints on the functioning metabolism, highlighting the existence of a light-driven respiration depending on the incident photon flux density [116].

Flux balance analysis (FBA) based on the network consisting of 484 metabolic reactions and 458 intracellular metabolites for *C. reinhardtii* indicates that aerobic heterotrophic growth on acetate has a low yield on carbon, while mixotrophically and autotrophically grown cells are significantly more carbon efficient [117]. A genome-scale extension for *C. reinhardtii* has been made with the network consisting of 1,080 genes, associated with 2,190 reactions and 1,068 metabolites (named *i*RC1080), that enables quantitative growth prediction for a given light source, resolving wavelength and photon flux. This offers insight into algae metabolism and potential for genetic engineering and efficient light source design [118]. Another comprehensive literature-based genome-scale model with the network of 866 ORFs, 1,862 metabolites, 2,249 gene-enzyme-reaction-association entries, and 1,725 reactions has been developed (named AlgaGEM), where it predicted observable metabolic effects under autotrophic, heterotrophic, and mixotrophic conditions, and predicts increased hydrogen production when cyclic electron flow is disrupted, and the physiological pathway for H_2 production, which identified new targets for further improvement of H_2 yield [119].

FBA approach has also been employed for cyanobacteria with the emphasis on the alleged glyoxylate shunt and the role of photorespiration in cellular growth and analyzed the diurnal light/dark cycles of the metabolism [120]. Genome-scale metabolic model of *Synechococcus elongates* PCC7942 (named *i*Syf715) has also been developed with the network of 851 reactions and 838 metabolites, and the applicability has been demonstrated for autotrophic growth conditions [121].

Metabolic modification for various biofuel and biochemical production

Algae have the potential for the genetic modification of their lipid pathways by upregulation of fatty acid biosynthesis or by downregulation of β-oxidation. By knocking out or modifying enzymes responsible for the synthesis of polyunsaturated lipids in the cell, it may be possible to dramatically increase the production of mono-unsaturated lipids [122]. Under optimal growth condition, the wild-type *Chlamydomonas* strains accumulate very low amount of oil (<1 μg per 10^6 cells) [45]. When cells are subjected to nitrogen starvation, oil content can be increased more than tenfold (up to 10 μg per 10^6 cells) [43-45]. Intracellular TAG amounts also fluctuate during the diurnal cycle because TAGs produced during the day provide a carbon and energy source for the night [123]. This is the major factor to yield loss in open pond microalgae cultivation. Researchers have not been able to achieve efficient homologous recombination in the nuclear genome of the commonly transformed laboratory algal strain *C. reinhardtii*, but this is not the case for the marine alga and biofuel candidate *Nannochloropsis* [124].

As a result of genetic engineering, some obligate photoautotrophs, formerly unable to partake in a sweet diet, have been given a taste of heterotrophy through the introduction of hexose transporters [125]. In the starchless mutant of *Chlamydomonas*, the flux through starch production redirects to lipid accumulation under nitrogen-starved condition. The strain should be designed to switch off completely for starch accumulation under nitrogen-starved condition. Blocking oil turnover processes might help increase the level of oil accumulated, as was observed in *Arabidopsis* leaves where the oil content was increased tenfold by knocking out a lipase gene [126].

Most metabolic engineering investigations have been made using such typical model organisms as *Synechocystis* sp. and *Synechococcus elongatus* sp. as well as *Aanabaena* sp., whereas much more complex genetic engineering is required for algae [16].

Bioethanol can be produced by introducing pyruvate decarboxylase (PDC) and alcohol dehydrogenase (ADH) genes from *Z. mobilis* into the shuttle vector and then transform *Synechosystis* sp. [127,128] (Table 1).

As shown in Figure 4, 2-ketoisovalerate can be converted to isobutyraldehyde by introducing ketoacid decarboxylase gene *kivd* from *Lactococcus lactis* into the genome of *S. elongatus* PCC7942 [61]. The flux to 2-ketoisovalerate can be improved by introducing *alsS* gene from *Bacillus subtilis* and the *ilvC* and *ilvD* genes from *E. coli* into the chromosome of *S. elongatus*. Carbon fixation can be improved by overexpression of RubisCO genes to increase the productivity of isobutyraldehyde. Iso-butanol can then be produced by introducing alcohol dehydrogenase YqhD from *Escherichia coli* [61].

2-Methyl-1-butanol (2 MB) can be produced in *S. elongatus* by expressing heterologous enzymes for citramalate pathway [129] to direct pyruvate toward isoleucine sysnthesis pathway, where 2-keto-3-methyvalerate is converted to 2 MB by Kivd and YqhD [130] (Table 1).

1-Butanol can be produced from AcCoA by anaerobic dark condition by engineered *S. elongatus* by introducing acetyl transferase (AtoB) from *E. coli*, 3-hydroxybutyryl-CoA dehydrogenase (Hbd) from *Clostridium acetobutylicum*, trans-2-enoyl-CoA reductase (Ter) from *Treponema denticola*, crotonase (Crt) from *C. acetobutylicum*, and bifunctional aldehyde/alcohol dehydrogenase (AdhE2) from *C. acetobutylicum* [131]. 1-Butanol can be also produced by aerobic culture of *S. elongates* PCC7942, where condensation of AcCoA is made by consuming ATP with CO_2 evolution [132], where ATP-dependent malonyl-CoA synthesis enzyme NphT7 was introduced, and NADH-dependent enzymes were replaced by NADPH-dependent enzymes in this strain. Butanol tolerance of *Synechocystis* can be improved by 150% by evolution by gradually increasing butanol concentration from 0.2% to 0.5% (*v*/*v*) [133] (Table 1).

Lactic acid has been used in the food and pharmaceutical industries and for biodegradable polymers [134], and this can be produced in *Synechosystis* after heterologous expression of LDH [135,136]. Since Pyk is inhibited under light condition in *Synechosystis*, heterologous expression of Pyk can enhance the pyruvate production and in turn enhance the lactate production, where Pyk is allosterically activated by fructose 1,6-bisphosphate (FBP) in the case of Pyk-F originated form *E. coli*, while the original Pyk does not show such characteristics [136]. Moreover, Ppc may be knocked down to direct the carbon flow from PEP towards lactate production via PYR, but the cell growth is depressed, since Ppc is also an important pathway for CO_2 fixation [136]. In most bacteria, LDH requires cofactor NADH, whereas NADPH is abundant in *Synechosystis* as mentioned before, and thus NADPH-dependent LDH may increase the lactate production, where this may be partly attained by introducing the LDH from *B. subtilis*, of which LDH co-utilizes NADH and NADPH [136,137] (Table 1).

Isoprene is a volatile compound and utilized in the synthesis of rubber etc., where isoprene is easily evaporated from the culture broth, and thus the toxicity to the cell can be relaxed by evaporation, where it can be trapped in the gas phase. Isoprene can be synthesized by *Synechocystis* sp. PCC6803 by introducing *lspS* gene from vine *Pueraria montana* and utilizing the naturally occurring methyl-erythritol-4-phosphate (MEP) pathway (Figure 4) [138] (Table 1).

Ethylene is another volatile compound, where this can be also produced by *Synechocystis* sp. PCC6803 by introducing ethylene forming pathway gene *efe* from *Pseudomonas syringae pv phaseolicola* (Figure 4) [139] (Table 1).

Fatty acid and fatty alcohol production can be made by *Synechocystis* sp. PCC6803 by overexpression of endogenous fatty acyl-ACP synthase gene *slr1609* (Figure 4) [140]. Only fatty acids can be produced and excreted outside of the cell by modification of *Synechocystis* sp. PCC6803 [141,142]. Fatty alcohols such as hexadecanol and octadecanol can be produced by introducing fatty acyl-CoA reductase genes from jojoba, which catalyze a fatty-acyl-ACP (Figure 4) [143] (Table 1).

Eucaryotic algae have also been considered for fatty acid production, where they can accumulate lipids up to about 70% of dry biomass [41,144]. The limitations of using algae are the complexity of the eukaryotic system and less available genetic tools, although some attempts have been done [145]. Some efforts are being made to

Table 1 Biofuel and biochemical production by cyanobacteria

Product	Species	Titer or productivity	Overexpressed or knockout gene (s)	Cultivation	Reference
Ethanol	*Synechococcus*	230 mg/L in 28 days	*pdc, adh*	Shake flask	[210]
	Synechocystis	552 mg/L in 6 days	*pdc, adh*	Photobioreactor	[127]
	Synechocystis	608 mg/L in 18 days	*pdc, adh*	Photobioreactor	[128]
Isobutyraldehyde	*Synechococcus*	1,100 mg/L in 8 days	*alsS, ilvC,D, kivd,rbcls*	Bottle with NaHCO3	[61]
Isobutanol	*Synechococcus*	18 mg/L	*kivd, yqhD*	Shake flask with NaHCO3	[61]
	synechococcus	450 mg/L in 6 days	*alsS, ilvC, D, kivd, yqhD*	Shake flask with NaHCO3	[61]
2 Methyl-1-butanol	*Synechococcus*	2 mg/L	*kivd, yqhD, cims*	Shake flask with NaHCO3	[61,130]
1-Butanol	*Synechococcus*	14.5 mg/L in 7 days	*hbd, crt, adhE2, ter, atoB*	Bottle under anoxic cond.	[131]
	Synechococcus	30 mg/L in 18 days	*ter, nphT7, bldh, yqhD, phaJ, B*	Shake flask	[132]
Fatty alcohol	*Synechocystis*	0.2 mg/L in 18 days	*far*	Photobaioreactor with 5% CO_2	[143]
	Synechocystis	0.02 mg/L/OD	*far,aas*	Shake flask	[140]
	Synechocystis	2.87 mg/gDCW	*Δsll0208, Δsll0209*	Flask	[211]
Fatty acids	*Synechocystis*	197 mg/L in 17 days	*tesA, accBCDA, fatB1, B2, tesA137*	1% CO_2 bubbling	[142]
Alka (e) nes	*Synechocystis*	0.162 mg/L/OD	*accBCDA*	Shake flask	[143]
	Synechocystis	2.3 mg/l/OD	*sll0208, sll0209*	Shake flask	[212]
Hydrogen	*Synechococcus*	2.8 μmol/h/mgChlorophyll-a	*hydEF, hydG, hydA*	Anaerobic condition	[153]
	Synechococcus	54 mmol/1,017 cells in 4 days	*Δldh*	Anoxic condition	[213]
L-Lactate	*Synechocystis*	0.0178 mmol/gDCW/h	*ldh, sth*	Shaking incubator	[214]
	Synechocystis	0.2512 mmol/gDCW/h	*pyk, ldh*	Shaking incubator	[136]
D-Lactate	*Synechocystis*	2.17 g/L in 24 days	*gldA, sth*	Photoautotropic with acetate	[215]
1,2-propanediol	*Synechococcus*	Approximately 150 mg/L	*mgsA, gldA, yqhD*	Shake flask	[216]
Isoprene	*Synechocystis*	50 μg/gDCW/d	*IspS*	Sealed culture	[138]
Ethylene	*Synechocystis*	26 μmol/gDCW/h	*efe* (RS1010)	Rotary shaker	[217]
	Synechocystis	111.6 μmol/gDCW/h	*efe* (slr068)	Rotary shaker	[217]
	Synechococcus	84.8 μmol/gDCW/h	*efe* (pUC303)	Flask	[218]
	Synechococcus	80.5 μmol/gDCW/h	*efe* (psbAI)	Flask	[219]
Acetone	*Synechocystis*	36.0 mg/L in 4 days	*ctfAB, adc, ΔphaCE, Δpta*	Flask	[148]
PHAs	Synechocystis	1.4 mg/100 mgDCW	*sigE*	Bubbled with 1% CO_2 in the air	[106]
	Synechocystis	533.4 mg/L in 21 days	*Δslr1829, Δslr1830*	Flask	[220]

identify stress conditions and key enzymes for fatty acid synthesis in the green algae *Haematococcus pluvialis* [146]. Effects of light conditions on fatty acid production were also investigated for *Nannochloropsis* [147].

Acetone can be used as the precursor for isopropanol. This can be produced in *Synechocystis* PCC6803 by introducing CoA transferase (CtfAB) and acetoacetate decarboxylase (Adc) from *C. acetobutylicum* for converting AcCoA to acetone (Figure 4) [148], where the PHB-forming pathway genes *phaCE* and the acetate-forming pathway gene *pta* may be disrupted. Alpha-olefin production can be also made by *Synechococcus* sp. PCC7008 [149] (Table 1).

Methane can be produced by co-culture of *C. reinhardtii* and methanogenic bacteria, where glyconate is produced from the former, while methane is produced from the latter by assimilating glyconate [150]. *Rhodobacter* are non-sulfur photosynthetic bacteria that produce hydrogen (H_2) from acetate etc. Hydrogen production may be enhanced by introducing aldehyde dehydrogenase (ALDH) gene from *Rhodospirillum rubrum* into *Rhodobacter sphaeroides* [151]. Hydrogen production can be enhanced by introducing exogenous hydrogenase into nitrogen fixing [152] and non-nitrogen fixing cyanobacteria [153]. Hydrogen production can be improved by introducing hydrogenase from *Clostridium thermocellum* into *Rhodopseudomonas*

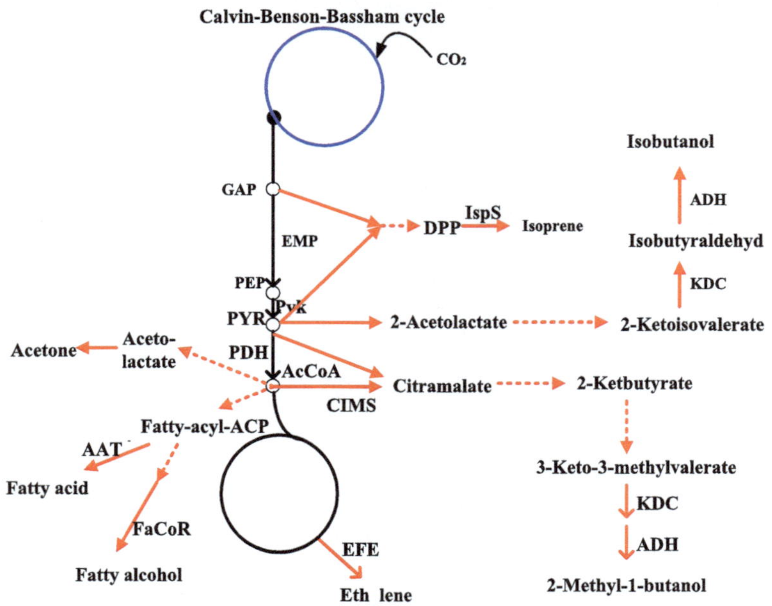

Figure 4 A variety of metabolic pathways for the production of biofuels and biochemicals.

palustris CGA009 and cultivated at 38°C by considering the outdoor bioreactors [152]. Hydrogen can be produced by introducing a hydrogenase HydA from *C. acetobutylicum* into *S. elongates* [153], and its production can be significantly improved in *Arthosporia* by continuously removing it from the culture broth [154].

Another important aspect of utilizing photosynthetic organisms is their ability of producing pharmaceuticals due to the reduction of ketones [155-157]. Microalgae are also attractive for their production of antioxidants, where *Fischerella ambigua* and *Chlorella vulgaris* show higher antioxidant activities [158]. Phenolic compounds have also antioxidant properties, where their production can be enhanced in *Spirulina platensis* by manipulating light intensities [159].

Phycocyanin is also attractive, where its production by *Arthospira* (*Spirulina*) *platensis* was investigated in marine environment [160]. The pigment sesquiterpene β-caryophyllene can be produced by *Synechocystis* sp. PCC6803 with the aid of a β-caryophyllene synthase gene from *Artemisia annua* [161], where this compound is used in the fragrance and cosmetic industry, and in natural remedies for its anti-inflammatory and anti-microbial properties. Lycopene is also important food additives and pigment, and can be produced by a purple non-sulfur bacterium, *R. rubrum*, by deletion of downstream phytoene desaturase gene *crtC* and *crtD* [162].

Cyanobacteria can excrete fructose, lactate, and glucose by introducing transport genes from *E .coli* [134]. Many cyanobacteria naturally produce sucrose as an osmotic response to their saline habitat together with manipulation of transport and secretion genes [163].

The biodegradable plastic such as PHB can be produced by *Rhodovulum sulfidophilum* P5 using inexpensive nitrogen and carbon sources [164]. PHB can be also produced in the filamentous cyanobacterium *Nostoc muscorum* under phosphate limitation by recombinant *Synechocystis* sp. PCC6803 [165] (Table 1). Glycogen can be produced in halophilic bacterium *A. palentensis* by manipulating growth condition [166]. Even ammonia can be produced by nitrogen-fixing cyanobacteria *Anabaena* sp. ATCC 33047 [167].

Cultivation and harvesting methods

Raceway pond

Microalgae can use sunlight more efficiently than other crop plants to produce oil [168]. The oil production capacity is almost one or two times higher than any other crop [169]. The open pond system is better for large-scale production. The main disadvantage of open pond systems is that by being open to the atmosphere, they lose water by evaporation at a rate similar to land crops and are also susceptible to contamination [122,170]. Some protozoa may contaminate the system and hamper the growth of microalgae.

An effective culture system may consist of the following criteria: (1) effective illumination area, (2) optimal gas-liquid transfer, (3) easy to operate, (4) low contamination level, (5) low capital and production cost, and (6) minimal land area requirement [171]. The system can be made of paddle wheel to avoid microalgae biomass sedimentation, and CO_2 may be sparged at the bottom of the raceway as carbon source [172].

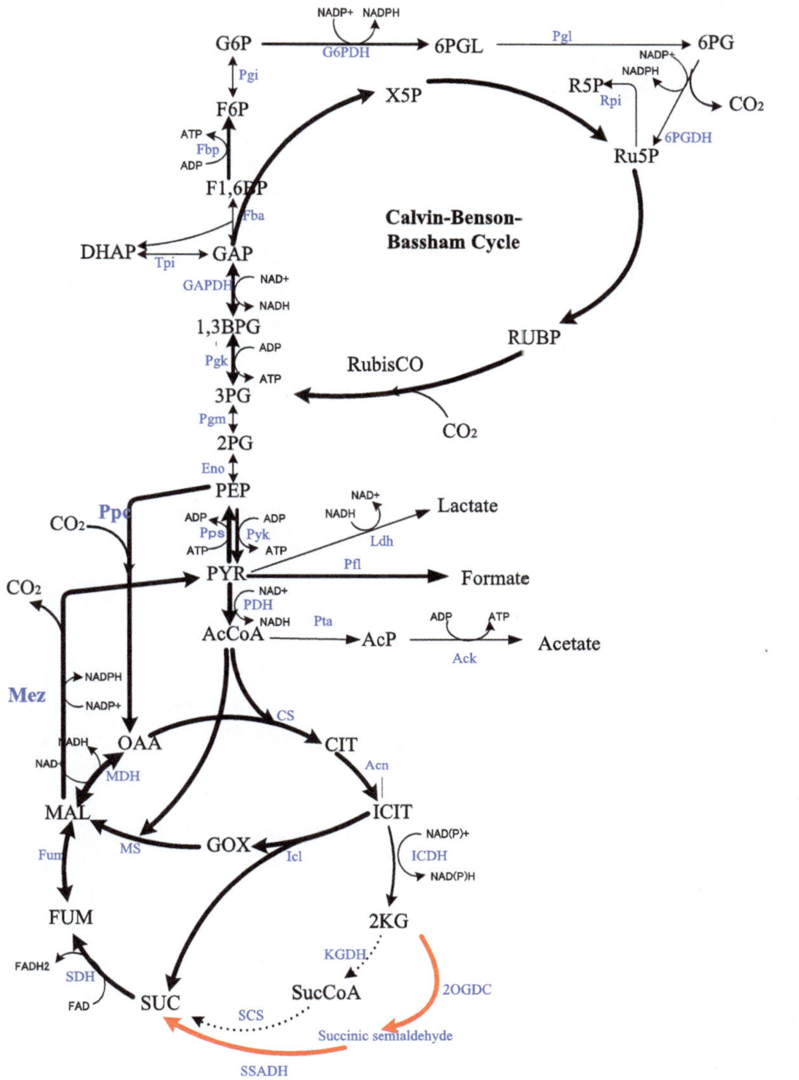

Figure 5 TCA cycle in cyanobateria.

Photobioreactor

Closed photobioreactor may be considered to overcome the limitations encountered in raceway pond [170]. There are several designs of closed photobioreactor such as air-lift tubular, flat plate, and vertical-column reactors and culture parameters such as nutrient levels, temperature, amount of inlet CO_2, etc. [173]. The reactor permits selective culture strain, in which optimum growth condition can be maintained to give high biomass and lipid productivity. The tubuler photobioreactor may be one of the most typical cultivation apparatus, where a vertical tubuler photobioreactor can increase the residence time of sparged gas, giving higher CO_2 utilization efficiency [174]. The higher intensity of light cannot reach to most of the cell in the large-scale photobioreactor. As a consequence, the metabolism also changes from light to dark condition. This phenomenon is undesirable for large-scale production.

Harvesting of algal biomass

The microalgae need to be separated from water to recover their biomass for downstream processing. There are several methods for microalgae harvest: (1) bulk harvesting- to separate microalgae from suspension such as natural gravity sedimentation, flocculation, and floatation, and (2) thickening to concentrate the microalgae slurry after bulk harvesting such as centrifugation and filtration [169].

Flocculation, the aggregation and sedimentation (or flocculation) of algal biomass, is also a very common primary harvesting method used to concentrate algae. Algal strains can also be engineered such that the addition of a polymer or a change in an environmental variable triggers flocculation [175,176].

Conventional flocculation method poses several disadvantages: (1) high dosage of multivalent slat is required to achieve satisfactory result, (2) it produces large quantity of sludge that increases the difficulty to dehydrate the biomass, and (3) flocculation efficiency is highly dependent on pH level [169,177]. By introduction of a coagulant that is positively charged into the culture medium, the negative charge surrounding the microalgae cells will be neutralized. At the same time, flocculant can be added to promote agglomeration by creating bridges between the neutralized cells to become dense flocs and settle down due to natural gravity [178]. Another possible method to harvest microalgae is through immobilization, in which microalgae are embedded in an entrapment matrix and continuously grow within the matrix. Only alginate gel entrapment method is feasible to immobilize microalgae so far [179]. Some of the advantages of using alginate gel are the requirement of only mild condition during immobilization process with negligible toxicity and high transparency [179,180]. Immobilized microalgae beads can be applied in diverse research areas such as for high-value product synthesis, organic pollutant removal, heavy metal removal, and toxicity measurement (biosensor) [180-186]. A few issues need to be addressed in immobilization of microalgae before the process can be upgraded to commercialization stage such as (1) stability, (2) leakage of microalgae cells, and (3) mass transfer limitation.

Downstream processing

Oil extraction

Effective lipid extraction is required particularly for microalgae with low lipid content, since the loss of the lipid during extraction process brings a serious problem for the production cost of microalgae biofuels [187]. The energy consumed in lipid extraction from dried microalgae biomass is a relatively small portion to the overall energy [172,188]. Various cell disruption methods are microwave application, sonication, bead beating, autoclaving [189,190], grinding, osmotic shock, homogenization, freeze drying [189], and 10% (*w/v*) NaCl addition [190,191].

Solvent extraction method

The solvent must be inexpensive, nontoxic, volatile, nonpolar, and it must selectively extract the lipid of the cell [189]. The potential of using co-solvent mixtures of ionic liquids and polar covalent molecules has been shown for lipid extraction [192]. The Soxhlet extraction method uses hexane, while the Bligh and Dyer's method uses mixture of chloroform and methanol as solvents to extract lipids [193]. The other solvents include benzene and ether, but hexane has gained more popularity as a chemical for solvent extraction, and it is also relatively inexpensive [194]. Although n-hexane is widely used to extract oil from various seed crops, it is inefficient to extract microalgae lipid. This is because microalgae lipid contains high concentration of unsaturated fatty acid, while n-hexane is a nonpolar solvent. Thus the selectivity of lipid towards the solvent is reduced [187]. Methanol and n-hexane are not sustainable, since both solvents are conventionally derived from nonrenewable fossil fuels. On the other hand, ethanol is a greener solvent, since it has a low toxicity level and can be derived from renewable source such as sugar-based plant (e.g., sugar cane and sweet sorghum) and lignocellulosic material (e.g., weed and corn stover) [170]. The ethanol, however, gives low extraction efficiency. Ultrasonication and microwave can be also used for cell disruption. The cell wall-less mutant is better for oil extraction.

Super critical fluid extraction

Several supercritical fluids are CO_2, ethane, methanol, ethanol, benzene, toluene, and water [195,196]. The basic principle of this technology is by achieving a certain phase (supercritical) that is beyond the critical point of a fluid, in which meniscus separating the liquid and vapor phase disappears, leaving only a single homogeneous phase [196]. Supercritical CO_2 has received much interest typically in extraction of pharmaceutical and health-related products from microalgae [197-200]. In fact, supercritical CO_2 offers several advantages in comparison with chemical solvent extraction such as (1) nontoxic and provide nonoxidizing environment to avoid degradation of extracts, (2) low critical temperature (around 31°C) which prevents thermal degradation of products, (3) high diffusivity and low surface tension which allow penetration of pores smaller than those accessible by chemical solvents, and (4) easy separation of CO_2 at ambient temperature after extraction [195,197,200].

Transesterification of oils

The most suitable catalyst for transesterification of oils with low free fatty acids (FFA) content is necessary. The presence of high free fatty acid content in microalgae lipid (more than 0.5% *w/w*) prevents the use of homogeneous base catalyst for transesterification reaction [201-203]. Alkaline metal alkoxides, even in small concentration of 0.5 mol %, are highly active catalysts [194]. Metal alkoxides (e.g., potassium methoxide) in methanol are better options than metal hydroxides (NaOH, KOH). In a short reaction time of about 30 min, they give high yields of about 98% [194]. They performed better in absence of water, which makes them inappropriate for industrial processes [204]. FFA will react with base catalyst to form soap leading to lower biodiesel yield and increase the difficulty to separate biodiesel from glycerol.

Acid catalyst (e.g., H_2SO_4) is not sensitive towards FFA level in oil, and thus esterificaiton (FFA is converted to alkyl ester) and transesterification can occur simultaneously. A chemically catalyzed transesterification process requires a high amount of energy, and separation of catalyst from the product is cost effective. Glycerol, produced as a byproduct of alcoholysis, readily adheres to the surface of immobilized lipase and decreases its enzyme activity, where glycerol removal is a complex process, which may hinder the large-scale operations [189].

Free fatty acids contained in extracted oil have to be removed prior to reaction in order to maintain activity of the alkaline catalysts. In the superheated method, free fatty acid and triglycerides are converted to fatty acid methyl ester directly. Academic and commercial enterprises have realized the importance of such research on this topic, and a direct transesterification method for total microalgal lipid content has produced appreciable levels of biodiesel even when there are undetectable levels of neutral lipids [205,206]. The advantages of non-catalytic alcoholysis reaction for the production of biodiesel are as follows: (1) the purification process to remove catalyst after reaction is not required, (2) the by-product (glycerol) can be directly utilized in other industry, (3) not only the triglycerides but also the free fatty acid might be converted into fatty acid methyl ester, (4) neutralization process for removal of free fatty acid is not required prior to the reaction process, and (5) the yield of the total system will be improved, and the cost required for the process will be reduced.

Improvement of product recovery

In a large-scale operation, the presence of microorganisms, medium composition, and process condition may cause emulsion formation, which lowers the product recovery efficiency. A better understanding on the mechanism of emulsion formation is necessary for the performance improvement based on nanotechnology as well as electrochemical properties such as ion, charge, viscosity, interface stabilization [207].

Conclusions

The fluctuation in global prices of crude oil, increasing threats to the environment by exhaust emissions, global warming, and threats of supply instabilities have adversely impacted the developing countries, more so to the petroleum-importing countries. The rising of sea level caused by green-house gas (CO_2) also threatens the most populated areas of the world. It is important to find a safe alternative fuel to relieve the escalating energy crisis and to protect the environment. Photosynthetic microorganisms have emerged as one of the most promising sources for biodiesel production.

Although algae and cyanobacteria have been paid recent attention from the point of view of sustainable biosynthesis as well as biofuel and biochemical production, the cell growth rate is significantly lower as compared to the typical biofuel-producing microorganisms such as *E. coli* and yeast. It is, therefore, strongly desirable to design microbial cell factories by means of a synthetic biology approach with in-depth understanding of the metabolic regulation mechanism with the aid of a systems biology approach such as modeling.

Several attempts are being made for improving the efficiency of capturing light energy and CO_2 fixation as mentioned before. One of the reasons of the lower cell growth rate in cyanobacteria may be due to an incomplete TCA cycle lacking KGDH and succinyl CoA synthetase (SCS). Recent investigation on *Synecococcus* sp. PCC7002 indicates that the genes encoding αKG decarboxylase (or 2-oxoglutarate decarboxylase) and succinic semialdehyde dehydrogenase are present, where NADPH instead of NADH is produced without producing guanosine triphosphate (GTP) by substrate level phosphorylation (Figure 5) [208]. It is important to elucidate the nature of such TCA cycles in cyanobacteria and plants from the point of view of functional significance of the metabolic feature in a broader evolutionary context [209]. Further investigation is necessary to improve the cell growth rate with balanced energy generation and biosynthesis.

Competing interests
The authors declare that they have no competing interests.

Authors' contributions
DS and KS investigated the references and drafted the manuscript. Both authors read and approved the final manuscript.

Acknowledgements
The authors would like to acknowledge Sheuli Bala for her helping during typing the manuscript.

Author details
[1]Department of Microbiology, Primeasia University, Banani, Dhaka 1213, Bangladesh. [2]Kyushu Institute of Technology, Iizuka, Fukuoka 820-8502, Japan. [3]Institute of Advanced Bioscience, Keio University, Tsuruoka, Yamagata 997-0017, Japan.

References
1. Canadel JG, Schulze ED (2014) Global potential of biospheric carbon management for climate mitigation. Nature Commun 5:5282
2. Schmidt LD, Dauenhauer PJ (2007) Chemical engineering: hybrid routes to biofuels. Nature 447:914–915
3. Rodolfi L, Chini Zittelli G, Bassi N, Padovani G, Biondi N, Bonini G et al (2009) Microalgae for oil: Strain selection, induction of lipid synthesis and outdoor mass cultivation in a low-cost photobioreactor. Biotechnol Bioeng 102(1):100–112
4. Jones CS, Mayfield SP (2012) Algae biofuels: versatility for the future of bioenergy. Curr Opin Biotechnol 23(3):346–351
5. Scott SA, Davey MP, Dennis JS, Horst I, Howe CJ, Lea-Smith DJ et al (2010) Biodiesel from algae: challenges and prospects. Curr Opin Biotechnol 21(3):277–286
6. Smil V (2005) Energy at the Crossroads: Global perspectives and uncertainties. MIT press.
7. Larkum AWD, Ross IL, Kruse O, Hankamer B (2012) Selection, breeding and engineering of microalgae for bioenergy and biofuels production. Trends Biotechnol 30(4):198–205
8. Stephens AI (2010) Influence of nitrogen-limitation regime on the production by *Chlorella vulgaris* of lipids for biodiesel feedstocks. Biofuels 1:47–58

9. Larkum AWD (2010) Limitations and prospects of natural photosynthesis for bioenergy production. Curr Opin Biotechnol 21:271–276
10. Zhu XG, Long SP, Ort DR (2010) Improving photosynthetic efficiency for greater yield. Annu Rev Plant Biol 61:235–261
11. Beer C, Reichstein M, Tomelleri E, Ciais P, Jung M, Carvalhais N et al (2010) Terrestrial gross carbon dioxide uptake: global distribution and covariation with climate. Science 329:834–838
12. Yuan L, Grotewold E (2015) Metabolic engineering to enhance the value of plants as green factories. Metab Eng 27:83–91
13. Overmann J, Garcia-Pichel F (2006) The photosynthetic way of life. In: Dworkin M, Falkow S (eds) The procaryotes: A handbook on the biology of bacteria, vol 2. Springer, NewYork, pp 32–85
14. Shimizu K (2014) Biofuels and biochemical production by microbes, Nova Publ. Co.
15. Sheehan J (2009) Engineering direct conversion of CO_2 to biofuel. Nat Biotechnol 27:1128–1129
16. Savakis P, Hellingwerf KJ (2015) Engineering cyanobacteria for direct biofuel production from CO_2. Curr Opin Biotechnol 33:8–14
17. Yu J (2014) Bio-based products from solar energy and carbon dioxide. Trends Biotechnol 32(1):5–10
18. Khan SA, Hussain MZ, Prasad S, Banerjee UC (2009) Prospects of biodiesel production from microalgae in India. Renew Sustain Energy Rev 13(9):2361–2372
19. Chisti Y (2007) Biodiesel from microalgae. Biotechnol Adv 25:294–306
20. Mata TM, Martins AA, Caetano NS (2010) Microalgae for biodiesel production and other applications: a review. Renew Sustain Energy Rev 14:217–232
21. Georgianna DR, Mayfield SP (2012) Exploiting diversity and synthetic biology for the production of algal biofuels. Nature 488:329–335
22. Wijffels RH, Barbosa MJ (2010) An outlook on microalgal biofuels. Science 329:796–799
23. Boyle NR, Morgan JA (2009) Flux balance analysis of primary metabolism in *Chlamydomonas reinhardtii*. BMC Syst Biol 3(1):4
24. Chiu SY, Kao CY, Chen CH, Kuan TC, Ong SC, Lin CS (2008) Reduction of CO_2 by a high-density culture of *Chlorella sp.* in a semicontinuous photobioreactor. Bioresour Technol 99(9):3389–3396
25. Rosenberg JN, Mathias A, Korth K, Betenbaugh MJ, Oyler GA (2011) Microalgal biomass production and carbon dioxide sequestration from an integrated ethanol biorefinery in Iowa: A technical appraisal and economic feasibility evaluation. Biomass Bioenergy 35(9):3865–3876
26. Sydney EB, Sturm W, de Carvalho JC, Thomaz-Soccol V, Larroche C, Pandey A et al (2010) Potential carbon dioxide fixation by industrially important microalgae. Bioresour Technol 101(15):5892–5896
27. Yoo C, Jun SY, Lee JY, Ahn CY, Oh HM (2010) Selection of microalgae for lipid production under high levels carbon dioxide. Bioresour Technol 101(1):S71–S74
28. Zhao B, Zhang Y, Xiong K, Zhang Z, Hao X, Liu T (2011) Effect of cultivation mode on microalgal growth and CO_2 fixation. Chem Eng Res Des 89 (9):1758–1762
29. Rosenberg JN, Oyler GA, Wilkinson L, Betenbaugh MJ (2008) A green light for engineered algae: redirecting metabolism to fuel a biotechnology revolution. Curr Opin Biotechnol 19(5):430–436
30. O'Grady J, Schwender J, Shachar-Hill Y, Morgan JA (2012) Metabolic cartography: experimental quantification of metabolic fluxes from isotopic labelling studies. J Exp Bot 63(6):2293–2308
31. Dennis DT, Blakeley SD (2000) Carbohydrate metabolism. In: Buchanan BB, Gruissem W, Jones RL (eds) Biochemistry and Molecular Biology of Plants. American Society of Plant Physiologists, Rockville MD, pp 630–675
32. Sriram G, Fulton DB, Iyer VV, Peterson JM, Zhou R, Westgate ME et al (2004) Quantification of compartmented metabolic fluxes in developing soybean embryos by employing biosynthetically directed fractional ^{13}C labeling, two-dimensional [13C, 1H] nuclear magnetic resonance, and comprehensive isotopomer balancing. Plant Physiol 136(2):3043–3057
33. Sriram G, Iyer VV, Bruce Fulton D, Shanks JV (2007) Identification of hexose hydrolysis products in metabolic flux analytes: A case study of levulinic acid in plant protein hydrolysate. Metab Eng 9(5):442–451
34. Schwender J, Shachar-Hill Y, Ohlrogge JB (2006) Mitochondrial metabolism in developing embryos of Brassica napus. J Biol Chem 281(45):34040–34047
35. Allen DK, Shachar-Hill Y, Ohlrogge JB (2007) Compartment-specific labeling information in ^{13}C metabolic flux analysis of plants. Phytochem 68(16):2197–2210
36. Lonien J, Schwender J (2009) Analysis of metabolic flux phenotypes for two Arabidopsis mutants with severe impairment in seed storage lipid synthesis. Plant Physiol 151(3):1617–1634
37. Gerhardt R, Stitt M, Heldt HW (1983) Subcellular metabolite determination in spinach leaves through non-aqueous fractionation. Physiol Chem 364:1130–1131
38. Masakapalli SK, Le Lay P, Huddleston JE, Pollock NL, Kruger NJ, Ratcliffe RG (2010) Subcellular flux analysis of central metabolism in a heterotrophic Arabidopsis cell suspension using steady-state stable isotope labeling. Plant Physiol 152(2):602–619
39. Hu Q, Sommerfeld M, Jarvis E, Ghirardi M, Posewitz M, Seibert M et al (2008) Microalgal triacylglycerols as feedstocks for biofuel production: perspectives and advances. Plant J 54(4):621–639
40. Bigogno C, Khozin-Goldberg I, Boussiba S, Vonshak A, Cohen Z (2002) Lipid and fatty acid composition of the green oleaginous alga *Parietochloris incisa*, the richest plant source of arachidonic acid. Phytochem 60(5):497–503
41. Radakovits R, Jinkerson RE, Darzins A, Posewitz MC (2010) Genetic engineering of algae for enhanced biofuel production. Eukaryot Cell 9(4):486–501
42. Harris EH (2009) The *Chlamydomonas sourcebook:* introduction to *Chlamydomonas* and its laboratory use Vol. 1, Acad Press.
43. Wang ZT, Ullrich N, Joo S, Waffenschmidt S, Goodenough U (2009) Algal lipid bodies: stress induction, purification, and biochemical characterization in wild-type and starchless *Chlamydomonas reinhardtii*. Eukaryot Cell 8(12):1856–1868
44. Work VH, Radakovits R, Jinkerson RE, Meuser JE, Elliott LG, Vinyard DJ et al (2010) Increased lipid accumulation in the *Chlamydomonas reinhardtii* sta7-10 starchless isoamylase mutant and increased carbohydrate synthesis in complemented strains. Eukaryot Cell 9(8):1251–1261
45. Siaut M, Cuiné S, Cagnon C, Fessler B, Nguyen M, Carrier P et al (2011) Oil accumulation in the model green alga Chlamydomonas reinhardtii: characterization, variability between common laboratory strains and relationship with starch reserves. BMC Biotechnol 11(1):7
46. Moellering ER, Benning C (2010) RNA interference silencing of a major lipid droplet protein affects lipid droplet size in *Chlamydomonas reinhardtii*. Eukaryot Cell 9(1):97–106
47. Li Y, Han D, Hu G, Dauvillee D, Sommerfeld M, Ball S et al (2010) *Chlamydomonas* starchless mutant defective in ADP-glucose pyrophosphorylase hyper-accumulates triacylglycerol. Metab Eng 12(4):387–391
48. Merchant SS, Kropat J, Liu B, Shaw J, Warakanont J (2012) TAG, You're it! *Chlamydomonas* as a reference organism for understanding algal triacylglycerol accumulation. Curr Opin Biotechnol 23(3):352–363
49. Liu B, Benning C (2013) Lipid metabolism in microalgae distinguishes itself. Curr Opin Biotechnol 24(2):300–309
50. Cagnon C, Mirabella B, Nguyen HM, Beyly-Adriano A, Bouvet S, Cuiné S et al (2013) Development of a forward genetic screen to isolate oil mutants in the green microalga *Chlamydomonas reinhardtii*. Biotechnol Biofuels 6(1):178
51. Solovchenko AE, Khozin-Goldberg I, Didi-Cohen S, Cohen Z, Merzlyak MN (2008) Effects of light intensity and nitrogen starvation on growth, total fatty acids and arachidonic acid in the green microalga *Parietochloris incise*. J Appl Phycol 20(3):245–251
52. Harun R, Danquah MK, Forde GM (2010) Microalgal biomass as a fermentation feedstock for bioethanol production. J Chem Technol Biotechnol 85(2):199–203
53. Yacoby I, Pochekailov S, Toporik H, Ghirardi ML, King PW, Zhang S (2011) Photosynthetic electron partitioning between [FeFe]-hydrogenase and ferredoxin: NADP + –oxidoreductase (FNR) enzymes in vitro. PNAS USA 108(23):9396–9401
54. Park JH, Yoon JJ, Park HD, Kim YJ, Lim DJ, Kim SH (2011) Feasibility of biohydrogen production from *Gelidium amansii*. Int J Hydrogen Energy 36(21):13997–14003
55. Shi X, Jung KW, Kim DH, Ahn YT, Shin HS (2011) Direct fermentation of *Laminaria japonica* for biohydrogen production by anaerobic mixed cultures. Int J Hydrogen Energy 36(10):5857–5864
56. Peers G (2014) Increasing algal photosynthetic productivity by integrating ecophysiology with systems biology. Trends Biotechnol 32(11):551–555
57. Sage RF, Sage TL, Kocacinar F (2012) Photorespiration and evolution of C4 photosynthesis. Annu Rev Plant Biol 63:19–47
58. Wang L, Czedik-Eysenberg A, Mertz RA, Si Y, Tohge T, Nunes-Nesi A et al (2014) Comparative analysis of C_4 and C_3 photosynthesis in developing leaves of maize and rice. Nat Biotechnol 32(11):1158–1165

59. Ducat DC, Silver PA (2012) Improving carbon fixation pathways. Curr Opin Chem Bol 16:337–344
60. Savage DF, Afonso B, Chen AH, Silver PA (2010) Spatially ordered dynamics of the bacterial carbon fixation machinery. Science 327:1258–1261
61. Atsumi S, Higashide W, Liao JC (2009) Direct photosynthetic recycling of carbon dioxide to isobutyraldehyde. Nat Biotechnol 27:1177–1180
62. Genkov T, Meyer M, Griffiths H, Spreitzer RJ (2010) Functional hybrid rubisco enzymes with plant small subunits and algal large subunits: engineered *rbsC* cDNA for expression in *Chlamidomonas*. J Biol Chem 285:19833–19841
63. Banti V, Giuntoli B, Gonzali S, Loreti E, Magneschi L, Novi G et al (2013) Low oxygen response mechanisms in green organisms. Int J Mol Sci 14:1–30
64. Geigenberger P (2003) Response of plant metabolism to too little oxygen. Curr Opin Plant Biol 6:247–256
65. Veyel D, Erban A, Fehrle I, Kopka J, Schroda M (2014) Rationals and approaches for studying metabolism in eukaryotic microalgae. Metabolites 4:184–217
66. Mettler T, Mühlhaus T, Hemme D, Schöttler MA, Rupprecht J, Idoine A et al (2014) Systems analysis of the response of photosynthesis, metabolism and growth to an increase in irradiance in the photosynthetic model organism *Chlamydomonas reinhardtii*. Plant Cell 26:2310–2350
67. Yang C, Hua Q, Shimizu K (2000) Energetics and carbon metabolism during growth of microalgal cells under photoautotrophic, mixotrophic and cyclic light-autotrophic/dark-heterotrophic conditions. Biochem Eng J 6(2):87–102
68. Quintana N, van der Kooy F, van der Rhee MD, Voshol GP, Verpoorte R (2011) Renewable energy from Cyanobacteria: energy production optimization by metabolic pathway engineering. Appl Microbiol Biotechnol 91:471–490
69. Pearce J, Leach CK, Carr NG (1969) The incomplete tricarboxylic acid cycle in the blue-green alga *Anabaena variabilis*. J Gen Microbiol 55:371–378
70. Vazquez-Bermudex MF, Herrero A, Flores E (2000) Uptake of 2-oxoglutarate in Synechococcus strain transformed with the *Escherichia coli kgtP* gene. J Bacteriol 182:211–215
71. Pearce J, Carr NG (1967) The metabolism of acetate by the blue-green algae, Anabaena variabilis and Anacystis nidulans. J Gen Microbiol 49:301–313
72. Yang C, Hua Q, Shimizu K (2002) Quantitative analysis of intracellular metabolic fluxes using GC-MS and two-dimensional NMR spectroscopy. J Biosci Bioeng 93(1):78–87
73. Yang C, Hua Q, Shimizu K (2002) Metabolic flux analysis in *Synechocystis* using isotope distribution from ^{13}C-labeled glucose. Metab Eng 4(3):202–216
74. Yang C, Hua Q, Shimizu K (2002) Integration of the information from gene expression and metabolic fluxes for the analysis of the regulatory mechanisms in *Synechocystis*. Appl Microbiol Biotechnol 58(6):813–822
75. Schwender J (2008) Metabolic flux analysis as a tool in metabolic engineering of plants. Curr Opin Biotechnol 19:131–137
76. Wiechert W, Noh K (2005) From stationary to instationary metabolic flux analysis. Adv Biochem Eng Biotechnol 92:145–172
77. Young JD, Shastri AA, Stephanopoulos G, Morgan JA (2011) Mapping photoautotrophic metabolism with isotopically nonstatinary ^{13}C flux analysis. Metab Eng 13:656–665
78. Colman B, Norman EG (1997) Serine synthesis in cyanobacteria by a nonphotorespiratory pathway. Physiol Plant 100:133–136
79. Owittrim GW, Colman B (1988) Phosphoenolpyruvate carboxylase mediated carbon flow in a cyanobacterium. Biochem Cell Biol 66:93–99
80. Bricker TM, Zhang S, Laborde SM, Mayer II PR, Frankel LK, Moroney JV (2004) The malic enzyme is required for optimal photoautotrophic growth of *Synechocystis* sp. strain PCC6803 under continuous light but not under a diurnal light regimen. J Bacteriol 186(23): 8144-8148
81. StompM HJ, Voros L, Pick FR, Laamanen M, Haverkamp T, Stal LJ (2007) Colourful coexistence of red and green picocyanobacteria in lakes and seas. Ecol Lett 10:290–298
82. Griffiths MJ et al (2009) Lipid productivity as a key characteristic for choosing algal species for biodiesel production. J Appl Phycol 21:493–507
83. Klok AJ, Lamers PP, Martens DE, Draaisma RB, Wijffels RH (2014) Edible oils from microalgae: insights in TAG accumulation. Trends Biotechnol 32(10):521–528
84. Blaby IK, Glaesener AG, Mettler T, Fitz-Gibbon ST, Gallaher SD, Liu B et al (2013) Systems-level analysis of nitrogen starvation-induced modifications of carbon metabolism in a *Chlamydomonas reinhardtii* starchless mutant. Plant Cell 25:4305–4323
85. Flores E, Herrero A (2005) Nitrogen assimilation and nitrogen control in cyanobacteria. Biochem Soc Trans 33:164–167
86. Valladares A, Montesinos ML, Herrero A, Flores E (2002) An ABC-type, high-affinity urea permease identified in cyanobacteria. Mol Microbiol 43:703–715
87. Garcia-Fernandez JM, de Marsac NT, Diez J (2004) Streamlined regulation and gene loss as adaptive mechanisms in *Prochlorococcus* for optimized nitrogen utilization in oligotrophic environments. Microbiol Mol Biol Rev 68:630–638
88. Allen MM (1984) Cyanobacterial cell inclusions. Annu Rev Microbiol 38:1–25
89. Li H, Sherman DM, Bao S, Sherman LA (2001) Pattern of cyanophycin accumulation in nitrogen-fixing and non-nitrogen-fixing cyanobacteria anobacteria. Arch Microbiol 176:9–18
90. Grossman AR, Schaefer MR, Chiang GG, Collier JL (1993) The phycobilisome, a light-harvesting complex responsive to environmental conditions. Microbiol Rev 57:725–749
91. Dong HP, Williams E, Wang DZ, Xie ZX, Hsia RC, Jenck A et al (2013) Responses of *Nanochloropsis oceanica* IMET1 to long-term nitrogen starvation and recovery. Plant Physiol 162:1110–1126
92. Li H, Sherman LA (2002) Characterization of *Synechocystis* sp. Strain PCC6803 and deltanbl mutants under nitrogen-deficient conditions. Arch Microbiol 178:256–266
93. Baier K, Nicklisch S, Grundner C, Reinecke J, Lockau W (2001) Expression of two *nblA*-homologous genes required for phycobilisome degradation in nitrogen-starved *Synechocystis* sp. PCC6803. FEMS Microbiol Lett 195:35–39
94. Kiyota H, Yokota-Hirai M, Ikeuchi M (2014) NblA1/A2-dependent homeostasis of amino acid pools during nitrogen starvation in *Synechocystis* sp. PCC6803. Metabolites 4:517–531
95. Schwarz D, Orf I, Kopla J, Hagemann M (2014) Effects of inorganic carbon limitation on the metabolome of the *Synechocystis* sp. PCC6803 mutant defective in *glnB* encoding the central regulator PII of cyanocacterial C/N acclimation. Metabolites 4:232–247
96. Azuma M, Osanai T, Hirai MY, Tanaka K (2011) A response regulator Rre37 and an RNA polymerase sigma factor SigE represent two parallel pathways to activate sugar catabolism in a cyanobacterium *Synechocystis* sp. PCC 6803. Plant Cell Physiol 52:404–412
97. Ehira S, Ohmori M (2006) NrrA directly regulates expression of hetR during heterocyst differentiation in the cyanobacterium *Anabaena* sp. strain PCC 7120. J Bacteriol 188:8520–8525
98. Liu D, Yang C (2014) The nitrogen-regulated response regulator NrrA controls cyanophicin synthesis and glycogen catabolism in the cyanobacterium Synechocystis sp. PCC 6803. J Biol Chem 289(4):2055–2071
99. Krasikov V, von Wobeser AE, Dekker HL, Huisman J, Matthijs HC (2012) Time-series resolution of gradual nitrogen starvation and its impact on photosynthesis in the cyanobacterium *Synechocystis* PCC 6803. Physiol Plant 145:426–439
100. Markou G, Nerantzis E (2013) Microalgae for high-value compounds and biofuels production: A review with focus on cultivation under stress conditions. Biotechnol Adv 31:1532–1542
101. Miyake M, Kataoka K, Shirai M, Asada Y (1997) Control of poly-β-hydroxybutyrate synthase mediated by acetyl phosphate in cyanobacteria. J Bacteriol 179:5009–503
102. Hauf W, Schlebusch M, Huge J, Kopka J, Hagemann M, Forchhammer K (2013) Metabolic changes in *Synechocystis* PCC6803 upon nitrogen-starvation: Excess NADPH sustains polyhydroxybutyrate accumulation. Metabolites 3:101–118
103. Muro-Pastor AM, Herrero A, Flores E (2001) Nitrogen-regulated group 2 sigma factor from Synechocystis sp. Strain PCC6803 involved in survival under nitrogen stress. J Bacteriol 183:1090–1095
104. Osanai T, Kanesaki Y, Nakano T, Takahashi H, Asayama M, Shirai M et al (2005) Positive regulation of sugar catabolic pathways in the cyanobacterium *Synechocystis* sp. PCC 6803 by the group 2 sigma factor *sigE*. J Biol Chem 280:30653–30659
105. Osanai T et al (2011) Genetic engineering of group 2 sigma factor SigE widely activates expressions of sugar catabolic genes in *Synechocystis* sp. PCC 6803. J Biol Chem 286:30962–30971
106. Osanai T, Numata K, Oikawa A, Kuwahara A, Iijima H, Doi Y et al (2013) Increased bioplastic production with an RNA polymerase sigma factor SigE during nitrogen starvation in Synechocystis sp. PCC6803. DNA Res 20:525–535
107. Osanai T, Kuwahara A, Iijima H, Toyooka K, Sato M, Tanaka K et al (2013) Pleiotropic effect of *sigE* over-expression on cell morphorogy, photosynthesis, and hydrogen production in *Synechocystis* sp. PCC6803. Plant J 76:456–465

108. Wada N, Sakamoto T, Matsugo S (2013) Multiple roles of photosynthetic and sunscreen pigments in cyanobacteria focusing on the oxidative stress. Metabolites 3:463–483
109. Reijnders MJMF, van Heck RGA, Lam CMC, Scaife MA, Martins dos Santos VAP, Smith AG et al (2014) Green genes: bioinformatics and systems-biology innovations drive algal biotechnology. Trends Biotechnol 32(12):617–626
110. Poolman MG, Fell DA, Thomas S (2000) Modeling photosynthesis and its control. J Exp Bot 51:319–328
111. Poolman MG, Olcer H, Lloyd JC, Raines CA, Fell DA (2001) Computer modeling and experimental evidence for two steady states in the photosynthetic Carbin cycle. Eur J Biochem 268:2810–2816
112. Garcia-Camacho F, Sanchez-Miron A, Morina-Grima E, Camacho-Rubio F, Merchuck JC (2012) A mechanistic model of photosynthesis in microalgae including photoacclimation dynamics. J Theor Biol 304:1–15
113. Camacho-Rubio F, Garcia-Camacho F, Fernandez-Sevilla JM, Chisty Y, Grima E (2003) A mechanistic model of photosynthesis in microalgae. Biotechnol Bioeng 81:459–473
114. Gerin S, Mathy G, Franck F (2014) Modeling the dependence of respiration and photosynthesis upon light, acetate, carbon dioxide, nitrate and ammonium in *Chlamydomonas reinhardtii* using design of experiments and multiple regression. BMC Sys Biol 8:96
115. Kliphuis AMJ, Klok AJ, Martens DE, Lamers PP, Janssen M, Wijffels RH (2012) Metabolic modeling of *Chlamydomonas reinhardtii*: Energy requirements for autotrophic growth and maintenance. J Appl Phycol 24:253–266
116. Cogne G, Rugen M, Bockmayr A, Titica M, Dussap C-G, Cornet J-F et al (2011) A model-based method for investigating bioenergetics processes in autotrophically growing eukaryotic microalgae: Application to the green algal *Chlamydomonas reinhardtii*. Biotechnol Prog 27(3):631–640
117. Boyle NR, Morgan JA (2009) Flux balance analysis of primary metabolism in *Chlamydomonas reinhardtii*. BMC Sys Biol 3:4
118. Chang RL, Ghamsari L, Manichaikul A, Hom EFY, Balaji S, Fu W et al (2011) Metabolic network reconstruction of *Chlamydomonas* offers insight into light-driven algal metabolism. Mol Sys Biol 7:518
119. de Oliveira Dal'Molin CG, Quek L-E, Palfreyman RW, Nielsen LK (2011) AlgaGEM^a genome-scale metabolic reconstruction of algae based on the *Chalamydomonas reinhardtii* genome. BMC Genomics 12: (Suppl 4) S5
120. Knoop H, Grundel M, Zilliges Y, Lehmann R, Hoffmann S, Lockau W, Steuer R (2013). Flux balance analysis of cyanobacterial metabolism: The metabolic network of *Synechocyctis* sp. PCC6803. Plos Comp Biol;9(6): 21003081
121. Triana J, Montagud A, Siurana M, Fuente D, Urchueguia A, Gamermann D et al (2014) Generation and evaluation of a genome-scale metabolic network model of *Synechococcus elongatus* PCC7942. Metabolites 4:680–698
122. Schenk PM, Thomas-Hall SR, Stephens E, Marx UC, Mussgnug JH, Posten C et al (2008) Second generation biofuels: high-efficiency microalgae for biodiesel production. Bioenergy Res 1(1):20–43
123. Lacour T, Sciandra A, Talec A, Mayzaud P, Bernard O (2012) Disel variations of carbohydrates and neutral lipids in nitrogen-sufficient and nitrogen-starved cyclostat cultures of Isochrysis sp. 1. J Geophys Res 48(4):966–975
124. Kilian O, Benemann CS, Niyogi KK, Vick B (2011) High-efficiency homologous recombination in the oil-producing alga Nannochloropsis sp. PNAS USA 108(52):21265–21269
125. Singh A, Nigam PS, Murphy JD (2011) Renewable fuels from algae: an answer to debatable land based fuels. Bioresour Technol 102(1):10–16
126. James CN, Horn PJ, Case CR, Gidda SK, Zhang D, Mullen RT et al (2010) Disruption of the Arabidopsis CGI-58 homologue produces Chanarin–Dorfman-like lipid droplet accumulation in plants. PNAS USA 107(41):17833–17838
127. Dexter J, Fu PC (2009) Metabolic engineering of cyanobacteria for ethanol production. Energy Environ Sci 2:857–864
128. Dienst D, Georg J, Abts T, Jakorew L, Kuchmina E, Borner T, Wilde A, Duhring U, Enke H, Hess WR (2014) Transcriptomic response to prolonged ethanol production in the cyanobacterium *Synechosystis* sp. PCC6803. Biotechnol for Biofuels 7:21
129. Atsumi S, Liao JC (2008) Directed evolution of *Methanococcus jannaschii* citramalate synthase for biosynthesis of 1-propanol and 1-butanol by *Escherichia coli*. Appl Environ Microbiol 74:7802–7808
130. Shen CR, Liao JC (2012) Photosynthetic production of 2-methyl-1-butanol from CO_2 in cyanobacterium *Synechococcus elongatus* PCC 7942 and characterization of the native acetohydroxyacid synthase. Energy Environ Sci 5:9574
131. Lan EI, Liao JC (2011) Metabolic engineering of cyanobacteria for 1-butanol production from carbon dioxide. Metab Eng 13:353–363
132. Lan EI, Liao JC (2012) ATP drives direct photosynthetic production of 1-butanol in cyanobacteria. PNAS USA 109:6018–6023
133. Wang Y, Shi M, Niu X, Zhang X, Gao L, Chen L et al (2014) Metabolomic evolution of butanol tolerance in photosynthetic *Synechocystis* sp. PCC6803. Microb Cell Fact 13:151
134. Datta R, Henry M (2006) Recent advances in products, processes and technologies-a review. J Chem Technol Biotechnol 81:1119–1129
135. Niederholtmeyer H, Wolfstadter BT, Savage DF, Silver PA, Way JC (2010) Engineering cyanobacteria to synthesize and export hydrophilic products. Appl Environ Microbiol 76:3462–3466
136. Angermayr SA, van der Woude AD, Correddu D, Vreugdenhil A, Verrone V, Hellingwerf KJ (2014) Exploring metabolic engineering design principles for the photosynthetic production of lactic acid by *Synechosystis* sp. PCC6803. Biotechnol for Biofuels 7:99
137. Richter N, Zienert A, Hummel W (2011) A single-point mutation enables lactate dehydrogenage from *Bacillus subtilis* to utilize NAD^+ and $NADP^+$ as cofactor. Eng Lif Sci 11:26–36
138. Lindberg P, Park S, Melis A (2010) Engineering a platform for photosynthetic isoprene production in cyanobacteria, using *Synechocystis* as the model organism. Metab Eng 12:70–79
139. Ungerer J, Tao L, Davis M, Ghirardi M, Maness PC, Yu JP (2012) Sustained photosynthetic conversion of CO_2 to ethylene in recombinant cyanobacterium *Synechocystis 6*803. Energy Environ Sci 5:8998–9006
140. Gao Q, Wang W, Zhao H, Lu X (2012). Effects of fatty acid activation on photosynthetic production of fatty acid-based biofuels in *Synechocystis* sp. PCC6803. Biotechnol for Biofuels; 5:17
141. Liu X, Fallon S, Sheng J, Curtiss R 3rd (2011) CO_2-limitation-inducible Green Recovery of fatty acids from cyanobacterial biomass. PNAS USA 108:6905–6908
142. Liu X, Sheng J, Curtiss R 3rd (2011) Fatty acid production in genetically modified cyanobacteria. PNAS USA 108:6899–6904
143. Tan X, Yao L, Gao Q, Wang W, Qi F, Lu X (2011) Photosynthesis driven conversion of carbon dioxide to fatty alcohols and hydrocarbons in cyanobacteria. Metab Eng 13:169–176
144. Sivakumar G, Xu J, Thompson RW, Yang Y, Randol-Smith P, Weathers PJ (2012) Integrated green algal technology for bioremediation and biofue. Bioresour Technol 107:1–9
145. Rasala BA, Chao S-S, Pier M, Barrera DJ, Mayfield SP (2014) Enhanced genetic tools for engineering multigene traits into green algae. PLoS One 9(4):e94028
146. Lei A, Chen H, Shen G, Hu Z, Chen L, Wang J (2012) Expression of fatty acid synthesis genes and fatty acid accumulation in haematococcus pluvialis under different stressors. Biotechnol Biofuels 5:18
147. Anandarajah K, Mahendraperumal G, Sommerfeld M, Hu Q (2012) Characterization of microalga *Nannochloropsis* sp. mutants for improved production of biofuels. Appl Energy 96:371–377
148. Zhou J, Zhang H, Zhang Y, Li Y, Ma Y (2012) Designing and creating a modularized synthetic pathway in cyanobacterium *Synechocystis* enables production of acetone from carbon dioxide. Metab Eng 14:394–400
149. Mendez-Perez D, Begemann MB, Pfleger BF (2011) Modular synthase-encoding gene involved in alpha-olefin biosynthesis in Synechococcus sp. strain PCC 7002. Appl Environ Microbiol 77:4264–4267
150. Gunther A, Jakob T, Goss R, Konig S, Spindler D, Rabiger N et al (2012) Methane production from glycolate excreting algae as a new concept in the production of biofuels. Bioresour Technol 121:454–457
151. Kobayashi J, Hasegawa S, Ito K, Yoshimune K, Komoriya T, Asada A et al (2012) Expression of aldehyde dehydrogenase gene increases hydrogen production from low concentration of acetate by *Rhodobacter sphaeroides*. Int J Hydrogen Energy 37:9602–9609
152. Lo SC, Shih SH, Chang JJ, Wang CY, Huang CC (2012) Enhancement of photoheterotrophic biohydrogen production at elevated temperatures by the expression of a thermophilic clostridial hydrogenase. Appl Microbiol Biotechnol 95:969–977
153. Ducat DC, Sachdeva G, Silver PA (2011) Rewiring hydrogenase-dependent redox circuits in cyanobacteria. PNAS USA 108:3941–3946
154. Ananyev GM, Skizim NJ, Dismukes GC (2012) Enhancing biological hydrogen production from cyanobacteria by removal of excreted products. J Biotechnol 162:97–104

155. Havel J, Weuster-Botz D (2007) Cofactor regeneration in phototrophic cyanobacteria applied for asymmetric reduction of ketones. Appl Microbiol Biotechnol 75:1031–1037
156. Nakamura K, Tohi YR, Hamada KH (2000) Cyanobacterium-catalyzed asymmetric reduction of ketones. Tetrahedron Lett 41:6799–6802
157. Yang ZH, Luo L, Chang X, Zhou W, Chen GH, Zhao Y et al (2012) Production of chiral alcohols from prochiral ketones by microalgal photo-biocatalytic asymmetric reduction reaction. J Ind Microbiol Biotechnol 39:835–841
158. Hajimahmoodi M, Faramarzi MA, Mohammadi N, Soltani N, Oveisi MR, Nafissi-Varcheh N (2010) Evaluation of antioxidant properties and total phenolic contents of some strains of microalgae. J Appl Phycol 22:43–50
159. Kepekci RA, Saygideger SD (2012) Enhancement of phenolic compound production in Spirulina platensis by two-step batch mode cultivation. J Appl Phycol 24:897–905
160. Leema JT, Kirubagaran R, Vinithkumar NV, Dheenan PS, Karthikayulu S (2010) High value pigment production from Arthrospira (Spirulina) platensis cultured in seawater. Bioresour Technol 101:9221–9227
161. Reinsvold RE, Jinkerson RE, Radakovits R, Posewitz MC, Basu C (2011) The production of the sesquiterpene beta-caryophyllene in a transgenic strain of the cyanobacterium *Synechocystis*. J Plant Physiol 168:848–852
162. Wang GS, Grammel H, Abou-Aisha K, Sagesser R, Ghosh R (2012) High-level production of the industrial product lycopene by the photosynthetic bacterium *Rhodospirillum rubrum*. Appl Environ Microbiol 78:7205–7215
163. Ducat DC, Avelar-Rivas JA, Way JC, Silver PA (2012) Rerouting carbon flux to enhance photosynthetic productivity. Appl Environ Microbiol 78:2660–2668
164. Haase SM, Huchzermeyer B, Rath T (2012) PHB accumulation in *Nostoc muscorum* under different carbon stress situations. J Appl Phycol 24:157–162
165. Tyo KE, Jin YS, Espinoza FA, Stephanopoulos G (2009) Identification of gene disruptions for increased poly-3-hydroxybutyrate accumulation in *Synechocystis* PCC 6803. Biotechnol Prog 25:1236–1243
166. Aikawa S, Izumi Y, Matsuda F, Hasunuma T, Chang JS, Kondo A (2012) Synergistic enhancement of glycogen production in *Arthrospira platensis* by optimization of light intensity and nitrate supply. Bioresour Technol 108:211–21
167. Razon LF (2012) Life cycle energy and greenhouse gas profile of a process for the production of ammonium sulfate from nitrogen-fixing photosynthetic cyanobacteria. Bioresour Technol 107:339–346
168. Carlozzi P, Buccioni A, Minieri S, Pushparaj B, Piccardi R, Ena A et al (2010) Production of bio-fuels (hydrogen and lipids) through a photofermentation process. Bioresour Technol 101(9):3115–3120
169. Chen CY, Yeh KL, Aisyah R, Lee DJ, Chang JS (2011) Cultivation, photobioreactor design and harvesting of microalgae for biodiesel production: a critical review. Bioresour Technol 102(1):71–81
170. Lam MK, Lee KT (2012) Microalgae biofuels: a critical review of issues, problems and the way forward. Biotechnol Adv 30(3):673–690
171. Xu L, Weathers PJ, Xiong XR, Liu CZ (2009) Microalgal bioreactors: challenges and opportunities. Eng Life Sci 9(3):178–189
172. Stephenson AL, Kazamia E, Dennis JS, Howe CJ, Scott SA, Smith AG (2010) Life-cycle assessment of potential algal biodiesel production in the United Kingdom: a comparison of raceways and air-lift tubular bioreactors. Energy Fuel 24(7):4062–4077
173. Rosenberg J, Oyler G, Wilkinson L, Betenbaugh M (2008) A green light for engineered algae: redirecting metabolism to fuel a biotechnology revolution. Biotechnol 19:430–436
174. Ono E, Cuello JL (2004) Design parameters of solar concentration systems for CO_2 mitigating algal photobioreactors. Energy 29:1651–1657
175. Christenson L, Sims R (2011) Production and harvesting of microalgae for wastewater treatment, biofuels, and bioproducts. Biotechnol Adv 29(6):686–702
176. Wu Z, Zhu Y, Huang W, Zhang C, Li T, Zhang Y et al (2012) Evaluation of flocculation induced by pH increase for harvesting microalgae and reuse of flocculated medium. Bioresour Technol 110:496–502
177. Renault F, Sancey B, Badot PM, Crini G (2009) Chitosan for coagulation/flocculation processes–an eco-friendly approach. Eur Polymer J 45(5):1337–1348
178. de Godos I, Guzman HO, Soto R, García-Encina PA, Becares E, Muñoz R et al (2011) Coagulation/flocculation-based removal of algal–bacterial biomass from piggery wastewater treatment. Bioresour Technol 102(2):923–927
179. Moreno-Garrido I (2008) Microalgae immobilization: current techniques and uses. Bioresour Technol 99(10):3949–3964
180. Moreira SM, Moreira-Santos M, Guilhermino L, Ribeiro R (2006) Immobilization of the marine microalga *Phaeodactylum tricornutum* in alginate for in situ experiments: Bead stability and suitability. Enz Microbiol Technol 38(1):135–141
181. Ertuğrul S, Bakır M, Dönmez G (2008) Treatment of dye-rich wastewater by an immobilized thermophilic cyanobacterial strain: *Phormidium* sp. Ecol Eng 32(3):244–248
182. Guedri H, Durrieu C (2008) A self-assembled monolayers based conductometric algal whole cell biosensor for water monitoring. Microchim Acta 163(3–4):179–184
183. Mallick N (2006) Immobilization of microalgae. In Immobilization of Enzymes and Cells (pp. 373–391). Humana Press.
184. Pérez-Martínez C, Sánchez-Castillo P, Jiménez-Pérez MV (2010) Utilization of immobilized benthic algal species for N and P removal. J Appl Phycol 22(3):277–282
185. Ruiz-Marin A, Mendoza-Espinosa L (2008) Ammonia removal and biomass characteristics of alginate-immobilized Scenedesmus obliquus cultures treating real wastewater. Fresenius Environ Bull 17:1236–1241
186. Ruiz-Marin A, Mendoza-Espinosa LG, Stephenson T (2010) Growth and nutrient removal in free and immobilized green algae in batch and semi-continuous cultures treating real wastewater. Bioresour Technol 101(1):58–64
187. Ranjan A, Patil C, Moholkar VS (2010) Mechanistic assessment of microalgal lipid extraction. Ind Eng Chem Res 49(6):2979–2985
188. Sander K, Murthy GS (2010) Life cycle analysis of algae biodiesel. Int J Life Cycle Assess 15(7):704–714
189. Rawat I, Ranjith Kumar R, Mutanda T, Bux F (2011) Dual role of microalgae: phycoremediation of domestic wastewater and biomass production for sustainable biofuels production. Appl Energy 88(10):3411–3424
190. Amaro HM, Guedes A, Malcata FX (2011) Advances and perspectives in using microalgae to produce biodiesel. Appl Energy 88(10):3402–3410
191. Lee J-Y, Yoo C, Jun S-Y, Ahn C-Y, Oh H-M (2010) Comparison of several methods for effective lipid extraction from microalgae. Biores Technol 101:575–577
192. Young G, Nippgen F, Titterbraudt S, Cooney MJ (2010) Lipid extraction from biomass using co-solvent mixtures of ionic liquids and polar covalent molecules. Sep Purif Technol 72:118–121
193. Kim YH, Choi YK, Park J, Lee S, Yang YH, Kim HJ et al (2012) Ionic liquid-mediated extraction of lipids from algal biomass. Bioresour Technol 109:312–315
194. Pragya N, Pandey KK, Sahoo PK (2013) A review on harvesting, oil extraction and biofuels production technologies from microalgae. Renew Sustain Energy Rev 24:159–171
195. Mendes RL, Nobre BP, Cardoso MT, Pereira AP, Palavra AF (2003) Supercritical carbon dioxide extraction of compounds with pharmaceutical importance from microalgae. Inorg Chim Acta 356:328–334
196. Sawangkeaw R, Bunyakiat K, Ngamprasertsith S (2010) A review of laboratory-scale research on lipid conversion to biodiesel with supercritical methanol (2001–2009). J Supercrit Fluids 55(1):1–13
197. Jaime L, Mendiola JA, Ibáñez E, Martin-Álvarez PJ, Cifuentes A, Reglero G et al (2007) β-Carotene isomer composition of sub-and supercritical carbon dioxide extracts. Antioxidant activity measurement. J Agric Food Chem 55(26):10585–10590
198. Kitada K, Machmudah S, Sasaki M, Goto M, Nakashima Y, Kumamoto S et al (2009) Supercritical CO2 extraction of pigment components with pharmaceutical importance from Chlorella vulgaris. J Chem Technol Biotechnol 84(5):657–661
199. Macías-Sánchez MD, Mantell Serrano C, Rodríguez Rodríguez M, de la Martínez Ossa E, Lubián LM, Montero O (2008) Extraction of carotenoids and chlorophyll from microalgae with supercritical carbon dioxide and ethanol as cosolvent. J Sep Sci 31(8):1352–1362
200. Ota M, Watanabe H, Kato Y, Watanabe M, Sato Y, Smith RL et al (2009) Carotenoid production from Chlorococcum littorale in photoautotrophic cultures with downstream supercritical fluid processing. J Sep Sci 32(13):2327–2335
201. Ehimen EA, Sun ZF, Carrington CG (2010) Variables affecting the in situ transesterification of microalgae lipids. Fuel 89(3):677–684
202. Zhu LY, Zong MH, Wu H (2008) Efficient lipid production with *Trichosporon fermentans* and its use for biodiesel preparation. Bioresou Technol 99(16):7881–7885
203. Ziino M, Lo Curto RB, Salvo F, Signorino D, Chiofalo B, Giuffrida D (1999) Lipid composition of *Geotrichum candidum* single cell protein grown in continuous submerged culture. Bioresour Technol 67(1):7–11

204. Schuchardt U, Sercheli R, Vargas RM (1998) Transesterification of vegetable oils: a review. J Braz Chem Soc 9(3):199–210
205. Wahlen BD, Willis RM, Seefeldt LC (2011) Biodiesel production by simultaneous extraction and conversion of total lipids from microalgae, cyanobacteria, and wild mixed-cultures. Bioresour Technol 102(3):2724–2730
206. Du W, Xu YY, Zeng J, Liu DH (2004) Novozym 435-catalysed transesterification of crude soya bean oils for biodiesel production in a solvent-free medium. Biotechnol Appl Biochem 40(2):187–190
207. Heeres AS, Picone CSF, van der Wielen LAM, Cunha RL, Cuellar MC (2014) Microbial advanced biofuels production: overcoming emulsification challenges for large-scale operation. Trends Biotechnol 32(4):221–229
208. Zhang S, Bryant DA (2011) The tricarboxylic acid cycle in cyanobacteria. Science 334:1551–1553
209. Steinhauser D, Fernie AR, Araujo WL (2012) Unusual cyanobacterial TCA cycles: not broken just different. Trends Plant Sci 17(9):503–509
210. Deng M-D, Coleman JR (1999) Ethanol synthesis by genetic engineering in cyanobacteria. Appl Environ Microbiol 65(2):523–528
211. Yao L, Qi F, Tan X, Lu X (2014) Improved production of fatty alcohols in cyanobacteria by metabolic engineering. Biotechnol Biofuels 7:94
212. Wang W, Liu X, Lu X (2013) Engineering cyanobacteria to improve photosynthetic production of alka(e)nes. Biotechnol Biofuels 6:69
213. McNeely K, Xu Y, Bennette N, Bryant DA, Dismukes GC (2010) Redirecting reductant flux into hydrogen production via metabolic engineering of fermentative carbon metabolism in a cyanobacterium. Appl Environ Microbiol 76(15):5032–5038
214. Angelmyr SA, Paszota M, Hellingwerf KJ (2012) Engineering a cyanobacterial cell factory for production of lactic acid. Appl Environ Microbiol 78:7098–7106
215. Varman AM, Yu Y, You L, Tan YJ (2013) Photoautotrophic production of D-lactic acid in an engineered cyanobacterium. Microb Cell Fact 12:117
216. Li H, Liao JC (2013) Engineering a cyanobacterium as the catalyst for the photosynthetic conversion of CO_2 to 1,2-propanediol. Microb Cell Fact 12:4
217. Guerrero F, Carbonell V, Cossu M, Correddu D, Jones PR (2012) Ethylene synthesis and regulated expression of recombinant protein in *Synechocystis* sp. PCC 6803. PLoS One 7(11):e50470
218. Sakai M, Ogawa T, Matsuoka M, Fukuda H (1997) Photosynthetic conversion of carbon dioxide to ethylene by the recombinant cyanobacterium, *Synechococcus* sp. PCC 7942, which harbpors a gene for the ethylene-forming enzyme of *Pseudomonas syringae*. J Ferm Bioeng 84(5):4343–443
219. Takahama K, Matsuoka M, Nagahama K, Ogawa T (2003) Construction and analysis of a recombinant cyanobacterium expressing a chromosomally inserted gene for an ethylene-forming enzyme at the *psbAI* locus, J. Biosci Bioeng 95(3):302–305
220. Wang B, Pugh S, Nielsen DR, Zhang W, Meldrum DR (2013) Engineering cyanobacteria for photosynthetic production of 3-hydroxybutyrate directly from CO_2. Metab Eng 16:68–77

Effect of cell immobilization and pH on *Scheffersomyces stipitis* growth and fermentation capacity in rich and inhibitory media

Diogo Portugal-Nunes[1,2], Violeta Sànchez i Nogué[2], Susana R Pereira[1], Sandra C Craveiro[3], António J Calado[3] and Ana MRB Xavier[1*]

Abstract

Background: A wide range of value-added products can potentially be produced by bioprocessing hardwood spent sulfite liquors (HSSLs) that are by-products of pulp and paper industry with a high pentose sugar content. However, besides sugars, HSSLs contain considerable amounts of sulfonated lignin derivatives and acetic acid that inhibit the metabolic activity of most microorganisms. *Scheffersomyces stipitis* is a yeast with high capacity to ferment the pentose sugar xylose under appropriate microaerophilic conditions but it has limited tolerance to HSSL inhibitors. In the present study, cultivations of suspended and immobilized *S. stipitis* were compared in terms of growth capacity and by-product formation using rich medium and HSSL to investigate whether the immobilization of cells in calcium alginate beads could be a protection against inhibitors while favoring the presence of microaerophilic conditions.

Results: Whereas cell immobilization clearly favored the fermentative metabolism in rich medium, pH control was found to play a more important role than cell immobilization on the ethanol production efficiency from bio-detoxified HSSL (*bd*HSSL), leading to an improvement of 1.3-fold on the maximum ethanol productivity than using suspended cells. When immobilization and pH control were applied simultaneously, the ethanol yield improved by 1.3-fold with unchanged productivity, reaching 0.26 g ethanol.(g glucose + xylose)$^{-1}$. Analysis of the immobilized beads inside revealed that the cells had grown in the opposite direction of the cortex.

Conclusions: Immobilization and pH control at 5.5, when applied simultaneously, have a positive impact on the fermentative metabolism of *S. stipitis*, improving the ethanol production efficiency. For the first time light microscopic analysis of the beads suggested that the nutrient and mass transfer limitations played a more important role in the fermentation than a possible protective role against inhibitors.

Keywords: *Scheffersomyces stipitis*; Hardwood spent sulfite liquor; Cell immobilization; Light microscopy; Ca alginate beads; Xylose fermentation; Stress tolerance

Background

Spent sulfite liquors (SSLs) are by-products that result from the acidic wood cooking process containing high amounts of fermentable sugars [1,2]. SSLs are by-products with low-cost associated, being potential substrates for the production of several value-added compounds. Among them, bioethanol - a renewable, biodegradable and non-toxic biofuel - derived from SSLs might constitute a cost-effective and large-scale production alternative, with associated environmental benefits [1,3].

The chemical composition of SSLs depends on the wood species used in the pulping process. Typically, undiluted hardwood SSLs (HSSLs) from *Eucalyptus globulus* contain 35 to 45 g.L^{-1} sugars, mainly xylose [2,4]. This significant xylose content makes HSSL a suitable substrate for pentose-fermenting yeasts, such as *Scheffersomyces stipitis* [2,5]. However, up to 8.0 to 10.0 g.L^{-1} acetic acid and 59.0 to 78.8 g.L^{-1} lignosulfonates have also been reported and they both act as inhibitors of microbial metabolism. In fact, the wood cooking process

* Correspondence: abx@ua.pt
[1]CICECO-Aveiro Institute of Materials and Chemical Department, University of Aveiro, Campus Universitário de Santiago, 3810-193 Aveiro, Portugal
Full list of author information is available at the end of the article

uses harsh conditions (pH 1 to 2, 125°C to 135°C, 8 to 12 h), which leads to the formation of microbial inhibitory chemical compounds [2]. Notably aromatic, polyaromatic, phenolic, and aldehyde compounds can be generated from lignin and the hemicellulose fraction degradation, thus affecting the integrity of biological membranes and inhibiting yeast metabolism [6-8]. Several studies reported a synergistic effect on inhibition when the furan derivatives, weak acids and/or phenolic compounds were simultaneously present [9-12].

S. stipitis was described as having the highest native capacity for D-xylose fermentation of any known microbe [1,13]. This yeast induces fermentative activity in response to oxygen limitation and reaches the optimal activity with microaerophilic conditions (1% to 15% O_2) [14,15]. One of the major drawbacks of *S. stipitis* is the low tolerance towards inhibitors and high ethanol concentrations [16]. *S. stipitis* performances in untreated HSSLs are strongly affected, since growth occurs at a considerable lower level and the ethanol yields can only reach 30% to 50% of the theoretical maximum - 0.51 g ethanol.g sugars^{-1} [1,17].

Physico-chemical detoxification of HSSL, metabolic engineering and strain adaptation are possible strategies to improve the alcoholic fermentation efficiency in yeast species [4,7,18-20]. Biological detoxification of HSSL with the fungus *Paecilomyces variotii* was also found to improve the ethanol production by *S. stipitis* [4]. Finally, cell immobilization could be an advantageous approach to ferment lignocellulosic by-products with inhibitors, improving cell stability, activity and stress tolerance [21-23]. In the case of cell entrapment, the layer creates a microenvironment that acts as a protection for high ethanol concentrations and inhibitors limiting mass transfer phenomena. However, it might also affect the oxygen mass transfer. Facilitated product recovery, the possibility to reuse the biomass and the lower risk of microbial contamination are other advantages of cell immobilization that could compensate for the additional investment that this process involves [24].

The aim of the present study was to investigate the effects of cell immobilization and pH control on the performances of *S. stipitis* in rich medium and in biodetoxified HSSL (*bd*HSSL). Immobilization was done by Ca alginate entrapment in order to investigate if it could limit the oxygen available for the cells, and therefore favor the fermentative metabolism of *S. stipitis*, and/or confer a protection layer against the inhibitory compounds that were still present in the *bd*HSSL. Growth and by-product formation using suspended and calcium alginate immobilized cultures of *S. stipitis* were compared and light microscopy was used to analyze the inside of beads in order to observe the mass transfer effects.

Methods

Strain

S. stipitis NRRL-7124 was supplied by the Agricultural Research Service Culture Collection at National Center for Agricultural Utilization Research, USDA. Stock cultures were stored in 30% glycerol at −80°C. *S. stipitis* culture was grown for 48 h at 28°C and maintained at 4°C in YM (yeast medium) agar plates containing 10 g.L^{-1} glucose, 5 g.L^{-1} peptone, 3 g.L^{-1} malt extract, 3 g.L^{-1} yeast extract and 20 g.L^{-1} agar.

Hardwood spent sulfite liquor

HSSL (Caima SA- ALTRI Group, Constância, Portugal) was obtained from a magnesium-based E. globulus wood cooking process. HSSL was physico-chemically pretreated as described by Xavier et al. in 2010 [5]. After the pretreatment, HSSL contained 24.6 g.L^{-1} xylose, 8.5 g.L^{-1} mannose, 7.8 g.L^{-1} arabinose, 4.5 g.L^{-1} galactose and 2.3 g.L^{-1} glucose [5]. Besides these sugars, HSSL contained the following potential inhibitory compounds: 78.2 g.L^{-1} lignosulfonates, 9.6 g.L^{-1} acetic acid and traces of 5-hydroxymethyl-2-furaldehyde and 2-furaldehyde [5]. Therefore further HSSL bio-detoxification was performed with a filamentous fungus as reported by Pereira et al. in 2012 [4].

Yeast cultures

YMglc and YMxyl (10 g.L^{-1} glucose or xylose, 3.0 g.L^{-1} malt extract and 3.0 g.L^{-1} yeast extract) were used as pre-culture medium for the rich medium cultivations. YNBglc medium containing 20 g.L^{-1} glucose, 6.7 g.L^{-1} yeast nitrogen base without amino acids, 10.2 g.L^{-1} potassium hydrogen phthalate and 2.2 g.L^{-1} KOH was used as pre-culture medium for the *bd*HSSL cultivations. The inocula were prepared taking a single colony from the YM plate into YNBglc. Cells were grown until the culture reached the beginning of the stationary phase at 30°C and 180 rpm.

Cultivations

Rich medium cultivations

Pre-cultures were performed in 100 mL of YMglc or YMxyl in 250 mL shake flasks. Cultures using free or immobilized cells were carried out in duplicate in 200 mL of YMglc or YMxyl in 250 mL shake flasks with a starting OD_{620nm} of 0.5 to 0.6. In the case of immobilized cells, the cell wet pellet correspondent to the mentioned OD was gently mixed with 2.8% sodium alginate and after complete homogenization, the mixture was added to 2.0% $CaCl_2$ using sterile needles and a distance needle solution of 4 to 7 cm. After 20 min with gentle agitation, the beads were transferred into 0.5% $CaCl_2$ and process was repeated (20 min, gentle agitation). The beads were then sterile transferred to the cultivation medium. OD_{620nm}, pH and metabolites concentration were followed. Samples for high performance liquid chromatography (HPLC) were centrifuged and the supernatant stored at −20°C until further analysis.

Screening in bdHSSL

After pre-cultures, *S. stipitis* was inoculated to six different *bd*HSSL levels (0%, 20%, 40%, 60%, 80%, and 100%) at an initial OD_{620nm} of 0.5. The media were prepared using YNB buffer 10× (10.2 $g.L^{-1}$ potassium hydrogen phthalate, 2.2 $g.L^{-1}$ KOH), YNB 10× (6.7 $g.L^{-1}$ yeast nitrogen base), HSSL sugars solution 5× (123.0 $g.L^{-1}$ xylose, 42.5 $g.L^{-1}$ mannose, 39.0 $g.L^{-1}$ arabinose, 22.5 $g.L^{-1}$ galactose, 11.5 $g.L^{-1}$ glucose). Aerobic growth cultures were performed in 50 mL conical tubes containing 5 mL of media. Cells were grown at 30°C and 180 rpm and OD_{620nm} was followed over time. The pH was determined at the end of the experiments. All cultivations were performed in duplicate.

bdHSSL cultivations

Pre-cultures were performed in 50 mL conical tubes or 250 mL baffled flasks and were used to inoculate 100 mL of 60% *bd*HSSL at an initial OD_{620nm} of 0.5. Cells were grown at 28°C and at the beginning of the stationary phase a cell pellet was obtained in order to inoculate undiluted *bd*HSSL with 7.3×10^{-2} g $CWW.mL^{-1}$ (g cell wet weight. mL^{-1}), approximately 10^8 CFUs. mL^{-1} (colony forming units per mL), which corresponds to an amount of cells that bio-industries use to start bioprocesses. Usually, this process is done by cell dry weight measurement but in this case it would be impossible since the cells were reutilized to start the cultivation assays.

Besides the assays with suspended cells, *S. stipitis* cells were also immobilized by entrapment in calcium alginate gel beads (as described in the 'Rich medium cultivations' section). The same procedure in order to obtain the correct CWW to start the cultivation was used as for the suspended cultures. When immobilized cells were used, the *bd*HSSL was previously supplemented with 0.2% $CaSO_4$.

Cultivations with both suspended and immobilized cells were also performed controlling the pH. KOH 3 M or H_2SO_4 3 M were pumped by an automatic control system in order to maintain the pH at 5.5.

Batch cultivations were carried out, in duplicate, in 1,000 mL fermenters with a working volume of 400 mL, at 30°C, initial pH of 5.5, airflow of 0.3 vvm and 160 to 180 rpm. OD_{620nm} and pH were measured over time. Samples for HPLC were centrifuged and the supernatant stored at −20°C until further analysis.

HPLC analysis

Concentrations of glucose, xylose, ethanol, acetate, xylitol and glycerol were determined by using a Eurokat H 10 μm 300 × 8 column (Knauer, Berlin, Germany) with H_2SO_4 0.01 N at 0.4 $mL.min^{-1}$ as mobile phase at 40°C. The injected volume was 20 μL. HPLC (VWR-Hitachi LaChrom Elite® -Tokyo, Japan) system was equipped with a Pump L-2130 and auto-sampler L-2200 from Hitachi (Tokyo, Japan), oven Gecko 2000 (Hattersheim, Germany) set at 40°C and refraction index (RI) detector L-2490 (Hitachi-Tokyo, Japan).

S. stipitis beads analysis

The distribution of cells inside the calcium alginate beads was examined by direct microscopic observation of thin sections. The beads were fixed in 2.5% glutaraldehyde in cacodylate buffer 0.1 M for about 2 h, briefly washed in the same buffer and dehydrated through a graded ethanol series followed by propylene oxide. The beads were then embedded in Spurr's resin TAAB: (TAAB Laboratories Limited, Aldermaston, England) overnight with constant agitation (100 rpm). After polymerization for 8 h at 70°C, 1.5-μm thick sections were cut with glass knives on an EM UC6 ultramicrotome (Leica Microsystems, Wetzlar, Germany). The sections were stained with toluidine blue and examined by differential interference contrast microscopy on a Zeiss Axioplan 2 imaging light microscope (Zeiss, Jena, Germany) equipped with a DP70 Olympus camera (Shinjuku, Tokyo, Japan).

Results and discussion

Cultivations in rich medium

The effect of immobilization of *S. stipitis* in a calcium alginate matrix was first investigated in rich medium using glucose or xylose as sole substrate in shake flasks. The main goals were to analyze whether the immobilization would improve the ethanol production and/or limit substrate uptake and to study the differences in fermentation profile between hexose and pentose sugars under these conditions.

The maximum glucose uptake rate using suspended cells (Figure 1A) was 1.5-fold higher than using immobilized cells (Figure 1B), 0.34 and 0.23 g $glucose.L^{-1}.h^{-1}$, respectively. This suggested that the calcium alginate matrix limited the mass transfer so that the glucose transfer rate through the immobilizing matrix was slower than the glucose uptake rate of cells inside the beads. Thus, the limiting step was the mass transfer process and not the substrate uptake kinetics. In contrast, in the cultivations using YMxyl (Figure 1C,D) the immobilization did not result in any negative effect on the maximum uptake rate - 0.17 g $xylose.L^{-1}.h^{-1}$ for both suspended and immobilized cells. However *S. stipitis* xylose uptake rates (Figure 1C,D) were lower than those obtained for glucose (Figure 1A,B), since pentose metabolism is slower and less energetic than hexose metabolism [25]. Thus, in this case, the bottleneck was most probably not the substrate available inside the beads

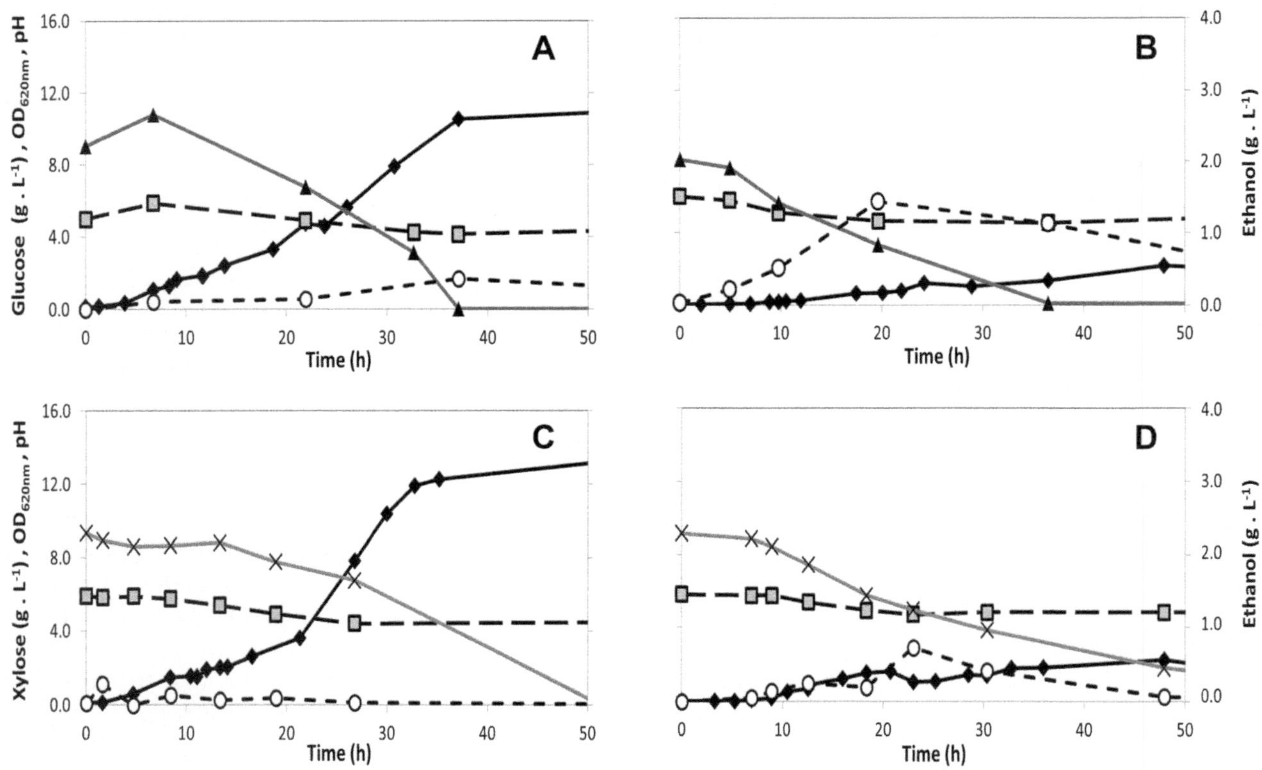

Figure 1 ***S. stipitis*** **growth and metabolites production in glucose or xylose rich medium.** *S. stipitis* growth and metabolites production in YMglc **(A,B)** and YMxyl **(C,D)** using suspended **(A,C)** and immobilized **(B,D)** cells. OD_{620nm} (filled diamond), pH (square), glucose (triangle), xylose (cross), and ethanol (circle) profiles are shown.

but the lower uptake rate of xylose by *S. stipitis* cells. The limiting step was the pentose uptake kinetics and not the substrate mass transfer. In the cultivations with suspended cells (Figure 1A,C), biomass reached approximately the same level on glucose and xylose and was the main cultivation product, which indicates that most of the carbon was directed towards biomass formation and not towards ethanol production due to the aerobic conditions provided by the shake flask set-up. Increased OD_{620nm} was also recorded in the other set-up (Figure 1B,D) despite the immobilization process. This means that some of the beads were disrupted by the stirring or because of cells growth. Alternatively, cells that bound to the outer layer of the matrix during the entrapment process may have been released in the medium by the agitation and aeration levels in the flasks.

When immobilized *S. stipitis* cells were used for the cultivation, there was a significant increase in ethanol production both from glucose and xylose (Figure 1B,D), resulting in higher titer, yield and volumetric productivity. With suspended cells, only 0.43 and 0.18 g ethanol.L^{-1} were produced after 50 h in YMglc and YMxyl, respectively, whereas maximum ethanol concentrations of 1.43 g ethanol.L^{-1} on glucose and 0.72 g ethanol.L^{-1} on xylose were achieved after 20 h with immobilized cells. After reaching its maximum, ethanol content decreased in both substrates, either because it was used as a carbon source or due to evaporation. This 3.3-fold and 4.0-fold higher titer with immobilized cells using glucose or xylose, respectively, suggests the existence of an O_2 limitation that is likely to arise from the immobilization matrix. The calcium alginate matrix might have limited the dissolved oxygen transfer and thus optimized the oxygen content inside the beads towards microaerophilic conditions that promote fermentative metabolism.

Impact of *bd*HSSL inhibitors on *S. stipitis* growth

Pereira et al. in 2012 demonstrated the inability of *S. stipitis* to grow in chemically treated HSSL, due to presence of weak acids such as acetic acid and gallic acid, among other inhibitory compounds. After the biodetoxification made by *P. variotii*, *S. stipitis* was able to grow in *bd*HSSL and ethanol production was obtained (0.24 g ethanol.g sugars^{-1}). However, there were still some remaining inhibitory compounds present on the *bd*HSSL, affecting the cell viability and performance: lignosulfonates, malic acid, propanoic acid, syringic acid, furoic acids and also acetic acid.

The impact of *bd*HSSL toxicity on *S. stipitis* performance was first evaluated by the determination of maximum growth rate (μ_{max}) at six different *bd*HSSL levels (0%, 20%, 40%, 60%, 80%, and 100%, Figure 2). As diluted *bd*HSSL was compensated for sugars and salts in order to avoid nutrient limitation, changes in μ_{max} were assumed to be caused by the presence of the inhibitory compounds only. Without *bd*HSSL, the obtained μ_{max}, 0.37 h^{-1}, was similar and consistent to previously reported values [5]. With 20% *bd*HSSL, μ_{max} decreased to 0.34 h^{-1}. Between 20% and 60%, μ_{max} slightly decreased despite the increase in inhibitors levels. Above 60%, the negative impact on growth was significant, with μ_{max} around 0.15 h^{-1}. Still *S. stipitis* could grow in 80% and 100% *bd*HSSL. From the screening assays, 60% *bd*HSSL was considered as a good compromise between the final biomass, μ_{max} and the ability to successfully resist to the presence of yeast inhibitors. Hence, 60% *bd*HSSL was used in the following experiments as a pre-adaptation step in order to enhance growth and fermentation efficiencies in *bd*HSSL.

Bioreactor cultivations in 100% *bd*HSSL using pre-adapted cells

Batch cultivations in bioreactor were used with 100% *bd*HSSL to evaluate the effects of (i) immobilization in a calcium alginate matrix and (ii) pH control in suspended and immobilized *S. stipitis* cultures. Several metabolites were analyzed by HPLC and the cultivation profiles are shown in Figure 3. Glycerol, furfural and HMF, whose concentrations were below the detection limit, are not discussed.

Using suspended *S. stipitis* (Figure 3A), pH remained stable during the 25 h of cultivation and then it started to slightly increase until 50 h. Glucose and xylose were not completely consumed until the end of the cultivation (t_F), remaining approximately 79% of the initial present sugars. The decrease in acetate might indicate that it was not present in a fully inhibitory concentration. The consumption of organic acids was unexpected, since there were monomeric sugars available. Ethanol concentration reached its maximum (1.1 $g.L^{-1}$) after 48 h, corresponding to a maximum ethanol yield ($Y_{E/S}$) of 0.20 ± 0.00 g ethanol.g^{-1} substrate and maximum ethanol productivity (P_E) of 27.6 ± 0.0 mg ethanol.$(L.h)^{-1}$ (Table 1). Biomass variation suggests that sugars, organic acids and even ethanol were used for growth and cell division. These results were probably arising from the presence of fully aerobic conditions in the bioreactor, since *S. stipitis* favors growth under excess of oxygen and ethanol production is maximized when microaerophilic conditions are present [14,15].

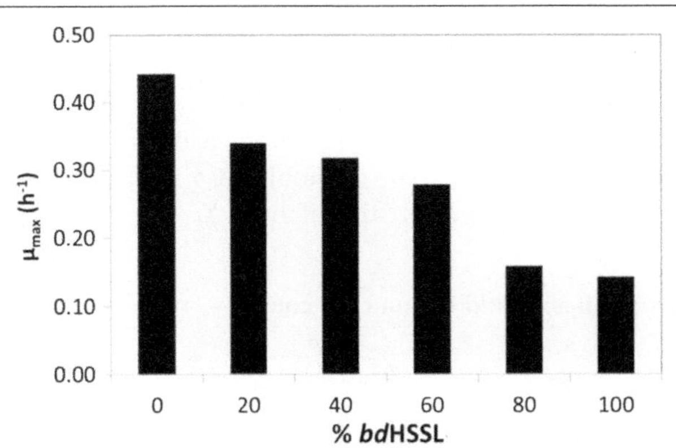

Figure 2 Maximum specific growth rate of *S. stipitis* in *bd*HSSL diluted to different percentages. Each % *bd*HSSL was compensated for sugars and salts, being the inhibitory compounds the only difference.

Effect of cell immobilization

Figure 3B shows the changes in *S. stipitis* cultivation profile caused by the immobilization. The pH remained stable until the 25 h and then slightly increased from 5.3 to 6.3 at t_F, resulting in a final higher pH than with suspended cells. Unexpectedly, cell immobilization without pH control did not improve ethanol production. In contrast, ethanol production was low and the maximum concentration (0.62 $g.L^{-1}$) was achieved after 42 h. The $Y_{E/S}$ was approximately the same as in suspended culture without pH control but the maximum productivity was reduced 3.6-fold.

During the experiment, the OD significantly increased. During *S. stipitis* entrapment process some cells bound to the outer layer of the matrix. The agitation and aeration levels on the bioreactors caused the release of those cells into the *bd*HSSL. Besides that, some Ca alginate beads were disrupted during the fermentation due to the shear stress on the bioreactors. Before this assay a first trial with immobilized cells in 100% *bd*HSSL was performed without the addition of $CaSO_4$. After 18 h, most of the beads were disrupted and alginate accumulated in the bottom of the bioreactors (data not shown). This result suggested that the anions present in *bd*HSSL bound to the calcium cations that were interacting previously with alginate. Since the amount of lignin sulfonated groups in the SSLs is commonly high, they could act as calcium sequestering agents in the media [26]. Therefore, a previous addition of $CaSO_4$ 0.2% (m/V) aimed at providing more Ca^{2+} in the media, avoiding sequestration of the calcium cations out of the calcium alginate matrix. The results showed that the beads remained stable during more time, but some degradation still occurred probably due to the depletion of $CaSO_4$. Thus, disruption of some calcium alginate barriers resulted in the release of *S. stipitis* cells from beads to the medium that can have created an artificial and continuous supply of fresh cells, these cells being also able to propagate as suspended forms.

Effect of pH control

*bd*HSSL is a highly buffered media, meaning that pH should not vary unless significant changes are applied.

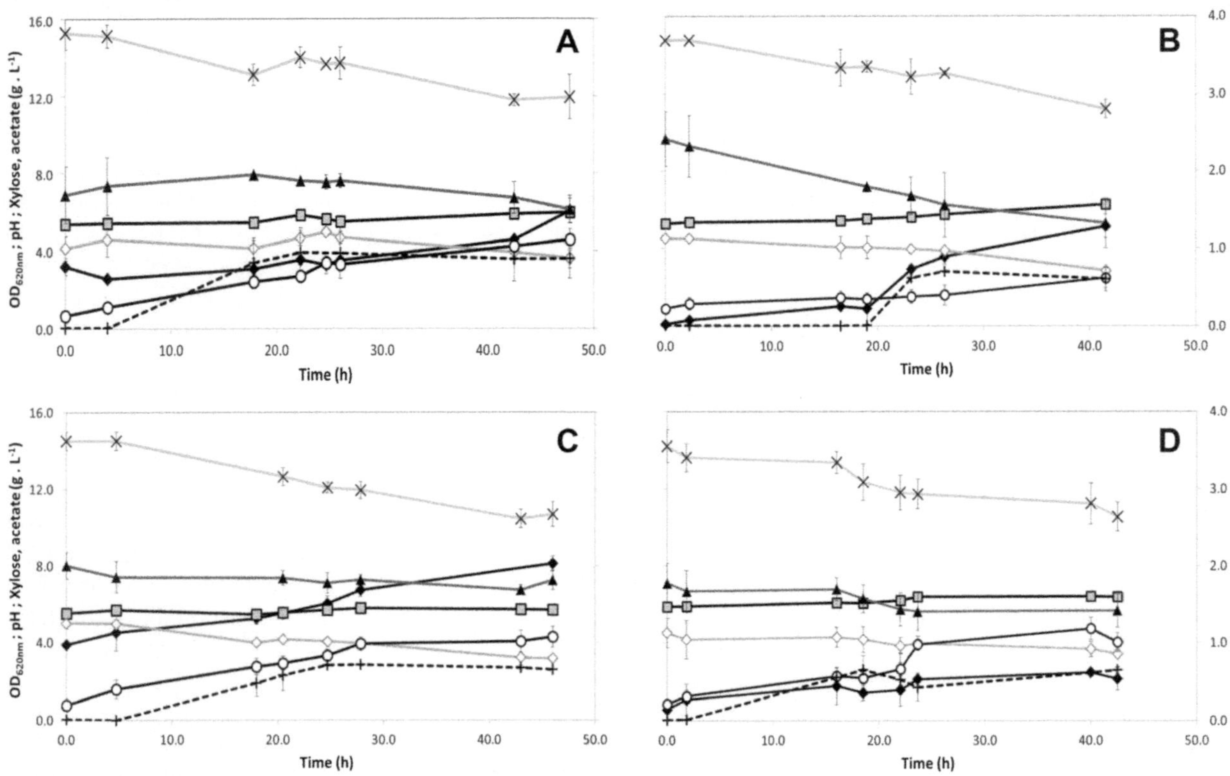

Figure 3 Effect of immobilization and pH control at 5.5 on *S. stipitis* performance in *bd*HSSL. Xylose (cross), acetate (open diamond), glucose (triangle), ethanol (circle) and xylitol (dashed line) concentrations using suspended *S. stipitis* **(A)**, immobilized *S. stipitis* **(B)**, suspended *S. stipitis* with pH control at 5.5 **(C)**, and immobilized *S. stipitis* with pH control at 5.5 **(D)**, in 100% *bd*HSSL. pH (square) and OD_{620nm} (filled diamond) values are also shown.

Since pH increased after 25 h on assays with both suspended and immobilized cells, cultivations with pH controlled at 5.5 were performed in order to evaluate its effect on *S. stipitis* cultivations. Figure 3C,D presents the time courses for the batch tests performed using suspended and immobilized *S. stipitis*, respectively, in 100% *bd*HSSL with pH control at 5.5. Using suspended cells, the pH maintenance led to a faster consumption of xylose and acetate (Figure 3C). More than 5 $g.L^{-1}$ of xylose was consumed in 50 h. Acetate started to be consumed after 20 h and the uptake rate was slightly higher than for suspended culture without pH control (Figure 3A). Maximum ethanol concentration (1.14 g $ethanol.L^{-1}$) was obtained at t_F. More interestingly, at 17 h the ethanol concentration reached 0.90 $g.L^{-1}$, which is 50% more than the obtained with suspended cells without pH control (0.60 $g.L^{-1}$). This means that with pH control the production of ethanol was more efficient and occurred at a higher rate (P_E was increased by 1.3-fold). The maximum specific ethanol productivity ($P_{E/X}$) and the maximum substrate uptake rate (r_S) were also significantly higher with pH control, reinforcing its positive impact on *S. stipitis* performance (Table 1).

As in suspended cultures, the positive effect of the pH control in fermentation efficiency was also found in the cultivation with immobilized cells and pH control at 5.5 (Figure 3D). Immobilization together with pH control resulted in the highest $Y_{E/S}$ in this study,

Table 1 *S. stipitis* ethanol production and substrates uptake in *bd*HSSL with and without pH control

S. stipitis	pH control	$Y_{E/s}$	P_E	$P_{E/X}$	r_S
Suspended	No	0.20 ± 0.00	27.6 ± 0.0	0.29 ± 0.03	1.04 ± 0.12
	Yes	0.20 ± 0.01	35.4 ± 1.3	0.39 ± 0.02	1.37 ± 0.39
Immobilized	No	0.19 ± 0.09	7.6 ± 1.4	NA	NA
	Yes	0.26 ± 0.06	27.5 ± 4.9	NA	NA

$Y_{E/S}$, maximum ethanol yield (g ethanol.(g consumed glucose + xylose)$^{-1}$); P_E, maximum ethanol productivity (mg ethanol.(L.h)$^{-1}$); $P_{E/X}$, maximum specific ethanol productivity (mg ethanol.(g CWW.h)$^{-1}$); r_S, maximum substrate uptake rate (mg consumed glucose + xylose.(g CWW.h)$^{-1}$); NA, not applicable.

0.26 ± 0.06 g ethanol.g^{-1} substrate. This represents a 1.4-fold improvement on the $Y_{E/S}$ when compared to the immobilized cultivations without pH control (Table 1). $P_{E/X}$ and r_S were not determined for the immobilization assays since the biomass could not be precisely measured due to the residues of the calcium alginate matrix present on the media.

The higher ethanol production suggests that the calcium alginate matrix limited the dissolved oxygen mass transfer and optimized the oxygen content inside the beads, being the *S. stipitis* cells in microaerophilic conditions. This limitation has not occurred in the previous trials, indicating than only under specific and constant pH together with immobilized cells it is possible to reach conditions that favor the fermentative metabolism on the bioreactor. Moreover, the disruption of the beads during time might have acted as an artificial and continuous supply of fresh cells in the bioreactor. In addition the physical barrier provided by the calcium alginate might have played an important role in terms of tolerance, since the cells were less exposed to the inhibitors (e.g. lignosulfonates) and harsh conditions.

Considering OD_{620nm} in the immobilization experiments, the biomass increased over time both without and with pH control but in a lower level in the last case. This indicates that pH control at 5.5 and immobilization caused a positive effect on the cell viability in *bd*HSSL. The addition of $CaSO_4$ to avoid the beads disruption was partially successful since, even with shear stress, there were beads that resisted during all the cultivation. The initial pH at 5.5 contributed also to an improved stability of the calcium alginate matrix, as a higher

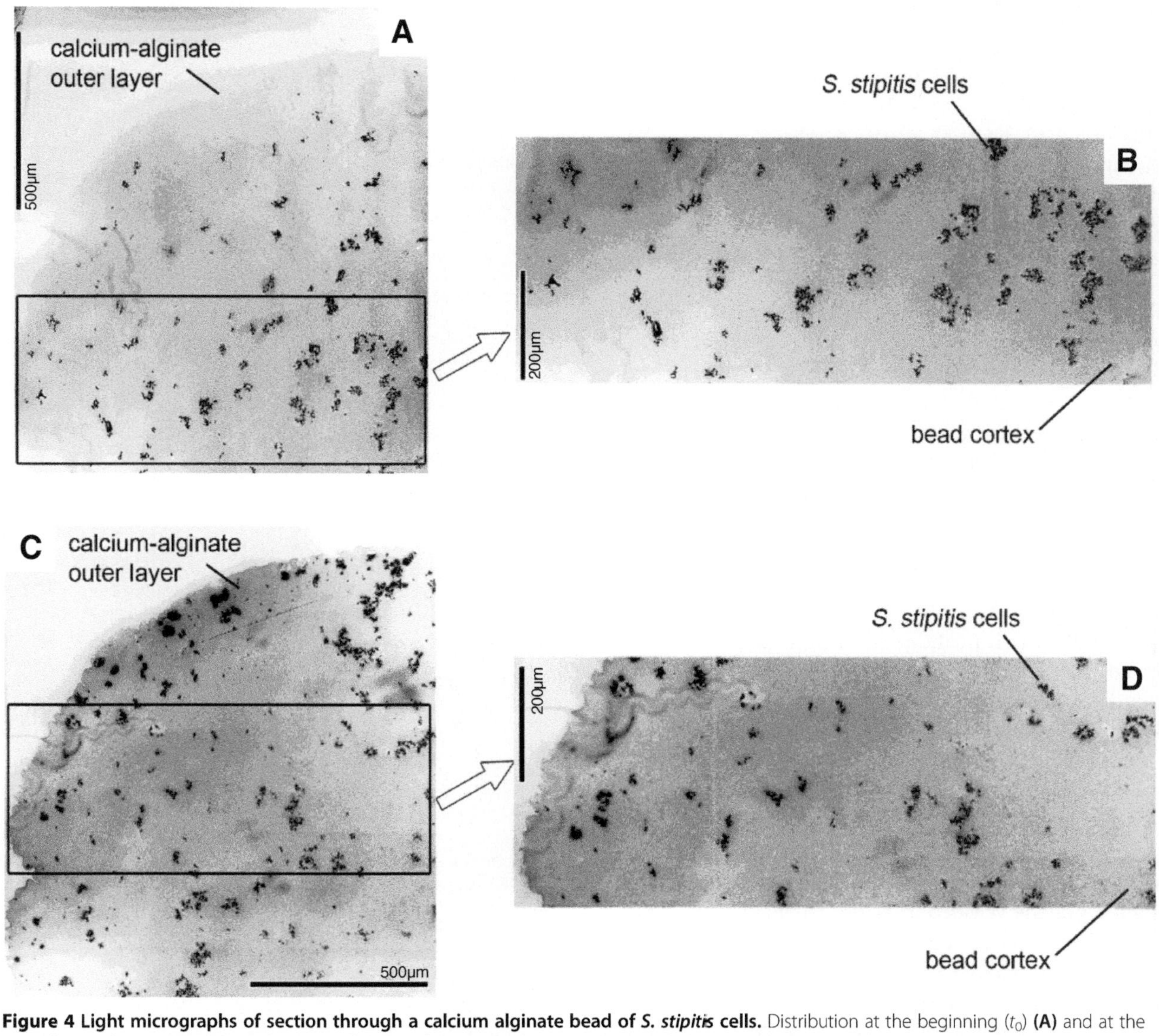

Figure 4 Light micrographs of section through a calcium alginate bead of *S. stipitis* cells. Distribution at the beginning (t_0) **(A)** and at the end of the cultivation (t_F) **(C)**. Higher magnification of the marked areas in **(A,C)** is shown in **(B,D)**, respectively.

number of calcium alginate beads that resisted to the disruption during cultivation with pH control was visible.

***S. stipitis* beads examination by light microscopy**

Sections of *S. stipitis* beads from t_0 and t_F were analyzed in light microscopy to evaluate the cells distribution and the effect of *bd*HSSL inside the calcium alginate matrix. In the immobilized cells from the assays with and without pH control no visible changes were detected on the microscope. The results showed on Figure 4 correspond to the fermentations with pH control at 5.5. At t_0 (Figure 4A) the cells were widely spread inside the bead, meaning that the immobilization process occurred as predicted. The shape of the beads was approximately rounded which should facilitate mass transfer. A perfect and uniform round shape and smaller beads size would maximize the efficiency of the transfer and avoid limitations that might occur because of the physical barrier, but this could only be analyzed with more accurate technical equipment [27]. At t_F there was a higher number of cells close to the calcium alginate outer layer (Figure 4C), indicating that the cells had grown/migrated in the opposite direction of the cortex. There, nutrients and carbon sources are more available and cells in the outer layer have more access to the substrates from the HSSL. However, it was expected that the *S. stipitis* cells would be more active mostly in the cortex area in order to be more protected from the inhibitory compounds. So this result seems to indicate that nutrient limitation has a greater impact than inhibitor stress under the studied conditions. Still the possibility of distinguishing viable cells amongst the total number could not be achieved with the utilized fixation and embedding processes. An improvement in this field or the use of another technique should allow the analysis of the areas where viable yeast cells are located in the end of the fermentation.

Conclusions

Cultivations in rich medium, both with glucose or xylose, showed that *S. stipitis* has the potential to be an ethanol producer using fermentative processes. Immobilization in calcium alginate matrix improved the ethanol production by creating aeration levels for *S. stipitis* close to those present in microaerophilic conditions. The screening assays showed differences in *S. stipitis* performance in six different levels of *bd*HSSL and 60% *bd*HSSL enabled a good compromise between the amount of inhibitors and the ability of the yeast to grow on a lignocellulosic-based by-product. Immobilization and pH 5.5, when applied simultaneously, led to a 1.3-fold increase of $Y_{E/S}$ when comparing to the suspended culture without pH control, reaching 0.26 g ethanol.(g glucose + xylose)$^{-1}$. Without pH control, immobilization seems to have a negative impact on the ethanol production, probably because it reduces the substrate uptake rate over the matrix. The process to perform the microtome cutting coupled with the microscopic observation allowed the analysis of the cell distribution inside the immobilized beads. To the best of our knowledge, it was the first time that this technique was used for this purpose. During cultivations, the cells have grown and/or migrated in opposite direction of the cortex, suggesting that the lack of nutrients and mass transfer limitations played a more important role than the inhibitors protection. This study contributed to a better knowledge of the influence that immobilization alone or together with pH can have on the metabolites distribution in *S. stipitis* cultivations. Further investigations under controlled microaerophilic conditions should be performed in order to evaluate if the pH control and immobilization have the same positive effects on ethanol production.

Abbreviations

*bd*HSSL: bio-detoxified HSSL; CDW: cell dry weight; glc: glucose; HSSLs: hardwood SSLs; SSLs: spent sulfite liquors; xyl: xylose; YM: yeast medium; YNB: yeast nitrogen base.

Competing interests

The authors declare that they have no competing interests.

Authors' contributions

DPN participated in the design of the study, performed the experimental work and wrote the manuscript. VSN and SRP participated in the design of the project, provided technical and scientific guidance and were involved in the manuscript draft. SCC participated in the design, analysis and interpretation of the *S. stipitis* beads data obtained through light microscopy. AJS helped on the interpretation of the data and figure manipulation. AMRBX conceived the idea, was part of the design of the study and provided scientific guidance and constructive feedback during the draft of the manuscript. All authors read and approved the final manuscript.

Acknowledgements

This work was partly funded in the scope of the project CICECO-Aveiro Institute of Materials (Ref. FCT UID/CTM/50011/2013), financed by national funds through the FCT/MEC and when applicable co-financed by FEDER under the PT2020 Partnership Agreement and also Swedish Energy Agency (Energimyndigheten). Marie F. Gorwa-Grauslund is gratefully acknowledged for scientific guidance and constructive discussions during the preparation of this manuscript. Dmitry Evtyugin is gratefully acknowledged for HSSL immobilization discussions. Eng. A. Prates from CAIMA - ALTRI Group, Indústria de Celulose S.A., Constância, Portugal is acknowledged for providing the HSSL. Authors also thank FCT for the PhD grant of S. R. Pereira (SFRH/BD/64552/2009). SCC was supported by a grant (SFRH/BPD/68537/2010) from the financing program 'QREN - POPH - Tipologia 4.1 - Formação Avançada' and by the European Social Funding (FSE) and the Portuguese Ministry of Education and Science (MEC).

Author details

[1]CICECO-Aveiro Institute of Materials and Chemical Department, University of Aveiro, Campus Universitário de Santiago, 3810-193 Aveiro, Portugal. [2]Division of Applied Microbiology, Department of Chemistry, Lund University, PO Box 124, SE-221 00 Lund, Sweden. [3]Department of Biology and GeoBioTec Research Unit, University of Aveiro, Campus Universitário de Santiago, 3810-193 Aveiro, Portugal.

References

1. Nigam JN (2001) Ethanol production from hardwood spent sulfite liquor using an adapted strain of *Pichia stipitis*. Journal of Industrial Microbiology & Biotechnology 26(3):145–150
2. Pereira SR, Portugal-Nunes DJ, Evtuguin DV, Serafim LS, Xavier AMRB (2013) Advances in ethanol production from hardwood spent sulphite liquors. Process Biochemistry 48(2):272–282. http://dx.doi.org/10.1016/j.procbio.2012.12.004
3. Jeffries TW, Grigoriev IV, Grimwood J, Laplaza JM, Aerts A, Salamov A, Schmutz J, Lindquist E, Dehal P, Shapiro H, Jin YS, Passoth V, Richardson PM (2007) Genome sequence of the lignocellulose-bioconverting and xylose-fermenting yeast *Pichia stipitis*. Nature Biotechnology 25(3):319–326
4. Pereira SR, Ivanuša Š, Evtuguin DV, Serafim LS, Xavier AMRB (2012) Biological treatment of eucalypt spent sulphite liquors: a way to boost the production of second generation bioethanol. Bioresource Technology 103(1):131–135. doi:10.1016/j.biortech.2011.09.095
5. Xavier AMRB, Correia MF, Pereira SR, Evtuguin DV (2010) Second-generation bioethanol from eucalypt sulphite spent liquor. Bioresource Technology 101(8):2755–2761. doi:10.1016/j.biortech.2009.11.092
6. Parajó JC, Domínguez H, Domínguez J (1998) Biotechnological production of xylitol. Part 3: operation in culture media made from lignocellulose hydrolysates. Bioresource Technology 66(1):25–40
7. Palmqvist E, Hahn-Hägerdal B (2000) Fermentation of lignocellulosic hydrolysates I: inhibition and detoxification. Bioresource Technology 74:17–24
8. Marques AP, Evtuguin DV SM, Amado FML, Prates A (2009) Structure of lignosulphonates from acidic magnesium-based sulphite pulping of *Eucalyptus globulus*. Journal of Wood Chemistry and Technology 29:337–357
9. Larsson S, Palmqvist E, Hahn-Hägerdal B, Tengborg C, Stenberg K, Zacchi G (1999) The generation of fermentation inhibitors during dilute acid hydrolysis of softwood. Enzyme and Microbial Technology 24:151–159
10. Palmqvist E, Grage H, Meinander NQ, Hahn-Hägerdal B (1999) Main and interaction effects of acetic acid, furfural, and p-hydroxybenzoic acid on growth and ethanol productivity of yeasts. Biotechnology and Bioengineering 63:46–55
11. Palmqvist E, Almeida JS, Hahn-Hägerdal B (1999) Influence of furfural on anaerobic glycolytic kinetics of *Saccharomyces cerevisiae* in batch culture. Biotechnology and Bioengineering 62(4):447–454
12. Ding MZ, Wang X, Yang Y, Yuan YJ (2011) Metabolomic study of interactive effects of phenol, furfural, and acetic acid on *Saccharomyces cerevisiae*. OMICS 15(10):647–653
13. Sanchez OJ, Cardona CA (2008) Trends in biotechnological production of fuel ethanol from different feedstocks. Bioresource Technology 99(13):5270–5295
14. Passoth V, Cohn M, Schafer B, Hahn-Hägerdal B, Klinner U (2003) Analysis of the hypoxia-induced ADH2 promoter of the respiratory yeast *Pichia stipitis* reveals a new mechanism for sensing of oxygen limitation in yeast. Yeast 20:39–51
15. Klinner U, Fluthgraf S, Freese S, Passoth V (2005) Aerobic induction of respirofermentative growth by decreasing oxygen tensions in the respiratory yeast *Pichia stipitis*. Applied Microbiology and Biotechnology 67:247–253
16. Huang C-F, Lin T-H, Guo G-L, Hwang W-S (2009) Enhanced ethanol production by fermentation of rice straw hydrolysate without detoxification using a newly adapted strain of *Pichia stipitis*. Bioresource Technology 100(17):3914–3920
17. Helle SS, Murray A, Lam J, Cameron DR, Duff SJB (2004) Xylose fermentation by genetically modified *Saccharomyces cerevisiae* 259ST in spent sulfite liquor. Bioresource Technology 92:163–171
18. Palmqvist E, Hahn-Hägerdal B, Szengyel Z, Zacchi G, Rèczey K (1997) Simultaneous detoxification and enzyme production of hemicellulose hydrolysates obtained after steam pretreatment. Enzyme and Microbial Technology 20:286–293
19. Jönsson LJ, Palmqvist E, Nilvebrant NO, Hahn-Hägerdal B (1998) Detoxification of wood hydrolysates with laccase and peroxidase from the white-rot fungus *Trametes versicolor*. Applied Microbiology and Biotechnology 49:691–697
20. Oliveira MED, Vaughan BE, Rykiel EJ, Jr (2005) Ethanol as fuel: energy, carbon dioxide balances and ecological footprint. Biosci 55(7):593–602
21. Desimone MF, Degrossi J, D'Aquino M, Diaz LE (2002) Ethanol tolerance in free and sol–gel immobilised *Saccharomyces cerevisiae*. Biotechnology Letters 24:1557–1559
22. Pourbafrani M, Talebnia F, Niklasson C, Taherzadeh MJ (2007) Protective effect of encapsulation in fermentation of limonene-contained media and orange peel hydrolyzate. International Journal of Molecular Sciences 8:777–787
23. Nikolić S, Mojović L, Rakin M, Pejin D (2009) Bioethanol production from corn meal by simultaneous enzymatic saccharification and fermentation with immobilized cells of *Saccharomyces cerevisiae* var. *ellipsoideus*. Fuel 88(9):1602–1607. http://dx.doi.org/10.1016/j.fuel.2008.12.019
24. Kourkoutas Y, Bekatrou A, Banat IM, Marchant R, Koutinas AA (2004) Immobilization technologies and support materials suitable in alcohol beverages production: a review. Food Microbiology 21:277–397
25. Zhang W, Mou H, Cui H, Zhang Y (2014) Isolation of xylose fermentation strains for ethanol production and xylose fermentation research. Chemical and Biochemical Engineering Quarterly 28(1):117–124
26. Varanasi P, Singh P, Auer M, Adams PD, Simmons BA, Singh S (2013) Survey of renewable chemicals produced from lignocellulosic biomass during ionic liquid pretreatment. Biotechnology for Biofuels 6(14):1–9
27. Sugiura S, Oda T, Izumida Y, Aoyagi Y, Satake M, Ochiai A, Ohkohchi N, Nakajima M (2005) Size control of calcium alginate beads containing living cells using micro-nozzle array. Biomaterials 26:3327–3331

Genotyping of amino acid-producing *Corynebacterium glutamicum* strains based on multi-locus sequence typing (MLST) scheme

Junjie Yang[1,2,4], Yimeng Kong[1,3] and Sheng Yang[1,2,4*]

Abstract

Background: Confusing parental information may hinder to dissect mechanisms of amino acid hyper-producing *Corynebacterium glutamicum* strains. Thus, an efficient method for genotyping of the *C. glutamicum* is heavily called.

Results: Multi-locus sequence typing (MLST) is currently the most popular molecular typing technique. But currently this method is not available for *C. glutamicum*. In this study, a MLST scheme was established based on sequences of seven housekeeping genes, for genotyping of *C. glutamicum*. The MLST method performed an efficient discrimination of 17 strains and helps to understand the population structure of this bacterium.

Conclusions: This work has expanded the MLST method to *C. glutamicum* and developed an efficient technique to discriminate strains of uncertain origin.

Keywords: *Corynebacterium glutamicum*; Multi-locus sequence typing; Amino acid producing; *Corynebacterium crenatum*; 16S rDNA

Background

Non-spore-forming Gram-positive bacteria *Corynebacterium glutamicum* are widely used in amino acid production industry with numerous ideal attributes [1,2]. Due to roughly identical 16S rDNA sequences, many independent strains previously regarded as *Brevibacterium* and *Corynebacterium* species in the Corynebacteriaceae family have been categorized into the *C. glutamicum* species, especially *Brevibacterium lactofermentum*, *Brevibacterium flavum*, and *Corynebacterium acetoacidophilum* [3-5]. Consequently, the *C. glutamicum* species dropped into a mixture of strains from diversified resources.

Clear genetic or phylogenetic information is occasionally missing after a long-term utilization and preservation. Confusing parental information may hinder to dissect mechanisms of amino acid hyper-production in these strains. For example, although genome sequences are now available for several glutamate-hyper-producing strains, e.g., SCgG1, SCgG2, and Z188, detail typing and parental information of these strains were still absent. To our knowledge, only an identification method by phage sensitivity has been reported, but this method is not very convenient because of the experimental procedure for handling phages [6]. Thus, an efficient method for genotyping of the *C. glutamicum* is heavily called for in this field.

Compared with other molecular typing techniques, multi-locus sequence typing (MLST) is currently the most popular one [7,8]. MLST relies on allelic variants in conserved genes to calculate phylogenetic relationship of strains. Multiple housekeeping genes (usually seven genes) are examined in the analysis as their sequences are constrained and variations are nearly neutral. Each strain is then assigned with an allelic profile or sequence type (ST). The MLST method excels other molecular

* Correspondence: syang@sibs.ac.cn
[1]Key Laboratory of Synthetic Biology, Institute of Plant Physiology and Ecology, Shanghai Institutes for Biological Sciences, Chinese Academy of Sciences, 300 Fenglin Road, Shanghai 200032, China
[2]Shanghai Research Center of Industrial Biotechnology, Shanghai 201201, China
Full list of author information is available at the end of the article

typing methods of being unambiguous and easily portable between laboratories.

To our knowledge, in the family of Corynebacteriaceae, MLST scheme is only developed in *Corynebacterium diphtheriae*. More than 70 STs were identified and fall into at least 11 groups of *C. diphtheriae*. Meanwhile, no ST is currently available for *C. glutamicum*. In order to have accurate typing techniques for *C. glutamicum*, we developed a MLST scheme, which would promote applied research of this amino acid-producing species.

Methods

The strains used in this study are listed in Table 1 [3-6,9-13]. Genome sequences of the strains used for MLST analysis include ATCC13032 [5,2,1], MB1 [14], R [15], ScCG1, ScCG2, ATCC14067 [16], S9114 [17], Z188, '*C. crenatum*' MT. Primers 27F 5′-AGAGTTTGATCMTGGCTCAG-3′ and 1492R 5′-TACGGYTACCTTGTTACGACTT-3′ were used to identify 16S rDNA sequences of the strains [3,4]. The retrieved 16S rDNA sequences are almost identical, with a similarity >99% compared with the sequence of the *C. glutamicum* ATCC 13032, indicating that all of the strains belong to *C. glutamicum* [3,4].

Referring to the genotyping scheme in *C. diphtheriae* [7], another species belonging to the same genus, seven housekeeping genes of the *C. glutamicum* strains were selected for analysis, including *atpA*, *dnaE*, *dnaK*, *fusA*, *rpoB*, *leuA*, and *odhA* (Table 2). All the loci are single copy in the genome sequences, according to the BLAST analysis. Considering the possibility to meet the case when only crude amino acid fermentation samples are available, which would limit the efficiency of PCR reactions, approximately 300~400 bp length of the PCR amplicon in housekeeping genes was designed for downstream experiments and analysis (Table 2).

Results and discussion

The PCR amplicon size ranges from 318 bp (*leuA*) to 402 bp (*fusA*), with a mean size of 369 bp (Table 2, Additional file 1: Figure S1 and Dataset S1). Each locus of housekeeping gene alleles for comparison and identification has a length ranging from 150 bp (*leuA*) to 243 bp (*dnaK*), with a mean length of 206 bp. All alleles for a given locus are of equal lengths and in a correct reading frame. The proportion of variation sites in each locus varies from 3% (*odhA*) to 17% (*leuA*) (Table 2). Primer and allele sequences are listed in Table 2 and Additional file 1: Dataset S2.

The dendrogram was drawn from allelic profile data using the UPGMA (unweighted pair group method with arithmetic mean) method. The online tool (http://pubmlst.org/) of the PHYLIP suite programs and Phylodendron [18] were implemented for tree generation, output, and display. The dendrogram demonstrates that eight groups of the strains could be classified according to their genetic distinctions (Figure 1). Except ST4/ST5 and ST9/ST10, which share six alleles respectively, most of STs share no more than three alleles with each other (Table 1), indicating that the strains could be well separated. The next section briefly describes the eight classified groups.

Group 1 (ST#1): The *C. glutamicum* type strain ATCC13032 is a member of this group. MB001 is a prophage-free variant of ATCC13032 with a 6% reduced genome [14]. Consistently, it has the same ST with ATCC13032.

Group 2 (ST#2): '*B. lactofermentum*' ATCC13869 represents an independent ST, as expected.

Group 3 (ST#3): '*C. acetoacidophilum*' ATCC13870 also represents an independent ST.

Group 4 (ST#4 and #5): Although '*B. flavum*' ATCC14067 and its derived strain ATCC21493 [13] contain different STs, in fact, there is only 1-bp difference (in *odhA*, resulting in Ala701Thr) among the seven analyzed genes. So, the classification is correct.

Group 5 (ST#6): *C. glutamicum* R [15] represents an independent ST.

Group 6 (ST#7): '*C. pekinense*' AS1.299 and its derived strain AS1.563 [9,10] contain the same ST and were consequently classified into the same group.

Group 7 (ST#8): *C. glutamicum* 617 (SIIM B1) also represents an independent ST.

Group 8 (ST#9 and #10): Several '*B. tianjinese*' and '*C. crenatum*' strains, including both 'wild-type' strains [11,12,19] and derived strains [12,17,20], represent the same ST (ST#9), so they are categorized in this group. The strain '*C. crenatum*' MT also fall into the group. It has a different ST (ST#10) from 'wild-type' strain AS1.542 with only one different locus.

Taken together, the MLST results reveal a correct typing of the related strains. We also tested a commercially obtained crude lysine sample and several amino acid-producing strains preserved in our laboratory. The crude lysine sample represented a MLST result same to 13032, so it would be produced by an ATCC 13032-derived strain. Three lysine-producing strains fall into group 1 (ST#1), group 2 (ST#2), and group 6 (ST#7), separately. A proline-producing strain falls into group 3 (ST#3). These results suggest that the eight groups defined in this study could encompass most of the amino acid-producing strains for laboratory use.

The aim of this work was to expand the MLST method to *C. glutamicum* and develop an efficient technique to discriminate strains of uncertain origin. After a long-term preservation, activation, and fermentation, parental information of the amino acid-producing strains is easily confused. The MLST scheme developed in our study addresses the problem

Table 1 Details and allelic profile of the strains used in this study

ST	Group	Strains	Synonym	Descriptions	Genome sequence[a]	Allelic profile						
						atpA	*dnaE*	*dnaK*	*fusA*	*leuA*	*odhA*	*rpoB*
		ATCC13032		*C. glutamicum* type strain	NC_003450, NC_006958							
1	1	MB001		Prophage-free variant of ATCC 13032 with a 6% reduced genome	NC_022040	1	1	1	1	1	1	1
		Crude sample		A crude lysine sample, commercially obtained								
2	2	ATCC13869	*B. lactofermentum*	'Wild-type *B. lactofermentum*'		1	2	2	4	2	4	4
3	3	ATCC13870	*C. acetoacidophilum*	'Wild-type *C. acetoacidophilum*'		4	6	5	5	6	1	1
4	4	ATCC14067	*B. flavum*	J. Bacteriol. 194 (3), 742-743 (2012)	AGQQ01000000	3	2	4	6	2	2	2
5	4	ATCC21493	*B. flavum*	Producing arginine, derived from ATCC 14067 (SIIM B234)		3	2	4	6	2	5	2
6	5	R		*C. glutamicum* isolated in Japan from a meadow soil sample	NC_009342	5	3	7	3	3	2	1
7	6	AS1.299	*C. pekinense*	'Wild-type *C. pekinense*,' producing glutamate (=CICC 10119, SIIM B3)		2	5	3	5	4	3	3
		AS1.563	*C. pekinense*	(=CICC 10178, SIIM B165) producing lysine, derived from AS1.299								
8	7	617		A glutamate-producing strain previously used in China (=CICC 10117, SIIM B1)		1	2	4	7	7	3	2
		T6-13	*B. tianjinese*	'Wild-type *B. tianjinese*' (=CICC 20182, SIIM B226)								
		SCgG1		Hyper-producing glutamate	NC_021351							
9	8	SCgG2		Hyper-producing glutamate	NC_021352	5	4	6	2	5	3	1
		S9114		A strain for industrial production of glutamate	AFYA01000000							
		Z188		Hyper-producing glutamate	AKXP01000000							
		AS1.542	*C. crenatum*	'Wild-type *C. crenatum*' (=CICC10124, SIIM B6)								
10	8	MT	*C. crenatum*	A mutant of AS1.542	AQPS01000000	6	4	6	2	5	3	1

[a]DDBJ/EMBL/GenBank accession number.

Table 2 Primer sequence and characteristics of the seven loci used in MLST

Gene	primer	Sequence (5'-3')	Amplicon size (bp)	Amplicon bps (from-to) in NC_006958.1	Allele size (bp)	Number of distinct alleles	Number of polymorphic sites
atpA	atpA_F	ATGTACCAGGGCAACCAC	390	1277256-1277645	240	6	13
	atpA_R	GGCAACCTTCTTCATACC					
dnaE	dnaE_F	GGCGAAGGATATTCCGTTG	366	2220485-2220120	225	6	23
	dnaE_R	ACTTCCCCATCGCGGTTG					
dnaK	dnaK_F	CACCTCACAGGAAATCTC	397	2959446-2959050	243	7	17
	dnaK_R	GGACTGGAACTTCTCTAC					
fusA	fusA_F	CAAGGCAGCTATCCGTAAGATG	402	524614-525015	237	7	11
	fusA_R	ATGTTACCAGCGTGTGCAACC					
leuA	leuA_F	CTGGTTCAGGCTCGTGAGCA	318	267591-267274	150	7	26
	leuA_R	GATCATTGGGTTCTCAGGA					
odhA	odhA_F	CTTTGGCTGGGTCATGGAAGG	327	1175355-1175681	186	5	5
	odhA_R	CTGGGCATCGTGCCAGAAAC					
rpoB	rpoB_F	TATGTGACCGCGGAGTTC	357	513278-513634	165	4	5
	rpoB_R	GAAACGCTCGGTGATCTG					

and proved to be an efficient way to distinguish them. According to our results, three glutamate-producing strains mentioned above (SCgG1, SCgG2, and Z188) would derive from T6-13 or AS1.542. Their common genetic origin serves a firm premise for downstream comparative genomic analysis to compare parental strains and their offsprings. According to our MLST result, a lysine-producing strain CICC20042 was found to show ST#9 (including T6-13, S9114, AS1.542, etc.) (detailed data not shown). But according to the CICC database, CICC20042 is derived from AS1.563 (ST#7, including AS1.299 and AS1.563). This suggests that confusion about the parental information might have happened.

Several strains previously regarded as *Brevibacterium* and *Corynebacterium* species had been classified to *C. glutamicum* [3,4]. Our analysis extends the conclusion and proposes that *C. glutamicum* should also contain several *C. pekinense*, *B. tianjinese*, and *C. crenatum* strains. The 'wild-type' strains of *B. tianjinese* and *C. crenatum* exhibit the same ST. Both of them are highly sensitive to specific phages [6], supporting a close relationship of these strains. A brief comparative genomic analysis of S9114 ([GenBank:AFYA01000000], derived from *B. tianjinese* T6-13) and '*C. crenatum*' MT [GenBank:AQPS01000000] discovered only hundreds of SNPs and InDels (unpublished data). Some of these SNPs and InDels may be responsible for the high glutamate yield. Further genomic analyses should be conducted in a broad range of *C. glutamicum* strains, especially the parental wild-type strains, to reveal details of their relationships as well as mechanisms for amino acid hyper-production.

Conclusions

This work has expanded the MLST method to C. glutamicum and developed an efficient technique to discriminate strains of uncertain origin.

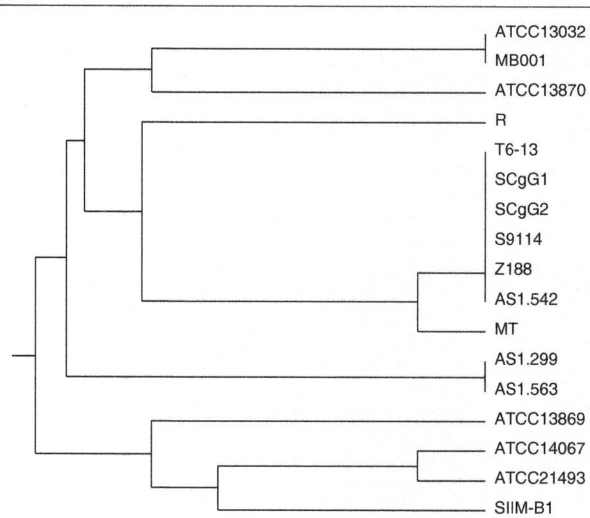

Figure 1 Dendrogram analysis of *Corynebacterium glutamicum* from allelic profile data. The dendrogram was generated with the UPGMA (unweighted pair group method with arithmetic mean) method based on the internal fragments of 7 housekeeping genes of the 17 strains. ATCC13032 and MB001, ATCC14067 and ATCC21493, and AS1.299 and AS1.563 were classified into the same groups, respectively. Several '*B. tianjinese*' and '*C. crenatum*' strains, including AS1.542, T6-13, S9114, etc., were also classified into the same group. Each of the other strains falls into an independent group separately.

Additional file

Additional file 1: Supplementary material. Figure S1. PCR fragments of seven genes: *atpA* 390 bp, *dnaE* 366 bp, *dnaK* 397 bp, *fusA* 402 bp, *rpoB* 357 bp, *leuA* 318 bp, and *odhA* 327 bp. Dataset S1. PCR fragment sequences of seven genes. Dataset S2. Allele sequences of seven genes.

Abbreviations

MLST: multi-locus sequence typing; ST: sequence type.

Competing interests

The authors declare that they have no competing interests.

Authors' contributions

JY and SY designed the research. JY carried out the experiments. JY and YK carried out the computational experiments. JY and YK wrote the manuscript. All authors read and approved the final manuscript.

Acknowledgements

We thank Prof. Ji-Bin Sun (Tianjin Institute of Industrial Biotechnology, Chinese Academy of Sciences) for providing the draft genome sequences of several strains. This work was supported by a grant from the National Basic Research Program (973 Program) of China (2014CB745100) and the National Key Technologies R&D Program of China (2012AA022101).

Author details

[1]Key Laboratory of Synthetic Biology, Institute of Plant Physiology and Ecology, Shanghai Institutes for Biological Sciences, Chinese Academy of Sciences, 300 Fenglin Road, Shanghai 200032, China. [2]Shanghai Research Center of Industrial Biotechnology, Shanghai 201201, China. [3]University of the Chinese Academy of Sciences, Beijing 100080, China. [4]Shanghai Collaborative Innovation Center for Biomanufacturing Technology, Shanghai 200237, China.

References

1. Ikeda M, Nakagawa S. The Corynebacterium glutamicum genome: features and impacts on biotechnological processes. Appl Microbiol Biotechnol. 2003;62(2–3):99–109. doi:10.1007/s00253-003-1328-1.
2. Kalinowski J, Bathe B, Bartels D, Bischoff N, Bott M, Burkovski A, et al. The complete Corynebacterium glutamicum ATCC 13032 genome sequence and its impact on the production of L-aspartate-derived amino acids and vitamins. J Biotechnol. 2003;104(1–3):5–25.
3. Liebl W, Ehrmann M, Ludwig W, Schleifer KH. Transfer of Brevibacterium divaricatum DSM 20297T, "Brevibacterium flavum" DSM 20411, "Brevibacterium lactofermentum" DSM 20412 and DSM 1412, and Corynebacterium glutamicum and their distinction by rRNA gene restriction patterns. Int J Syst Bacteriol. 1991;41(2):255–60.
4. Pascual C, Lawson PA, Farrow JA, Gimenez MN, Collins MD. Phylogenetic analysis of the genus Corynebacterium based on 16S rRNA gene sequences. Int J Syst Bacteriol. 1995;45(4):724–8.
5. Kinoshita S, Nakayama K, Akita S. Taxonomical study of glutamic acid accumulating bacteria, Micrococcus glutamicus nov. sp. Bull Agric Chem Soc Jpn. 1958;22:176–85.
6. Ge B-Z, Wang J-X, Zhu S-J, Si X-D. Identification of glutamic acid producing strains by phages (in Chinese with English abstract). Virol Sin. 1991;6(3):256–9.
7. Bolt F, Cassiday P, Tondella ML, Dezoysa A, Efstratiou A, Sing A, et al. Multilocus sequence typing identifies evidence for recombination and two distinct lineages of Corynebacterium diphtheriae. J Clin Microbiol. 2010;48(11):4177–85. doi:10.1128/JCM. 00274-10.
8. Maiden MC, van Rensburg MJ, Bray JE, Earle SG, Ford SA, Jolley KA, et al. MLST revisited: the gene-by-gene approach to bacterial genomics. Nat Rev Microbiol. 2013;11(10):728–36. doi:10.1038/nrmicro3093.
9. Chen Q, Zhang Z-Y, Li L-G. A new L-glutamic acid-producing species of Corynebacterium (in Chinese with English abstract). Acta Microbiol Sinica (Wei Sheng Wu Xue Bao). 1973;13(1):1–6.
10. Institute_of_Microbiology_Chinese_Academy_of_Sciences, Hangzhou_Glutamate_Factory. Study on the production of lysine by auxotrophic mutant of Corynebacterium pekinense AS1.299 (in Chinese). Microbiol China (Wei Sheng Wu Xue Tong Bao). 1974;1(1):7–11.
11. Chen Q, Li L-G. Studies on L-glutamic acid producing bacteria AS 1.542. I. Identification of strain AS 1.542 (in Chinese with English abstract). Acta Microbiol Sinica (Wei Sheng Wu Xue Bao). 1975;15(2):119–24.
12. Zhang K, Liu Y. Studies on glutamate dehydrogenase from Brevibacterium tianjinese T6-13 (in Chinese with English abstract). Acta Microbiol Sinica (Wei Sheng Wu Xue Bao). 1991;31(4):281–6.
13. Onoda T, Yoshinaga F, Kubota K, Kamijo H, Okumura S (1973) Method of producing l-arginine by microorganism. US Patent 3723249, patent date 27/03/1973
14. Baumgart M, Unthan S, Ruckert C, Sivalingam J, Grunberger A, Kalinowski J, et al. Construction of a prophage-free variant of Corynebacterium glutamicum ATCC 13032 for use as a platform strain for basic research and industrial biotechnology. Appl Environ Microbiol. 2013;79(19):6006–15. doi:10.1128/AEM. 01634-13.
15. Yukawa H, Omumasaba CA, Nonaka H, Kos P, Okai N, Suzuki N, et al. Comparative analysis of the Corynebacterium glutamicum group and complete genome sequence of strain R. Microbiology. 2007;153(Pt 4):1042–58. doi:10.1099/mic.0.2006/003657-0.
16. Lv Y, Liao J, Wu Z, Han S, Lin Y, Zheng S. Genome sequence of Corynebacterium glutamicum ATCC 14067, which provides insight into amino acid biosynthesis in coryneform bacteria. J Bacteriol. 2012;194(3):742–3. doi:10.1128/JB.06514-11.
17. Lv Y, Wu Z, Han S, Lin Y, Zheng S. Genome sequence of Corynebacterium glutamicum S9114, a strain for industrial production of glutamate. J Bacteriol. 2011;193(21):6096–7. doi:10.1128/JB.06074-11.
18. Jolley KA, Maiden MC. BIGSdb: scalable analysis of bacterial genome variation at the population level. BMC Bioinformatics. 2010;11:595. doi:10.1186/1471-2105-11-595.
19. Xu M, Rao Z, Dou W, Xu Z. The role of ARGR repressor regulation on L-arginine production in Corynebacterium crenatum. Appl Biochem Biotechnol. 2013;170(3):587–97. doi:10.1007/s12010-013-0212-4.
20. Yun F, Zhou W (1994) Breeding and application of a strain of high glutamic acid yielding bacterium S9114 (in Chinese with English abstract). J South China University of Technol (Natural Science). 1994;22(1):56-62.

17

Adaptation of *Pseudomonas* sp. AKS2 in biofilm on low-density polyethylene surface: an effective strategy for efficient survival and polymer degradation

Prosun Tribedi, Anirban Das Gupta and Alok K Sil*

Abstract

Background: *Pseudomonas* sp. AKS2 can efficiently degrade low-density polyethylene (LDPE). It has been shown that this degradation of LDPE by AKS2 is correlated to its ability to form biofilm on the polymer surface. However, the underlying mechanism of this biofilm-mediated degradation remains unclear. Since bioremediation potential of an organism is related to its adaptability in a given environment, we hypothesized that AKS2 cells undergo successful adaptation in biofilm on LDPE, which leads to higher level of LDPE degradation. To verify this, the current study investigated a number of parameters of AKS2 cells in biofilm that are known to be involved in adaptation process.

Results: Successful adaptation always develops a viable microbial population. So we examined the viability of AKS2 cells in biofilm. We observed the presence of viable population in the biofilm. To gain an insight, the growth of AKS2 cells in biofilm on LDPE at different time points was examined. Results showed a better reproductive competence and more colonization for AKS2 biofilm cells than planktonic cells, indicating the increased fitness of AKS2 biofilm cells than their planktonic counterpart. Towards understanding fitness, we determined the hydrolytic activity, different carbon source utilization potentials, functional diversity and homogeneity of AKS2 biofilm cells. Results showed increased hydrolytic activity (approximately 31%), higher metabolic potential, higher functional diversity (approximately 27%) and homogeneity for biofilm-harvested cells than planktonic cells. We also examined cellular surface hydrophobicity, which is important for cellular attachment to LDPE surface. Consistent with the above results, the cell surface hydrophobicity of biofilm-harvested AKS2 cells was found to be higher (approximately 26%) compared to that of their planktonic counterpart. All these results demonstrated the occurrence of physiological as well as structural adaptations of AKS2 cells in biofilm on LDPE surface that resulted in better attachment, better utilization of polymer and better growth of AKS2 cells, leading to the development of a stable colony on LDPE surface.

Conclusions: The present study shows that AKS2 cells in biofilm on LDPE surface undergo successful adaptation that leads to enhanced LDPE degradation, and thus, it helps us to understand the underlying mechanism of biofilm-mediated polymer degradation process by AKS2 cells.

Keywords: Polyethylene-based plastic material; Bioremediation; *Pseudomonas*; Biofilm; Adaptation

* Correspondence: alokksil7@gmail.com
Department of Microbiology, University of Calcutta, 35 B.C. Road, Kolkata 700019, India

Background

In the modern era, plastic-based materials have a variety of domestic and industrial applications. However, the widespread use of this non-biodegradable material poses a major threat to the environment. An example of a widely used non-biodegradable polymer is polyethylene. Although there are reports of microbial degradation of polyethylene, the rate is very slow [1,2]. Moreover, this microbial degradation requires the pre-oxidation of polyethylene, either by physical or chemical treatment [1-4]. Previously, we reported that *Pseudomonas* sp. AKS2 can degrade 5% ± 1% of low-density polyethylene (LDPE) in just 45 days, without any prior oxidation of polyethylene [5]. This report also documented that AKS2 developed biofilm on polyethylene surface efficiently, and there was a linear correlation between this biofilm formation and the ability to degrade polyethylene [5]. However, the underlying mechanism of this biofilm-mediated LDPE degradation by AKS2 remains unclear.

Biofilm represents a complex association of microorganisms in a given habitat [6], and its formation is a bacterial survival response to hostile environment [7]. Microorganisms are known to be capable of altering their structural and physiological activities through biofilm formation as this allows survival under varied environmental conditions. Such alteration in activities for better survival in a given habitat is known as adaptability.

In general, adaptation is a biological process by which an organism becomes more competent to live in a given habitat [8]. Following adaptation, microorganisms exhibit different physiological and structural activities compared to their non-adapted counterparts [9]. Thus, adaptation provides a kind of biological insurance for an organism to encounter varying environments in a given ecological niche. The adaptive traits may be structural, behavioural or physiological. Structural adaptation includes variations in shape and size of the organism. The alteration of membrane fluidity, by both psychrophilic and thermophilic bacteria, is an example of structural adaptation. This type of modulation in membrane fluidity serves as a protection against harsh temperatures. While behavioural adaptations are composed of inherited behaviour chains, physiological adaptations allow the organism to perform special functions for the adjustment of cellular growth and development, regulation of temperature, etc. Bacterial secretion of exo-polysaccharides for their attachment to a solid surface is an example of physiological adaptation.

Existing literature documents that bioremediation potential of an organism is related to its adaptability in a given environment [10]. Adapted *Rhodococcus erythropolis* DCL14 cells were shown to degrade alkanes and alcohols at higher rate compared to their non-adapted counterparts [11]. Thus, we hypothesized that successful adaptation of AKS2 population in biofilm on LDPE surface resulted in its sustained retention at this site, which resulted in enhanced polymer degradation.

To verify this hypothesis, we investigated the various parameters of AKS2 structure and physiology such as fitness, metabolic potential, functional diversity and homogeneity that are relevant to the adaptation. We observed increased fitness, higher levels of functional diversity and homogeneity, and increased cell surface hydrophobicity for biofilm adapted cells compared to the planktonic cells, indicating the physiological and structural adaptations of AKS2 cells in biofilm. Thus, these results demonstrate that successful adaptation of AKS2 cells in biofilm on LDPE surface resulted in enhanced degradation of LDPE.

Methods

Bacterial strain and culture condition

Pseudomonas sp. AKS2 was previously isolated from Kolkata municipal solid waste dumping ground soil (Kolkata, India) [12]. It is a potential degrader of polyethylene succinate (PES) [12] and LDPE [5]. This isolate was grown in 100 ml of sterile basal media containing 300 mg of sterile LDPE films at 30°C for different time periods as per the requirement of the experiments. Basal media were prepared as described previously [5]. Commercially available LDPE was used in all the experiments. LDPE films were made additive free by washing with 70% ethanol. The surface area of each polyethylene film used was 5 × 4 cm.

Dual staining for the assessment of AKS2 viability in biofilm

To determine the viability of AKS2 cells in biofilm on polyethylene surface after the incubation, LDPE films were removed from the conditioned medium and the adhered bacterial population, if any, were stained with 4 μg ml^{-1} acridine orange for 15 min [5]. Thereafter, LDPE films were washed with sterile Milli-Q water (Millipore Corporation, Billerica, MA, USA) and further treated with 4 μg ml^{-1} ethidium bromide for another 15 min. LDPE films were again washed with sterile Milli-Q water. Thereafter, dried films were observed under a fluorescence microscope (Olympus IX 71, Olympus Corporation, Tokyo, Japan).

Measurement of AKS2 fitness

Microbial fitness represents the ease of reproduction of an organism in a given environment. In order to examine the fitness of AKS2 biofilm cells, we compared the colonization and reproduction potential of biofilm-harvested cells with that of planktonic AKS2 cells. For this experiment, AKS2 cells were grown for 30 days in media containing LDPE films as sole C-source. After the incubation, cells that adhered to LDPE films were extracted. These cells represent biofilm-harvested cells. To compare the colonization and reproduction potential, biofilm-harvested and planktonic AKS2 cells were inoculated separately in equal numbers (approximately 10^4 cells) into 100 ml of basal media containing 300 mg of LDPE film as sole C-source and incubated at 30°C for different lengths of time. After incubation for 5 and 10 days, LDPE films were taken out from the growth media and examined under a fluorescence microscope after staining with acridine orange as described in the previous section. All experiments were performed in triplicate.

Fluorescein diacetate hydrolysis assay

To examine the hydrolytic activity of bacterial cells either harvested from biofilm or planktonic condition, fluorescein diacetate (FDA) hydrolysis assay was performed [13]. Briefly, equal numbers of cells (approximately 10^6 cells) taken from the respective origin were separately added to 1.5 ml of 60 mM sodium phosphate buffer, pH 7.6. FDA solution was added to it to attain a final concentration of 10 μg ml^{-1}. The flask was then shaken at 30°C for 30 min. These samples were then centrifuged at 6,000 rpm for 5 min, and the absorbance of the supernatant was measured by a spectrophotometer (V-630, Jasco, Tokyo, Japan) at 494 nm. Samples without FDA served as a control.

Evaluation of bacterial cell surface hydrophobicity

Cell surface hydrophobicity of biofilm-harvested cells and planktonic cells was measured by bacterial adhesion to hydrocarbon (BATH) assay [14]. For this purpose, equal numbers of biofilm-harvested and planktonic AKS2 cells (approximately 10^6 cells) were added separately to several tubes containing increasing volumes (ranging from 0 to 0.2 ml) of n-hexadecane. Tubes were then shaken for 10 min and allowed to stand for 15 min to complete the phase separation. The OD_{400} of the aqueous suspensions was measured. Cell surface hydrophobicity was calculated using the following formula:

$$\text{Cell surface hydrophobicity (in \%)} = 100 \times (\text{Initial OD}-\text{Final OD})/\text{Initial OD}$$

Physiological profiles of AKS2 population

The patterns of potential carbon source utilization by biofilm-harvested AKS2 cells and their planktonic counterpart were assessed using BiOLOG-ECO plates (Biolog, Bremen, Germany) containing triplicates of 31 different environmentally relevant carbon sources [15]. To perform the experiment, 150 μl of either biofilm-harvested cells or planktonic cells containing approximately 10^3 CFU was separately added into each well of the BiOLOG-ECO plates. The plates were incubated at 30°C for 72 h, and the absorbance of each well was recorded at 590 nm. Microbial metabolic activity in each microplate, expressed as average well colour development (AWCD), was determined as follows: $AWCD = \Sigma Absorbance_i/31$, where $Absorbance_i$ is the absorbance at 590 nm from each well. The Shannon diversity index (H), an indicator of functional diversity, was calculated using the following equation: $H = -\Sigma pi \ln pi$, where pi is the ratio of the activity on each substrate ($Absorbance_i$) to the sum of activities of all substrates ($\Sigma Absorbance_i$) and assuming absorbance at 590 nm of 0.25 as threshold for positive response [16]. The corresponding Lorenz curve that provides a graphical depiction of the information contained in the Shannon diversity index was plotted. Thereafter, this curve is used to derive the Gini coefficient (G), which is a measure of functional inequality, using the formula:

$$G = 1-2\int_0^1 L\, dF$$

where L is the Lorenz curve and F is the standardized cumulative distribution of the standardized population.

Cluster analysis

For cluster analysis, data from the richness tests using BiOLOG-ECO plates were collected from either biofilm-harvested cells or planktonic cells of AKS2. The similarity matrix was generated by Euclidean distances, which were used to build a dendrogram with the unweighted pair group mean averages (UPGMA) algorithm wherein the linkage was single. Cluster analysis was performed by using the software Minitab 16.

Statistical analysis

Experimental results were subjected to statistical analysis of one-way analysis of variance (ANOVA) in order to evaluate statistically significant differences among samples. Mean values were compared at different levels of significance using the software Minitab 16. All experiments were performed in triplicate.

Results

AKS2 cells exhibit increased fitness for their growth on LDPE surface

In a habitat, an efficient adaptation of an organism should lead to the development of a viable population. Therefore, to verify the hypothesis that AKS2 cells undergo adaptation in biofilm, we examined the viability of AKS2 cells in biofilm on LDPE surface by performing dual staining with ethidium bromide and acridine orange. Acridine orange stains both the dead and viable cells, whereas ethidium bromide selectively stains the dead cells as it cannot pass through the intact membrane of living cells. Consistent with our previously published report [5], we observed AKS2 biofilm formation from day 30 onwards and a considerable weight loss of LDPE film (5%) after incubation with AKS2 for 45 days (Additional file 1: Figure S1 and Additional file 2: Figure S2). Dual staining of LDPE film obtained after 30 days of incubation with AKS2 showed a large number of green cells (stained with acridine orange) in contrast to a few red cells (stained with ethidium bromide), indicating that majority of AKS2 cells present in biofilm were live (Figure 1). This result demonstrated the development of viable AKS2 population on polyethylene surface signifying an adaptation of AKS2 cells in biofilm on LDPE surface.

To better understand the adaptation, we examined the fitness of AKS2 cells in biofilm. Fitness is the ability of an organism to survive as well as reproduce in a given environment, and thus, successful adaptation is represented by better fitness. To examine the fitness of AKS2 cells in biofilm, we compared their colonization and reproduction potential on LDPE surface with that of planktonic AKS2 cells (see 'Methods' for details). For this purpose, biofilm-harvested and planktonic AKS2 cells were inoculated separately in equal numbers (approximately 10^4 cells) into basal media containing LDPE film as sole C-source and incubated for different lengths of time. The LDPE films were then examined under a fluorescence microscope after staining with acridine orange to monitor the extent of AKS2 attachment to LDPE. The result showed greater adherence to the LDPE film by biofilm-harvested cells than planktonic cells (Figure 2A). We also compared the reproduction efficiency between biofilm-harvested cells and planktonic cells of AKS2 and observed an approximately 2.5-fold increase in cell number for biofilm-harvested cells from day 5 to day 10 (Figure 2B). For planktonic cells, the corresponding fold increase over the same period of time was only approximately 1.4-fold (Figure 2B). These results indicate that biofilm-harvested cells have higher reproduction ability than their planktonic counterparts. Thus, there is a significant increase in the fitness of biofilm-harvested AKS2 cells compared to that of their planktonic counterpart.

Biofilm-harvested AKS2 cells exhibit increased functional diversity and metabolic activity

In an ecological niche, metabolic functional diversity influences productivity of an organism by increasing their ability to utilize a greater variety of nutrients [17]. Thus, it is possible that in the biofilm, AKS2 cells increase metabolic functional diversity, which in turn enhances their LDPE degradation capability. To verify this, we compared the functional diversity of biofilm-harvested and planktonic AKS2 cells by determining the Shannon diversity index from the utilization spectrum of 31 different eco-sensitive carbon sources in BiOLOG-ECO plates [18,19]. We observed that the Shannon diversity index is significantly higher (approximately 27%) for biofilm-harvested cells than for planktonic cells (Figure 3A). To further validate the result, we compared metabolic potentials of biofilm-harvested cells and planktonic cells. Metabolic potential of an organism was examined by measuring the AWCD of BiOLOG-ECO plate [18]. As expected, the biofilm-harvested cells of AKS2 showed higher level of AWCD compared to the planktonic cells (Figure 3B). Thus, the results demonstrated increased functional diversity and metabolic potential of biofilm-harvested AKS2 cells.

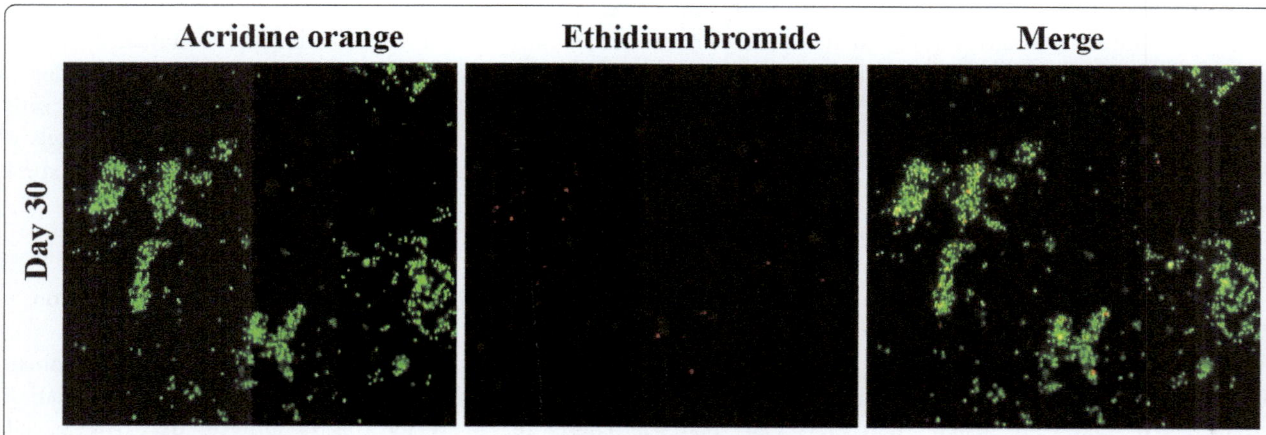

Figure 1 AKS2 develops viable microbial population in biofilm on polyethylene surface. LDPE films recovered from the conditioned media after 30 days of incubation were stained with acridine orange and ethidium bromide and observed under a fluorescence microscope. The figure is a representative of images obtained from 20 different fields and from three independent experiments.

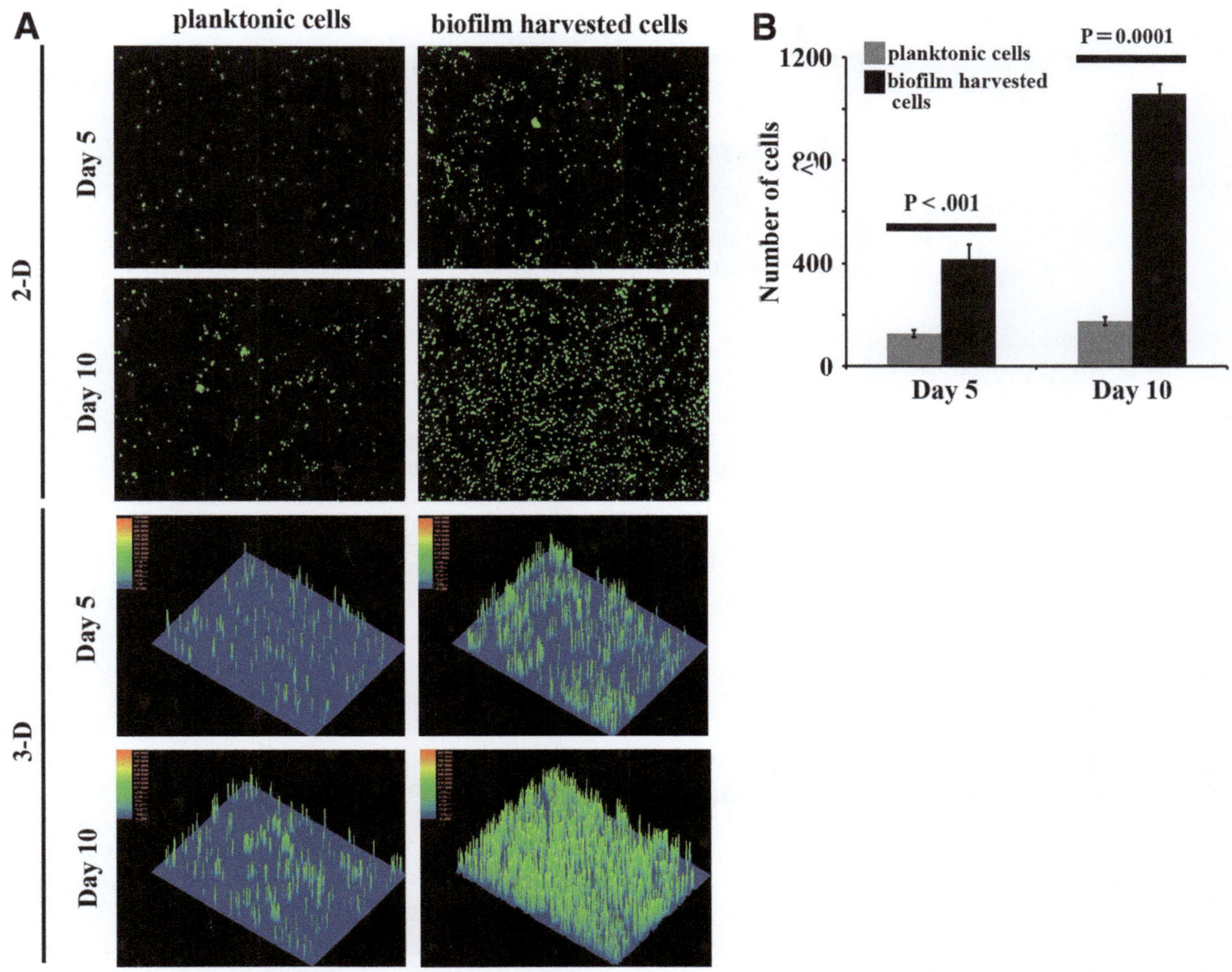

Figure 2 Colonization and reproduction efficiency comparison between biofilm-harvested and planktonic AKS2 cells. Equal numbers (approximately 10^4 CFU) of biofilm-harvested cells and planktonic cells of AKS2 were separately inoculated in basal media containing sterile LDPE films and incubated at 30°C for different time points. After the incubation, polyethylene films were recovered from the conditioned media, stained with acridine orange and observed under a fluorescence microscope. Microbial colonization efficiency **(A)** and reproduction efficiency **(B)** were examined. The figure is representative of images obtained from 20 different fields and from three independent experiments. Statistical significance between the groups of reproduction efficiency was evaluated by ANOVA.

Viable microbial cells produce a large array of hydrolytic enzymes, which can cleave FDA to produce fluorescein that can be detected spectrophotometrically, and this assay is widely used to measure cell metabolic activity [18,20]. FDA is sensitive to esterase and lipase activity as it contains an ester linkage. Therefore, to verify the metabolic activity, we compared the FDA hydrolysis activity of biofilm-harvested AKS2 cells with the corresponding activity of AKS2 planktonic cells. The result showed higher level (approximately 31%) of FDA hydrolysis activity in the extract obtained from biofilm-harvested AKS2 cells than in the extract from the planktonic cells (Figure 4). This result indicates that biofilm-harvested AKS2 cells harbour higher level of hydrolytic enzymes, which ensures the enhanced metabolic activity for biofilm cells than for planktonic cells.

AKS2 cells exhibited increased functional homogeneity in biofilm

Functional homogeneity or evenness of a microbial population contributes towards the development of a stable colony, and therefore, we determined the functional homogeneity of biofilm-harvested cells. To examine it, we plotted the Lorenz curves deduced from different carbon source utilization patterns by AKS2 under different conditions. Lorenz curve is the graphical representation of the degree of inequality in a population. The Lorenz curve for biofilm-harvested cells was found to be closer to the line of equality compared to their non-adapted planktonic counterpart (Figure 5A). The closer the curve to the line of equality, the higher the evenness will be, i.e., the system is more homogenously distributed. Thus, this result indicates higher

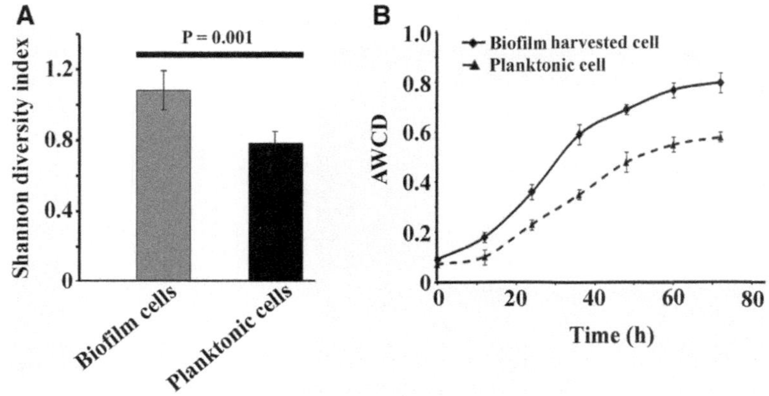

Figure 3 Functional diversity and metabolic activity profile of biofilm-harvested and planktonic AKS2 cells. Equal numbers (approximately 10^3 CFU) of either biofilm-harvested or planktonic cells of AKS2 were separately added to each well of BiOLOG-ECO plate and incubated for 3 days at 30°C. Absorbance at 590 nm of each well was recorded at different time points. Shannon diversity index **(A)** and average well colour development **(B)** were derived from well colour absorbance and plotted. Three replicates have been used for each experiment, and the result represents the average of these three replicates. Error bars indicate standard deviation (±SD). Statistical significance between the groups was evaluated by ANOVA.

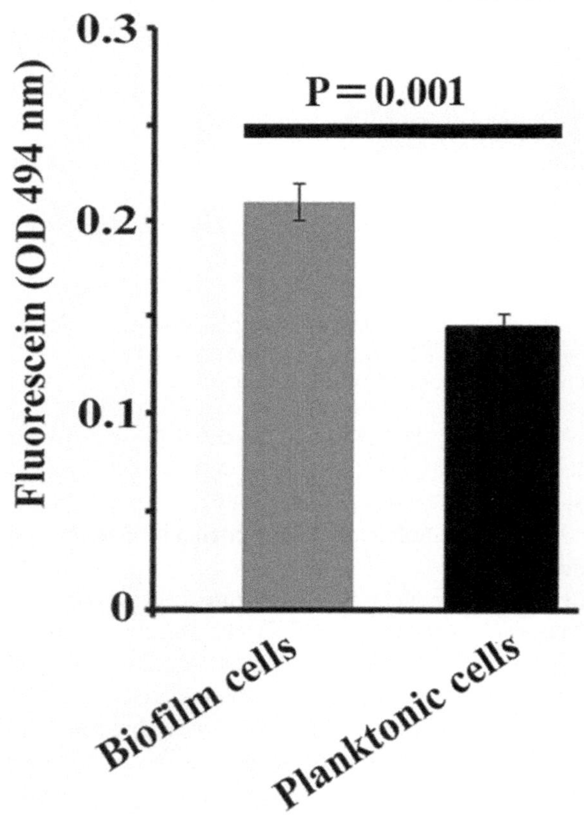

Figure 4 Comparison of hydrolytic activity. Equal numbers (approximately 10^6 CFU) of either biofilm-harvested or planktonic cells of AKS2 were separately examined for esterase activity by FDA hydrolysis. Three replicates have been used for each experiment, and the result represents the average of these three replicates. Error bars indicate standard deviation (±SD). Statistical significance between the groups was evaluated by ANOVA.

level of functional homogeneity for biofilm-harvested cells than planktonic cells. The subsequent analysis of Gini coefficient, a widely used inequality coefficient, showed a reduction in the coefficient value for biofilm-harvested cells compared to the planktonic cells (Figure 5A). Since evenness of a system varies inversely with the Gini coefficient, this result again indicates that AKS2 cells in biofilm have a higher degree of evenness with respect to their ability to utilize different C-sources than their planktonic counterpart. To lend support to this observation, we have also performed metabolic cluster analysis. AKS2 cells harvested from biofilm showed distinct and different patterns of metabolic cluster from their planktonic form (Figure 5B). Unlike planktonic cells, biofilm-harvested cells showed a large cluster (represented by a red colour) where most of the carbon sources were utilized with maximum similarity (Figure 5B). This result further confirms greater functional homogeneity or evenness for AKS2 cells in biofilm than for planktonic cells. Again, the increased evenness or functional homogeneity of AKS2 cells in biofilm contributes to better degradation of the polymer. Taken together, the increased levels of functional diversity, metabolic activity specially the hydrolytic activity and functional homogeneity indicate the occurrence of physiological adaptation of AKS2 cells on LDPE surface.

Biofilm-harvested AKS2 cells exhibit higher level of cell surface hydrophobicity

The cell surface hydrophobicity significantly contributes to cellular attachment to polymer surface, and thus, it plays an important role towards biofilm formation and polymer degradation as it brings the substrate (LDPE polymer) in close proximity to the enzyme [5,21,22].

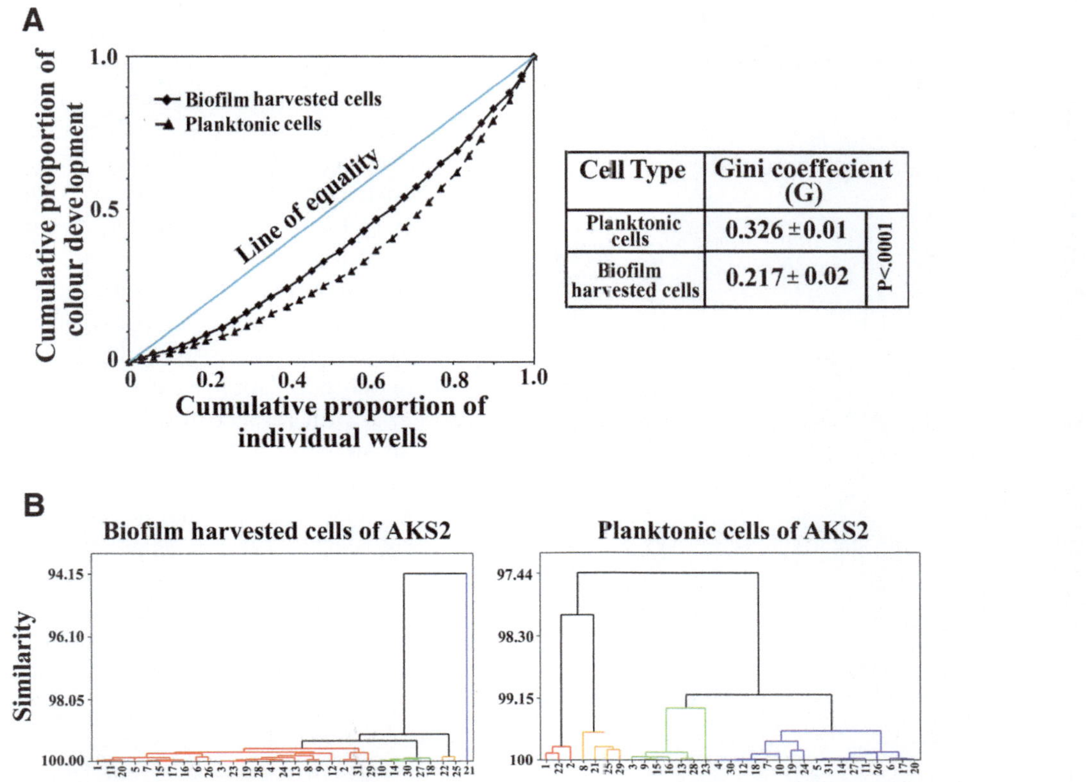

Figure 5 Biofilm-harvested AKS2 cells exhibit higher evenness. Equal numbers (approximately 10^3 CFU) of either biofilm-harvested or planktonic cells of AKS2 were separately added to each well of BiOLOG-ECO plate and incubated for 3 days at 30°C. Absorbance at 590 nm of each well was recorded at different time points. **(A)** Lorenz curves and Gini coefficient. **(B)** UPGMA cluster analysis. Statistical significance between the groups of Gini coefficient was evaluated by ANOVA.

Therefore, we examined the cell surface hydrophobicity of biofilm-harvested AKS2 cells and compared the same with planktonic cells. The result showed significantly higher (approximately 26%) level of surface hydrophobicity in biofilm-harvested cells than planktonic AKS2 cells (Figure 6). It indicates the occurrence of structural adaptation of AKS2 cells in biofilm on LDPE surface.

Polyethylene surface modulates the adaptation of AKS2

It is possible that LDPE surface may contribute towards the observed adaptation of AKS2 cells on its surface as abiotic components are known to play an important role in the stability of an ecosystem [23]. Towards understanding the role of LDPE in this adaptation, we compared functional diversity and evenness of AKS2 biofilm cells on this polymer with those of AKS2 biofilm cells formed on another polymer, polyethylene succinate (PES), as these two components have direct correlation with adaptation. We observed a significant difference in functional diversity and evenness between biofilm cells taken from each polymer (Figure 7A,B). Since we started this experiment with an equal number of AKS2 cells for both the polymers, the different levels of functional diversity can only be attributed to the difference in the polymeric surface. Thus, this result clearly suggests a possible role of each polymer in the adaptation process of AKS2 in a given habitat.

Discussion

The present study investigated structural and physiological properties of AKS2 cells during LDPE degradation. It has been reported that adapted marine microorganisms can survive in extremely unfavourable environmental conditions containing high concentrations of pollutants and toxic substances like heavy metals, hydrocarbons, xenobiotics and other recalcitrant compounds by forming biofilm [10]. Similarly, LDPE degradation by AKS2 was also found to be increased concomitantly with the increased biofilm formation [5]. Though biofilm was shown to enhance bioremediation, the high cell density inside a biofilm is known to cause a stressful environment [24,25]. The ability of an organism to adapt to the different microenvironments in biofilm is an important survival strategy against this environmental stress [26,27]. In addition, a previous report has documented that under stressful conditions, microorganisms undergo phenotypic diversification to enhance their adaptive potential [28]. Towards understanding the

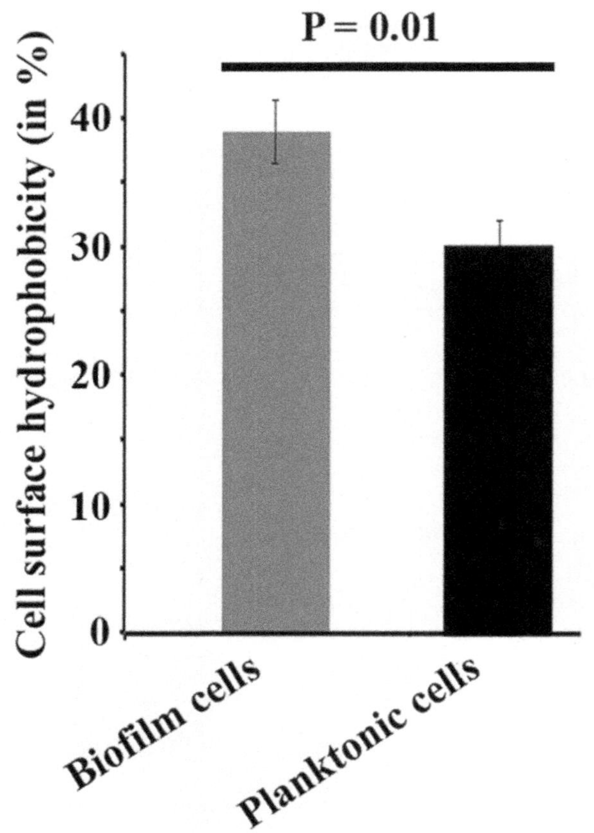

Figure 6 Comparison of cell surface hydrophobicity. Equal numbers of either biofilm-harvested cells or planktonic cells of AKS2 were separately examined for cell surface hydrophobicity by BATH assay. Three replicates have been used for each experiment, and the result represents the average of these three replicates. Error bars indicate standard deviation (±SD). Statistical significance between the groups was evaluated by ANOVA.

adaptation, we observed that biofilm cells on polyethylene surface exhibited higher metabolic activity, in particular hydrolytic activity, compared to planktonic cells. This increased metabolic activity may help these cells to degrade and utilize the polymer to establish a sustainable population. Biofilm-harvested cells also exhibited increased functional diversity. This indicates that microorganisms are undergoing phenotypic diversifications, which leads to better adaptation. Moreover, biofilm-harvested cells exhibit lower Gini coefficient which is an indicator of increased functional homogeneity. Therefore, majority of the individual AKS2 cells within the population have similar levels of metabolic potential with respect to utilization of a wide range of carbon sources. This enables the individual microorganisms to efficiently grow and establish a stable population in biofilm as the possibility of intra-species competition is greatly reduced. It has been documented that *Rhodococcus* tolerates extreme conditions by structural adaptations such as the modification of cell membrane and the alteration in cell surface hydrophobicity [11]. We also observed increased cell surface hydrophobicity of biofilm-harvested AKS2 cells compared to their planktonic counterpart. This increased cell surface hydrophobicity enables AKS2 cells to attach to the hydrophobic LDPE surface more efficiently compared to planktonic cells. The emergence of this trait demonstrated the occurrence of structural adaptation in biofilm of AKS2 cells on polyethylene surface. Taken together, the increased metabolic potential, higher level of functional diversity and homogeneity, and the increased surface hydrophobicity resulted in better colonization and higher reproduction potential of biofilm-harvested AKS2 cells.

Phenotypic alteration by an organism in an imposed condition has been considered an important strategy for better adaptation [29]. In our previous study, we observed that AKS2 cells in biofilm on LDPE surface

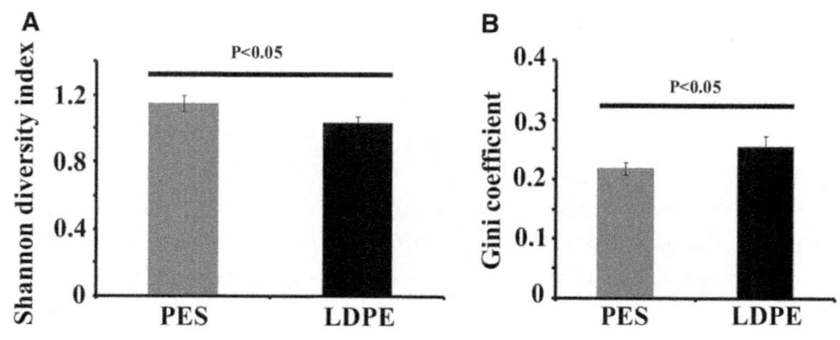

Figure 7 Functional diversity of AKS2 cells on LDPE and PES surface. Equal numbers (approximately 10^3 CFU) of biofilm-harvested cells of AKS2, taken from both LDPE and PES surfaces, were aseptically separately added to each well of BIOLOG-ECO plate and incubated for 3 days at 30°C. Absorbance at 590 nm of each well was recorded at different time points. Shannon diversity index **(A)** and Gini coefficient **(B)** were derived from well colour absorbance at 590 nm and plotted. Three replicates have been used for each experiment, and the result represents the average of these three replicates. Error bars indicate standard deviation (±SD). Statistical significance between the groups was evaluated by ANOVA.

exhibited a significant alteration in their shape and size compared to the planktonic form [5]. In biofilm, AKS2 cells become more round shaped and smaller in size than planktonic cells [5]. This structural adaptation may provide AKS2 cells a better access to the available nutrients in its surroundings. In the same study, we also observed that AKS2 cells adhering to LDPE secreted exo-polysaccharides for their better attachment and formation of biofilm on the polymer [5]. Again, it suggests the occurrence of physiological adaptations of AKS2 cells on LDPE surface. Collectively, these results demonstrated that AKS2 cells have adapted successfully in biofilm on LDPE surface resulting in a viable population and better degradation of the LDPE polymer.

The stability of an ecosystem depends on the balanced interactions between biotic and abiotic components [23]. These biotic and abiotic components are connected together through nutrient cycles and energy flows [30]. Thus, the abiotic surface of an ecosystem is likely to play a significant role towards adaptation of an organism in a given ecological niche. Towards this, our results showed different levels of functional diversity and evenness of the same organism, AKS2 cells, for two different polymers: LDPE and PES. Thus, the result demonstrates the possible involvement of LDPE towards adaptation of AKS2 cells in biofilm on LDPE surface.

Conclusions

Towards understanding biofilm-mediated LDPE degradation by AKS2, we verified the adaptation of AKS2 cells in biofilm by examining their viability and fitness. The results showed a viable population of AKS2 cells in biofilm with increased fitness compared to their planktonic counterpart. Further investigation revealed higher metabolic potential, higher functional diversity and homogeneity, and higher level of surface hydrophobicity for the biofilm-harvested AKS2 cells than planktonic cells. All these physiological and structural properties are known to be connected to the adaptability of an organism, and thus, these observations strongly support the view that AKS2 cells have adapted successfully in biofilm and thus developed a viable and stable population which resulted in enhanced polymer degradation. Thus, the current study deciphers the underlying mechanism of LDPE degradation by AKS2 biofilm cells with an enhanced rate.

In conclusion, the current study demonstrates structural and physiological adaptation of AKS2 cells in biofilm on polyethylene surface wherein the nature of the polymer plays an important role. This adaptation leads to enhanced LDPE degradation through biofilm formation.

Additional files

Additional file 1: Figure S1. Scanning electron micrograph. The micrograph shows the formation of biofilm by AKS2 on LDPE surface.

Additional file 2: Figure S2. Bar diagram. The bar diagram shows the extent of LDPE degradation by AKS2.

Abbreviations

ANOVA: Analysis of variance; AWCD: Average well colour development; BATH: Bacterial adhesion to hydrocarbon; CFU: Colony-forming unit; C-source: Carbon source; FDA: Fluorescein diacetate; *G*: Gini coefficient; *H*: Shannon diversity index; *L*: Lorenz curve; LDPE: Low-density polyethylene; OD: Optical density; PES: Polyethylene succinate; UPGMA: Unweighted pair group mean averages.

Competing interests

The authors declare that they have no competing interests.

Authors' contributions

PT and AKS conceived the idea and designed the experiments. PT performed all the experimental works. PT, AD and AKS interpreted the results and wrote the manuscript. All authors read and approved the final manuscript.

Acknowledgements

We thank Dr. Srimonti Sarkar for critical reading of the manuscript. This work is supported by a grant in aid from the Department of Biotechnology, Government of West Bengal, India (Sanction no. 555-BT (Estt)/RD-21/11).

References

1. Roy PK, Titus S, Surekha P, Tulsi E, Deshmukh C, Rajagopal C (2008) Degradation of abiotically aged LDPE films containing pro-oxidant by bacterial consortium. Polym Degrad Stab 93:1917–1922
2. Chatterjee S, Roy B, Roy D, Banerjee R (2010) Enzyme-mediated biodegradation of heat treated commercial polyethylene by *Staphylococcal* species. Polym Degrad Stab 95:195–200
3. Albertsson AC, Erlandsson B, Hakkarainen M, Karlsson S (1998) Molecular weight changes and polymeric matrix changes correlated with the formation of degradation products in biodegraded polyethylene. J Environ Polym Degrad 6:187–195
4. Volke-Sepulveda T, Saucedo-Castaneda G, Gutierrez-Rojas M, Manzur A, Favela-Torres E (2002) Thermally treated low density polyethylene biodegradation by *Penicillium pinophilum* and *Aspergillus niger*. J Appl Polym Sci 83:305–314
5. Tribedi P, Sil AK (2013) Low-density polyethylene degradation by *Pseudomonas* sp. AKS2 biofilm. Environ Sci Pollut Res Int 20:4146–4153
6. Cvitkovitch DG, Li YH, Ellen RP (2003) Quorum sensing and biofilm formation in streptococcal infections. J Clin Investig 112:1626–1632
7. Kim J, Kim HS, Han S, Lee JY, Oh JE, Chung S, Park HD (2013) Hydrodynamic effects on bacterial biofilm development in a microfluidic environment. Lab Chip 13:1846–1849
8. Dobzhansky T, Hecht MK, Steere WC (1968) On some fundamental concepts of evolutionary biology. In: Evolutionary biology volume 2 (1st edition). Appleton-Century-Crofts, New York, pp 1–34
9. Li YH, Hanna MN, Svensater G, Ellen RP, Cvitkovitch DG (2001) Cell density modulates acid adaptation in *Streptococcus mutans*: implications for survival in biofilms. J Bacteriol 183:6875–6884
10. Dash HR, Mangwani N, Chakraborty J, Kumari S, Das S (2013) Marine bacteria: potential candidates for enhanced bioremediation. Appl Microbiol Biotechnol 97:561–571
11. de Carvalho CCCR (2012) Adaptation of *Rhodococcus erythropolis* cells for growth and bioremediation under extreme conditions. Res Microbiol 163:125–136

12. Tribedi P, Sarkar S, Mukherjee K, Sil AK (2012) Isolation of a novel *Pseudomonas* sp. from soil that can efficiently degrade polyethylene succinate. Environ Sci Pollut Res Int 19:2115–2124
13. Chrzanowski TH, Crotty RD, Hubbard JG, Welch RP (1984) Applicability of the fluorescein diacetate method of detecting active bacteria in freshwater. Microb Ecol 10(2):179–185
14. Rosenberg M, Perry A, Bayer EA, Gutnick DL, Rosenberg E, Ofek I (1981) Adherence of *Acinetobacter calcoaceticus* RAG-1 to human epithelial cells and to hexadecane. Infect Immun 33:29–33
15. Choi KH, Dobbs FC (1999) Comparison of two kinds of BiOLOG microplates (GN and ECO) in their ability to distinguish among aquatic microbial communities. J Microbiol Method 36:203–213
16. Garland JL (1997) Analysis and interpretation of community-level physiological profiles in microbial ecology. FEMS Microbiol Ecol 24:289–300
17. Tilman D (2001) Functional Diversity. pp. 109-120. In: Encyclopedia of Biodiversity. Volume 3 (Levin, S.A., ed.). Academic Press, San Diego, 870 pp.
18. Teng Y, Luo Y, Sun M, Liu Z, Li Z, Christie P (2010) Effect of bioaugmentation by *Paracoccus* sp. strain HPD-2 on the soil microbial community and removal of polycyclic aromatic hydrocarbons from an aged contaminated soil. Bioresour Technol 101:3437–3443
19. Tribedi P, Sil AK (2013) Bioaugmentation of polyethylene succinate-contaminated soil with *Pseudomonas* sp. AKS2 results in increased microbial activity and better polymer degradation. Environ Sci Pollut Res Int 20:1318–1326
20. Killham K, Staddon WJ (2002) Bioindicators and sensors of soil health and the application of geostatistics. In: Burns RG, Dick R (eds) Enzymes in the environment: activity, ecology and applications. Marcel Dekker, New York, pp 391–405
21. Gilan(Orr) I, Hadar Y, Sivan A (2004) Colonization, biofilm formation and biodegradation of polyethylene by a strain of *Rhodococcus ruber*. Appl Microbiol Biotechnol 65:97–104
22. Balasubramanian V, Natarajan K, Hemambika B, Ramesh N, Sumathi CS, Kottaimuthu R, Rajash KV (2010) High-density polyethylene (HDPE)-degrading potential bacteria from marine ecosystem of Gulf of Mannar, India. Lett Appl Microbiol 51:205–211
23. Chapin FS, Pamela AM, Harold AM (2002) Principles of terrestrial ecosystem ecology. Springer, New York, ISBN 0-387-95443-0
24. deBeer D, Stoodley P, Roe F, Lewandowski Z (1994) Effects of biofilm structure on oxygen distribution and mass transport. Biotechnol Bioeng 43:1131–1138
25. Stoodley P, Sauer K, Davies DG, Costerton JW (2002) Biofilms as complex differentiated communities. Annu Rev Microbiol 56:187–209
26. Aertsen A, Michiels CW (2004) Stress and how bacteria cope with death and survival. Crit Rev Microbiol 30:263–273
27. Boles BR, Thoendel M, Singh PK (2004) Self-generated diversity produces "insurance effects" in biofilm communities. Proc Natl Acad Sci U S A 101:16630–16635
28. Koh KS, Lam KW, Alhede M, Queck SY, Labbate M, Kjelleberg S, Rice SA (2007) Phenotypic diversification and adaptation of *Serratia marcescens* MG1 biofilm-derived morphotypes. J Bacteriol 189:119–130
29. Price TD, Qvarnstrom A, Irwin DE (2003) The role of phenotypic plasticity in driving genetic evolution. Proc R Soc Lond B 270:1433–1440
30. Odum EP (1971) Fundamentals of ecology, 3rd edn. Saunders, New York. ISBN 0534420664

Improved yield of α-L-arabinofuranosidase by newly isolated *Aspergillus niger* ADH-11 and synergistic effect of crude enzyme on saccharification of maize stover

Harshvadan Patel[1], Digantkumar Chapla[2], Jyoti Divecha[3] and Amita Shah[1*]

Abstract

Background: In the view of depleting resources and ever-increasing price of crude oil, there is an urge for the development of alternative sources to solve the issue of fuel in the coming years. Lignocellulosic biomass is considered to be the most potential alternative resources for fossil fuel. Bioconversion of cellulosic and hemicellulosic components into fermentable sugars is the key step in producing fuel ethanol from lignocellulose. The enzymatic hydrolysis of lignocellulosic biomass needs a highly balanced composition of cellulases and hemicellulases. Commercial enzymes are usually poor in accessory hemicellulolytic enzymes like α-L-arabinofuranosidase. The addition of such accessory enzymes in combination with cellulase or hemicellulase plays a vital role in improving the total yield of fuel ethanol by enhancing the saccharification yield.

Results: The newly isolated fungal strain *Aspergillus niger* ADH-11 produced a maximum of 22.14 U/g of α-L-arabinofuranosidase under solid-state fermentation using wheat bran as the substrate and modified Mandels-Weber medium at 30°C after 180 h of incubation. The optimization of various fermentation parameters was performed by response surface methodology employing Plackett-Burman design followed by Box-Behnken design. The yield of α-L-arabinofuranosidase was enhanced by 2.34-fold after executing statistical optimization of various fermentative parameters. Crude α-L-arabinofuranosidase was found to be highly stable for 3 h at its optimum temperature (55°C) and pH (4.0). The assessment of the crude enzyme extract in saccharification of alkali-treated maize stover revealed that the supplementation of crude α-L-arabinofuranosidase to commercial cellulase and crude xylanase mixture increased the saccharification yield up to 730 mg/g of maize stover.

Conclusions: The newly isolated *A. niger* ADH-11 was found to be a potential producer of α-L-arabinofuranosidase. The crude enzyme was active at low pH and high temperature which makes it suitable for various industrial applications such as enzymatic saccharification of lignocellulosic biomass. The supplementation of α-L-arabinofuranosidase enzyme to commercial cellulases and hemicellulases improves the bioconversion of lignocellulosic biomass to a greater extent.

Keywords: *Aspergillus niger* ADH-11; α-L-Arabinofuranosidase; Solid-state fermentation; Response surface methodology; Maize stover

* Correspondence: arshah02@yahoo.com
[1]BRD School of Biosciences, Sardar Patel Maidan Satellite Campus, Sardar Patel University, PO Box No. 39, Vallabh Vidyanagar 388120, Gujarat, India
Full list of author information is available at the end of the article

Background

Hemicelluloses are the second most abundant renewable source on the earth after celluloses representing up to 30 to 35% of the total dry weight of the carbohydrate polymers, out of which xylan is the major constituent [1,2]. Xylan-rich agroindustrial wastes are among the most important biomass sources in the world, representing an annual generation of 40 million tons causing considerable damage to economic activities in the agroindustrial sector and environment as well [3]. As xylan is the complex structure, its complete degradation requires the action of accessory hemicellulolytic enzymes out of which α-L-arabinofuranosidase is one of the most important enzymes [2,4].

α-L-Arabinofuranosidase is an exo-type enzyme that generally catalyzes the cleavage of the terminal α-L-arabinofuranosyl residues of arabinoxylan, arabinan, and arabinogalactan present in the hemicellulose structure. The arabinose side chains on hemicelluloses participate in cross-linking within the plant cell wall structure. L-Arabinose substitutions on xylan strongly inhibit the action of xylan-degrading enzymes, thus preventing the complete degradation of the polymer to its basic xylose units [5]. It is now well established that the action of α-L-arabinofuranosidase alone or in the combination with other hemicellulolytic enzymes is inevitable for enzymatic bioconversion of lignocellulosic biomass in to sugars which can be further fermented to ethanol. This enzyme is also finds a wide range of applications in food, feed, paper pulp, and pharmaceutical industries.

The production of α-L-arabinofuranosidase is widespread among bacteria and fungi. However, filamentous fungi are more attractive than bacteria as potential producers of these enzymes as fungi secrete higher enzyme levels in the culture medium than bacteria [6]. α-L-Arabinofuranosidase production has been reported by many fungal strains, such as *Trichoderma reseei* (142 nkat/mg), *Aspergillus awamori* (22.0 U/mg), *Aspergillus nidulans* (30.0 mIU/ml), *Aspergillus terreus* (6.0 nkat/ml), *Penicillium purpurogenum* (1.0 U/ml), *Aspergillus oryzae* (0.02 U/mg), and *Penicillium chrysogenum* (0.34 U/mg) [7-13].

Solid-state fermentation (SSF) is a microbial process in which solid material is used as the substrate which may provide necessary nutrients and serves as an anchorage for the cell. SSF can be of special interest in those processes where the crude fermented product may be used directly as the enzyme source [14]. Fungi grow well on moist substrates in the absence of free-flowing water [15] and produce large amount of extracellular enzymes like cellulases, hemicellulases, and ligninases. The production of enzymes using fungi under SSF is many a times higher than submerged fermentation (SmF), since SSF processes reproduce the natural living conditions of such fungi [16]. The use of SSF for production of enzymes has many economic advantages over SmF like no need for complex and sophisticated machinery, easy product recovery, simple and inexpensive substrate for the fermentation, low energy demand, high volumetric productivity, and often a high yield of product [17]. Response surface methodology has been extensively used in optimizing various fermentative parameters for enzyme production using solid-state fermentation [18-21].

Considering the potential and future prospects of α-L-arabinofuranosidase, the present study was aimed at production of α-L-arabinofuranosidase under SSF by a newly isolated fungal strain of *Aspergillus niger* ADH-11 employing response surface methodology. The properties of crude α-L-arabinofuranosidase were also evaluated to predict its end application, and the crude enzyme was supplemented to cocktails containing cellulase and xylanase to check synergistic action during saccharification of maize stover.

Methods

Materials

All the reagents, media, and chemicals used under study were of analytical grade purchased from Qualigens Fine Chemicals Pvt Ltd, HiMedia Laboratories Pvt Ltd, Merck & Co., and Loba Chemie Pvt Ltd from Mumbai, Maharashtra, India. *p*-Nitrophenyl-α-L-arabinofuranoside was obtained from Sigma-Aldrich, St. Louis, MO, USA. Wheat straw, rice straw, corn cobs, and maize bran were provided by a local farmer, whereas wheat bran and rice bran were procured from local market. The raw materials were washed thoroughly with water, dried at 80°C, and stored at room temperature in air tight plastic bags until use. The commercial cellulase used in this study was provided by MAPs Enzyme Limited (Ahmedabad, Gujarat, India).

Microbial strain

The α-L-arabinofuranosidase-producing fungal strain was isolated from decaying custard apple and it was identified as *A. niger* ADH-11 by the Agharkar Research Institute (ARI) Pune, India on the basis of its molecular characteristics. The sequence of the strain was submitted to NCBI gene bank with accession no. KF026012. The culture was grown on potato dextrose agar (PDA) slant and store at 4°C. It was subcultured every 15 days.

α-L-Arabinofuranosidase production under SSF using different agro residues by *A. niger* ADH-11

The α-L-arabinofuranosidase production was carried out in 250-ml Erlenmeyer flasks containing 5 g of washed dried agro residues (wheat bran, wheat straw, rice bran, rice straw, corn cobs, and maize bran) moistened with 10 ml of modified Mandels-Weber medium containing the following (g/l): urea, 0.3; ammonium sulfate, 1.4; KH_2PO_4, 0.3; $CaCl_2$, 0.3; $MgSO_4.7H_2O$, 0.3; proteose peptone. 1.0 and (mg/l) $MnSO_4.7H_2O$. 1.6; $ZnSO_4.7H_2O$, 1.4; $CoCl_2$, 2; Tween 80, 0.1%; and initial pH 5.3. The medium and the substrate were sterilized separately at 121°C for 15 min at 15 lbs. The inoculum was prepared using 7 days old *A. niger* ADH-11 slants. The spore suspension was prepared by adding 5 ml of 1% Tween 80 to each slant. The spore count was carried out using Neuber's chamber. The medium and the substrate were mixed at the time of inoculation at 10^8 spore/ml and incubated at 30°C under static condition. The contents of the flasks were mixed intermittently (twice a day), and the crude enzyme was extracted from each flask at regular interval of time (at every 24 h).

Enzyme extraction

The contents from each flask were extracted using a minimum quantity (30 ml) of 50 mM sodium citrate buffer (pH 5.3) along with 0.2 ml of Tween 80. The flasks were kept at 30°C with 150 rpm for 30 min for thorough mixing of the contents, followed by filtration through a wet muslin cloth with thorough squeezing, and the filtrate was centrifuged at 8,000 rpm for 30 min. The clear supernatant thus obtained was used as a crude enzyme for further studies.

Enzyme assays

The α-L-arabinofuranosidase activity was determined according to the method reported by Yanai and Sato [22] with some modifications. The reaction mixture consisting of 1 mM *p*-nitrophenyl-α-L-arabinofuranoside in 50 mM sodium citrate buffer (pH 5.3) was incubated with enzyme at 50°C for 10 min in total volume of 0.5 ml. The reaction was terminated by adding 1 ml of 2 M sodium carbonate solution. The amount of *p*-nitrophenol released was determined by measuring absorbance at 410 nm. One unit of α-L-arabinofuranosidase activity is defined as amount of enzyme required to release 1 μmol of *p*-nitrophenol per minute under assay condition. The xylanase activity was measured using 1% birch wood xylan solution as reported earlier [23]. The substrate was prepared in 50 mM sodium citrate buffer of pH 5.3. The enzyme reaction was carried out at 50°C for 10 min. After incubation, the reaction was terminated by adding 1 ml of DNS reagent. The enzyme activity was determined by measuring the release of reducing sugar using xylose as standard [24]. One unit of xylanase activity is defined as the amount of enzyme releasing 1 μmol of xylose per minute under assay condition. The filter paper activity was measured according to IUPAC recommendation employing filter paper (Whatman no. 1) as a substrate [25]. The release of reducing sugars in 60 min at 55°C and pH 4.8 (50 mM sodium citrate buffer) was measured as glucose equivalent using DNS method. One unit of filter paper activity is defined as amount of enzyme releasing 1 μmol of glucose per minute under assay condition.

Protein estimation

The soluble protein was estimated by Folin's method using bovine serum albumin as a standard [26].

Optimization of α-L-arabinofuranosidase production using response surface methodology

The optimization of the physicochemical parameters for α-L-arabinofuranosidase production was performed in two stages. Initially, ten variables were considered for screening using Plakett-Burman design (PBD) to identify the variables, which significantly influenced α-L-arabinofuranosidase production and in the second stage the significant variables screened from PBD were further optimized using a Box-Behnken design (BBD).

Screening of parameters affecting α-L-arabinofuranosidase production by PBD

In this study, urea, ammonium sulfate, proteose peptone, yeast extract, KH_2PO_4, $CaCl_2$, $MgSO_4$, pH, moisture ratio, and fermentation time were selected as the independent variables. Each variable was set at two levels, higher and lower (Table 1). The experimental design is given in Table 2. The significance of regression coefficients was tested by *T* test for α-L-arabinofuranosidase production (Table 3).

Optimization of significant parameters for α-L-arabinofuranosidase production by BBD

BBD involves full factorial search by observing simultaneous, systematic, and efficient variation of important components on the fermentation process. Urea, $MgSO_4$, $CaCl_2$, pH, ammonium sulfate, and fermentation time were selected as individual variable for the production of α-L-arabinofuranosidase. BBD in six factors having three center runs (with a total of 54 experimental runs) was

Table 1 Actual and coded level of variables tested with Plakett-Burman design

	Code	−1	1
Urea	X1	0.1	5
$MgSO_4$	X2	0.1	3
$CaCl_2$	X3	0.1	3
pH	X4	3	8
Ammonium sulfate	X5	1	10
Fermentation time	X6	48	216
Protease peptone	X7	1	5
Yeast extract	X8	1	5
KH_2PO_4	X9	0.1	3
Moisture ratio	X10	1:1	1:3

Table 2 PBD matrix for the screening of variables influencing α-L-arabinofuranosidase production

Run	X1	X2	X3	X4	X5	X5	X7	X8	X9	X10	α-L-Arabinofuranosidase (U/g)
1	0.1	0.1	0.1	3	1	48	1	1	0.1	1	8.75
2	5	0.1	0.1	8	10	216	1	5	0.1	3	35.41
3	5	0.1	3	3	1	216	5	5	0.1	1	35.01
4	0.1	3	3	3	10	48	5	5	0.1	3	9.4
5	0.1	3	3	8	1	48	1	5	3	3	4.97
6	5	0.1	0.1	3	10	48	1	5	3	1	12.35
7	0.1	3	3	8	10	216	1	1	0.1	1	19.85
8	0.1	0.1	0.1	8	1	216	5	5	3	1	35.08
9	0.1	0.1	0.1	3	10	216	5	1	3	3	30.41
10	5	3	3	3	1	216	1	1	3	3	32.15
11	5	3	3	8	10	48	5	1	3	1	10.79
12	5	0.1	0.1	8	1	48	5	1	0.1	3	11.81

used for the optimization of α-L-arabinofuranosidase production.

The design allowed to evaluate the main and interactive effects of urea (X1), $MgSO_4$ (X2), $CaCl_2$ (X3), pH (X4), ammonium sulfate (X5), and fermentation time (X6) for α-L-arabinofuranosidase yield (U/g). The α-L-arabinofuranosidase yield (U/g) corresponding to the combined effects of six variables was studied in their specific range as shown in Table 4. The temperature was kept constant at 30°C throughout the experiment. All flasks were analyzed for α-L-arabinofuranosidase yield at specific time intervals as planned in BBD. The plan of BBD in the coded levels of six independent variables is shown in Table 5.

For statistical calculation the independent variables were coded as follows:

$$x_i = (X_i - X_o)/\delta X_i \quad (1)$$

Where X_i is the experimental value of the variable; X_o is the midpoint of X_i, δX_i is the step change in X_i, and x_i is the coded value for X_i, $i = 1 - 6$.

The α-L-arabinofuranosidase yield was fitted using response surface methodology applying Equation 2 and was analyzed using MINITAB 16.0 version:

$$Y = \beta_o + \beta_i x_i + \sum \beta_{ii} x_i^2 + \sum \beta_{ij} x_i x_i \quad (2)$$

where Y is the predicted response variable, β_o, β_i, β_{ii}, β_{ij} are fixed regression coefficients of the model, x_i, x_j (i = 1, 2, 3, 4, 5 and 6, $i \neq j$, $i < j$ = 1, 2, 3, 4, 5, 6) represent independent variables in the form of original values.

Table 3 Regression coefficient for α-L-arabinofuranosidase production

Term	Effect	Coefficient	Standard error Coefficient	*T* value	*P* value
Constant		26.02	1.579	16.48	0.039
Urea	15.88	7.94	1.579	5.03	0.025
Ammonium sulfate	−12.63	−6.32	1.579	−4.00	0.050
Protease peptone	4.21	2.10	1.579	1.33	0.410
Yeast extract	4.12	2.06	1.579	1.30	0.417
KH_2PO_4	−0.12	−0.06	1.579	−0.04	0.059
$CaCl_2$	−28.81	−14.40	4.737	−3.04	0.002
$MgSO_4$	36.24	18.12	5.470	3.31	0.047
pH	−12.73	−6.37	1.579	−4.03	0.050
Moisture Ratio	−0.65	−0.33	1.579	−0.21	0.870
Fermentation time	32.68	16.34	1.579	10.35	0.047

Interpretation and data analysis

The results of the experimental design were analyzed and interpreted using Minitab 16 (Minitab Inc, State College, PA, USA) statistical software. The prediction of fermentation parameters and generation of response contour plot by the model were also done by the same software. Analysis of variance (ANOVA) was used to establish the significance of the model parameters.

Table 4 Actual and coded level of variables tested with Box-Behnken design for α-L-arabinofuranosidase production

Process variable	Coded level of variable		
	−1	0	+1
Urea (X1)	0.1	2.55	5
$MgSO_4$ (X2)	0.1	1.55	3
$CaCl_2$ (X3)	0.1	1.55	3
pH (X4)	3	5.5	8
Ammonium sulfate (X5)	1	5.5	10
Fermentation time (X6)	48	132	216

Table 5 Full factorial BBD matrix for α-L-arabinofuranosidase production by *Aspergillus niger* ADH-11

	Urea	$MgSO_4$	$CaCl_2$	pH	Ammonium sulfate	Fermentation time	α-L-Arabinofuranosidase	
Run	X1	X2	X3	X4	X5	X6	Predicted activity (U/g)	Experimental activity (U/g)
1	−1	−1	0	−1	0	0	24.803	23.750
2	1	−1	0	−1	0	0	23.506	21.940
3	−1	1	0	−1	0	0	21.359	24.070
4	1	1	0	−1	0	0	22.517	20.990
5	−1	−1	0	1	0	0	21.465	22.510
6	1	−1	0	1	0	0	22.089	19.860
7	−1	1	0	1	0	0	20.756	21.840
8	1	1	0	1	0	0	23.835	25.370
9	0	−1	−1	0	−1	0	18.686	20.850
10	0	1	−1	0	−1	0	18.814	20.140
11	0	−1	1	0	−1	0	19.025	19.720
12	0	1	1	0	−1	0	14.183	12.240
13	0	−1	−1	0	1	0	16.684	17.410
14	0	1	−1	0	1	0	19.828	20.350
15	0	−1	1	0	1	0	21.293	18.750
16	0	1	1	0	1	0	19.467	18.520
17	0	0	−1	−1	0	−1	20.493	17.850
18	0	0	1	−1	0	−1	17.673	19.390
19	0	0	−1	1	0	−1	12.448	13.680
20	0	0	1	1	0	−1	16.328	15.430
21	0	0	−1	−1	0	1	22.543	23.930
22	0	0	1	−1	0	1	18.641	16.920
23	0	0	−1	1	0	1	21.868	20.640
24	0	0	1	1	0	1	24.666	26.820
25	−1	0	0	−1	−1	0	16.175	16.160
26	1	0	0	−1	−1	0	17.414	17.340
27	−1	0	0	1	−1	0	14.703	12.370
28	1	0	0	1	−1	0	17.861	17.410
29	−1	0	0	−1	1	0	19.621	19.590
30	1	0	0	−1	1	0	18.245	21.060
31	−1	0	0	1	1	0	17.154	17.710
32	1	0	0	1	1	0	17.697	17.230
33	0	−1	0	0	−1	−1	15.256	15.060
34	0	1	0	0	−1	−1	15.484	14.150
35	0	−1	0	0	1	−1	15.089	14.820
36	0	1	0	0	1	−1	18.333	16.440
37	0	−1	0	0	−1	1	22.735	23.410
38	0	1	0	0	−1	1	17.793	19.280
39	0	−1	0	0	1	1	23.168	25.720
40	0	1	0	0	1	1	21.242	20.220
41	−1	0	−1	0	0	−1	16.569	15.160
42	1	0	−1	0	0	−1	16.174	18.470
43	−1	0	1	0	0	−1	13.391	15.960

Table 5 Full factorial BBD matrix for α-L-arabinofuranosidase production by *Aspergillus niger* ADH-11 *(Continued)*

44	1	0	1	0	0	−1	20.412	21.240
45	−1	0	−1	0	0	1	24.726	23.410
46	1	0	−1	0	0	1	19.487	16.430
47	−1	0	1	0	0	1	20.467	18.660
48	1	0	1	0	0	1	22.643	24.540
49	0	0	0	0	0	0	18.998	21.490
50	0	0	0	0	0	0	18.998	16.630
51	0	0	0	0	0	0	18.998	20.180
52	0	0	0	0	0	0	18.998	17.460
53	0	0	0	0	0	0	18.998	18.670
54	0	0	0	0	0	0	18.998	19.560

Wheat bran was used as the substrate at 30°C under SSF.

Effect of temperature and pH on α-L-arabinofuranosidase activity

The optimum temperature for α-L-arabinofuranosidase was determined by assaying relative activity at different temperatures ranging from 40°C to 75°C. The optimum pH for α-L-arabinofuranosidase was determined by assaying relative activity at different pH (3 to 7) using 50 mM sodium citrate buffer for pH 3, 4, 5, and 50 mM sodium phosphate buffer for pH 6 and 7.

Temperature and pH stability of α-L-arabinofuranosidase

To determine the thermal stability of α-L-arabinofuranosidase, the enzyme solution was treated at different temperatures (45°C, 55°C, and 65°C) in 50 mM sodium citrate buffer (pH 4.0) in a temperature-controlled water bath and the residual activity was measured at different time intervals (60 min) up to 3 h. To determine the pH stability of α-L-arabinofuranosidase, the enzyme solution was appropriately diluted at different pH buffers (4, 5, 6 and 7) and left to room temperature and the residual activity was measured at different time intervals (60 min) up to 3 h.

Pretreatments of maize stover

The substrate (maize stover) was washed, dried, and sieved to get uniform particle size before its use. The maize stover was given two separate pretreatments. In one case, pretreatment was given by preparing 15% slurry of maize stover in 4% NaOH solution and was incubated at 30°C for 12 h. The second pretreatment was given by 15% ammonia solution in the ratio of (1:4.5 (*w*/*v*)) to prepare the slurry. The slurry was autoclaved at 121°C and 15 lbs for 1 h. The substrate was washed completely with distilled water until neutral and dried in oven at 80°C until moisture was evaporated.

Enzymatic hydrolysis of maize stover

The enzymatic saccharification of pretreated and untreated maize stover was carried out using crude α-L-arabinofuranosidase from *A. niger* ADH-11, crude xylanase from *Aspergillus foetidus* MTCC 4898 [18], and commercial cellulase individually. The crude enzyme from *A. niger* ADH-11, crude enzyme from *A. foetidus* MTCC 4898, and commercial cellulase also have other cellulolytic and xylanolytic enzymes. Crude α-L-arabinofuranosidase was used at 4.0 U/g, crude xylanase was used at 300 U/g, while commercial cellulase was used at 5.0 FPU/g during saccharification of maize stover. The enzymatic hydrolysis was performed in 150-ml screw cap Erlenmeyer flasks containing 2.5% pretreated and untreated maize stover and diluted enzyme, as mentioned earlier, in 50 mM sodium citrate buffer (pH 4.8) containing 0.1% Tween 80 in a final volume of 40 ml. Controls were kept for each reaction in which the active enzyme was replaced with heat-inactivated enzyme. The reaction system was fortified with 10 mg% sodium azide. The reaction was carried out at 50°C in water bath with mild shaking. The samples were withdrawn every 4 h and incubated in boiling water bath (100°C) to inactivate the enzyme; the reaction mixture was then filtered through wet muslin cloth by thorough squeezing and centrifuged to collect the clear supernatant. This supernatant was used for further analysis to estimate total reducing sugar by DNS method.

Enzymatic hydrolysis of pretreated maize stover using commercial cellulase supplemented with crude xylanase and crude α-L-arabinofuranosidase

The enzymatic hydrolysis of alkali (NaOH)-treated maize stover was performed using commercial cellulase (5.0 FPU/g), crude xylanase from *A. foetidus* MTCC 4898 (300 U/g), and crude α-L-arabinofuranosidase from *A. niger* ADH-11 (4.0 U/g) individually and cocktail of all. The reaction system was same as described in 'Enzymatic hydrolysis of maize stover' section. The hydrolysed products were analyzed by high-performance liquid chromatography (HPLC; Phenomenex, Rezex ROA-organic acid H+, column; Phenomenex Inc, Torrance, CA, USA). The degree of synergy or synergy is defined as 'the ratio of the yield of product released by enzymes when used together to the sum of yield of these products when the enzymes are used separately in the same amounts as they were employed in the mixture'.

Results and discussion

α-L-Arabinofuranosidase production using different agro residues under SSF

The production of α-L-arabinofuranosidase was carried out using various agro residues like wheat bran, wheat straw, rice bran, rice straw, corn cobs, and maize bran as a sole carbon source and Mandels-Weber medium as a moistening agent at 30°C under SSF. It was observed that all six substrates supported production of α-L-arabinofuranosidase but wheat bran was found to be the most appropriate substrate under SSF, yielding a maximum yield of α-L-arabinofuranosidase up to 9.45 U/g after 192 h of incubation (Figure 1); hence, wheat bran was selected as the substrate for production of α-L-arabinofuranosidase. The biochemical composition of wheat bran indicates that it contains predominantly non-starch carbohydrate polymers (approximately 58%), starch (approximately 19%), and crude protein (approximately 18%). The non-starch carbohydrate polymers are being primarily approximately 70% arabinoxylans, approximately 24% cellulose, and approximately 6% β-(1,3) (1,4)-glucan [27]. The presence of higher amount of arabinoxylans in wheat bran may have induced α-L-arabinofuranosidase production. Relatively very few attempts have been reported on the production of these enzymes under SSF. Khandeparker et al. [28] reported wheat bran was the best substrate for α-L-arabinofuranosidase production by *Arthrobacter* sp. MTCC 5214 under SSF.

Optimization of α-L-arabinofuranosidase production under SSF using response surface methodology

Response surface methodology has been successfully applied for the optimization of fermentation medium components, conditions for enzymatic hydrolysis, and fermentation processes [29-32]. It predicts the maximum enzyme production among the selected range of variables and also studies the interaction among independent variables.

In the present study, the effect of urea, ammonium sulfate, proteose peptone, yeast extract, KH_2PO_4, $CaCl_2$, $MgSO_4$, pH, moisture ratio, and fermentation time was evaluated on the basis of α-L-arabinofuranosidase production using PBD (Table 1). Among these variables, urea, ammonium sulfate, KH_2PO_4, $CaCl_2$, $MgSO_4$, pH, and fermentation time were identified as the most significant and contributing variables (Table 3) for α-L-arabinofuranosidase production (Figure 2a). These parameters were further analyzed as variables using BBD for better production of α-L-arabinofuranosidase.

The result of 54 run BBD in six variables urea, ammonium sulfate, KH_2PO_4, $CaCl_2$, $MgSO_4$, pH, and fermentation time for the optimization of α-L-arabinofuranosidase production is shown in Table 5.

The experimental results suggest that the variables selected for the fermentation process had a strong effect on α-L-arabinofuranosidase production. On the basis of these experimental values, statistical testing was carried out using Minitab 16.0. The analysis of the model was tested by Fisher's F test and Student's T test. ANOVA of α-L-arabinofuranosidase production showed that the response surface model was significant ($P = 0.002$) (Table 6) due to the square portion of the regression model. A P value below 0.05 indicates that the test parameters are significant. In common, the larger the magnitude of T and smaller the value of P, more significant is the consequent coefficient term [33]. The fitted second-order response surface model as specified by Equation 2 for α-L-arabinofuranosidase yield in coded process variables is as follows:

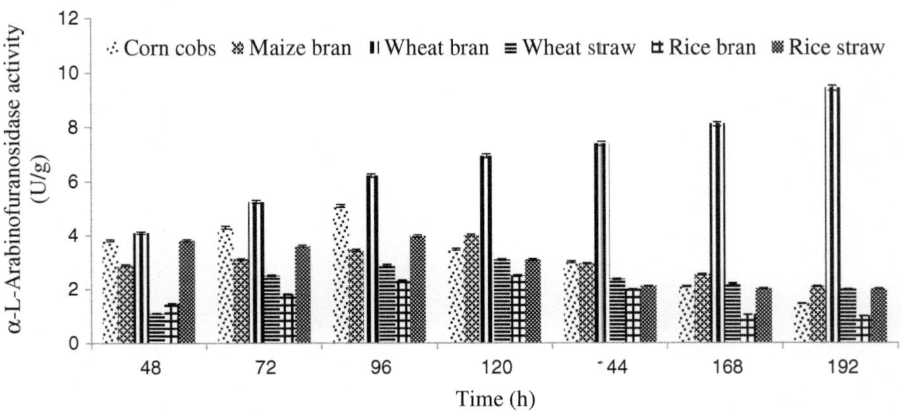

Figure 1 α-L-Arabinofuranosidase production using agro-residues under solid-state fermentation at 30°C.

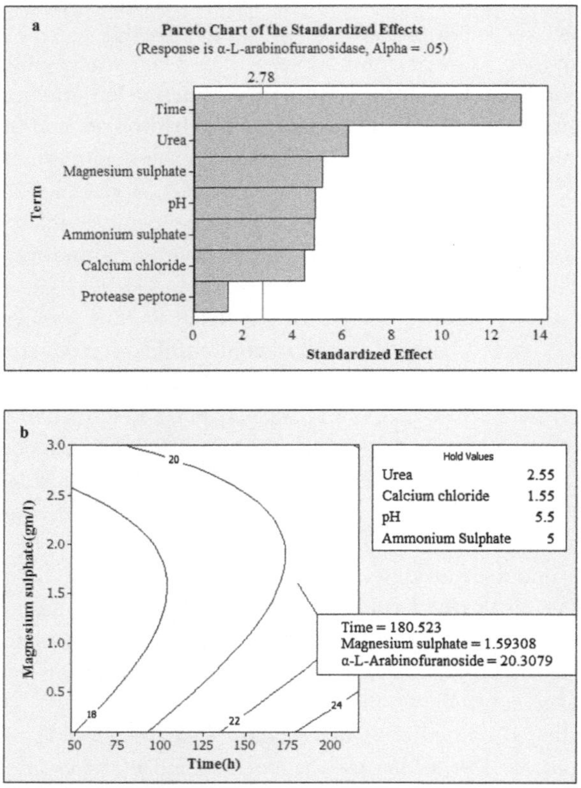

Figure 2 Pareto chart and contour plot. (a) Pareto chart showing effect of media components on α-L-arabinofuranosidase production. **(b)** Contour plot showing interaction of $MgSO_4$ and incubation time on α-L-arabinofuranosidase production at hold values of urea, $CaCl_2$, ammonium sulfate and pH.

$$\begin{aligned} Y = {} & 18.9983 + 0.4454x_1 - 0.4446x_2 - 0.0054x_3 \\ & - 0.5050x_4 + 0.8204x_5 + 2.5971x_6 + 0.5010x_1^2 \\ & + 2.4422x_2^2 - 0.2028x_3^2 + 5997x_4^2 - 2.7403x_5^2 \\ & - 0.0628x_6^2 + 0.6138x_1x_2 + 1.8538x_1x_3 \\ & - 1.2113x_1x_6 + 0.7537x_2x_5 - 1.2925x_2x_6 \\ & + 0.6838x_3x_4 + 1.0675x_3x_5 - 0.2487x_4x_6 \\ & + 0.1500x_5x_6 \end{aligned} \quad (3)$$

where Y is α-L-arabinofuranosidase yield square root of the predicted response, and x_1, x_2, x_3, x_4, x_5, and x_6 are the coded values of urea, $MgSO_4$, $CaCl_2$, pH, ammonium sulfate, and fermentation time, respectively. A comparison of the experimentally obtained values of the enzyme with the predicted values indicated that these data are in reasonable agreement with predictive values (Table 5).

The corresponding P value showed that the independent variable $MgSO_4$ (X2) and fermentation time (X6) had significant effect (0.003, 0.000) on α-L-arabinofuranosidase yield (Table 7). A significant interaction was also observed between urea (X1) and $CaCl_2$ (X3) and between pH (X4) and ammonium sulfate (X5) (0.034, 0.035) contributing to the response at significant level for α-L-arabinofuranoside production (Table 7). The R^2 value provides a measure of variability in the observed response value that can be explained by the experimental factors and their interactions. The coefficient R^2 of α-L-arabinofuranoside was 76.4%.

The fitted response for the above regression model was plotted in contour plot (Figure 2b) for the pairs of interactive variables while keeping other variables at their predicted optimum level. Among the variables, $MgSO_4$ (X2), $CaCl_2$ (X3), pH (X4), and fermentation time (X6) were the most significantly linear and had a positive effect on enzyme production. The significant interaction was observed between $MgSO_4$ (X2) and fermentation time (X6) for α-L-arabinofuranosidase production. Increasing the fermentation time (X6) up to certain level may lead to an increase in α-L-arabinofuranosidase yield, while increasing the $MgSO_4$ (X2) up to 2.55 g/l may lead to a drastic decrease in the α-L-arabinofuranosidase yield. $MgSO_4$ is considered as a good stimulator of mycelial growth which decreases the dormancy of the spores and can affect the enzyme yields [34]. Guerfali et al. [35] performed PBD followed by CCD and found that $MgSO_4$ played an important role in α-L-arabinofuranosidase production.

The application of RSM with BBD predicted that the maximum α-L-arabinofuranosidase production will be up to 20.30 U/g at decoded values of process parameters as $MgSO_4$ 1.59, urea 2.55, $CaCl_2$ 1.55, and ammonium

Table 6 ANOVA for the response surface quadratic model for α-L-arabinofuranosidase production

Source	Degree of freedom	Sequential sums of squares	Adjusted sums of squares	Adjusted mean squares	*F* value	*P* value
Regression	27	464.93	464.93	17.220	3.12	0.002
Linear	6	193.24	193.24	32.207	5.84	0.001
Square	6	123.23	123.23	20.539	3.73	0.008
Interaction	15	148.46	148.46	9.897	1.80	0.092
Residual error	26	143.28	143.28	5.511		
Lack-of-fit	21	127.27	127.27	6.061	1.89	0.249
Pure error	5	16.00	16.00	3.201		
Total	53	608.21				

Table 7 Regression coefficient for α-L-arabinofuranosidase production

Term	Coefficient	Standard error coefficient	*T* value	*P* value
Constant	18.9983	0.9583	19.824	0.000
Urea	0.4454	0.4792	0.930	0.361
$MgSO_4$	−0.4246	0.4792	−0.886	0.384
$CaCl_2$	−0.0054	0.4792	−0.011	0.991
pH	−0.5050	0.4792	−1.054	0.302
Ammonium sulfate	0.8204	0.4792	1.712	0.099
Fermentation time	2.5997	0.4792	5.420	0.000
Urea × Urea	0.5010	0.7319	0.684	0.500
$MgSO_4$ × $MgSO_4$	2.4422	0.7319	3.337	0.003
$CaCl_2$ × $CaCl_2$	−0.2028	0.7319	−0.277	0.784
pH × pH	0.5997	0.7319	0.819	0.420
Ammonium sulfate × Ammonium sulfate	−2.7403	0.7319	−3.744	0.001
Fermentation time × Fermentation time	−0.0628	0.7319	−0.086	0.932
Urea × $MgSO_4$	0.6138	0.8300	0.739	0.466
Urea × $CaCl_2$	1.8538	0.8300	2.234	0.034
Urea × pH	0.4800	0.5869	0.818	0.421
Urea × Ammonium sulfate	−0.6538	0.8300	−0.788	0.438
Urea × Fermentation time	−1.2113	0.8300	−1.459	0.156
$MgSO_4$ × $CaCl_2$	−1.2425	0.8300	−1.497	0.146
$MgSO_4$ × pH	0.6838	0.8300	0.824	0.418
$MgSO_4$ × Ammonium sulfate	0.7537	0.5869	1.284	0.210
$MgSO_4$ × Fermentation time	−1.2925	0.8300	−1.557	0.131
$CaCl_2$ × pH	1.6750	0.8300	2.018	0.054
Ammonium sulfate × Fermentation time	1.0675	0.8300	1.286	0.210
$CaCl_2$ × Ammonium sulfate	−0.2706	0.5869	−0.461	0.649
$CaCl_2$ × Fermentation time	−0.2487	0.8300	−0.300	0.767
pH × Ammonium sulfate	1.8425	0.8300	2.220	0.035
pH × Fermentation time	0.1500	0.8300	0.181	0.858

sulfate 5 g/l with initial pH 5.5 after incubation 180 h of fermentation time.

Validation of the experimental model

A repeat fermentation for α-L-arabinofuranosidase production was carried out under optimal conditions to validate the parameters predicted by the model. The α-L-arabinofuranosidase production under optimized parameters *viz.* $MgSO_4$ 1.59, urea 2.55, $CaCl_2$ 1.55, and ammonium sulfate 5 g/l at pH 5.5 and fermentation time of 180 h yielded α-L-arabinofuranosidase activity of 22.14 U/g (3.16 U/ml). This was significantly higher than the predicted value (20.30) by the model. Crude enzyme extract also contained 140 U/g (20.0 U/ml) endo-xylanase, 70.0 U/g (10.0 U/ml) β-xylosidase, and 0.14 U/g (0.02 U/ml) filter paper activity. Thus, it was proved that the statistical optimization increased α-L-arabinofuranosidase production by 2.34-fold as compared to their initial production which was 9.45 U/g (1.35 U/ml) before statistical optimization. Khandeparker [28] reported a maximum 3 U/g of α-L-arabinofuranosidase production after 120 h of incubation by *Arthrobacter* sp. MTCC 5214 under SSF. Guerfali et al. [35] reported a maximum 0.6 U/ml of α-L-arabinofuranosidase production after 120 h of incubation by *Talaromyces thermophilus* using response surface methodology under submerged fermentation. Hence, newly isolated *A. niger* ADH-11 was found to be a strong producer of α-L-arabinofuranosidase; however, the fermentation time was higher as compared to other reports.

Effect of temperature and pH on α-L-arabinofuranosidase activity

The influence of temperature on α-L-arabinofuranosidase activity was evaluated in the range of 40°C to 75°C. The results revealed that the optimum temperature for α-L-arabinofuranosidase activity was 55°C (Figure 3a). α-L-

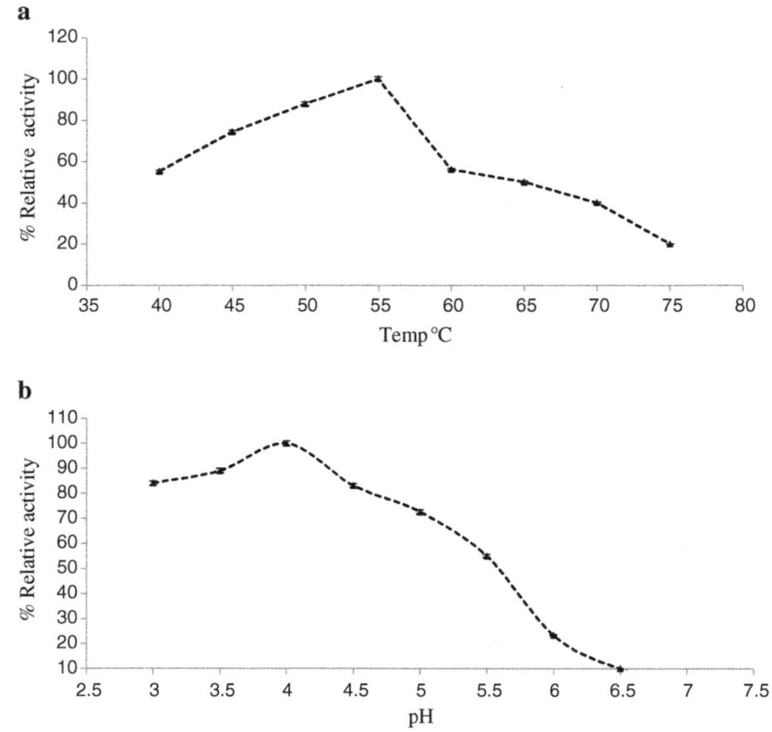

Figure 3 Effect of temperature (a) and pH (b) on α-L-arabinofuranosidase activity.

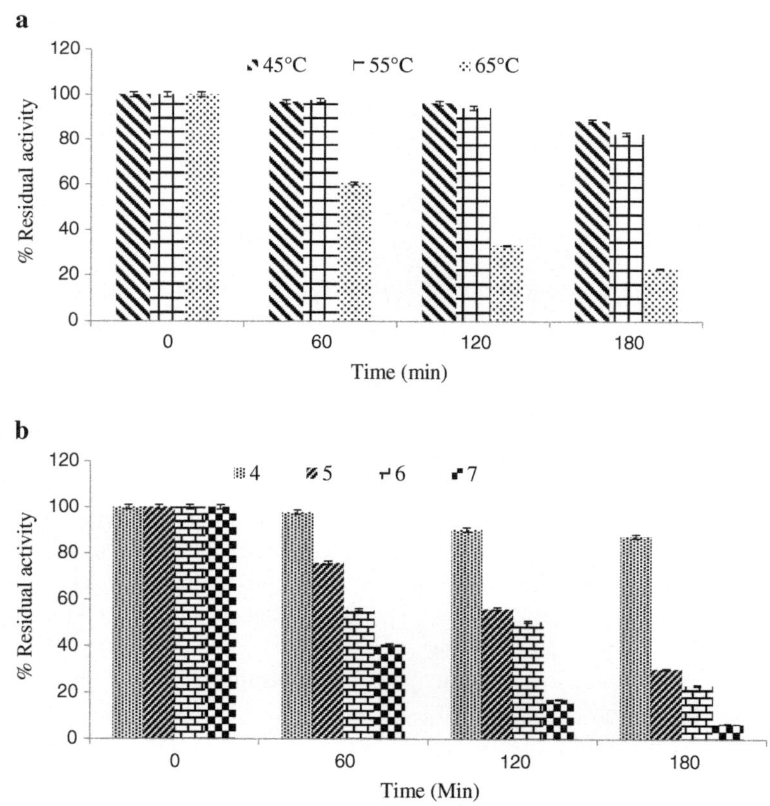

Figure 4 Thermal stability (a) and pH stability (b) of α-L-arabinofuranosidase.

Arabinofuranosidase activity was reduced by only 13% and 44% at 50°C and 60°C, respectively. The result was similar to many α-L-arabinofuranosidases from different strains of fungi [36]. Guerfali et al. [35] also reported that the optimum temperature of α-L-arabinofuranosidase from *Talaromyces thermophilus* was 55°C.

The α-L-arabinofuranosidase activity at various pH (3 to 7) was measured using *p*-nitrophenyl-α-L-arabinofuranoside as a substrate at 55°C. The optimum pH for α-L-arabinofuranosidase activity was found to be at 4 (Figure 3b). The enzyme was remarkably active even at pH 3.5 and 4.5, with loss of only 11% and 17% activities. Kaneko et al. [37] also found the optimum pH as 4.0 for α-L-arabinofuranosidase from *A. niger* 5-16. Most of the fungal α-L-arabinofuranosidases exhibited acidic pH in the range of 4.0 to 5.0. The enzyme activity is markedly affected by variation in pH outside its optimum pH. This may be due to substrate binding and catalysis, which are often affected by charge distribution on both substrate and enzyme molecules [38].

Temperature and pH stability of α-L-arabinofuranosidase

The α-L-arabinofuranosidase from *A. niger* ADH-11 was found to be highly stable in the range of 45°C to 55°C (Figure 4a). At 55°C the enzyme retained 82.38% of its activity after 3 h. Yan et al. [39] reported that α-L-arabinofuranosidase from *Chaetomium* sp. was stable at 55°C for 30 min. Guais et al. [40] found that α-L-arabinofuranosidase activity was dropped by 50% at 60°C after 12 min. The α-L-arabinofuranosidase could retain its activity up to 87.37% at pH 4 after incubation of 3 h (Figure 4b). The activity of α-L-arabinofuranosidase was gradually

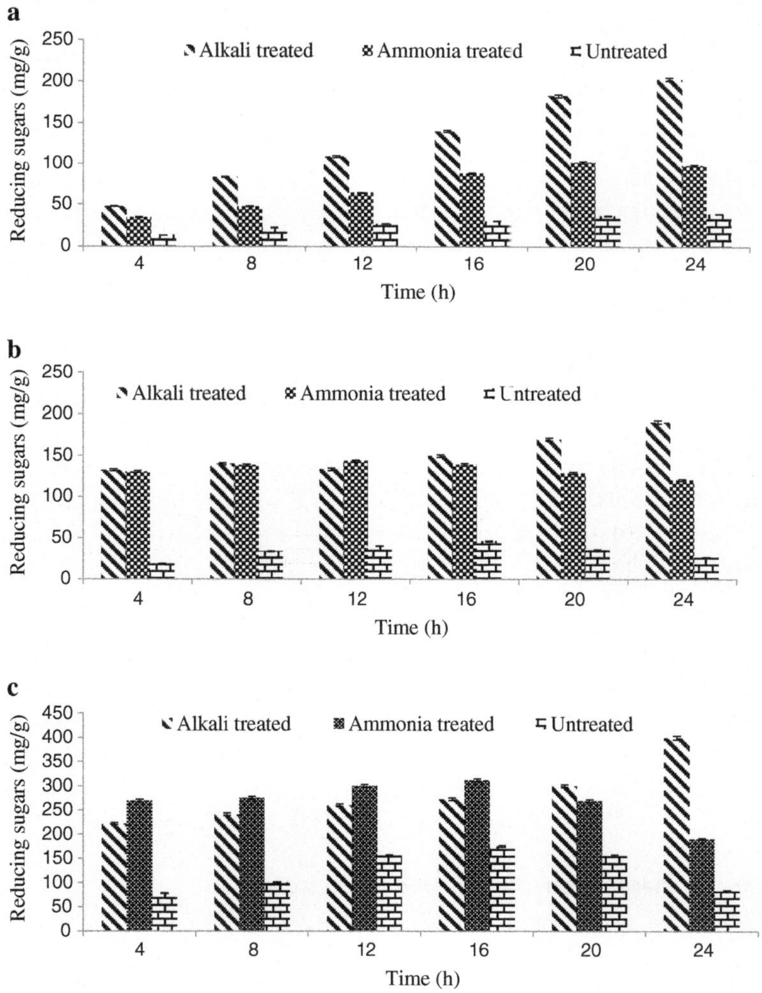

Figure 5 Enzymatic hydrolysis of untreated and alkali (NaOH)- and ammonia treated maize stover using different enzymes. (a) Enzymatic hydrolysis of untreated and alkali- and ammonia-treated maize stover using 4.06 U/g crude α-L-arabinofuranosidase at 50°C. **(b)** Enzymatic hydrolysis of untreated and alkali-, and ammonia-treated maize stover using 300 U/g crude xylanase at 50°C. **(c)** Enzymatic hydrolysis of untreated and alkali- and ammonia-treated maize stover using 5.0 FPU/g commercial cellulase at 50°C.

Table 8 Composition of enzymes in each cocktail

Enzyme	β-Xylosidase (U/ml)	α-L-Arabinofuranosidase (U/ml)	Feruloyl esterase (U/ml)	Xylanase (U/ml)	Filter paper activity (U/ml)	Endoglucanase (U/ml)	β-Glucosidase (U/ml)
Crude α-L-arabinofuranosidase[a]	10.0	3.16	1.0	20.0	0.02	5.56	0.18
Crude xylanase[b]	0.2	ND	ND	375.0	0.12	4.5	0.20
Commercial cellulase[c]	ND	ND	ND	300.0	43.9	2,798.0	43.0

[a]Crude α-L-arabinofuranosidase from *A. niger* ADH-11; [b]Crude xylanase from *A. foetidus* MTCC 4898; [c]Commercial cellulase MAPs 450. ND, not detected.

reduced to 30.21% and 23.07% at pH 5 and 6 after 3 h, respectively, while only 6.59% activity was retained at pH 7 after 3 h. Filho et al. [41] reported that α-L-arabinofuranosidases I and II from *Penicillium capsulatum* retained 50% activity at pH 4.0 after 9 and 17.5 min, respectively. Yan et al. [39] reported that α-L-arabinofuranosidase from *Chaetomium* sp. retained 80% activity at pH 4.0 after 30 min. In comparison to the above reports, the crude α-L-arabinofuranosidase from *A. niger* ADH-11 was found to be more stable at high temperature and low pH, so it can be suitable for saccharification of lignocellulosic biomass along with commercial cellulases.

Enzymatic hydrolysis of maize stover

The major conversion steps in biochemical production of cellulosic biofuels are pretreatment and enzymatic hydrolysis. Together, these steps work to breakdown the cellulose and hemicelluloses found in the cell walls of plants (biomass) to simple sugars which can then can be fermented to ethanol. Maize stover is currently the largest waste biomass in the India, consisting of more than one-third of the total solid waste produced. Maize stover typically contains 70% cellulose and hemicellulose and 15% to 20% lignin. Ammonia and NaOH-pretreated maize stover were used to test the ability of the crude α-L-arabinofuranosidase, crude xylanase, and commercial cellulase for the production of fermentable sugar by saccharification. Maximum reducing sugars of 203.6 mg/g were produced using crude α-L-arabinofuranosidase, while crude xylanase and commercial cellulase yielded 190.7 and 400.0 mg/g respectively using NaOH-pretreated maize stover (Figure 5a,b,c). A higher yield of reducing sugars was achieved when the substrate was pretreated with NaOH as compared to ammonia.

The maximum bioconversion of lignocellulosic biomass requires action of not only efficient cellulolytic enzymes but also main chain depolymerizing and debranching hemicellulolytic enzymes at appropriate levels. It is evident from Table 8 that the commercial enzyme used in this study was deficient in β-xylosidase, α-L-arabinofuranosidase, and feruloyl esterase and hence, the indigenously produced crude α-L-arabinofuranosidase was supplemented to form a balanced enzyme cocktail for saccharification of alkali-pretreated maize stover. There was a remarkable increase in β-xylosidase, α-L-arabinofuranosidase, feruloyl esterase, endo-xylanase, and β-glucosidase loading upon supplementation of crude α-L-arabinofuranosidase to commercial cellulase-crude xylanase cocktail (Table 9). As shown in Figure 6, the cocktail produced maximum 730.1 mg/g of reducing sugar after 24 h which was higher as compared to each enzyme used alone (crude α-L-arabinofuranosidase 203.6 mg/g, crude xylanase 134.4 mg/g, and commercial cellulase 294.1 mg/g, respectively) for saccharification. The co-action of crude α-L-arabinofuranosidase with commercial cellulase-crude xylanase produced a superior saccharification yield. The HPLC analysis of enzymatic hydrolysates revealed synergistic rise in glucose, xylose, and arabinose (degree of synergy 1.21, 1.00, and 1.33, respectively). The levels of glucose, xylose, and arabinose were 5.57, 4.99, and 0.64 mg/ml when the cocktail was used (Table 10). Thus, the supplementation of crude α-L-arabinofuranosidase (4.0 U/g) with commercial cellulase and crude xylanase increased the glucose yield up to 176.8% and xylose yield up to 166.8% (Table 10). Although supplementation of crude xylanase with commercial cellulase increased the

Table 9 Enzyme loading for saccharification of pretreated maize stover

Enzyme	β-Xylosidase (U/g)	α-L-Arabinofuranosidase (U/g)	Feruloyl esterase (U/g)	Xylanase (U/g)	Filter paper activity (U/g)	Endoglucanase (U/g)	β-Glucosidase (U/g)
Crude α-L-arabinofuranosidase[a]	12.7	4.0	1.27	25.4	0.025	7.06	0.22
Crude xylanase[b]	0.16	ND	ND	300	0.09	3.6	0.16
Commercial cellulase[c]	ND	ND	ND	33.0	5.0	307.7	4.78

[a]Crude α-L-arabinofuranosidase from *A. niger* ADH-11; [b]Crude xylanase from *A. foetidus* MTCC 4898; [c]Commercial cellulase MAPs 450. ND, not detected.

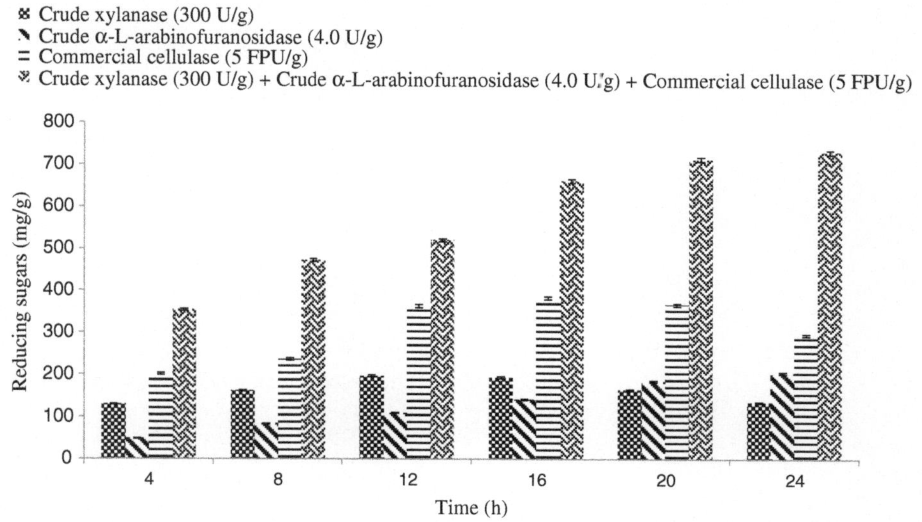

Figure 6 Enzymatic hydrolysis of alkali-treated maize stover using individual and cocktail of different enzymes. Enzymatic hydrolysis of alkali-treated maize stover using individual and cocktails of crude α-L-arabinofuranosidase, crude xylanase, and commercial cellulases at 50°C.

loading of endoglucanase, β-glucosidase, and endoxylanase (Table 9), the yield of glucose, xylose, and arabinose was lesser compared to the cocktail of crude α-L-arabinofuranosidase (4.0 U/g) with commercial cellulase (5 FPU/g) and crude xylanase (300 U/g) cocktail. These results clearly suggest that the higher yield of reducing sugars obtained by this cocktail was due to higher amount of α-L-arabinofuranosidase and β-xylosidase enzymes which were contributed by the crude α-L-arabinofuranosidase produced by *A. niger* ADH-11. The presence of α-L-arabinofuranosidase plays an important role by removing arabinose from the side chains of xylan, and β-xylosidase plays a crucial role in reducing the end product inhibition of xylanases by hydrolyzing xylobiose and hence facilitates better bioconversion of maize stover. A similar observation was reported by Delabona et al. [42]. Their studies on sugarcane bagasse strongly suggested that the supplementation of α-L-arabinofuranosidase with commercial cellulase resulted in greater sugar release. The present study clearly demonstrated that the crude α-L-arabinofuranosidase-rich enzyme obtained from *A. niger* ADH-11 can be used to enhance the efficiency of commercial cellulase for saccharification of biomass.

Table 10 Monomer sugar composition of each cocktail used for enzymatic saccharification of NaOH-treated maize stover

Enzyme cocktails	Glucose (mg/ml)	Xylose (mg/ml)	Arabinose (mg/ml)
Commercial cellulase (5 FPU/g)	2.98	1.69	ND
Crude α-L-arabinofuranosidase (4.0 U/g)	0.69	1.59	0.48
Crude xylanase (300 U/g)	0.93	1.70	ND
Commercial cellulase (5 FPU/g) + Crude xylanase (300 U/g)	3.15	2.99	ND
Commercial cellulase (5 FPU/g) + Crude α-L-arabinofuranosidase (4.0 U/g) + Crude xylanase (300 U/g)	5.57	4.99	0.64

ND, not detected.

Conclusions

The present investigation showed that newly isolated *A. niger* ADH-11 is a potential α-L-arabinofuranosidase producer. The statistical optimization for α-L-arabinofuranosidase production by SSF was highly advantageous as enzyme production was enhanced by 2.34-fold. To the best of our knowledge, this report describes the highest level of α-L-arabinofuranosidase production under SSF. The crude enzyme was highly active and stable at 55°C and pH 4.0, which makes it suitable for application in enzymatic saccharification of biomass. Moreover, it was evident from the present investigation that the supplementation of crude α-L-arabinofuranosidase can play significant role in the enzymatic hydrolysis of maize stover.

Competing interests
The authors declare that they have no competing interests.

Authors' contributions
HP designed and performed the experiments along with the data analysis and organized the manuscript draft. JD designed and analyzed the statistical data for optimization of enzyme production. DC helped in designing the experiments and preparing the manuscript draft. AS supervised the complete study along with experimentation, interpretation, and manuscript preparation. All authors read and approved the final manuscript.

Acknowledgements
The authors gratefully acknowledge Gujarat State Biotechnology Mission (GSBTM), Gandhinagar, Gujarat, India for providing the research grants to support this work. The authors are also thankful to MAPs Enzyme Limited, India for providing the cellulase enzyme.

Author details
[1]BRD School of Biosciences, Sardar Patel Maidan Satellite Campus, Sardar Patel University, PO Box No. 39, Vallabh Vidyanagar 388120, Gujarat, India. [2]Department of Microbiology, Shree Ramkrishna Institute of Computer Education and Applied Sciences - MTB College Campus, Athwalines, Surat 395001, Gujarat, India. [3]Department of Statistics, Sardar Patel University, Vallabh Vidyanagar 388120, Gujarat, India.

References

1. Aristidou A, Penttila M (2000) Metabolic engineering applications to renewable resource utilization. Curr Opin Biotechnol 39:1931–1938
2. Beg QK, Kapoor M, Mahajan L, Hoondal GS (2001) Microbial xylanases and their industrial applications: a review. Appl Microbiol Biotechnol 56:326–338
3. Cano A, Palet C (2007) Xylooligosaccharide recovery from agricultural biomass waste treatment with enzymatic polymeric membranes and characterization of products with MALDI-TOF-MS. J Membr Sci 291:96–105
4. Collins T, Gerday C, Feller G (2005) Xylanases, xylanase families and extremophilic xylanases. FEMS Microbiol Rev 29:3–23
5. Kaji A (1984) L-Arabinosidases. Adv Carbohydr Chem Biochem 42:383–394
6. Okafor UA, Okochi VI, Onyegeme-Okerenta BM, Nwodo-Chinedu S (2007) Xylanase production by *Aspergillus niger* ANL 301 using agro-wastes. Afr J Biotechnol 6(14):1710–1714
7. Potanen K (1988) An α-L-arabinofuranosidase of *Trichoderma reesei*. J Biotechnol 7:271–282
8. Kormelink FJM, MJF S-V I, Wood TM, Voragen AGJ (1991) Purification and characterization of a (1,4)-β-arabinoxylan arabinofuranohydrolase from *Aspergillus awamori*. Appl Microbiol Biotechnol 35:753–758
9. Ramon D, Veen PVD, Visser J (1993) Arabinan degrading enzymes from *Aspergillus nidulans*: induction and purification. FEMS Microbiol Lett 113:15–22
10. Lounteri E, Siika-aho M, Tenkanen M, Viikari L (1995) Purification and characterization of three α-arabinosidases from *Aspergillus terreus*. J Biotechnol 38:279–291
11. Ioannes PD, Peirano A, Steiner J, Eyzaguirre J (2000) An α-L-arabinofuranosidase from *Penicillium purpurogenum*: production, purification and properties. J Biotechnol 76:253–258
12. Hashimoto T, Nakata Y (2003) Synergistic degradation of arabinoxylan with α-L-arabinofuranosidase, xylanase and β-xylosidase from soy sauce koji mold, *Aspergillus oryzae*, in high salt condition. J Biosci Bioeng 95(2):164–169
13. Sakamoto T, Kawasaki H (2003) Purification and properties of two type-B α-L-arabinofuranosidases produced by *Penicillium chrysogenium*. Biochem Biophys Acta 1621:204–210
14. Tengerdy RP (1996) Solid state fermentation for enzyme production. In: Pandey A (ed) Advances in biotechnology. Educational Publishers and Distributors, New Delhi, pp 13–16
15. Shah A, Madamwar D (2005) Xylanase production under solid-state fermentation and its characterization by an isolated strain of *Aspergillus foetidus* in India. World J Microbiol Biotechnol 21:233–243
16. Barrios-Gonzalez J (2012) Solid-state fermentation: physiology of solid medium, its molecular basis and application. Process Biochem 47:175–185
17. Pandey A, Soccol CR, Mitchell D (2000) New development in solid state fermentation: I-bioprocesses and products. Process Biochem 35:1153–1169
18. Chapla D, Divecha J, Madamwar D, Shah A (2010) Utilization of agro-industrial waste for xylanase production by *Aspergillus foetidus* MTCC 4898 under solid state fermentation and its application in saccharification. Biochem Eng J 49:361–369
19. Trivedi S, Divecha J, Shah A (2012) Optimization of inulinase production by a newly isolated *Aspergillus tubingensis* CR16 using low cost substrate. Carbo Poly 90:483–490
20. Li Y, Jiang H, Xu Y, Zhang X (2008) Optimization of nutrient components for enhance phenazine-1-carboxylic acid production by gacA-inactivated *Pseudomonas* sp. M18G using response surface method. Appl Microbiol Biotechnol 77:1207–1217
21. Min BJ, Park YS, Kang SW, Song YS, Lee JH, Park C, Kim CW, Kim SW (2007) Statistical optimization of medium components for the production of xylanase by *Aspergillus niger* KK2 in submerged cultivation. Biotechnol Bioprocess Eng 12:302–307
22. Yanai T, Sato M (2000) Purification and characterization of a novel α-L-arabinofuranosidase from *Pichia capsulata* X91. Biosci Biotechnol Biochem 64:1181–1188
23. Bailey MJ, Biely P, Poutanen K (1992) Interlaboratory testing of methods for assay of xylanase activity. J Biotechnol 23:257–270
24. Miller GL (1959) Use of dinitrosalicylic acid reagent for determination of reducing sugar. Anal Chem 31:426–428
25. Ghosh T (1994) Measurement of cellulase activities, Commission on Biotechnology. IUPAC., pp 1–12
26. Lowry OH, Rosebrough NJ, Farr AL, Randall RJ (1951) Protein measurement with the folin phenol reagent. J Biol Chem 31:426–428
27. Sun X, Liu Z, Qu Y, Li X (2008) The effect of wheat bran composition on the production of biomass hydrolyzing enzymes by *Penicillium decumbens*. Appl Biochem Biotechnol 146:119–128
28. Khandeparker R, Numan MTH, Mukherjee B, Satwekar A, Bhosle NB (2008) Purification and characterization of α-L-arabinofuranosidase from *Arthrobacter* sp. MTCC 5214 in solid state fermentation. Process Biochem 43:707–712
29. Narra M, Dixit G, Divecha J, Madamwar D, Shah A (2012) Production of cellulases by solid state fermentation with *Aspergillus terreus* and enzymatic hydrolysis of mild alkali-treated rice straw. Bioresour Technol 121:355–361
30. Ho SH, Anh TV, Koo YM (2013) Optimization of lipase-catalyzed synthesis of caffeic acid phenethyl ester in ionic liquid by response surface methodology. Bioprocess Biosys Eng 36:799–807
31. Katapodis P, Christakopoulou V, Kekos D, Christakopoulou P (2007) Optimization of xylanase production by *Chaetomum thermophilum* in wheat straw using response surface methodology. Biochem Eng J 35:136–141
32. Li Y, Cui F, Liu Z, Xu Y, Zhao H (2007) Improvement of xylanase production by *Penicillium oxalicum* ZH-30 using response surface methodology. Enzym Microbial Technol 40:1381–1388
33. Montgomery DC (1991) Design and analysis of experiments, 7th edn. Wiley, New York. ISBN ISBN-978-0-470-12866-4
34. Kirillova IP, Agre N, Kalakutskii LB (1975) Effect of the composition of the medium and magnesium and calcium ions on the germination of spores of *Thermoactinomyces vulgaris*. Mikrobiologiia 44(6):1034–1040
35. Guerfali M, Chaabouni M, Gargouri A, Belgith H (2010) Improvement of α-L-arabinofuranosidase production by *Talaromyces thermophilus* and agro-industrial residues saccharification. Appl Microbiol Biotechnol 85:1361–1372
36. Saha BC (2000) α-L-Arabinofuranosidase: biochemistry, molecular biology and application in biotechnology. Biotechnol Adv 18:403–423
37. Kaneko S, Shimasaki T, Kusakabe I (1993) Purification and some properties of intracellular α-L-arabinofuranosidase from *Aspergillus niger* 5-16. Biosci Biotechnol Biochem 57:1161–1165
38. Shah AR, Madamwar D (2005) Xylanase production by a newly isolated *Aspergillus foetidus* strain and its characterization. Process Biochem 40:1763–1771
39. Yan Q, Tang L, Yang S, Zhou P, Zhang S, Jiang Z (2012) Purification and characterization of a novel thermostable α-L-arabinofuranosidase (α-L-AFase) from *Chaetomium sp.* Process Biochem 47:472–478
40. Guais O, Tourrasse O, Dourdoigne M, Parrou JL, Francois JM (2010) Characterization of the family GH54 α-L-arabinofuranosidases in *Penicillium funiculosum*, including a novel protein bearing a cellulose-binding domain. Appl Microbiol Biotechnol 87:1007–1021
41. Filho EXF, Puls J, Coughlan MP (1996) Purification and characterization of two arabinofuranosidases from solid state culture of the fungus *Penicillium capsulatum*. Appl Environ Microbiol 62(1):168–172
42. Delabona PS, Cota J, Hoffmam ZB, Paixao DA, Farinas CS, Cairo JP, Lima DJ, Squina FM, Ruller R, Pradella JG (2013) Understanding the cellulolytic system of *Trichoderma harzianum* P49P11 and enhancing saccharification of pretreated sugarcane bagasse by supplementation with pectinase and α-L-arabinofuranosidase. Bioresor Technol 131:500–507

Fabrication of silver nanoparticles from leaf extract of *Butea monosperma* (Flame of Forest) and their inhibitory effect on bloom-forming cyanobacteria

Venkatesh Chaturvedi[1] and Pradeep Verma[2*]

Abstract

Background: Silver nanoparticles (SNPs) are used extensively in areas such as medicine, catalysis, electronics, environmental science, and biotechnology. Therefore, facile synthesis of SNPs from an eco-friendly, inexpensive source is a prerequisite. In the present study, fabrication of SNPs from the leaf extract of *Butea monosperma* (Flame of Forest) has been performed. SNPs were synthesized from 1% leaf extract solution and characterized by ultraviolet-visible (UV-vis) spectroscopy and transmission electron microscopy (TEM). The mechanism of SNP formation was studied by Fourier transform infrared (FTIR), and anti-algal properties of SNPs on selected toxic cyanobacteria were evaluated.

Results: TEM analysis indicated that size distribution of SNPs was under 5 to 30 nm. FTIR analysis indicated the role of amide I and II linkages present in protein in the reduction of silver ions. SNPs showed potent anti-algal properties on two cyanobacteria, namely, *Anabaena* spp. and *Cylindrospermum* spp. At a concentration of 800 μg/ml of SNPs, maximum anti-algal activity was observed in both cyanobacteria.

Conclusions: This study clearly demonstrates that small-sized, stable SNPs can be synthesized from the leaf extract of *B. monosperma*. SNPs can be effectively employed for removal of toxic cyanobacteria.

Keywords: Silver nanoparticles (SNPs); *Butea monosperma*; Cyanobacteria; Anti-algal

Background

Nanotechnology is an emerging branch of science which deals with synthesis and characterization of various metallic and non-metallic nanoparticles of different compositions, sizes, and shapes [1]. Nanoparticles are molecular aggregates ranging from 1 to 100 nm in diameter [2,3]. Nanoparticles are being extensively used in areas such as medicine, catalysis, electronics, environmental science, and biotechnology [4]. Among various types of nanoparticles available, silver nanoparticles (SNPs) are the most commonly employed as they are simple to produce following reduction of silver nitrate [5]. Further, silver has been known for its excellent antimicrobial properties and employed as an antiseptic agent in the treatment of wounds and burns [1,2]. There are many physical and chemical methods for the synthesis of SNPs, such as chemical and photochemical reactions [6], thermal decomposition of metal compounds [7], microwave-assisted reduction [8], and laser-mediated reduction [9]. These methods have several demerits specifically high cost and toxicity in the biological system [10]. Recently, biological synthesis of SNPs by employing different plant extracts is considered as an inexpensive and eco-friendly method for large-scale production of SNPs [10,11]. Most of the plants employed for the reduction process also exhibit medicinal properties [12,13]. *Butea monosperma* Lam. Kuntze (Fabaceae) is a plant, which is native to India and is called Flame of Forest because it forms orange-colored flowers during the summer season (April to June). This plant is of great medicinal

* Correspondence: vermaprad@yahoo.com
[2]Department of Microbiology, Central University of Rajasthan, N.H. 8 Bandarsindri, Kishangarh, Ajmer, Rajasthan 305801, India
Full list of author information is available at the end of the article

importance to the native people. Leaves of this plant exhibit anti-inflammatory activity and are also used to control giardiasis caused by the protozoa *Giardia lamblia* [14]. To date, there is no report on the synthesis of SNPs using this plant.

Cyanobacteria are primitive photosynthetic bacteria, which are considered as an excellent source of pigments, vitamins, polysaccharides, proteins, pharmaceuticals, and other biologically active compounds [15-17]. However, there are certain cyanobacteria that are responsible for the production of algal blooms. Algal blooms cause depletion of dissolved oxygen in water leading to anoxia and subsequently death of fishes. These cyanobacteria also produce various toxins (microcystin, saxitoxin, etc.) which are extremely toxic to mammals including humans [18,19]. Therefore, there is urgent need to identify chemicals and metabolites which inhibit algal bloom formation [18]. Recent studies have demonstrated that silver nanoparticles can inhibit bloom formation by the toxin-producing cyanobacteria *Microcystis aeruginosa* [19]. However, there is no report on the impact of SNPs on other toxic cyanobacteria. In the present study, an attempt has been made to synthesize and characterize SNPs from the leaf extract of *B. monosperma* owing to its medicinal properties. Further, the anti-algal potential of SNPs on two toxic bloom-forming cyanobacteria, namely, *Anabaena* spp. and *Cylindrospermum* spp., was also evaluated.

Methods

Synthesis of SNPs

Fresh leaves of *B. monosperma* growing in forests of Bilaspur, Chhattisgarh, India, were collected (Figure 1). Fresh leaves were washed extensively with water followed by final wash with distilled water. The veins of the leaves were removed and cut into small pieces (1 cm^2). One gram (1 g) of leaf pieces was suspended in 100 ml of double distilled water and heated on a water bath for 20 min. After heating, the extract was filtered using filter paper (Whatman filter paper no. 1) and stored at 4°C until further use. For synthesis of SNPs, different dilutions of the leaf extract in double distilled water were prepared, and to it, 0.01 M of $AgNO_3$ solution was added. The solution was incubated in a rotary shaker at 100 rpm for 2 to 3 h in the dark.

Figure 1 The photograph of *Butea monosperma* (Flame of Forest) plant.

Characterization of SNPs by UV-vis spectroscopy and TEM

Synthesis of SNPs was monitored using a double-beam spectrophotometer (Systronics 2203, Systronics, Ahmedabad, India). The absorption spectrum in the range of 200 to 800 nm of the 0.5% leaf extract solution and the 1% leaf extract solution containing 0.01 mM $AgNO_3$ was recorded. The presence of SNPs was confirmed by observing a peak at 410 to 430 nm, corresponding to the surface plasmon resonance of SNPs. The 1% leaf extract solution mixed with 100 ml of Millipore water (Millipore Corporation, Billerica, MA, USA) was employed as a positive control. The particle size and distribution of SNPs were determined by transmission electron microscopy (TEM) analysis. For TEM analysis, a drop of the solution containing 1% leaf extract solution supplemented with 0.01 mM $AgNO_3$ was employed. TEM was performed using a Morgagni 268D transmission electron microscope (FEI Electron Optics, Hillsboro, OR, USA). Size distribution of SNPs from the TEM image was calculated by the software ImageJ.

Organism and culture conditions

In the present study, the two organisms employed were *Anabaena* spp. and *Cylindrospermum* spp. (blue-green freshwater cyanophyte). Both the cyanobacterial strains were obtained from the Department of Life Sciences, Guru Ghasidas Vishwavidyalaya, Bilaspur (Chhattisgarh). For maintenance of cultures, BG-11 medium [20] was employed. The cultures used were maintained in BG-11 medium in a culture room at 26°C to 30°C with a 16/8-h photoperiod under 3,000 lx from a cool white light. The cultures were hand shaken daily at least twice to keep the organisms in a homogeneous state.

Growth measurements

Growth of the organisms was monitored spectrophotometrically (double-beam spectrophotometer, Systronics 2203) at regular intervals with increase in absorbance at 660 nm [20].

Pigment extraction from cyanobacteria

Sample preparation

Culture sample (5.0 ml) was withdrawn, and cells were centrifuged at 4,800 rpm for 5 min. The supernatant obtained was removed, and the cell pellet was then resuspended in 5.0 ml of distilled water to remove any salts that

Figure 2 Synthesis of SNPs using different leaf extract concentrations. **(A)** Control flask containing 0.01 M $AgNO_3$. **(B)** Visible color change in 0.01 M $AgNO_3$ solution after addition of 0.5% leaf extract solution. **(C)** Visible color change in 0.01 M $AgNO_3$ solution after addition of 1.0% leaf extract solution.

could have been retained with the biomass, and again centrifuged [21].

Extraction and quantification

The solvent absolute methanol (99%) was used for pigment extraction. Briefly, cells were suspended in 5.0 ml of solvent followed by vortexing for 15 s and incubated for 20 min. In the absence of other cell wall disruption methods, the cells were centrifuged at 4,000 rpm for 5 min. The absorbance of the supernatant obtained was read spectrophotometrically at the wavelength corresponding to the equations [21].

Chlorophyll-a and carotenoid determination

For chlorophyll-a and carotenoid determination, the pellet obtained was kept and the supernatant was discarded, and to the pellet, 5.0 ml of 25% absolute methanol was added and it was incubated at 50°C for 1 h. The sample was then centrifuged at 4,000 rpm for 5 min, and the clear supernatant obtained was used for measuring optical density in an ultraviolet-visible (UV-vis) spectrophotometer at 663 and 480 nm, respectively [22]. Protein determination was performed using the modified Lowry method [23].

Anti-algal properties of SNPs

After synthesis of SNPs, the solution was centrifuged at 12,000 rpm for 30 min at 30°C, and the nanoparticle pellet was collected, washed twice with 5 ml of Millipore water, and dried in a hot air oven. For toxicity studies, a stock solution of SNPs (5 mg/ml) was prepared. SNPs were suspended by sonicating the solution at 40 kHz, with a pulse rate of 40 for 10 min. The stock was stored in the dark at 4°C prior to use. Cells were taken (50 ml from the 7-day-old

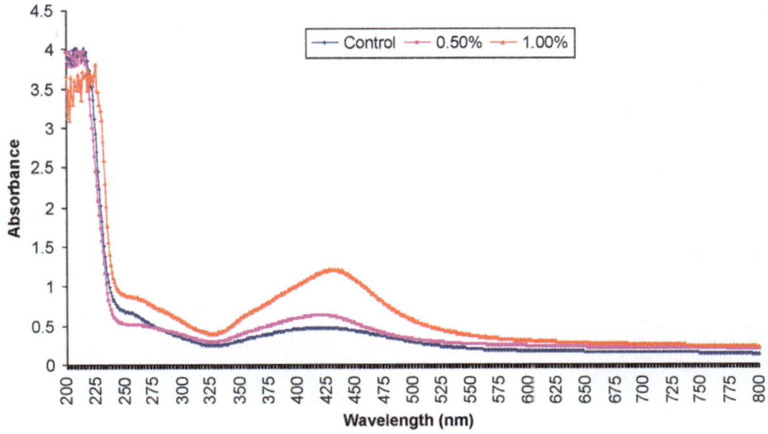

Figure 3 UV-vis spectrum of leaf extract and $AgNO_3$ solutions. The figure shows the UV-vis spectrum of different concentrations of the leaf extract solution (0.5% and 1%) in the presence of 0.01 M $AgNO_3$ solution after 3 h of incubation.

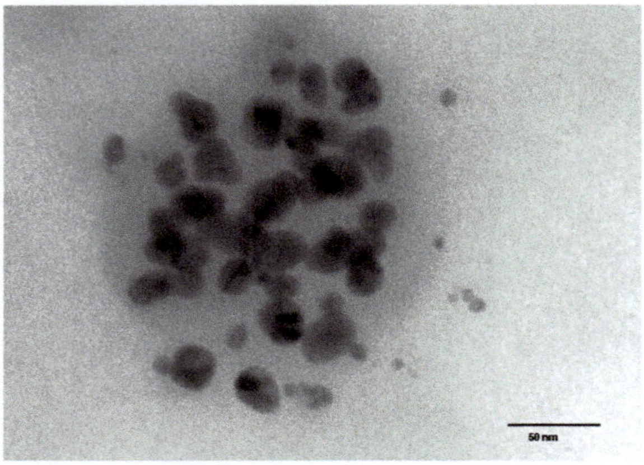

Figure 4 TEM image of synthesized SNPs using *Butea monosperma* (Flame of Forest), 1% leaf extract solution.

culture) and transferred aseptically into four sets of 100-ml Erlenmeyer flasks for each strain. Different concentrations of SNPs (400, 600, and 800 μg/ml) were added into each flask except for one, which was kept as control, and flasks were incubated in a culture room at 26°C to 30°C with a 16/8-h photoperiod under 3,000 lx from a cool white light for 15 days. Periodic analysis was done (0th, 4th, 8th, and 12th day) by harvesting the cells from each set of flasks, followed by evaluation of its effect on growth and protein and pigment content of the cyanobacterial cells. All the experiments were performed in triplicate, and results were represented as mean ± standard deviation. The results were compared by one-way ANOVA followed by the Tukey-Kramer comparison test.

Results and discussion

Synthesis of SNPs

There are several reports on the synthesis of SNPs from plants having high medicinal and ornamental value, such as *Aloe vera* [24], *Cinnamomum camphora* [25], *Azhadirachta indica* [26], and *Ocimum sanctum* [27]. *B. monosperma* is an ornamental and medicinal plant, native to mountainous regions of India. Almost all the parts of the plant have medicinal value. The leaves show diuretic and

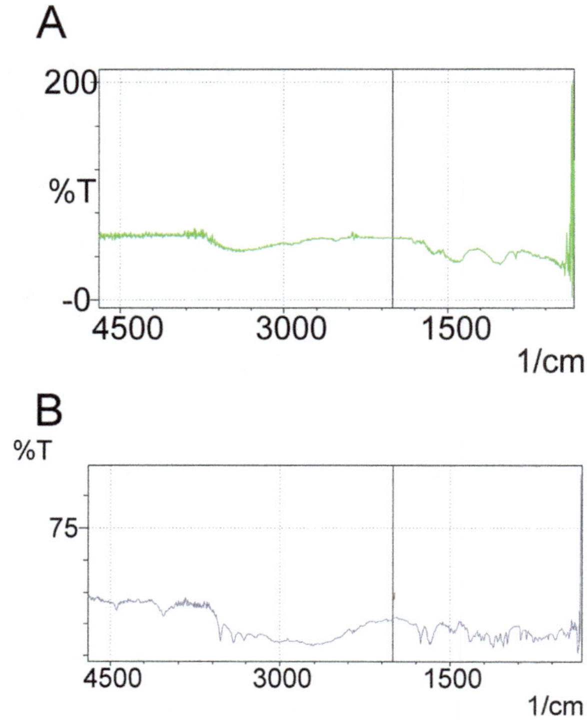

Figure 6 FTIR spectrum of leaf extract (A) and 1% leaf extract in the presence of 0.01 M $AgNO_3$ (B).

anti-ovulatory properties [14]. In this study, the synthesis of SNPs was carried out by employing the leaf extract of *B. monosperma*. The method was optimized by varying the concentration of the leaf extract (0.5%, 1.0%) with a fixed concentration (0.01 M) of $AgNO_3$. After the reaction, the synthesis was confirmed by a dark brown color of the solution. Studies have shown that during reduction of silver ions by the leaf extract, the solution first turns yellow and in later stages turns dark brown. The dark brown color represents the most stable form of SNPs, whereas the gray color represents instability and aggregation of SNPs [1,10]. In this study, at a concentration of 0.05% leaf extract, the color of the solution was yellow but it turned grayish after some time (Figure 2B). At a concentration of 1.0% leaf extract, the solution was yellow in the beginning and turned

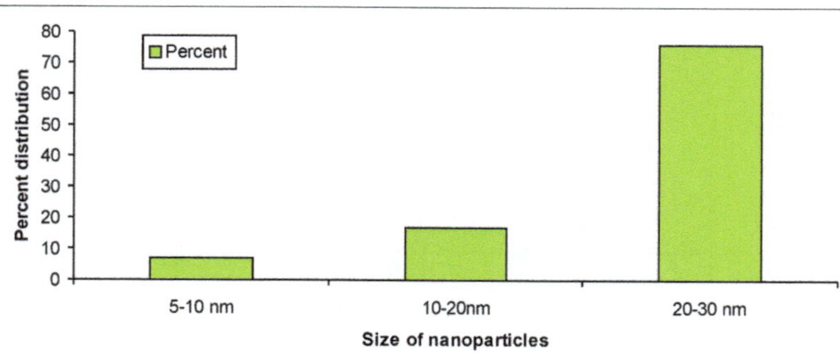

Figure 5 Size distribution of SNPs.

dark brown on standing (Figure 2C). This demonstrates that stable SNPs are produced at this concentration.

Characterization of SNPs

Chemically synthesized silver nanoparticles were characterized by UV-vis spectroscopy and TEM. The UV-visible spectrum of the leaf extract and $AgNO_3$ solutions is shown in Figure 3. It is clear from the spectrum that at 0.5% leaf extract solution, the peak at 420 to 430 nm was low and broad, suggesting instability of the nanoparticles. At a concentration of 1.0% leaf extract solution, the peak was comparatively high and less broad, suggesting the stability and small size of the silver nanoparticles [28,29]. High absorbance indicates higher concentration of SNPs. At 0.5% leaf extract concentration, the peak was observed at 420 nm. At 1.0% leaf extract concentration, the peak shifted to 430 nm. This redshift has been attributed to the increase in the size of nanoparticles. The greater is the shift, the larger is the size of nanoparticles [30]. However, in the present study, the shift was not very high (420 to 430 nm), and it indicates the small size of SNPs.

In order to further confirm SNP formation, TEM was performed. The photograph of TEM is shown in Figure 4. It is clear from the figure that nearly spherical SNPs were formed. From the TEM images, particle size and distribution were calculated. The results are depicted in Figure 5. It was observed that majority of the nanoparticles (76%) were in the range of 20 to 30 nm, 17% were in the range of 10 to 20 nm, and 7% were in the range of 5 to 10 nm. These results clearly represent that the size distribution of nanoparticles falls under 5 to 30 nm, with the major portion under the 20- to 30-nm range. By employing this method, the size distribution of SNPs obtained was smaller when compared to previous reports, *viz.* 44 to 225 nm using *Pergularia daemia* plant latex [31] and 71 to 110 nm using mangrove plant *Avicennia marina* [32].

FTIR analysis of SNPs

To elucidate the role of various functional groups present in the leaf extract in the synthesis of SNPs, Fourier transform infrared (FTIR) analysis of the leaf extract and

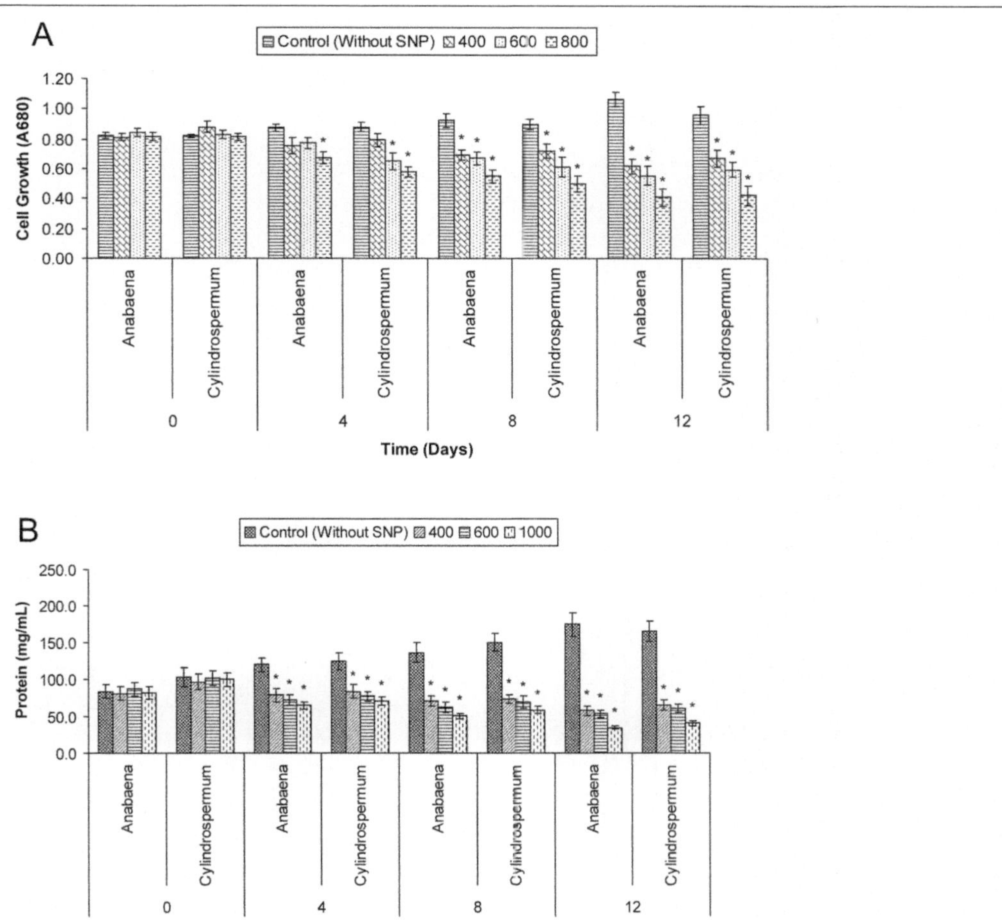

Figure 7 Inhibitory activity of different concentration of SNPs (400, 600, and 800 μg/ml) on *Anabaena* spp. and *Cylindrospermum* spp. **(A)** Cell growth and **(B)** protein. The asterisk denotes significance at $p \leq 0.01$.

$AgNO_3$ was carried out (Figure 6A,B). It is clear from the spectrum that in the presence of $AgNO_3$, broadening of peaks in the 1,680 cm^{-1} and 3,450 cm^{-1} regions suggests a plausible role of amide I and II linkages present in protein in the reduction of silver ions. The broadening of these peaks was not evident in the FTIR spectrum of the leaf extract. This result is in agreement with previous reports which suggest the involvement of proteins present in the plant extract in the reduction of silver ions [29].

Evaluation of anti-algal properties of SNPs on cyanobacteria

Cyanobacterial blooms are defined as intense growth of cyanobacteria in water bodies [18]. Algal blooms occur in natural waters when cyanobacteria grow quickly in water due to the presence of high amounts of nitrogen and phosphorus present due to eutrophication. A dense algal bloom can block sunlight and take up oxygen in the water, killing aquatic plants and animals. Some cyanobacteria produce cyanotoxins that severely affect animals and humans. Some common bloom-forming cyanobacteria are *Anabaena*, *Cylindrospermopsis*, *Nodularia*, *Oscillatoria*, *Microcystis*, and *Planktothrix* [33,34]. The current scenario demands development of efficient methods to control growth of these cyanobacteria. Various chemical methods have been employed such as treatment with copper sulfate and ozone, chlorination, or potassium permanganate. These methods have shown little efficacy; they are costly and their application to water bodies is sometimes practically not feasible [35]. In the present investigation, use of SNPs to control growth of cyanobacteria has been evaluated. There are numerous reports on the antimicrobial nature of SNPs [36]. The antimicrobial nature of silver/silver ions can be attributed to their ability to bind and denature proteins causing

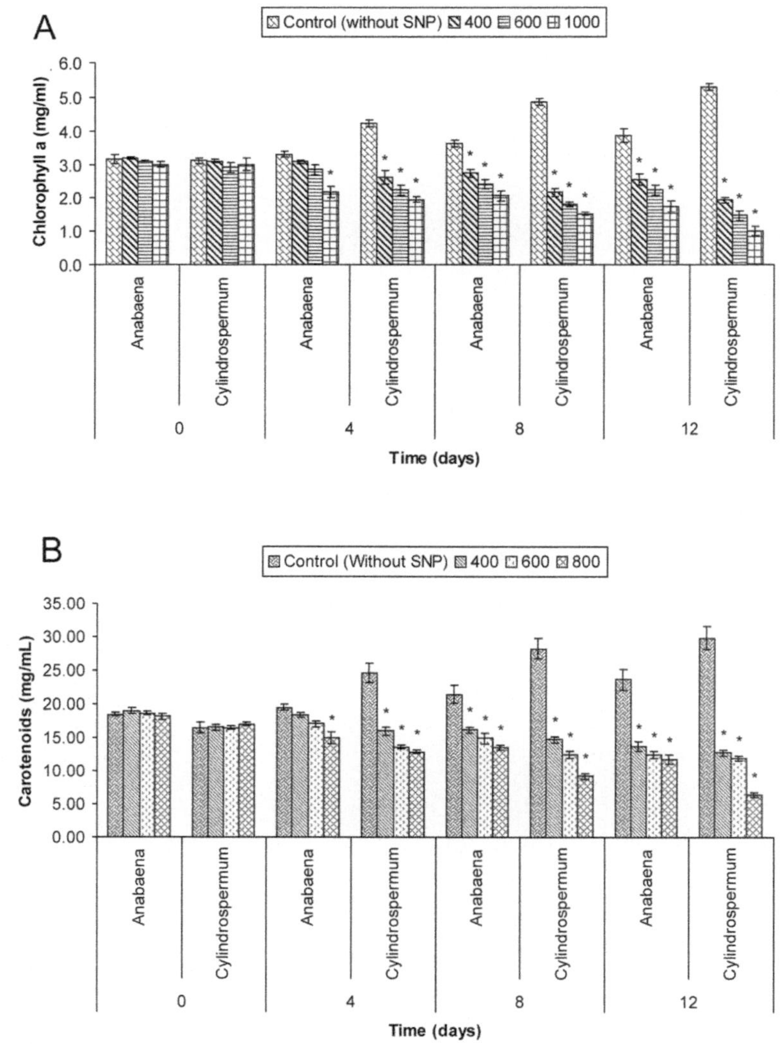

Figure 8 Inhibitory activity of different concentration of SNPs (400, 600, and 800 µg/ml) on *Anabaena* spp. and *Cylindrospermum* spp. **(A)** Chlorophyll-a and **(B)** carotenoid. The asterisk denotes significance at $p \leq 0.01$.

structural changes in the cell wall and nuclear membrane, leading to cell death. Silver ions can also inhibit bacterial replication by binding to DNA [2]. The anti-algal activity of SNPs on the toxic bloom-forming cyanobacteria *M. aeruginosa* has been reported (park). However, studies pertaining to the effect of SNPs on other cyanobacteria are still in the primitive stage. In the present study, the anti-algal activity of SNPs on two cyanobacteria, namely, *Anabaena* spp. and *Cylindrospermum* spp., was evaluated. The different concentrations of silver nanoparticles, i.e., 400, 600, and 800 μg/ml, were employed to assess the anti-algal property in the two cyanobacteria. Pre-grown cultures of cyanobacteria were incubated with different concentrations of SNPs. A similar culture without SNPs served as control. After a regular time interval of 4 days, growth, protein concentration, and chlorophyll-a and carotenoid content were estimated and compared (Figures 7 and 8). It is clearly evident from the results that dose-dependent reductions in the above mentioned parameters were observed in the two cyanobacteria. In both the cyanobacteria, significant inhibition ($p \leq 0.01$) was observed at a concentration of 800 μg/ml of SNPs. When toxicity of both cyanobacteria was compared, it was observed that *Anabaena* spp. was more sensitive towards SNPs than *Cylindrospermum* spp. The anti-algal activity remarkably increased with increasing concentrations of SNPs. The results were in agreement with the results obtained by Park et al. [19] where SNPs showed a dose-dependent inhibition on the harmful cyanobacterium *M. aeruginosa*. Yamanka et al. [3] proposed two hypotheses to explain the antimicrobial (here in this case, it is anti-algal) activity of SNPs. Initially, a cell in contact with SNPs takes up nanoparticles inside the cell, which inhibit various enzymes and biological processes causing severe damage to the cell. Secondly, reactive oxygen species (ROS) are also produced, possibly through the inhibition of respiratory enzymes, and attack the cell causing cell death.

Conclusions

In this study, SNPs using the leaf extract of *B. monosperma* (Flame of Forest) were synthesized. It was observed that at a concentration of 1% leaf extract, synthesis of stable, small-sized SNPs took place. TEM analysis indicated that the size distribution of SNPs was under 5 to 30 nm, which is smaller as compared to previous reports. SNPs showed strong anti-algal properties on two cyanobacteria, namely, *Anabaena* spp. and *Cylindrospermum* spp. A direct dose-dependent relationship between concentration of SNPs and cell death was observed. SNPs have been previously shown to be antimicrobial in nature. The antimicrobial activity of SNPs on different bacteria and fungi has been reported. Results of the present study clearly demonstrate that SNPs can be used for control of toxic cyanobacteria. However, this is preliminary study where toxicity was assessed in laboratory conditions. It is essential to test the feasibility of this method in natural environments and to develop a method for the application of SNPs to water bodies.

Competing interests
The authors declare that they have no competing interests.

Authors' contributions
VC carried out the synthesis and characterized the anti-algal properties of SNPS on cyanobacteria. PV designed the whole study and prepared the manuscript. Both authors read and approved the final manuscript.

Acknowledgements
PV is thankful to DBT for providing the financial support (Grant No. BT/304/NE/TBP/2012).

Author details
[1]School of Biotechnology, Banaras Hindu University, Varanasi, UP 221005, India. [2]Department of Microbiology, Central University of Rajasthan, N.H. 8 Bandarsindri, Kishangarh, Ajmer, Rajasthan 305801, India.

References
1. Lengke FM, Fleet EM, Southam G (2007) Biosynthesis of silver nanoparticles by filamentous cyanobacteria a from a silver(I) nitrate complex. Langmuir 23:2694–2699
2. Prabhu S, Poulose EK (2012) Silver nanoparticles: mechanism of antimicrobial action, synthesis, medical applications, and toxicity effects. Int Nano Lett 2:32
3. Yamanaka M, Hara K, Kudo J (2005) Synergistic bactericidal activity of silver nanoparticles and ciprofloxacin against phytopathogens. Appl Environ Microbiol 71(11):7589–7593
4. Sadowski Z (2010) Biosynthesis and Application of Silver and Gold Nanoparticles, In: David Pozo Perez (Ed.), Silver Nanoparticles, Rijeka, Croatia, InTech, pp 257-276
5. Rai M, Yadav A, Gade A (2009) Silver nanoparticles as a new generation of antimicrobials. Biotech Adv 2:776–83
6. An C, Peng S, Sun Y (2010) Facile synthesis of sunlight-driven AgCl:Ag plasmonic nanophotocatalyst. Adv Mater 22(23):2570–2574
7. Wang RC, Gao YS, Chen SJ (2009) Simple synthesis and size-dependent surface enhanced Raman scattering of Ag nanostructures on TiO2 by thermal decomposition of silver nitrate at low temperature. Nanotechnology 20 (37):375605
8. Horikoshi S, Abe H, Torigoe K, Abe M, Serpone N (2010) Access to small size distributions of nanoparticles by microwave-assisted synthesis. Formation of Ag nanoparticles in aqueous carboxymethylcellulose solutions in batch and continuous-flow reactors. Nanoscale 2(8):1441–1447
9. Peng Z, Spliethoff B, Tesche B, Walther T, Kleinermanns K (2006) Laser-assisted synthesis of Au-Ag alloy nanoparticles in solution. J Phys Chem B 2110 (6):2549–2554
10. Murugan K, Senthilkumar B, Senbagam D, Al-Sohaibani S (2014) Biosynthesis of silver nanoparticles using Acacia leucophloea extract and their antibacterial activity. Int J Nanomedicine 9:2431–2438
11. Rahimi-Nasrabadi M, Pourmortazavi SM, Shandiz SA, Ahmadi F, Batooli H (2014) Green synthesis of silver nanoparticles using Eucalyptus leucoxylon leaves extract and evaluating the antioxidant activities of extract. Nat Prod Res 27:1–6
12. Rajasekharreddy P, Rani PU (2014) Biofabrication of Ag nanoparticles using Sterculia foetida L. seed extract and their toxic potential against mosquito vectors and HeLa cancer cells. Mater Sci Eng C Mater Biol Appl 39:203–212
13. Suresh G, Gunasekar PH, Kokila D, Prabhu D, Dinesh D, Ravichandran N, Ramesh B, Koodalingam A, Vijaiyan Siva G (2014) Green synthesis of silver nanoparticles using Delphinium denudatum root extract exhibits antibacterial and mosquito larvicidal activities. Spectrochim Acta A Mol Biomol Spectrosc 127:61–66
14. Burlia DA, Khadeb AB (2007) A comprehensive review on Butea monosperma (Lam.) Kuntze. Pharmacogn Rev 1:333–7

15. Pulz O, Gross W (2004) Valuable products from biotechnology of microalgae. Appl Microbiol Biotechnol 65:155–165
16. Rastogi RP, Richa SRP, Singh SP, Hader DP (2010) Photoprotective compounds from marine organisms. J Ind Microbiol Biotechnol 37:537–558
17. Singh SP, Kumari S, Rastogi RP, Singh KL, Richa SRP (2010) Photoprotective and biotechnological potentials of cyanobacterial sheath pigment, scytonemin. Afr J Biotechnol 9(5):580–588
18. Humpage AR, Hardy SJ, Moore EJ, Frosscio SM, Falcones IR (2000) Microcystins (cyanobacterial toxin) in drinking water enhances the growth of aberrant cryptic foci in the mouse colon. J Toxicol Environ Health part A 61:155–165
19. Park MH, Kim KH, Lee HH, Kim JS, Hwang SJ (2010) Selective inhibitory potential of silver nanoparticles on the harmful cyanobacterium Microcystis aeruginosa. Biotechnol Lett 32(3):423–428
20. Chaturvedi V, Chandravanshi M, Rahangdale M, Verma P (2013) An integrated approach of using polystyrene foam as an attachment system for growth of mixed culture of cyanobacteria with concomitant treatment of copper mine waste water. J Waste Management doi:10.1155/2013/282798
21. Henriques M, Silva A, Rocha J (2007) Extraction and Quantification of Pigments from a Marine Microalga: A Simple and Reproducible Method. In: Mendez-Vilas A. (Ed.), Communicating Current Research and Educational Topics and Trends in Applied Microbiology, Badajoz, Spain, Formatex, pp. 586–593.
22. Jodlowska S, Latala A (2011) The comparison of spectrophotometric method and high-performance liquid chromatography in photosynthetic pigments analysis. OnLine J Bio Sci 11(2):63–69
23. Markwell MA, Haas SM, Tolbert NE, Bieber LL (1981) Protein determination in membrane and lipoprotein samples: manual and automated procedures. Methods Enz 72:296–303
24. Chandran SP, Chaudhary M, Pasricha R, Ahmad A, Sastry M (2006) Synthesis of gold nanotriangles and silver nanoparticles using Aloe vera plant extract. Biotechnol Prog 22:577–583
25. Huang Q, Li D, Sun Y, Lu Y, Su X, Yang H, Wang Y, Wang W, Shao N, Hong J, Chen C (2007) Biosynthesis of silver and gold nanoparticles by novel sundried Cinnamomum camphora leaf. Nanotechnology 18:105104
26. Lalitha A, Subbaiya R, Ponmurugan P (2013) Green synthesis of silver nanoparticles from leaf extract Azhadirachta indica and to study its anti-bacterial and antioxidant property. Int J Curr Microbiol App Sci 2:228–235
27. Singhal G, Bhavesh R, Kasariya K, Sharma AR, Singh RP (2011) Biosynthesis of silver nanoparticles using Ocimum sanctum (Tulsi) leaf extract and screening its antimicrobial activity. J Nanopar Res 13:2981–2988
28. Mohanta YK, Behera SK (2014) Biosynthesis, characterization and antimicrobial activity of silver nanoparticles by Streptomyces sp. SS2. Bioprocess Biosyst Eng. doi:10.1007/s00449-014-1205-6
29. Ajitha B, Ashok Kumar Reddy Y, Sreedhara Reddy P (2014) Biosynthesis of silver nanoparticles using Plectranthus amboinicus leaf extract and its antimicrobial activity. Spectrochim Acta A Mol Biomol Spectrosc 128:257–262
30. Ashkarran AA, Bayat A (2013) Surface plasmon resonance of metal nanostructures as a complementary technique for microscopic size measurement. Int Nano Lett 3:50
31. Patil CD, Borase HP, Patil SV, Salunkhe RB, Salunke BK (2012) Larvicidal activity of silver nanoparticles synthesized using Pergularia daemia plant latex against Aedes aegypti and Anopheles stephensi and nontarget fish Poecillia reticulate. Parasitol Res 111(2):555–562
32. Gnanadesigan M, Anand M, Ravikumar S, Maruthupandy M, Syed Ali M, Vijayakumar V, Kumaraguru AK (2011) Antibacterial potential of biosynthesized silver nanoparticles using Avicennia marina mangrove plant. Appl Nanosci 2:143–147
33. Carmichael WW (2001) Health effects of toxin producing cyanobacteria: the cyanoHABs. Hum Ecol Risk Assess 7:1393–1407
34. Codd GA, Morrison LF, Metcalf JS (2005) Cyanobacterial toxins: risk management for health protection. Tox Appl Phar 203:264–272
35. Westrick J, Szlag D, Southwell B, Sinclair J (2010) A review of cyanobacteria and cyanotoxins removal/inactivation in drinking water treatment. Anal Bional Chem 397:1705–1714
36. Banerjee P, Satapathy M, Mukhopahayay A, Das P (2014) Leaf extract mediated green synthesis of silver nanoparticles from widely available Indian plants: synthesis, characterization, antimicrobial property and toxicity analysis. Biores Bioproc 1:3

Improving the stability of glutamate fermentation by *Corynebacterium glutamicum* via supplementing sorbitol or glycerol

Yan Cao[1*], Zhen-ni He[2], Zhong-ping Shi[2] and Mpofu Enock[3]

Abstract

Background: *Corynebacterium glutamicum* is widely used in glutamate fermentation. The fermentation feature of the strain varies sometimes. These variations may lead to the reduction in the ability of the strain to resist environmental changes and to synthesize glutamate, resulting in abnormal glutamate fermentations.

Results: In the abnormal glutamate fermentations, glutamate accumulation stopped after glucose feeding and the final glutamate concentration was at a lower level (50 to 60 g/L). The r_{NAD^+}/r_{NADH} ratio was lower than that in normal batch which was reflected by lower oxidation-reduction potential (ORP) value. The abnormal fermentation performance was improved when glucose was co-fed with sorbitol/glycerol at a weight ratio of 5:1 or adding 10 to 15 g/L of sorbitol/glycerol in the initial medium. Under these conditions, glutamate synthesis continued after substrate (s) feeding and final glutamate concentration was restored to normal levels (≥72 g/L). r_{NAD^+}/r_{NADH} ratio, ORP, and pyruvate dehydrogenase (PDH), isocitrate dehydrogenase (ICDH), and cytochrome c oxidase (CcO) activities were maintained at higher levels.

Conclusions: Sorbitol and glycerol were not used as carbon sources for the fermentation. They were considered as effective protective agents to increase cells' resistance ability against environmental changes and maintain key enzymes activities.

Keywords: Enzyme activity; Fermentative stability; Glutamate fermentation; Oxidation-reduction potential; Protective substance; r_{NAD^+}/r_{NADH}

Background

L-Glutamate is mainly used as a flavor enhancer in food industry and nutrient in pharmaceutical industry. The annual production has exceeded 2.2 million tons by fermentation with *Corynebacterium glutamicum* [1]. The biosynthetic pathway of glutamate includes complex enzymatic reactions, as shown in Figure 1a. The enzymes effectively convert substrate (such as glucose) into glutamate only when they are well coordinated with coenzymes NAD^+ or NADH. Intracellular levels of NAD^+ and NADH significantly affect the catalytic efficiency of the enzymes [2]. The ratio of NAD^+/NADH in vivo is a key factor affecting energy transfer and redox state of the cells, and the optimal value at different stages during glutamate fermentation usually varied [3]. The metabolic flux distribution can be altered by variation of NAD^+/NADH ratio or r_{NAD^+}/r_{NADH} ratio in glutamate fermentation [4].

NAD^+/NADH ratio is indirectly reflected by oxidation-reduction potential (ORP) [5,6], which represented the redox state of the cells. The optimal ORP range corresponding to different fermentation processes is different. For example, maximum lysine yield was obtained when ORP was controlled between the range of −230 and −210 mV, while the preferable ORP range was −275 to −225 mV in homoserine and valine fermentations [6,7]. The redox state of cells is changed if certain auxiliary substances (such as sorbitol and glycerol) are supplemented and intracellular NAD^+/NADH ratio can be varied correspondingly [8]. These auxiliary substrates are usually non-repressive

* Correspondence: caoyan_115@163.com
[1]National University of Singapore (Suzhou) Research Institute, 377 Linquan Street, Suzhou, Jiangsu Province, China
Full list of author information is available at the end of the article

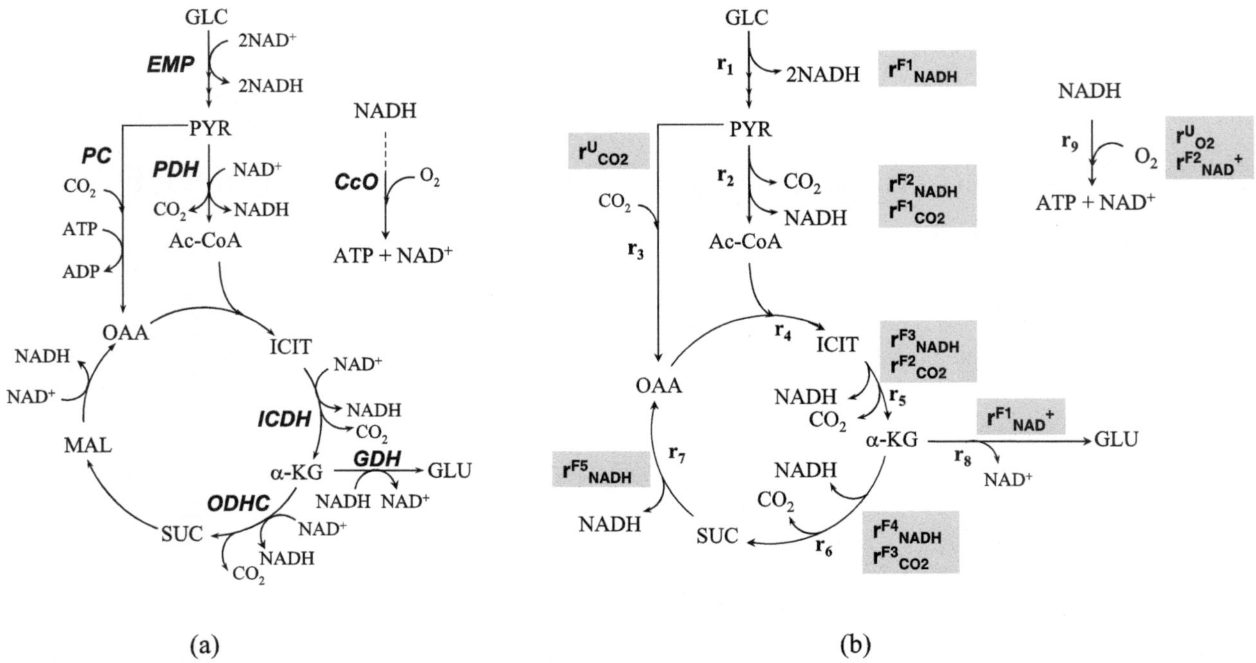

Figure 1 Simplified metabolic pathway of glutamate synthesis and generation/consumption of NAD$^+$ and NADH. (a) Metabolic pathway of glutamate synthesis; **(b)** simplified pathway of generation/consumption of NAD$^+$ and NADH. GLC, glucose; PYR, pyruvate; Ac-CoA, acetyl coenzyme A; ICIT, isocitrate; α-KG, α-ketoglutarate; SUC, succinate; MAL, malate; OAA, oxaloacetate; PDH, pyruvate dehydrogenase; PC, pyruvate carboxylase; ICDH, isocitrate dehydrogrnase; ODHC, α-ketoglutarate dehydrogenase; GDH, glutamate dehydrogenase; CcO, cytochrome c oxidase.

carbon sources. They can protect cells against stress in their living environment, enhance cell viability, and reduce the metabolic burden [9-12]. It was reported that the production of alkaline polygalacturonate lyase and lipase increased by 1.85-fold and 8.7-fold, respectively, when the strategy of methanol/sorbitol co-feeding was adopted [10,13]. Arruda and Felipe found that xylitol productivity could be increased by 35% when glycerol was added in the medium [14]. Therefore, fermentation with mixed carbon sources was considered as an effective way to enhance the targeted metabolite productions.

C. glutamicum used in industry is usually stored at 4°C for a short time and replaced regularly. In this way, production fluctuation resulting from strain change can be avoided. However, sometimes the fermentation characteristics of the strain vary, resulting in decreased glutamate synthesis ability and resistance to environmental changes. In such a case, fermentation performance becomes abnormal and glutamate production also ends at a very low level. Glutamate production fluctuated in a large range, and fermentation stability decreased greatly. Frequent rejuvenation is a common solution to this problem. But it is a costly, time-consuming, and troublesome procedure. Furthermore, glutamate is a low value-added product. It is more economical to adopt a simple way with low operation cost to maintain the fermentative stability. The strategy of feeding mixed substrates has been applied in other fermentations. This was an effective method to decrease cell mortality, maintain the enzyme activity, and promote targeted metabolite production [9-11]. Similar studies with regard to glutamate fermentation are also important, but few have been reported. In this study, sorbitol or glycerol was either co-fed with glucose or added in the initial medium, aiming at protecting the cells against environmental change/stress and stabilizing glutamate productions. Meanwhile, the theoretical mechanism was interpreted. The results gained in this study will provide some useful information and reference to the glutamate fermentation industry in terms of stabilizing the glutamate production.

Methods

Strain and culture condition

C. glutamicum ATCC13032 was used in this study. The seed microorganism was grown in a shaker at 32°C and 200 r/min for 8 to 10 h in liquid medium containing (in g/L) glucose 25, K_2HPO_4 1.5, $MgSO_4$ 0.6, $MnSO_4$ 0.005, $FeSO_4$ 0.005, corn slurry 25, and urea 2.5 (separately sterilized). Initial pH was adjusted to 7.0 to 7.2. The medium for jar fermentation contained (in g/L) glucose 140, K_2HPO_4 1.0, $MgSO_4$ 0.6, $MnSO_4$ 0.002, $FeSO_4$ 0.002, thiamine 5.0×10^{-5}, corn slurry 15, and urea 3.0 (separately sterilized).

The fed-batch fermentation was implemented in a 5-L fermentor (BIOTECH-5BG, Baoxing Co., Shanghai, China) equipped with on-line DO/pH/ORP electrodes. Initial medium volume was 3 L, and air aeration rate was 1.33 vvm. The temperature was maintained at 32°C during the entire fermentation period (about 34 to 36 h). pH was maintained at 7.0 to 7.2 by automatically pumping in 25% (*w*/*v*) ammonia water. Dissolved oxygen (DO) was controlled at 20% of air saturation by manually adjusting agitation rate. Concentrated glucose (50%, *w*/*v*) was fed when glucose concentration was lower than 20 g/L. Sorbitol or glycerol was supplemented by the following two methods:

Method #1: 50% (*w*/*v*) sorbitol or glycerol solution was co-fed with the addition of concentrated glucose solution. The feeding ratio of 1:5 (*w*/*w*) sorbitol/glycerol versus glucose was applied.

Method #2: 5 to 15 g /L sorbitol or glycerol was added into the initial medium before inoculation. Only glucose was fed when glucose concentration was lower than 20 g/L.

Analytical and measurement methods

Cell concentration was assayed by spectrophotometer at 620 nm (OD_{620}). Glucose and glutamate concentrations were measured by a biosensor (SBA-40C, Shandong Science Academy, Jinan, China). The concentrations of sorbitol and glycerol were analyzed by HPLC (Hitachi Chromaster Organizer, Hitachi, Ltd, Chiyoda-ku, Japan) equipped with an ion exclusion column (Aminex HPX-87H, 300 mm × 7.8 mm, Bio-Rad, Hercules, CA, USA) and a differential refractive index detector at 30°C. The mobile phase was 0.005 mol/L H_2SO_4 at a flow rate of 0.6 mL/min [10]. Two electronic balances (JA1102, Haikang Co., Shanghai, China) were connected to the computer and used to monitor the feeding amount of glucose and sorbitol or glycerol solution. O_2 and CO_2 partial pressure in exhaust gas were on-line measured by a gas analyzer (LKM2000, Lokas Co., Daejeon, Korea). CO_2 evolution rate (CER), O_2 uptake rate (OUR), and respiratory quotient (RQ) were then calculated by standard formula.

Enzyme activity assay

The activities of pyruvate dehydrogenase (PDH) and isocitrate dehydrogenase (ICDH) were analyzed by the methods reported [15,16]. Cytochrome c oxidase (CcO) was assayed using the kit for bacteria (Genmed Scientifics Inc., Wilmington, DE, USA). The enzyme activity was expressed as U/mg-DCW, where 1 U was the quantity of dry cell converting 1 μmol NAD^+ per minute. The relative enzymatic activity (REA) was used for comparison and interpretation. REA before feeding glucose/mixed carbon sources (18 h) was set as the unit (1), and REA after feeding glucose/mixed carbon sources (26 h) was described by Eq. (1).

$$REA_{aft}(k) = \frac{E_{aft}(k)}{E_{bef}(k)} \qquad (1)$$

Where $E_{bef}(k)$ and $E_{aft}(k)$ referred to the activities of *k*-th enzyme (PDH, ICDH, and CcO) before and after glucose/mixed carbon sources feeding.

Modeling and calculation of r_{NAD^+}/r_{NADH} ratio

The glutamate synthesis pathway was depicted according to the map reported [16], shown in Figure 1a. NADH and NAD^+ were generated or consumed to run the entire fermentation, and they were closely associated with glucose consumption, glutamate synthesis, CO_2 release, and O_2 consumption rates. Therefore, the generation or consumption rates of NADH and NAD^+ could be determined by a couple of measurable reaction rates, such as glucose consumption rate (r_{GLC}), glutamate formation rate (r_{GLU}), CER, and OUR. The metabolic pathways of NADH and NAD^+ were simplified as Figure 1b based on the following assumptions:

(1) The main products were glutamate and CO_2, because the concentrations of other byproducts (lactate, acetate, and other amino acids) were very low.

(2) Pentose phosphate (PP) pathway was ignored because it was not related to generation/consumption of NADH or NAD^+.

(3) The metabolic flux into the two reaction branches at pyruvate node followed the ideal condition, namely $r_2 = r_3$ in Figure 1b [16]. Therefore, the glyoxylate shuttle was ignored.

(4) The intermediate carbon metabolites were in pseudo-steady-state, and the net accumulation of them was 0. But it was not applied for NADH and NAD^+. r_{NAD^+}/r_{NADH} ratio was closely and positively associated with the NAD^+/NADH ratio, while r_{NAD^+} actually represented NADH consumption rate ($r^{(C)}_{NADH}$) and r_{NADH} represented NADH formation rate ($r^{(F)}_{NADH}$). NADH formation rate ($r^{(F)}_{NADH}$) could differ with its consumption rate ($r^{(C)}_{NADH}$), which led the variation in r_{NAD^+}/r_{NADH} ratio.

(5) Glutamate fermentation was a non-growth associated process; the cell concentration in production phase basically stayed at a constant level or declined slightly. So, we used the volume reaction rate to replace the specific reaction rate for convenience purpose.

The simplified metabolic pathway (Figure 1b) contains nine reactions shown in the Appendix, which covers the basic reactions occurring in EMP pathway, tricarboxylic acid (TCA) cycle, CO_2 fixing reaction, respiratory chain, and glutamate synthesis. According to the assumptions and simplifications above, the rates of all the reactions were coupled as follows.

$$r_1 = r_{GLC} \quad (2)$$

$$r_2 = r_3 = r_4 = r_5 = 0.5r_1 \quad (3)$$

$$r_6 = r_7 = r_5 - r_8 \quad (4)$$

$$r_8 = r_{GLU} \quad (5)$$

$$r_9 = 2r_{O2} = 2\ \mathrm{OUR} \quad (6)$$

The generation/consumption rates of NADH and NAD^+ as well as r_{NAD^+}/r_{NADH} ratio at a specified time t could be calculated by the equations of (7) to (16).

$$r^{F1}_{NADH}(t) = 2r_1(t) = 2r_{GLC}(t) \quad (7)$$

$$r^{F2}_{NADH}(t) = r_2(t) = r^{F1}_{CO_2}(t) \quad (8)$$

$$r^{F3}_{NADH}(t) = r_5(t) = r^{F2}_{CO_2}(t) \quad (9)$$

$$r^{F4}_{NADH}(t) = r_6(t) = r^{F3}_{CO_2}(t) \quad (10)$$

$$r^{F5}_{NADH}(t) = r_7(t) = r_6(t) \quad (11)$$

$$r^{F1}_{NAD^+}(t) = r_{GLU}(t) \quad (12)$$

$$r^{F2}_{NAD^+}(t) = r_9(t) = 2r^{U}_{O_2}(t) = 2\mathrm{OUR}(t) \quad (13)$$

$$r^{U}_{CO_2}(t) = r_3(t) \quad (14)$$

$$\mathrm{CER}(t) = r^{F1}_{CO_2}(t) + r^{F2}_{CO_2}(t) + r^{F3}_{CO_2}(t) - r^{U}_{CO_2}(t) \quad (15)$$

Where $r^{F}_{NAD^+}(t)$ is the NAD^+ formation rate (mmol/L/h), $r^{F}_{NADH}(t)$ NADH formation rate (mmol/L/h), $r^{U}_{O2}(t)$ O_2 uptake rate (mmol/L/h), $r^{F}_{CO2}(t)$ CO_2 evolution rate (mmol/L/h), $r^{U}_{CO2}(t)$ CO_2 uptake rate (mmol/L/h), $r_{GLC}(t)$ glucose consumption rate (mmol/L/h), and $r_{GLU}(t)$ glutamate production rate (mmol/L/h).

Results and discussion

Fermentation performance of normal and abnormal batches

Final glutamate production and cell concentration were two factors reflecting the fermentation performance. In 'normal' fermentation, final glutamate concentration and the maximum cell concentration (OD_{620}) were more than 70 g/L and 50, respectively. Otherwise, fermentations were categorized into 'abnormal.' Glutamate fermentation was non-growth associated, and glutamate was accumulated when cell growth had almost ceased (after 10 h). At this moment, about 1/3 glucose in the initial medium was consumed. Additional glucose was supplemented during the main glutamate production phase at 18 to 20 h. In the normal batch, glutamate concentration still increased after glucose was fed, although glutamate accumulation rate became slow. However, glutamate production stopped after glucose feeding in the abnormal batch, and the final glutamate concentration was around 50 g/L. In this case, cell growth and glutamate production before glucose feeding were almost the same as those in the normal fermentation (Figure 2a,b). There was no significant difference in the changing trend of RQ between the normal and abnormal batches. However, the changing pattern of ORP in the abnormal batch differed from that of normal batch significantly (Figure 2c,d). ORP was maintained in the normal range (−75 to −85 mV) before 20 h (glucose was fed at 18 h) and decreased slowly to a lower level (−120 mV) after 20 h. Generally, r_{NAD^+}/r_{NADH} ratio was closely and positively associated with the NAD^+/NADH ratio. It was shown that in vivo r_{NAD^+}/r_{NADH} ratio was at a lower level

$$\begin{aligned}
\frac{r_{NAD^+}}{r_{NADH}}(t) &= \frac{r^{F}_{NAD^+}(t)}{r^{F}_{NADH}(t)} = \frac{r^{F1}_{NAD^+}(t) + r^{F2}_{NAD^+}(t)}{r^{F1}_{NADH}(t) + r^{F2}_{NADH}(t) + r^{F3}_{NADH}(t) + r^{F4}_{NADH}(t) + r^{F5}_{NADH}(t)} \\
&= \frac{r_{GLU}(t) + 2\mathrm{OUR}(t)}{2r_{GLU}(t) + r^{F1}_{CO_2}(t) + r^{F2}_{CO_2}(t) + r^{F3}_{CO_2}(t) + r_7} \\
&= \frac{r_{GLU}(t) + 2\mathrm{OUR}(t)}{2r_{GLU}(t) + \mathrm{CER}(t) + r_3 + r_7} = \frac{r_{GLU}(t) + 2\mathrm{OUR}(t)}{2r_{GLU}(t) + \mathrm{CER}(t) + 2r_3 - r_8} \\
&= \frac{r_{GLU}(t) + 2\mathrm{OUR}(t)}{3r_{GLU}(t) + \mathrm{CER}(t) - r_{GLU}(t)}
\end{aligned} \quad (16)$$

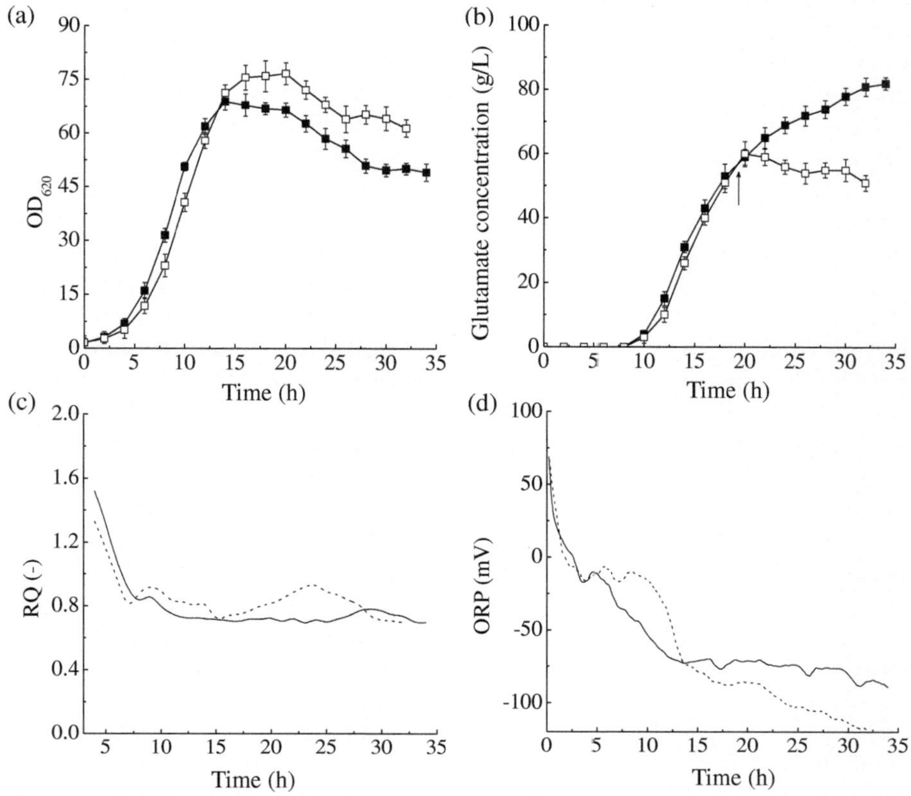

Figure 2 Comparison of fermentation parameters for normal and abnormal batches. Black square and solid line, normal batch; white square and dotted line, abnormal batch without sorbitol or glycerol; arrow, glucose fed. **a**: Time courses of cell concentration (OD_{620}) in different operation conditions; **b**: Time courses of glutamate concentration in different operation conditions; **c**: Time courses of RQ in different operation conditions; **d**: Time courses of ORP in different operation conditions.

and NADH was excessive in abnormal fermentation after 20 h, which was verified by the lower r_{NAD^+}/r_{NADH} ratio calculated in Table 1. Therefore, the improper r_{NAD^+}/r_{NADH} ratio after glucose feeding might be the reason for the non-accumulation of glutamate in the abnormal batch [17].

The abnormal performance appeared after glucose was fed, and it was not contaminated. It was speculated that abnormal fermentation was due to the change of the characteristics of strain which led to the following results: (1) glutamate production ability decreased and (2) fermentation environment changed after glucose was fed and the strain failed to adapt to the change, resulting in intracellular abnormal metabolism and stoppage of glutamate synthesis. It has been reported that some osmoregulators (such as trehalose or betaine) are either

Table 1 r_{NAD^+}/r_{NADH} ratio at different instants under different operation conditions

Time (h)	Fermentation batches					
	Batch #1	Batch #2	Batch #3	Batch #4	Batch #5	Batch #6
14	1.11	0.85	1.10	0.78	1.03	0.90
16	1.11	0.92	1.13	0.85	0.94	0.94
18	1.02	0.83	1.15	0.83	0.98	0.90
20	0.94	0.66	1.10	0.89	0.93	0.88
22	0.93	0.58	1.08	0.84	0.97	0.87
24	0.89	0.56	1.15	0.95	0.89	0.91
26	0.93	0.59	1.09	0.96	0.94	0.86
28	0.72	0.64	0.95	0.82	0.76	0.82

Batch #1, normal batch (control); *batch #2*, abnormal batch without sorbitol or glycerol; *batch #3*, sorbitol co-fed with glucose; *batch #4*, 15 g/L sorbitol added in the initial medium; *batch #5*, glycerol co-fed with glucose; *batch #6*, 10 g/L glycerol added in the initial medium.

produced by the microorganisms or taken up from the medium in lysine production by *C. glutamicum* in response to a hyperosmotic shock [18,19]. They can protect cells against environmental shock/stress. Sorbitol has the same effect. When glutamate accumulation ceased and the apparent fermentation parameters (OUR, CER, etc.) declined after glucose feeding for a period of 2 h, then 'abnormal' fermentation status was concluded. Twelve and 2 g/L sorbitol were added at 26 and 32 h, respectively; glutamate concentration increased gradually and reached 72 g/L at 36 h with a 2-h fermentation period extend, as shown in Figure 3.

Therefore, the abnormal fermentation was due to the decrease of resistance ability in response to the environmental alterations. In addition, the permeability of cell membrane increased to secrete glutamate extensively in the production phase, and the glucose addition easily brought about shock or stress in the living environment. The carbon flux distribution in vivo was adjusted, and consequently, the metabolism of NAD^+ and NADH was changed. On the other hand, PDH and ICDH required NAD^+ as the coenzyme and less NAD^+ amount restricted the enzymes' catalytic actions. As a result, the metabolic flux was redistributed and a series of abnormal effects arose.

Cells might be more tolerant to the environmental change if some osmoregulators, such as trehalose or betaine, were added before or at the same time when the environmental shock/stress occurred. However, trehalose and betaine are expensive. Sorbitol and glycerol were cheaper, and they were also efficient environmental shock/stress protective reagents. Hence, fermentation performance when co-feeding sorbitol/glycerol with glucose or adding sorbitol/glycerol in the initial medium was studied.

Fermentation performance in presence of sorbitol

The fermentation performance in presence of sorbitol is shown in Figure 4. The cell growth patterns did not change much despite the sorbitol supplement methods adopted. Glutamate production did not cease and final glutamate concentration reached 73 and 77 g/L at 34 h, respectively, when co-feeding sorbitol with glucose or adding sorbitol in the initial medium. No difference in RQ before 20 h in the batches with/without sorbitol was observed. RQ decreased continuously after 20 h in the presence of sorbitol (RQ was about 0.4 at 34 h). Lower RQ was favorable for glutamate accumulation and indicated that less glucose proceeded beyond the α-KG node in the TCA cycle [20]. In these cases, less NADH was accumulated in vivo and it was desirable for maintaining cellular activities [21]. Less NADH accumulation implied that $r^{(F)}_{NADH}$ is less than $r^{(C)}_{NADH}$, leading to a higher r_{NAD^+}/r_{NADH} ratio as well as a higher ORP levels. Under the condition of co-feeding sorbitol with glucose, ORP was maintained at a normal level (−75 to −85 mV) after feeding the mixed carbon sources. When 15 g/L sorbitol was added in the initial medium, ORP always decreased in the production phase, but it was much higher than that in the abnormal batch without sorbitol. r_{NAD^+}/r_{NADH} ratio was more than 0.8 after 20 h when co-feeding sorbitol with glucose or adding sorbitol in the initial medium, as shown in Table 1. The deterioration of the fermentation performance was reversed in the presence of sorbitol. This might be because the resistance to environmental change was enhanced and NADH consumption was returned to normal. The r_{NAD^+}/r_{NADH} ratio was returned to normal level in the presence of sorbitol, and higher ORP values supported this fact indirectly.

Fermentation performance in presence of glycerol

The fermentation performance when co-feeding glycerol with glucose or adding glycerol in the initial medium is shown in Figure 5. Similar fermentation curves and performance as those in the presence of sorbitol were obtained. The final glutamate concentration also reached

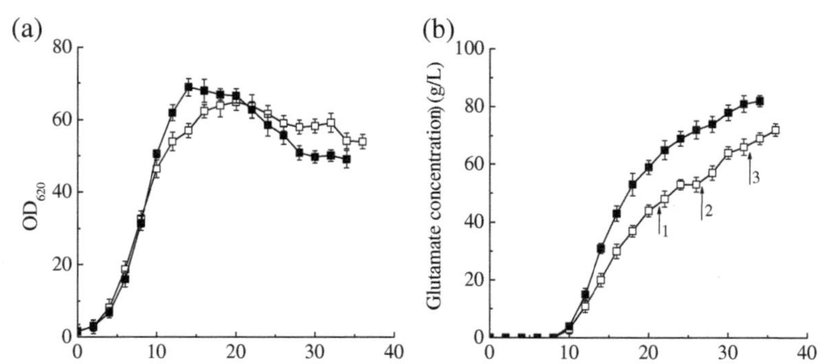

Figure 3 Changing patterns of cell and glutamate concentrations under the condition of supplementing sorbitol when glutamate accumulation ceased. Black square, normal batch; white square, abnormal batch with addition of sorbitol; arrow 1, glucose fed; arrow 2 and arrow 3, adding 12 and 2 g/L sorbitol, respectively. **a**: Time courses of cell concentration (OD_{620}) in different operation conditions; **b**: Time courses of glutamate concentration in different operation conditions.

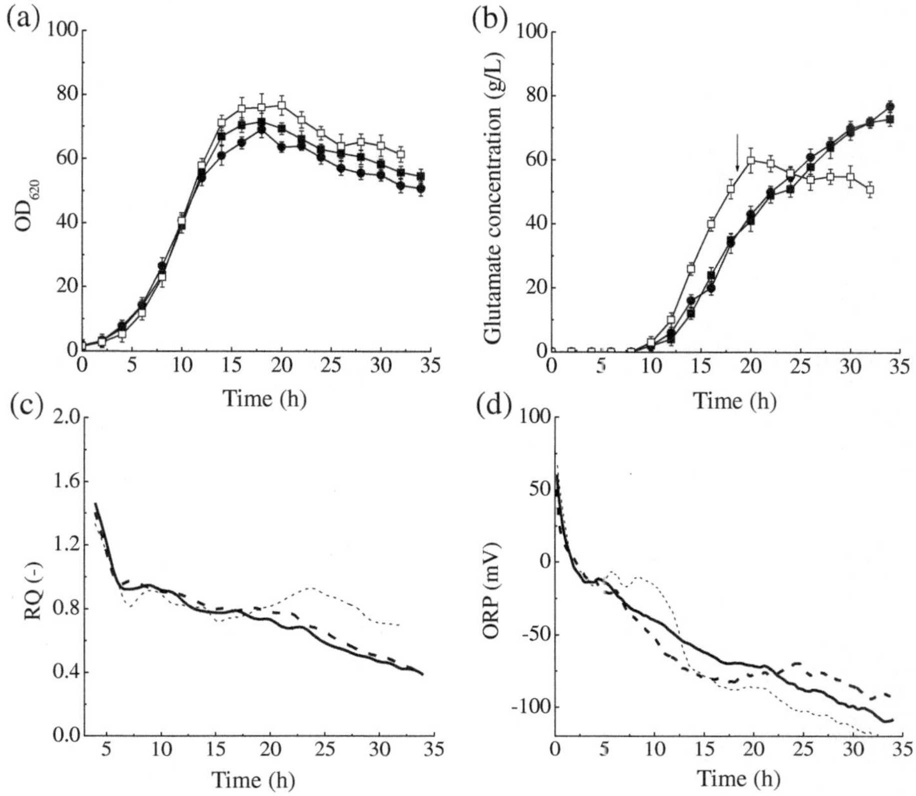

Figure 4 Comparison of fermentation parameters with/without sorbitol. White square and dotted line, abnormal batch without sorbitol; black square and dashed line, sorbitol co-fed with glucose; black circle and solid line, 15 g/L sorbitol added in the initial medium; arrow, glucose fed. **a**: Time courses of cell concentration (OD_{620}) in different operation conditions; **b**: Time courses of glutamate concentration in different operation conditions; **c**: Time courses of RQ in different operation conditions; **d**: Time courses of ORP in different operation conditions.

normal levels at 34 h (72 and 76 g/L, respectively) when co-feeding glycerol with glucose or adding glycerol in the initial medium. The improvements in glutamate production were also closely associated with higher r_{NAD^+}/r_{NADH} ratio (more than 0.8) as shown in Table 1, which was also reflected by higher ORP values (Figure 5) and lower RQ (0.6 to 0.7) in the late production phase.

From the results above, it could be concluded that abnormal glutamate fermentations could be restored to normal by supplementing the media with sorbitol or glycerol, especially when sorbitol or glycerol was added in the initial medium. It has been reported that sorbitol and glycerol could be assimilated by yeast and *Escherichia coli* to increase the targeted product yield by effectively providing the required energy [22,23]. Furthermore, both sorbitol and glycerol were used as effective protective agents of cell viability and enzymes due to their hygroscopicity, freezing tolerance, and oxidation resistance. The major role that sorbitol and glycerol played in the restoration of abnormal glutamate fermentation was analyzed subsequently.

Investigation of the role of sorbitol and glycerol during glutamate fermentation

The concentrations of sorbitol and glycerol were assayed, and results are shown in Figure 6. The results indicated that sorbitol and glycerol were hardly utilized by *C. glutamicum* when co-feeding them with glucose or adding them in the initial medium. When sorbitol or glycerol was co-fed with glucose, sorbitol and glycerol did not reduce after feeding; instead, they were gradually accumulated in the broth. While only a small portion of sorbitol or glycerol could be consumed when they were added in the initial medium. In summary, sorbitol and glycerol were not assimilated by *C. glutamicum* but functioned as protectants to improve the tolerance of the strain in response to the disturbance of the living environment.

The glutamate biosynthesis pathway is composed of many enzymatic reactions in which citrate synthase (CS), PDH, and ICDH are the key enzymes directing carbon flux towards TCA cycle (Figure 1a). Their activities are repressed by excessive NADH. Lower r_{NAD^+}/r_{NADH} ratio indicates that NADH was more than NAD^+ in the

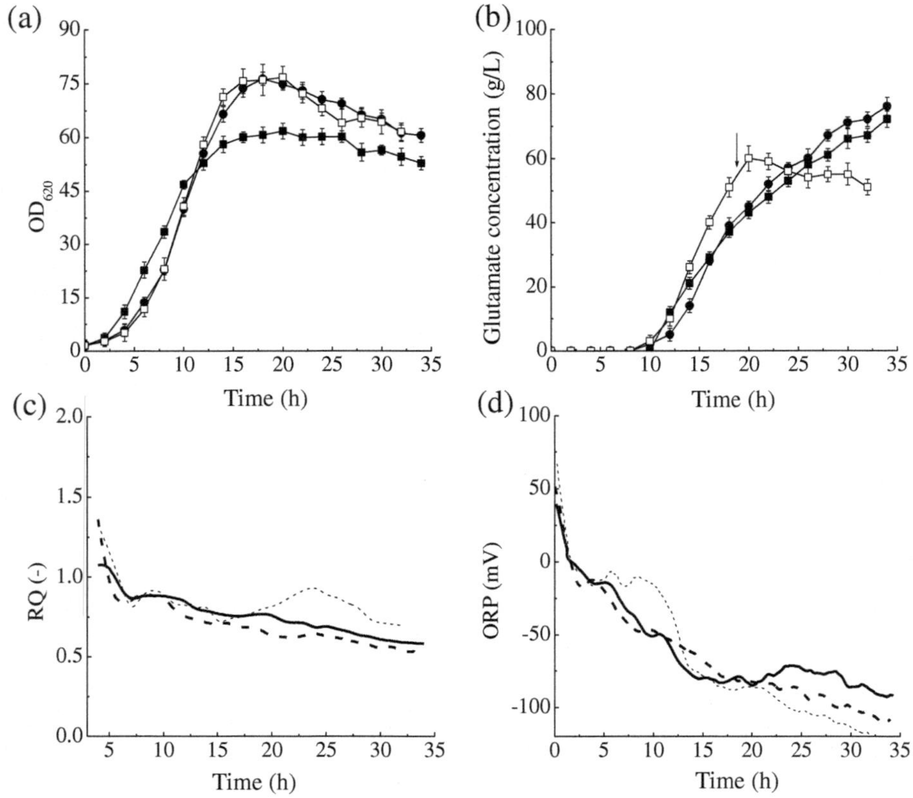

Figure 5 Comparison of fermentation parameters with/without glycerol. White square and dotted line, abnormal batch without glycerol; black square and dashed line, glycerol co-fed with glucose; black square and solid line, 10 g/L glycerol added in the initial medium; arrow, glucose fed. **a**: Time courses of cell concentration (OD_{620}) in different operation conditions; **b**: Time courses of glutamate concentration in different operation conditions; **c**: Time courses of RQ in different operation conditions; **d**: Time courses of ORP in different operation conditions.

cytoplasm. PDH and ICDH activities could be limited by the insufficiency of NAD^+. CcO is the key enzyme catalyzing the transformation of proton (H^+) from NADH to O_2 through the respiratory chain [24]. During this period, NADH was consumed and NAD^+ was regenerated. The activity of CS was difficult to measure, so the activities of PDH, ICDH, and CcO before and after glucose addition (also mixed substrates) under different operation conditions were analyzed. The relative activity of each enzyme before feeding glucose (also mixed substrates, 18 h) was defined as 1. The REA after feeding (26 h) are shown in Table 2. In the abnormal fermentation without sorbitol or

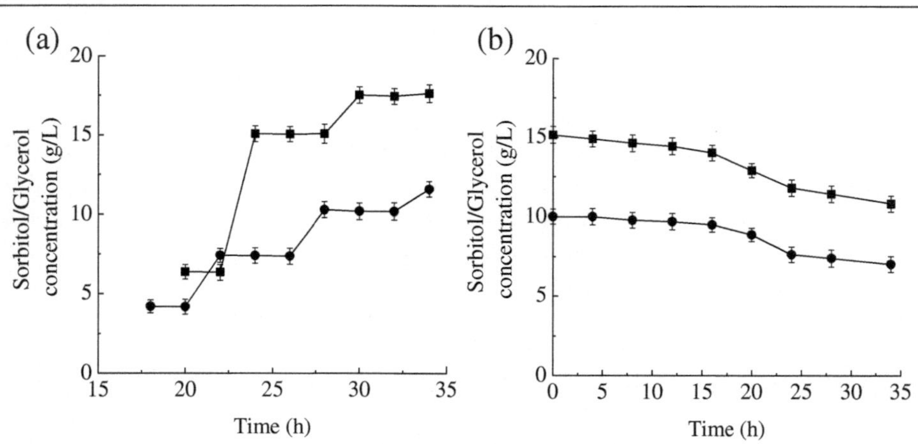

Figure 6 Time courses of sorbitol and glycerol concentrations under different supplementing conditions. (a) Sorbitol or glycerol co-fed with glucose; **(b)** sorbitol or glycerol added in the initial medium. Black square, sorbitol; black circle, glycerol.

Table 2 Relative enzymatic activities of PDH, ICDH, and CcO after substrate(s) feeding (26 h) in different batches

Key enzyme	Fermentation batches					
	Batch #1	Batch #2	Batch #3	Batch #4	Batch #5	Batch #6
PDH	0.84 ± 0.057	0.58 ± 0.043	0.72 ± 0.052	0.72 ± 0.036	0.76 ± 0.078	0.72 ± 0.061
ICDH	1.18 ± 0.051	0.72 ± 0.039	0.97 ± 0.061	0.92 ± 0.032	1.24 ± 0.076	1.27 ± 0.092
CcO	0.50 ± 0.063	0.00	1.50 ± 0.089	0.79 ± 0.043	0.75 ± 0.057	0.25 ± 0.04

Batch #1, normal batch (control); *batch #2*, abnormal batch without sorbitol or glycerol; *batch #3*, sorbitol co-fed with glucose; *batch #4*, 15 g/L sorbitol added in the initial medium; *batch #5*, glycerol co-fed with glucose; *batch #6*, 10 g/L glycerol added in the initial medium.

glycerol, the activities of PDH and ICDH decreased by 42 and 28%, respectively, and CcO was inactive after glucose feeding. In the presences of sorbitol or glycerol, the inactivation of these enzymes was relieved. The activity of PDH decreased by 20 to 30%, and ICDH activity was not significantly affected. The activity of CcO was almost maintained at a higher level. It was concluded that the repression of the key enzymes directing glucose into glutamate synthesis was relieved by addition of sorbitol or glycerol. Accompanied by the recovery of NAD^+ regeneration, the metabolic flux was shifted into the normal pathway of glutamate synthesis. Consequently, glutamate accumulated continuously after glucose feeding and a failed fermentation could be avoided.

Sorbitol and glycerol served as shield materials during glutamate fermentation. The activity of key enzymes could be properly maintained when supplementing sorbitol or glycerol. The r_{NAD^+}/r_{NADH} ratio was increased, and ORP was maintained around the normal range. The stability of glutamate fermentation was improved efficiently by adding sorbitol or glycerol, and the improvement was more obvious when sorbitol/glycerol was added in the initial medium.

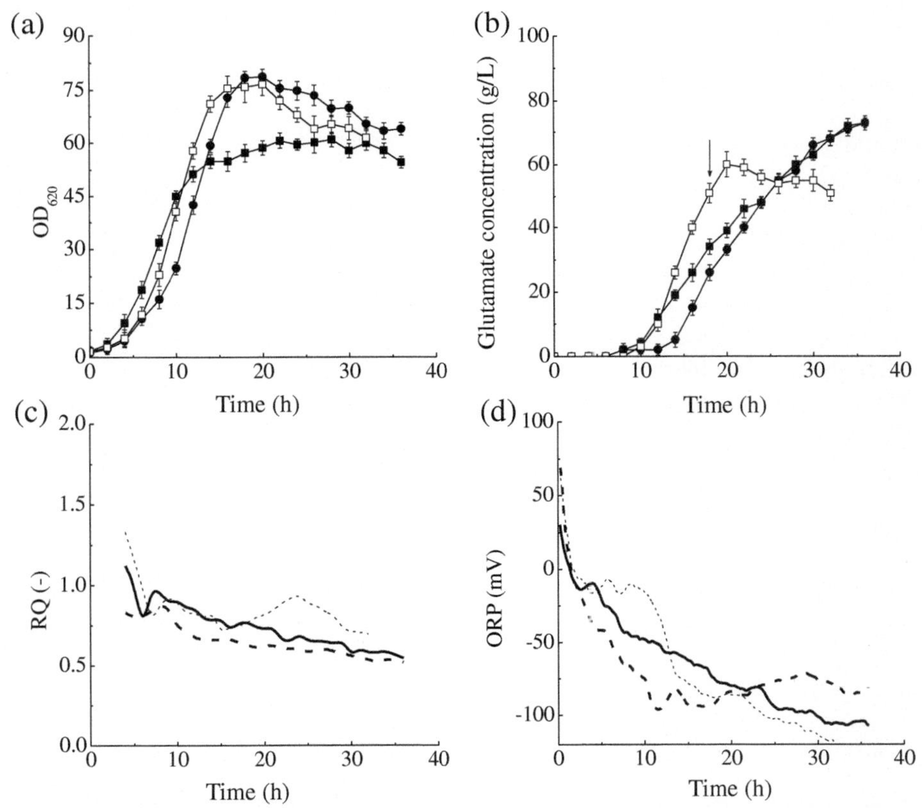

Figure 7 Glutamate fermentation performance when less amount of sorbitol or glycerol was added in the initial medium. White square and dotted line, abnormal batch without sorbitol or glycerol; black square and dashed line, 5 g/L sorbitol added in the initial medium; black circle and solid line, 4 g/L glycerol added in the initial medium; arrow, glucose fed. **a**: Time courses of cell concentration (OD_{620}) in different operation conditions; **b**: Time courses of glutamate concentration in different operation conditions; **c**: Time courses of RQ in different operation conditions; **d**: Time courses of ORP in different operation conditions.

Feasibility analysis in industry

A likely failed fermentation could be restored to normal when co-feeding sorbitol or glycerol with glucose or adding them in the initial medium. However, glutamate is a low value-added product, and the supplementing amount of sorbitol or glycerol should be minimized to save the raw-material cost in industry. Therefore, the cell growth and glutamate production were analyzed with less sorbitol or glycerol (4 to 5 g/L) addition in the initial medium, as shown in Figure 7. In these cases, cell growth was not affected and glutamate synthesis did not stop after feeding glucose and glutamate concentration ended at 74 g/L at 36 h. It was verified again that sorbitol and glycerol functioned mainly as protectants, and their protective effect strengthened with increasing concentration of shield materials [8].

Conclusions

The fermentation features of *C. glutamicum* changed during preservation process and glutamate accumulation stopped after glucose feeding, leading to an abnormal fermentation. This abnormal fermentation performance could be restored to normal by co-feeding sorbitol or glycerol with glucose or adding them in the initial medium. Restoration was more effective when sorbitol or glycerol was added in the initial medium. Glutamate fermentation stability was also improved efficiently. In these cases, sorbitol and glycerol were used as protective agents. When sorbitol or glycerol was added, the adaptive capability of cells to environmental change was promoted and the activities of PDH/ICDH/CcO could be maintained. The usage efficiency of NADH was improved, and r_{NAD^+}/r_{NADH} ratio increased to normal level which was reflected by higher ORP value. These results provided theoretical basis and feasibility for stabilizing glutamate fermentation in its industrial production.

Appendix

Simplified metabolic reactions in glutamate fermentation by *C. glutamicum*:

r_1: $GLC + NAD^+ \rightarrow 2PYR + 2NADH$

r_2: $PYR + NAD^+ \rightarrow Ac\text{-}CoA + NADH + CO_2$

r_3: $PYR + CO_2 + ATP \rightarrow OAA + ADP$

r_4: $OAA + Ac\text{-}CoA \rightarrow ICIT$

r_5: $ICIT + NAD^+ \rightarrow \alpha\text{-}KG + NADH + CO_2$

r_6: $\alpha\text{-}KG + NAD^+ \rightarrow SUC + CO_2 + NADH$

r_7: $SUC + NAD^+ \rightarrow OAA + NADH$

r_8: $\alpha\text{-}KG + NH_4^+ + NADH \rightarrow GLU + NAD^+$

r_9: $NADH + 0.5O_2 + Pi \rightarrow NAD^+ + ATP$

Competing interests

The authors declare that they have no competing interests.

Authors' contributions

YC carried out the experiments, performed the statistical analysis, and drafted the manuscript. Z-NH was involved in performing the experiments. ME helped carry out the experiments and revised the manuscript. Z-PS conceived the idea, participated in its design and coordination, and helped in drafting of the manuscript. All authors read and approved the final manuscript.

Acknowledgements

The authors thank the financial sponsors from the National High-Tech Program (#2006AA020301) and Major State Basic Research Development Program (#2007CB714303) of China.

Author details

[1]National University of Singapore (Suzhou) Research Institute, 377 Linquan Street, Suzhou, Jiangsu Province, China. [2]School of Biotechnology, Jiangnan University, 1800 Lihu Road, Wuxi, Jiangsu Province, China. [3]Department of Food Processing Technology, Harare Institute of Technology, 1505 Ganges Road, P.O. Box BE 277, Belvedere, Harare, Zimbabwe.

References

1. Sano C (2009) History of glutamate production. Am J Clin Nutr 90:728S–732S
2. Ying WH (2006) NAD(+) and NADH in cellular functions and cell death. Front Biosci 11:3129–3148
3. de Graef MR, Alexeeva S, Snoep JL, Teixeira de Mattos MJ (1999) The steady-state internal redox state (NADH/NAD) reflects the external redox state and is correlated with catabolic adaptation in *Escherichia coli*. J Bacteriol 181:2351–2357
4. Berrios-Rivera SJ, Bennett GN, San KY (2002) The effect of increasing NADH availability on the redistribution of metabolic fluxes in *Escherichia coli* chemostat cultures. Metab Eng 4:230–237
5. Kastner JR, Eiteman MA, Lee SA (2003) Effect of redox potential on stationary-phase xylitol fermentations using *Candida tropicalis*. Appl Microbiol Biot 63:96–100
6. Li J, Jiang M, Chen KQ, Ye Q, Shang LA, Wei P, Ying HJ, Chang HN (2010) Effect of redox potential regulation on succinic acid production by *Actinobacillus succinogenes*. Bioproc Biosyst Eng 33:911–920
7. Radjai MK, Hatch RT, Cadman TW (1984) Optimization of amino acid production by automatic self tuning digital control of redox potential. Biotechnol Bioeng Symp 14:657–679
8. Bhatnagar A, Srivastava SK (1992) Aldose reductase: congenial and injurious profiles of an enigmatic enzyme. Biochem Med Metab Biol 48:91–121
9. Azizi A, Ranjbar B, Khajeh K, Ghodselahi T, Hoornam S, Mobasheri H, Ganjalikhany MR (2011) Effects of trehalose and sorbitol on the activity and structure of *Pseudomonas cepacia* lipase: spectroscopic insight. Int J Biol Macromol 49:652–656
10. Wang Z, Wang Y, Zhang D, Li J, Hua Z, Du G, Chen J (2010) Enhancement of cell viability and alkaline polygalacturonate lyase production by sorbitol co-feeding with methanol in *Pichia pastoris* fermentation. Bioresource Technol 101:1318–1323
11. Liu Y, Zhang YG, Zhang RB, Zhang F, Zhu J (2011) Glycerol/glucose co-fermentation: one more proficient process to produce propionic pcid by *Propionibacterium acidipropionici*. Curr Microbiol 62:152–158
12. John GSM, Gayathiri M, Rose C, Mandal AB (2012) Osmotic shock augments ethanol stress in *Saccharomyces cerevisiae* MTCC 2918. Curr Microbiol 64:100–105
13. Ramon R, Ferrer P, Valero F (2007) Sorbitol co-feeding reduces metabolic burden caused by the overexpression of a rhizopus oryzae lipase in *Pichia pastoris*. J Biotechnol 130:39–46
14. Arruda PV, Felipe MGA (2009) Role of glycerol addition on xylose-to-xylitol bioconversion by *Candida guilliermondii*. Curr Microbiol 58:274–278
15. Popova O, Ismailov S, Popova T, Dietz KJ, Golldack D (2002) Salt-induced expression of NADP-dependent isocitrate dehydrogenase and ferredoxin-dependent glutamate synthase in *Mesembryanthemum crystallinum*. Planta 215:906–913
16. Hasegawa T, Hashimoto KI, Kawasaki H, Nakamatsu T (2008) Changes in enzyme activities at the pyruvate node in glutamate-overproducing *Corynebacterium glutamicum*. J Biosci Bioeng 105:12–19

17. Gourdon P, Lindley ND (1999) Metabolic analysis of glutamate production by *Corynebacterium glutamicum*. Metab Eng 1:224–231
18. Skjerdal OT, Sletta H, Flenstad SG, Josefsen KD, Levine DW, Ellingsen TE (1996) Changes in intracellular composition in response to hyperosmotic stress of NaCl, sucrose or glutamic acid in Brevibacterium lactofermentum and *Corynebacterium glutamicum*. Appl Microbiol Biotechnol 44:635–642
19. Park SM, Sinskey AJ, Stephanopoulos G (1997) Metabolic and physiological studies of *Corynebacterium glutamicum* mutants. Biotechnol Bioeng 55:864–879
20. Xiao J, Shi ZP, Gao P, Feng HJ, Duan ZY, Mao ZG (2006) On-line optimization of glutamate production based on balanced metabolic control by RQ. Bioproc Biosyst Eng 29:109–117
21. Savinell JM, Palsson BO (1992) Network analysis of intermediary metabolism using linear optimization. I. Development of mathematical formalism. J Theor Biol 154:421–454
22. Lin H, Bennett GN, San KY (2005) Effect of carbon sources differing in oxidation state and transport route on succinate production in metabolically engineered *Escherichia coli*. J Ind Microbiol Biotechnol 32:87–93
23. Murarka A, Dharmadi Y, Yazdani SS, Gonzalez R (2007) Fermentative utilization of glycerol by *Escherichia coli* and its implications for the production of fuels and chemicals. Appl Environ Microb 74:1124–1135
24. Farver O, Grell E, Ludwig B, Michel H, Pecht I (2006) Rates and equilibrium of CuA to heme a electron transfer in paracoccus denitrificans cytochrome c oxidase. Biophys J 90:2131–2137

Assembly of cellulases with synthetic protein scaffolds *in vitro*

Ting Yu, Xin Gao, Yuhong Ren* and Dongzhi Wei

Abstract

Background: Enzymatic cascades in metabolic pathways are spatially organized in such a way as to facilitate the flow of substrates. The construction of artificial cellulase complexes that mimic natural multienzyme assemblies can potentially enhance the capacity for cellulose hydrolysis. In this study, an artificial cellulase complex was constructed by tethering three cellulases to a synthetic protein scaffold.

Results: Three pairs of interacting proteins were selected and characterized. The artificial protein scaffolds were constructed by fusing three interacting proteins. Cellulases were tethered to these synthetic scaffolds in different orders. The optimal assembly resulted in a 1.5-fold higher hydrolysis of cellulose than that achieved by unassembled cellulases.

Conclusions: A novel artificial protein scaffold was constructed and used to assemble three cellulases. The resultant increase in enzymatic activity suggests that this can be used as a strategy for enhancing the biocatalytic capacity of enzyme cascades.

Keywords: Multienzyme; Assembly; Scaffold; Cellulose

Background

Multienzyme pathways in living systems comprise cascades in which enzymes are tethered together into assemblies that facilitate substrate flow between components, limit the diffusion of intermediate metabolic products, and increase the yield from sequential reactions [1,2]. There have been various attempts to produce multienzyme assemblies *in vitro*, including by gene fusion, protein or DNA scaffold construction, and chemical modification [3]. Although the simplest way is by enzyme fusion, this often results in loss of enzymatic activity or formation of inclusion bodies [4], while chemical modification can also impair enzymatic activity [5]. Moreover, the high cost of generating DNA scaffolds makes this approach infeasible on a large scale [6,7].

Protein scaffolds are an attractive strategy for bringing together enzymes. A 77-fold enhancement in product concentration was observed in an assembly of three mevalonate biosynthetic enzymes with a protein scaffold composed of metazoan signaling proteins [8]. In another study, a self-assembled enzyme complex using cellulosome achieved a 13.4-fold increase in reaction rate [9], while a proliferating cell nuclear antigen-based assembly of P450 with ferredoxin and ferredoxin reductase showed high catalytic activity [10]. The protein scaffolds used in these studies were limited to metazoan signaling proteins and cellulosome components [11-14], which may not be amenable to all types of enzyme cascades. As such, there is an ongoing need for novel and different types of protein scaffolds.

Cellulose is the most abundant renewable resource on Earth and plays a significant role in biofuel production [15]. Cellulose is broken down into oligosaccharides by endoglucanase (EG) and exoglucanase (CBH) before β-glucosidase (BGL) hydrolyzes cellobiose into glucose [16]. EG, CBH, and BGL are enzymes in the cellulose degradation pathway that act synergistically; coexpressing these enzymes improves the efficiency of cellulose degradation [17]. It was hypothesized that a highly ordered assembly containing the three cellulases would enhance their activities and thereby increase cellulose hydrolysis. Therefore, CelccA (EG), CelccE (CBH), and Cel2454 (BGL) were selected as a model system for protein scaffold-mediated assembly strategy.

In this study, considering the correct expression and assembly of proteins, three pairs of interacting proteins

* Correspondence: yhren@ecust.edu.cn
State Key Laboratory of Bioreactor Engineering, East China University of Science and Technology, 130 Meilong Rd, Shanghai 200237, People's Republic of China

(IPA/IPa, IPB/IPb, and IPC/IPc) from *E. coli* K12 [18] were selected to construct a protein scaffold for three cellulases - namely, CelccA (EG), CelccE (CBH), and Cel2454 (BGL) (Figure 1). The enzymes were arranged in different orders and the effect on activity was investigated. The multienzyme complex in best order was characterized in detail and compared with the unassembled components.

Methods

Strains and medium

Escherichia coli DH5α was used for cloning, and *E. coli* BL21 (DE3) was used for protein expression. Cells were cultured in Luria-Bertani (LB) medium (10.0 g/l tryptone, 5.0 g/l yeast extract, 10.0 g/l NaCl) supplemented with either 100 mg/l ampicillin or 50 mg/l kanamycin.

Plasmid construction

PCR primer sequences are listed in Table 1. All plasmid sequences were validated by sequencing. To generate the pET21a-IPA plasmid, the *IPA* gene (GenBank accession no. EU899769.1) was amplified from *E. coli* K12 genomic DNA using the IPA-F/IPA-R primer pair. The PCR product was digested with *Nde*I/*Xho*I and ligated into pET21a digested with the same restriction enzymes. Similarly, plasmids pET21a-IPa, pET21a-IPB, pET21a-IPb, pET21a-IPC, and pET21a-IPc were constructed by amplifying the DNA sequences of *IPa*, *IPB*, *IPb*, *IPC*, and *IPc* (GenBank accession nos. EU895384.1, EU8971IPC.1, AB9163IPA.1, EU893769.1, and EU891369.1, respectively) using the primer pairs IPa-F/IPa-R, IPB-F/IPB-R, IPb-F/IPb-R, IPC-F/IPC-R, and IPc-F/IPc-R, respectively. Plasmids pET28a-celCCA, pET28a-celCCE, and pET28a-cel2454 were constructed by amplifying the DNA sequences of celCCA, celCCE, and cel2454, respectively (GenBank accession nos. M93096.1, M87018.2, and ACL76783.1, respectively), from the genomic DNA of *Clostridium cellulolyticum* DSM 5812 using the primer pairs A-F/A-R, E-F/E-R, and 4-F/4-R, respectively [19,20].

Plasmid pET28a-IPaA had an expression cassette containing IPa and celCCA at the N and C termini, respectively. A DNA fragment encoding IPa was amplified with the primer pair IPaA-F/IPaA-R and cloned into pET28a-celCCA by Seamless Cloning [21]. Plasmid pET28a-AIPa containing celCCA and IPa at the N and C termini, respectively, was generated with the primer pair AIPa-F/AIPa-R. Plasmid pET28a-IPbE containing IPb and celCCE at the N and C termini, respectively, was generated with primer pair IPbE-F/IPbE-R. Plasmid pET28a-EIPb containing celCCE and IPb at the N and C termini, respectively, was constructed with primer pair EIPb-F/EIPb-R. Plasmid pET28a-IPc4 containing IPc and cel2454 at the N and C termini, respectively, was generated with primer pair IPc4-F/IPc4-R. Plasmid pET28a-4IPc containing cel2454 and IPc at the N and C termini, respectively, was constructed with primer pair 4IPc-F/4IPc-R.

Plasmid pET21a-ScafBAC had an expression cassette containing IPA flanked by IPB and IPC at the N and C termini, respectively. The DNA fragments encoding IPB, IPA, and IPC were amplified with the primer pairs ScafBAC-F1/IPB-R, ScafBAC-F2/IPA-R, and ScafBAC-F3/ScafBAC-R, respectively. The three fragments were cloned into pET21a by Seamless Cloning. Plasmid pET21a-ScafABC was constructed using primer pairs ScafABC-F1/IPA-R, ScafABC-F2/IPB-R, and ScafABC-F3/ScafABC-R, and plasmid pET21a-ScafBCA was generated using primer pairs ScafBAC-F1/IPB-R, ScafABC-F3/IPC-R, and ScafBCA-F/ScafBCA-R.

Protein expression and purification

Recombinant proteins were precultured overnight at 37°C in LB medium supplemented with appropriate antibiotics. The cultures were inoculated in fresh LB medium containing antibiotics and incubated at 37°C until the optical density at 600 nm reached 0.8. The cultures were then cooled to 18°C, and isopropyl-β-D-thiogalactopyranoside was added to a final concentration of 0.1 mM. After 20 h, cells were harvested by centrifugation for 10 min at 8,000 rpm and 4°C, resuspended in 20 mM phosphate-buffered saline (PBS; pH 7.0), and disrupted by sonication on ice. Cellular debris was removed by centrifugation for 40 min at 11,000 rpm. Proteins were purified using a HisTrapFF column (GE Healthcare, Waukesha, WI, USA), and protein concentration was determined by the Bradford method.

Biolayer interferometry

Binding affinities between proteins were measured by biolayer interferometry (Octet QKe; Fortebio, Menlo Park, CA, USA), which detects changes in mass (protein density)

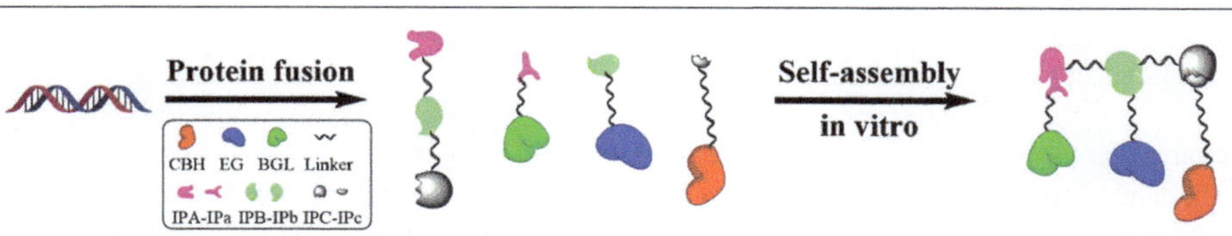

Figure 1 Schematic representation for assembly of cellulases on a protein scaffold. Three cellulases (CBH, EG, BGL) and three pairs of interaction proteins (IPA-IPa, IPB-IPb, IPC-IPc) were used for the assembly.

Table 1 Primers used in this work

Name	Sequence
IPA-F	GGAATTCCATATGATGCGAATTCTGGCTATC
IPA-R	CCGCTCGAGTTCTTTGCCCGGAAGTTTC
IPa-F	GGAATTCCATATGATGCGTGTACTGGGTATTG
IPa-R	CCGCTCGAGCGCAGCCGGTAACTCCGCC
IPB-F	GGAATTCCATATGATGAACGATGAAATGAAAG
IPB-R	CCGCTCGAGGGCTTTATTCTGACGCCAC
IPb-F	GGAATTCCATATGATGGCAAGATATTTGGG
IPb-R	CCGCTCGAGCTTGGAGTAAAGCTCGAC
IPC-F	GGAATTCCATATGATGACTGACAAAATGC
IPC-R	CCGCTCGAGCGCTTCAATGGCAGCAC
IPc-F	GGAATTCCATATGATGGCAACACGAATTG
IPc-R	CCGCTCGAGTGGAATCAGCAGGCTGG
A-F	GGAATTCCATATGTATGATGCTTCACTTATTCCG
A-R	CCGCTCGAGGTTGCTTGGAAGCTTAC
E-F	GTGCCGCGCGGCAGCCATATGATAGGACAAGCATTTGC
E-R	GTGGTGGTGGTGGTGCTCGAGAGTTGGAGGAGTCACTGACCC
4-F	GGAATTCCATATGCAATACGATCAGATAGATAAA
4-R	CCGCTCGAGCAGAGCAAGAGCTATAG
IPaA-F	GTGCCGCGCGGCAGCATGCGTGTACTGGGTATTGAAACTTC
IPaA-R	AAGTGAAGCATCATACGATCCGCCACCGCCAGAGCCACCTCCGCCCGCAGCCGGTAACTCCG
AIPa-F	AAGCTTCCAAGCAACGGCGGAGGTGGCTCTGGCGGTGGCGGATCGCGTGTACTGGGTATTG
AIPa-R	GTGGTGGTGGTGGTGCGCAGCCGGTAACTCCGCCAGCGGC
IPbE-F	GTGCCGCGCGGCAGCATGGCAAGATATTTGGGTCCTAAGC
IPbE-R	AAATGCTTGTCCTATCGATCCGCCACCGCCAGAGCCACCTCCGCCCTTGGAGTAAAGCTCG
EIPb-F	GTGACTCCTCCAACTGGCGGAGGTGGCTCTGGCGGTGGCGGCAAGATATTTGGGTCCTAAGC
EIPb-R	GTGGTGGTGGTGGTGCTTGGAGTAAAGCTCGACGATCAGGTG
IPc4-F	GTGCCGCGCGGCAGCATGACTGACAAAATGCAAAGTTTAG
IPc4-R	TATCTGATCGTATTGCGATCCGCCACCGCCAGAGCCACCTCCGCCCGCTTCAATGGCAGCAC
4IPc-F	ATAGCTCTTGCTCTGGGCGGAGGTGGCTCTGGCGGTGGCGGATCGACTGACAAAATGCAAAG
4IPc-R	GTGGTGGTGGTGGTGCGCTTCAATGGCAGCACGCAATTTTTTC
ScafBAC-F1	TAAGAAGGAGATATAAACGATGAAATGAAAGGTAAAAGCGGC
ScafBAC-F2	CGTCAGAATAAAGCCGGCGGAGGTGGCTCTGGCGGTGGCGGATCGCGAATTCTGGCTATCG
ScafBAC-F3	CTTCCGGGCAAAGAAGGCGGAGGTGGCTCTGGCGGTGGCGGATCGACTGACAAAATGCAAAGTT
ScafBAC-R	GTGGTGGTGGTGGTGCTCGAGCGCTTCAATGGCAGCACGC
ScafABC-F1	TAAGAAGGAGATATACGAATTCTGGCTATCGATACCGCGACAGAGGCCTGCTC
ScafABC-F2	CTTCCGGGCAAAGAAGGCGGAGGTGGCTCTGGCGGTGGCGGATCGAACGATGAAATGAAAGG
ScafABC-F3	CGTCAGAATAAAGCCGGCGGAGGTGGCTCTGGCGGTGGCGGATCGACTGACAAAATGCAAAG
ScafBCA-F	GCTGCCATTGAAGCGGGCGGAGGTGGCTCTGGCGGTGGCGGATCGCGAATTCTGGCTATCG
ScafBCA-R	GTGGTGGTGGTGGTGTTCTTTGCCCGGAAGTTTCTTCCATGC

on a biosensor; changes in the reflected interference wave pattern between the sample and an internal reference layer result in a phase shift that can be followed in real time in both kinetic and quantitative modes [22]. All experiments were performed in kinetic buffer (20 mM PBS, pH 7.0; 1 mg/ml bovine serum albumin (BSA), and 0.02% Tween 20). One of the proteins (1 μM) was biotinylated by incubating with 2 μl biotinyl N-hydroxysuccinimide ester for 1 h at room temperature, with excess biotin removed using a desalting column. The biotinylated protein was loaded onto the streptavidin biosensor by incubating for 240 s. The immobilized protein was equilibrated with kinetic buffer for 180 s, and the corresponding protein (1 μM) was associated to the biotinylated protein by incubating for 800 s. Dissociation was measured for 800 s in kinetic buffer. For each assay, a control experiment was carried out using BSA. Binding affinity was independent of which protein was loaded onto the streptavidin biosensor [23-26].

Enzyme assays

The BGL activity was measured by incubating 135 μl of 2.5 mM p-nitrophenyl β-D-glucopyranoside solution in 20 mM sodium phosphate buffer (pH 7.0) with 7.5 μg of pure enzyme solution at 37°C for 30 min. The reaction was terminated by adding 70 μl of 0.4 M Na_2CO_3 and the absorbance at 420 nm was measured. One unit of enzyme was defined as the activity producing 1 μmol of p-nitrophenol per min under the assay conditions. EG/CBH activity was measured by incubating 90 μl of 1.5% (wt/vol) carboxymethyl cellulose (CMC) in 20 mM sodium phosphate buffer (pH 7.0) with 5 μg of pure enzyme solution at 37°C for 30 min. A 100-μl volume of sample was mixed with 150 μl of 3,5-dinitrosalicylic acid reagent, and after boiling for 10 min, the absorbance at 540 nm was measured. One unit of enzymatic activity was defined as the amount of enzyme required to produce a 1 μmol reduction sugar per min under the assay conditions.

The activity of free or assembled enzymes was assayed in the presence of 0.75% (wt/vol) CMC at 37°C in 20 mM sodium phosphate buffer (pH 7.0). The reduction sugars were measured as described above. Glucose concentration was determined using an SBA biosensor analyzer (Biology Institute of Shandong Academy of Sciences, Jinan, China).

Results and discussion

Selection and characterization of interacting proteins

To synthesize the protein scaffold, three pairs of interacting proteins from *E. coli* K12 were expressed in *E. coli* BL21 (DE3). The six proteins were purified by Ni-chelating affinity chromatography (Figure 2A), and protein-protein affinity was measured by biolayer interferometry [23-26], in which the obtained values reflect the amount of protein bound to the biosensor. The response values (Figure 2B,C, D) confirmed their binding, and the K_D calculated by Fortebio Data Analysis software version 7.0 for IPA-IPa (7.36×10^{-8} M), IPB-IPb (1.39×10^{-12} M), and IPC-IPc (5.27×10^{-8} M) demonstrated their high affinity, suggesting that they are suitable for constructing protein scaffolds.

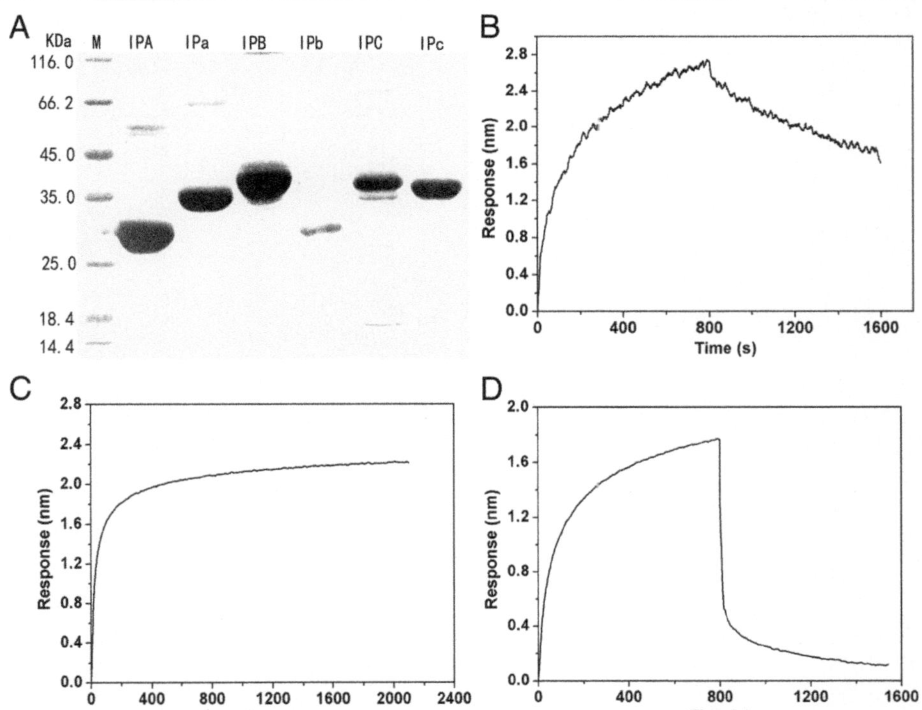

Figure 2 Purification of interaction proteins and their binding affinity. (A) SDS-PAGE analysis of purified interaction proteins. Biolayer interferometry sensorgrams of **(B)** IPA-IPa, **(C)** IPB-IPb, **(D)** IPC-IPc. IPA, IPB, and IPC were immobilized to the streptavidin-coated biosensors, respectively. Then IPa, IPb, and IPc were bound to the corresponding proteins, respectively. The figure shows the response of association and dissociation. All the response value was normalized by minusing the value of control.

Fusion of cellulases and interacting proteins

The three cellulases were fused with the interacting proteins IPa, IPb, and IPc, which endowed them with an interaction interface. To minimize the effect of fusion on enzymatic activity, six fusion proteins were generated by attaching the interacting proteins to the C or N terminus of cellulases. The commonly used glycine-serine linker $(G_4S)_2$ was used to link two proteins and thereby reduce folding interference [27]. The resultant constructs were expressed in *E. coli* with a 6-histidine tag at the N terminus and purified by Ni-chelating affinity chromatography. IPa could only be attached to the N terminus of celCCA (IPaA) (Figure 3A), given that partial proteolysis occurred at the C terminus [19]. Similarly, the ideal celCCE and IPb fusion was obtained by attaching IPb to the N terminus of celCCE (IPbE). For the cel2454-IPc fusion, the activity of IPc4 was slightly higher than that of 4IPc (Figure 3D), and the former was therefore chosen for the assembly. The three enzyme fusions that were ultimately selected retained almost all of the specific activity of the respective unfused proteins (Figure 3B,C,D), indicating that protein structure was not appreciably compromised by the fusion.

Construction of scaffolds and their effect on assembled enzyme complexes

To determine the appropriate order of cellulases within assemblies, three protein scaffolds were constructed by fusing the three interacting proteins (IPA, IPB, and IPC) in different orders and assembling three tri-enzyme complexes by mixing the scaffolds and cellulases in a 1:1:1:1 stoichiometry for 1 h. The reduction sugars from CMC hydrolysis catalyzed by the three tri-enzyme complexes were assayed. The complex assembled with ScafBAC had the highest sugar reduction (Figure 4), indicating that the synergistic effect was related to the location of the enzyme within the complex. Given that the activity of EG is much higher than that of CBH or BGL, oligosaccharides and cellobioses produced by EG can be immediately catalyzed by CBH and BGL when EG is between CBH and BGL.

Characterization of the tri-enzyme complex assembled by ScafBAC

The fused cellulases and protein scaffold were purified using a Ni-sepharose column, and the correct formation of the protein assembly as well as the affinity of each cellulase fused to the protein scaffold was verified by biolayer interferometry [28] (Figure 5). The response of biolayer interferometry is dependent on the amount of proteins bound to the biosensor which indicates the binding of proteins. Although different binding responses were observed for the three cellulase fusions, they were all attached to the scaffold. Moreover, the binding responses were similar to values obtained for the interacting proteins, indicating that the binding affinity was not affected by the fusion.

The potential for synergistic effects on enzyme cascades is the most interesting property of protein assemblies. The

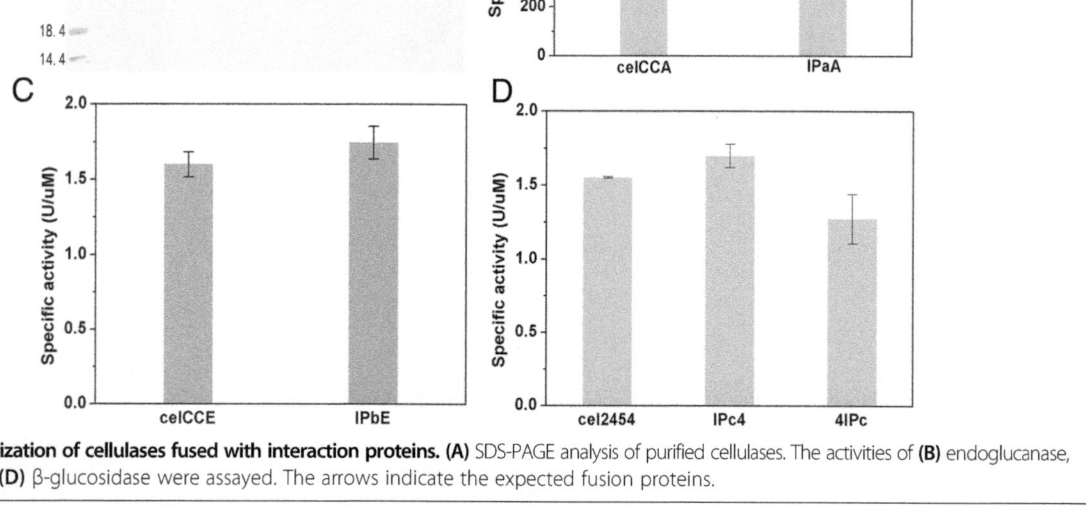

Figure 3 Characterization of cellulases fused with interaction proteins. (A) SDS-PAGE analysis of purified cellulases. The activities of **(B)** endoglucanase, **(C)** exoglucanase, **(D)** β-glucosidase were assayed. The arrows indicate the expected fusion proteins.

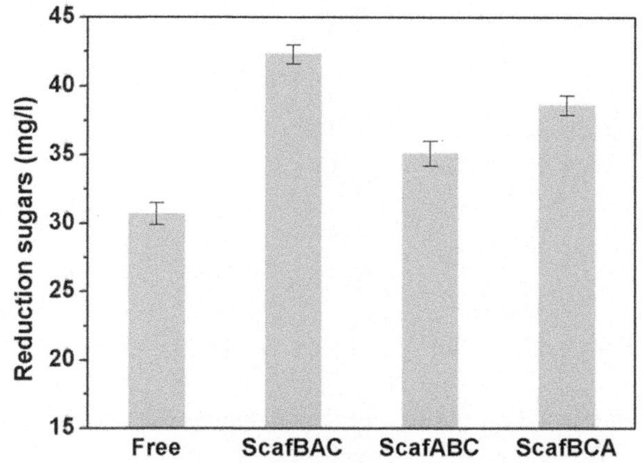

Figure 4 Reducing sugar production by free cellulases or three multienzyme complexes assembling with different scaffolds. 1.5 g/l of CMC was catalyzed by 0.6 μM of enzyme for 30 min. 'Free' means three fusion cellulases without scaffolds.

capacity for CMC hydrolysis was compared between the tri-enzyme complex and free cellulases. Catalytic activity was evaluated by glucose production using a SBA biosensor analyzer. After a 4-h reaction, a 1.5-fold enhancement in activity was observed for the tri-enzyme complex as compared to the corresponding free enzymes (Figure 6), suggesting the channeling of intermediates within the complex; that is, cellobiose generated by EG and CBH was immediately transferred to the adjacent BGL for glucose production, thereby resulting in enhanced cellulose hydrolysis.

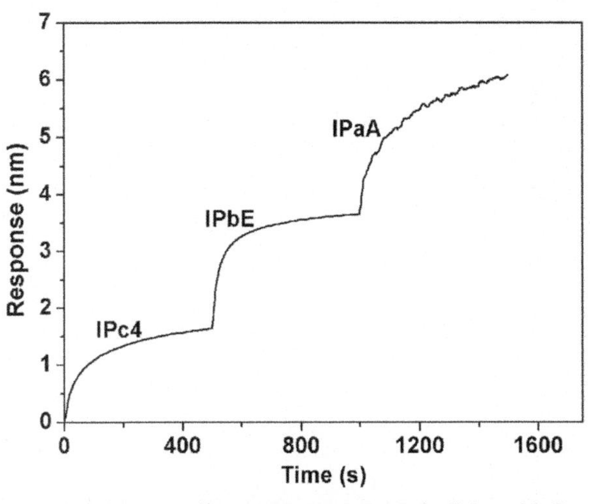

Figure 5 Biolayer interferometry sensorgrams of cellulases binding to biotinylated scaffold on streptavidin-coated biosensors. Biotinylated scaffold was immobilized to the streptavidin-coated biosensors. Then IPc4, IPbE, and IPaA were bound to the scaffold. The figure shows the response of association with IPc4, IPbE, and IPaA. All the response value was normalized by minusing the value of control.

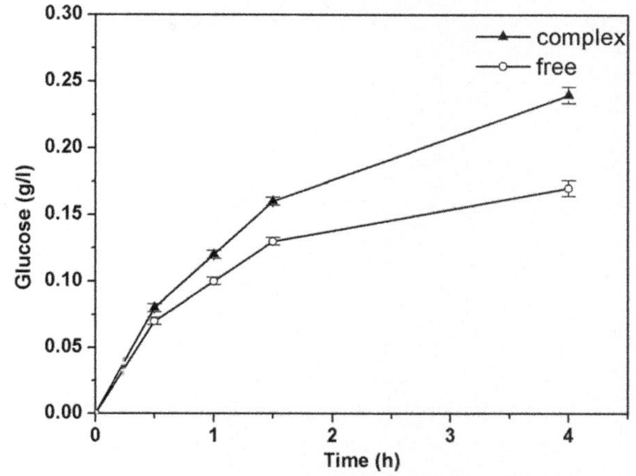

Figure 6 Glucose production of catalyzed by multienzyme complex or by equal amount of free cellulases. 7.5 g/l of CMC was catalyzed by 0.6 μM of enzyme. 'Free' means three fusion cellulases without scaffolds.

Conclusions

An artificial tri-enzyme complex was constructed by assembling three cellulases with a novel protein scaffold composed of interacting proteins. The effect of the order of cellulase within the scaffolds on the catalytic efficiency was determined. Moreover, the complex had higher catalytic activity than the individual components. These results suggest that this novel protein scaffold can serve as a powerful tool for facilitating multienzyme cascades.

Competing interests

The authors declare that they have no competing interests.

Authors' contributions

TY carried out the experiment and drafted the manuscript. XG participated in the design of the study and revised the manuscript. YR and DW conceived of the study, participated in its design and coordination. All authors read and approved the final manuscript.

Acknowledgements

This work supported by National Special Fund for State Key Laboratory of Bioreactor Engineering (2060204), and 'the Fundamental Research Funds for the Central Universities', People's Republic of China.

References

1. Ricca E, Brucher B, Schrittwieser JH (2011) Multi-enzymatic cascade reactions: overview and perspectives. Advan Synthesis Catalysis 353(13):2239–2262
2. Conrado RJ, Varner JD, DeLisa MP (2008) Engineering the spatial organization of metabolic enzymes: mimicking nature's synergy. Curr Opin Biotechnol 19(5):492–499
3. Schoffelen S, van Hest JCM (2012) Multi-enzyme systems: bringing enzymes together in vitro. Soft Matter 8(6):1736–1746
4. Adlakha N, Sawant S, Anil A, Lali A, Yazdani SS (2012) Specific fusion of beta-1,4-endoglucanase and beta-1,4-glucosidase enhances cellulolytic activity and helps in channeling of intermediates. Appl Environ Microbiol 78 (20):7447–7454
5. Schoffelen S, Beekwilder J, Debets MF, Bosch D, van Hest JC (2013) Construction of a multifunctional enzyme complex via the strain-promoted azide-alkyne cycloaddition. Bioconjug Chem 24(6):987–996

6. Fu J, Liu M, Liu Y, Woodbury NW, Yan H (2012) Interenzyme substrate diffusion for an enzyme cascade organized on spatially addressable DNA nanostructures. J Am Chem Soc 134(12):5516–5519
7. Sun Q, Madan B, Tsai SL, DeLisa MP, Chen W (2014) Creation of artificial cellulosomes on DNA scaffolds by zinc finger protein-guided assembly for efficient cellulose hydrolysis. Chem Commun (Camb) 50(12):1423–1425
8. Dueber JE, Wu GC, Malmirchegini GR, Moon TS, Petzold CJ, Ullal AV, Prather KL, Keasling JD (2009) Synthetic protein scaffolds provide modular control over metabolic flux. Nat Biotechnol 27(8):753–759
9. You C, Myung S, Zhang YH (2012) Facilitated substrate channeling in a self-assembled trifunctional enzyme complex. Angew Chem Int Ed Engl 51 (35):8787–8790
10. Hirakawa H, Nagamune T (2010) Molecular assembly of P450 with ferredoxin and ferredoxin reductase by fusion to PCNA. ChemBioChem 11 (11):1517–1520
11. Liu F, Banta S, Chen W (2013) Functional assembly of a multi-enzyme methanol oxidation cascade on a surface-displayed trifunctional scaffold for enhanced NADH production. Chem Commun (Camb) 49(36):3766–3768
12. Tsai SL, DaSilva NA, Chen W (2013) Functional display of complex cellulosomes on the yeast surface via adaptive assembly. ACS Synth Biol 2(1):14–21
13. Borne R, Bayer EA, Pages S, Perret S, Fierobe HP (2013) Unraveling enzyme discrimination during cellulosome assembly independent of cohesin-dockerin affinity. FEBS J 280(22):5764–5779
14. Fan LH, Zhang ZJ, Yu XY, Xue YX, Tan TW (2012) Self-surface assembly of cellulosomes with two miniscaffoldins on Saccharomyces cerevisiae for cellulosic ethanol production. Proc Natl Acad Sci U S A 109(33):13260–13265
15. Geddes CC, Nieves IU, Ingram LO (2011) Advances in ethanol production. Curr Opin Biotechnol 22(3):312–319
16. Rizk M, Antranikian G, Elleuche S (2012) End-to-end gene fusions and their impact on the production of multifunctional biomass degrading enzymes. Biochem Biophys Res Commun 428(1):1–5
17. Liu M, Yu H (2012) Co-production of a whole cellulase system in Escherichia coli. Biochem Eng J 69:204–210
18. Levchenko A, Hu P, Janga SC, Babu M, Díaz-Mejía JJ, Butland G, Yang W, Pogoutse O, Guo X, Phanse S, Wong P, Chandran S, Christopoulos C, Nazarians-Armavil A, Nasseri NK, Musso G, Ali M, Nazemof N, Eroukova V, Golshani A, Paccanaro A, Greenblatt JF, Moreno-Hagelsieb G, Emili A (2009) Global functional atlas of Escherichia coli encompassing previously uncharacterized proteins. PLoS Biol 7(4):e1000096
19. Fierobe HP, Gaudin C, Belaich A, Loutfi M, Faure E, Bagnara C, Baty D, Belaich JP (1991) Characterization of Endoglucanase A from Clostridium cellulolyticum. J Bacteriol 173(24):7956–7962
20. Gaudin C, Belaich A, Champ S, Belaich JP (2000) CelE, a multidomain cellulase from Clostridium cellulolyticum: a key enzyme in the cellulosome. J Bacteriol 182:1910–1915
21. You C, Zhang XZ, Zhang YH (2012) Simple cloning via direct transformation of PCR product (DNA multimer) to Escherichia coli and Bacillus subtilis. Appl Environ Microbiol 78(5):1593–1595
22. Naik S, Kumru OS, Cullom M, Telikepalli SN, Lindboe E, Roop TL, Joshi SB, Amin D, Gao P, Middaugh CR, Volkin DB, Fisher MT (2014) Probing structurally altered and aggregated states of therapeutically relevant proteins using GroEL coupled to bio-layer interferometry. Protein Sci 23(10):1461–1478
23. Fierobe HP, Mingardon F, Mechaly A, Belaich A, Rincon MT, Pages S, Lamed R, Tardif C, Belaich JP, Bayer EA (2005) Action of designer cellulosomes on homogeneous versus complex substrates: controlled incorporation of three distinct enzymes into a defined trifunctional scaffolding. J Biol Chem 280(16):16325–16334
24. Wei ZH, Chen H, Zhang C, Ye BC (2014) FRET-based system for probing protein-protein interactions between sigma(R) and RsrA from Streptomyces coelicolor in response to the redox environment. PloS One 9(3):e92330
25. Prischi F, Konarev PV, Iannuzzi C, Pastore C, Adinolfi S, Martin SR, Svergun DI, Pastore A (2010) Structural bases for the interaction of frataxin with the central components of iron-sulphur cluster assembly. Nat Commun 1(95):1–10
26. Maun HR, Wen X, Lingel A, de Sauvage FJ, Lazarus RA, Scales SJ, Hymowitz SG (2010) Hedgehog pathway antagonist 5E1 binds hedgehog at the pseudo-active site. J Biol Chem 285(34):26570–26580
27. Lu P, Feng MG (2008) Bifunctional enhancement of a beta-glucanase-xylanase fusion enzyme by optimization of peptide linkers. Appl Microbiol Biotechnol 79(4):579–587
28. Smith MC, Scaglione KM, Assimon VA, Patury S, Thompson AD, Dickey CA, Southworth DR, Paulson HL, Gestwicki JE, Zuiderweg ER (2013) The E3 ubiquitin ligase CHIP and the molecular chaperone Hsc70 form a dynamic, tethered complex. Biochemistry 52(32):5354–5364

Cellulose-degrading enzymes from *Aspergillus terreus* D34 and enzymatic saccharification of mild-alkali and dilute-acid pretreated lignocellulosic biomass residues

Adepu K Kumar* and Bhumika S Parikh

Abstract

Background: Production of cellulose-degrading enzymes from *Aspergillus terreus* D34 using different growth substrates was studied under solid-state cultivation. We have tested two lignocellulosic biomass residues *viz.*, rice straw (RS) and sugarcane bagasse (BG), both separately and in combinations, and crystalline cellulose as a sole source of carbon for cellulase production. We also demonstrated different cellulase cocktail formulations and enzymatic saccharification studies on mild-alkali and dilute-acid pretreated RS- and BG-biomass residues.

Results: Substrate-specific non-denaturing native gels showed two exoglucanases, four endoglucanases, three β-glucosidases, and four xylanases in the microbial culture extract of RS-grown cells. While in the BG-culture extract, two exoglucanases, five endoglucanases, three β-glucosidases, and four xylanases were detected. Similarly, in crystalline cellulose-grown culture extract, three exoglucanases, four endoglucanases, one β-glucosidase, and two xylanases were detected. However, the cellulase compositions were highly varied with the culture extracts obtained from the mixed biomass (RSBG) growth substrate. We found that few enzymes were specifically induced while others were repressed in RSBG-grown cultures. Enzymatic saccharification resulted in the production of maximum reducing sugars of 0.733 and 0.498 g g^{-1} with mild-alkali treated RS- and BG-biomass residues with saccharification yields reaching up to 82.8% ± 1.0% and 59.3% ± 1.7%, respectively.

Conclusions: The cellulase activities, namely FPase, CMCase, avicelase, β-glucosidase, and endoxylanase, were significantly higher in the BG-grown culture extract. Optimization of microbial growth carbon sources produced an efficient cellulase enzyme cocktail mixture with an approximately twofold higher total cellulase (FPase) activity that drastically reduced the required amount of enzyme (in terms of unit volumes) for enzymatic hydrolysis studies.

Keywords: *A. terreus*; Multiple expression; Cellulases; Lignocellulosic biomass; Saccharification

Background

Lignocellulosic plant biomass is the most abundant renewable bioenergy resource available on Earth. It will play a major role in the near future as an alternative bioresource to the currently existing non-renewable and exhaustive petroleum products for bioenergy applications. Globally, large quantities of agro-industrial wastes and crop residues are available in plenty and efficient utilization and microbial biodegradation of these biomaterials will lead to several processes of great economic value [1]. Particularly in Asian countries, surplus amounts of lignocellulosic residues are available annually without much usage, namely rice straw (667.6 million tons) and sugarcane bagasse (74.8 million tons). These could be used as a potential bioresource material for cellulosic ethanol production [2,3]. However, large amounts of these agricultural residues are still frequently burnt in the fields causing severe environmental pollution problems and may foresee in utilizing these biomass residues for various potential industrial applications, for example, production of microbial enzymes and liquid biofuels, etc.

* Correspondence: kiranbio@gmail.com
Bioconversion Technology Division, Sardar Patel Renewable Energy Research Institute, Near BVM Engineering College, Vallabh Vidyanagar, Post box No. 2, 388 120 Gujarat, India

The major components of lignocellulosic biomass residues are celluloses, hemicelluloses, and lignin. The cellulosic fractions could be broken down into simple reducing sugars either by enzymatic or chemical processes for fuel ethanol production [4,5]. A broad range of cellulolytic and hemicellulolytic enzymes belonging to the glycoside hydrolase (GH) super family are produced by a variety of bacteria and fungi and are primarily involved in the breakdown of complex biomass residues into simple reducing sugars [6]. The major cellulase enzymes involved in lignocellulosic bioconversion processes are the following: (i) exoglucanases (1,4-β-cellobiohydrolase; avicelase; E.C: 3.2.1.4) which act in a processive manner on the reducing and non-reducing ends of cellulose polysaccharide chains, liberating either cellobiose or glucose as major end products; (ii) endoglucanases (carboxymethyl cellulase; CMCase; E.C: 3.2.1.74) which randomly cut β-1,4 bonds of cellulose chains generating new ends; and (iii) β-glucosidases (E.C: 3.2.1.21) which specifically cut the cellobiose dimer units producing glucose monomers. Along with these enzymes, other key enzymes belonging to hemicellulases, mainly endoxylanases (E.C: 3.2.1.8), are known to act on the xylan portion of the biomass residue into pentoses, where xylan constitutes nearly 30% to 35% of total dry weight of the biomass.

Among several hypercellulolytic fungal strains, filamentous fungi such as *Trichoderma* and *Aspergillus* sps are considered as the major microbial sources for production of cellulose-degrading enzymes [7,8]. The genome sequencing of *T. reesei* revealed two cellobiohydrolases (CBHI and CBHII) and eight endoglucanases (EG) [9], among which only two major proteins (EGI and EGII) are usually secreted in notable quantities [10]. *Trichoderma* sps is a well-known commercial cellulase producer but lacks sufficient β-glucosidase activity in the enzyme mixture. Although up to 12 β-glucosidases have been identified in *T. reesei* genome, most of them are found to be intracellular with low expression levels [9]. In contrary, *Aspergillus* sps produces high amounts of β-glucosidase in the extracellular medium and is commonly used for commercial production of this enzyme. *A. terreus* NIH 2624 genome is known, and it contains genes encoding 5 exoglucanases, 22 endoglucanases, 18 β-glucosidases, and 7 xylanases and other genes with conserved domains that encode several putative cellulose-degrading enzymes and strongly establishes that multiple cellulase genes are present in *A. terreus*. Since expression of endoglucanases and cellobiohydrolases is very poor from *Aspergillus* sps, much research is focused towards β-glucosidase and xylanase production [11].

The process of lignocellulosic biomass bioconversion to ethanol involves three major steps, namely pretreatment, saccharification, and fermentation. Keeping in view of voluminous research reports on mild-alkali and dilute-acid pretreatments of lignocellulosic biomass residues and microbial fermentation processes [12-14], the detailed description on the two steps is not described in detail in this report and further researchers are suggested to refer recent reviews [15,16]. Saccharification of the pretreated lignocellulosic biomass depends mainly on the hydrolytic efficiency of each individual cellulase enzyme component present in the multi-cellulase complex and their synergistic action towards the breakdown of insoluble celluloses and hemicelluloses into simple reducing sugars [6]. In continuation to our earlier published work on the production of cellulases from *A. terreus* D34 and enzymatic hydrolysis on mild-alkali pretreated RS biomass [12,13], here we report the detailed zymographic identification studies of multiple cellulase hemicellulase enzymes and microbial growth substrate optimization for low-cost production of cellulases using cheap lignocellulosic biomass residues. In addition, enzymatic saccharification studies on mild-alkali and dilute-acid pretreated RS- and BG-biomass residues with different cellulase cocktails are also reported.

Methods

Chemicals and raw materials

Precut, washed, and dried RS- and BG-biomass residues and microcrystalline cellulose (Sigma-Aldrich, St. Louis, MO, USA) were used as substrates for microbial growth and cellulase production. Celluclast 1.5 L and Novozyme 188, Avicel, carboxymethyl cellulose (CMC), *p*-nitrophenyl-β-D-glucopyranoside (PNPG), 4-methylumbelliferyl-β-D-cellobioside (MUC), 4-methylumbelliferyl-β-D-glucopyranoside (MUG), birchwood xylan, glucose, xylose, and cellobiose were purchased from Sigma-Aldrich (St. Louis, MO, USA) and were used for activity staining and enzyme assay measurements. The reagents for polyacrylamide gel electrophoresis (PAGE) analysis were purchased from Bio-Rad, Gurgaon, Haryana, India. All other chemicals and reagents were of analytical grade quality and were obtained from commercial sources.

Lignocellulosic biomass preparation

RS- and BG-biomass residues were procured from local agricultural fields and sugarcane refineries near Anand district, Gujarat. Initially, the biomass residues were extensively washed with water and then dried in sunlight for 3 to 4 days until the moisture content was reduced <5% (*w*/*w*). The dried biomass residues were then cut into small size (5 to 10 mm) using hammer mill. The precut biomasses were directly used without any further processing for microbial growth and enzyme production.

Microbial culture and growth conditions

We have recently reported the isolation of lignocellulose-degrading filamentous fungus *A. terreus* D34; microbial growth conditions and identification and complete details were given previously [12,13]. In brief, the filamentous fungus was grown for 7 days at 45°C on sterile agar slants comprising 2% (*w*/*v*) potato dextrose agar (PDA) and subcultured on fresh agar slants every 15 days. For long-term storage, spore suspensions were made in 0.1% (*w*/*v*) Triton X-100 and then glycerol stock cultures were prepared and stored at −80°C until further use.

Enzyme production in solid-state cultivation

Solid-state cultivation was carried out separately in 250 mL Erlenmeyer flasks containing RS- and BG-biomass residues and in combinations as a sole source of carbon for microbial growth. Crystalline cellulose was used as a synthetic growth substrate, and comparative analysis were performed with the naturally available lignocellulosic biomass residues. The RS to BG combinations tested were the following: 100:0, 25:75, 50:50, 75:25, and 0:100, where 100 correlate to 5 g dry weight of the respective biomass. All other growth parameters were maintained similar to our recent published report on cellulase production [12,13]. The incubated flasks were removed periodically every 24 h for 7 days, and enzyme activities and protein concentrations were measured.

Crude cellulase extraction and ammonium sulfate precipitation

The culture extract was separated from the microbial cells by the addition of 40 mL of 50 mM sodium citrate buffer (pH 4.8) to the solid-state grown cells and incubated at 45°C for 1 h under mild stirring conditions (100 rpm) to remove loosely bound enzymes from the biomass residue. The solid residual slurry was separated by vacuum filtration, and then the filtrate was subjected to centrifugation at 10,000 xg for 10 min for separation of the suspended particles. The so-obtained supernatant was further filtered through 0.45 μm glass fiber filters under vacuum pressure (Millipore system, Millipore India Ltd, Bengaluru, Karnataka, India) to obtain a clear enzyme solution. Finally, the proteins in the clarified culture extract were subjected to ammonium sulfate precipitation (85% to 90% (*w*/*v*)), respectively. The precipitated proteins were resolubilized in 50 mM sodium citrate buffer (pH 4.8) and then desalted in the above buffer by membrane dialysis (10 kDa cutoff) for 3 to 4 h with frequent buffer exchange. The desalted crude cellulases were used for enzymatic saccharification and zymographic experiments.

Determination of cellulase activities

Total cellulase activity (FPase) was measured by standard filter paper assay method following the NREL protocol [17]. In brief, 50 mg of Whatman filter paper (grade I) incubated with 1 mL of diluted enzyme at pH 5.0 for 60 min at 50°C. The concentration of reducing sugars formed was determined by dinitrosalicylic acid (DNSA) method [18]. Avicelase (exoglucanase) activity was measured by incubating 0.5 ml of crude enzyme dilute with 2.5% (*w*/*v*) Avicel in 50 mM citrate buffer (pH 5.0) for 60 min at 50°C. One unit of FPase activity and avicelase activity is measured as the amount of enzyme required to release 1 μg of glucose per min. Endoglucanase (CMCase) activity was measured by incubating 0.5 mL of diluted enzyme with 0.5 mL of 2% (*w*/*v*) CMC in 50 mM citrate buffer (pH 5.0) for 30 min at 50°C. Finally, xylanase activity was measured by incubating 0.5 mL of diluted enzyme with 0.5 mL of 1% (*w*/*v*) birchwood xylan in 50 mM citrate buffer (pH 5.0) for 30 min at 50°C. The concentration of reducing sugars formed in the above assay experiments were determined as described above. The amount of enzyme required for liberating either 1 μg of glucose or 1 μg of xylose per minute under the above assay conditions was considered as one unit of CMCase or xylanase activity. β-Glucosidase assay was performed by incubating 0.2 mL of enzyme dilute and 0.2 mL of 0.01 M PNPG and 1.6 mL of 50 mM citrate buffer (pH 5.0) for 30 min at 50°C. Then, the reaction was stopped by the addition of 4 mL of 0.2 M NaOH-glycine buffer (pH 10.6) [19]. The amount of enzyme required for liberating 1 μg p-nitrophenol per minute was considered as one unit of β-glucosidase activity. All the assay experiments were carried out either voiding enzyme or substrate and were used as control. Enzyme activities were also determined for commercial enzymes (Celluclast 1.5 L and Novozyme 188) following the similar method described above. FPase and endoglucanase activities were measured for Celluclast 1.5 L, and β-glucosidase activity was measured for Novozyme 188. The mean values from triplicate experiments were calculated and presented here.

Native PAGE for zymographic analysis

Native PAGE using concentrated enzyme solution from RS-, BG-, and cellulose-grown culture extracts was performed. Zymogram analysis for the identification of avicelase, CMCase, xylanase, and β-glucosidase isoforms was done. CMCase and xylanase active protein bands were detected separately on a 10% (*w*/*v*) separating gel with 2% (*w*/*v*) CMC and 2% (*w*/*v*) birchwood xylan substrates, respectively. Congo red dye staining method was used for the detection of zone of clearance by the enzymatic action of CMCase and xylanase proteins. Briefly, gel was loading with 30 μg of crude enzyme sample obtained from RS-, BG-, RSBG-, and cellulose-grown cells. After running electrophoresis, the gel was incubated in 50 mM sodium citrate buffer (pH 5.0) for 60 min at 50°C and then stained with 0.1% (*w*/*v*) Congo red solution for 20 min. The gels are washed with 1 M NaCl to visualize the clear bands corresponding to the cellulase and hemicellulase active proteins. The developed gels were visualized and digitally imaged using gel documentation system (G-BOX, Syngene, Bengaluru, Karnataka, India). Whereas, avicelase and *β*-glucosidase enzyme proteins were identified separately with post-incubation of the gel either with 10 mM MUC or 10 mM MUG for 15 to 60 min at 37°C and was then viewed directly under UV light using gel documentation system. It is to be noted that the highly active proteins were initially detected as a clear band as soon as the dye reagent was added and were turned into dark color immediately after addition of developing agent. Whereas, at this staining period the less active protein bands were detected as clear white bands against dark background. Based on the mobility of the protein bands on the gel, the different enzyme active bands were labeled separately on each gel from top to bottom.

Pretreatment of lignocellulosic biomass residues

Mild-alkali and dilute-acid pretreatments were performed separately with 0.2 kg of the respective dried biomass. Alkali pretreatments were done at room temperature with 0.5% NaOH at a solid-to-liquid ratio of 1:20 for 24 h following the optimized conditions that was reported recently by us [12,13], while acid pretreatments were conducted at 121°C for 10, 30, and 60 min by incubating the biomass residues with 1.5% H_2SO_4 at a solid to liquid ratio of 1:10. After both pretreatments, the biomass residues were neutralized to pH 7.0 and washed thoroughly with distilled water until the residual glucose concentrations were <0.1 g L^{-1}. The pretreated biomass residues were stored in air-tight containers and kept at 4°C until use.

Enzymatic saccharification

All the enzymatic hydrolysis experiments were performed at 10% (*w*/*v*) solid loading at 45°C for 48 h in 50 mL screw-capped bottles. Based on our recently reported enzyme optimization studies, an enzyme loading of 9 FPU g^{-1} was selected in the current investigation [12,13]. All other experimental parameters were similar as detailed previously, unless specified. Control experiments were performed voiding both substrate and enzyme in the reaction mixtures. Enzymatic saccharification was also performed using commercial enzymes at an enzyme loading of 9 FPU of Celluclast 1.5 L and 100 U of Novozyme 188 per gram of pretreated biomass. After enzymatic saccharification, the liquefied fraction was separated and clarified from the residual biomass by filtration method using glass fiber membrane filters. Reducing sugars were analyzed by both HPLC and DNSA methods. Saccharification yields were calculated as described by Taniguchi [20].

Enzyme formulation and cellulase cocktail preparation

Before the experimental setup, the crude cellulases obtained from RS- and BG-grown cells were physically premixed to a final 9 FPU g^{-1} in a ratio of RS25:BG75, RS50:BG50, and RS75:BG25, respectively. Saccharification experiments were also performed with three different formulated cellulase enzyme cocktails. All other parameters of saccharification process were not altered and maintained similar as described in the above section.

Analytical methods

Protein concentration was determined using BSA as standard following the method described by Bradford [21]. The contents of cellulose, hemicellulose, and lignin in the untreated and pretreated biomass residues were determined following the method described by Goering and Van Soest [22]. All the assay experiments were carried out in triplicates, and the mean values are given. Statistical analysis of the enzyme activities and saccharification yields was performed using ANOVA software, and the significance of difference in each data was denoted in superscripts by special characters. All the data presented in this study were the mean of three or more experiments with a variation within 10%.

Results and discussion

Aspergillus sps were reported to utilize a broad range of lignocellulosic biomass residues for microbial growth and production of cellulose-degrading enzymes [11]. Also, the cellulase enzymes produced from *A. terreus* were highly active at low pH conditions and thermotolerant; thus it is a potential microbial source for hyperproduction of a broad range of cellulases for cellulose breakdown, especially towards enzymatic saccharification processes and bioethanol production. These cellulase active proteins form a synergetic network of complex biochemical mechanisms.

Microbial cellulase production

The cellulase and hemicellulase enzyme activities in the tested growth carbon substrates were found to gradually increase with the incubation time and reached maximum on the 7th day. Similarly, secretion of extracellular soluble proteins was also linearly increased with the growth incubation time irrespective of growth carbon substrate, and maximum concentration of proteins were found in the BG- (32.8 ± 2.2 g L^{-1}), followed by RS- (22.9 ± 1.8 g L^{-1}) and cellulose- (20.2 ± 1.3 g L^{-1}) grown cultures, respectively.

Filter paper activity

Total cellulase (FPase) activity (FPU g^{-1}) in the crude extract was found to be maximum in BG-grown culture (5.74 ± 0.3), followed by RS- (4.62 ± 0.4) and cellulose- (1.96 ± 0.3) grown cultures, respectively (Figure 1). Nearly a threefold increase in FPase activity in BG-grown culture extract and approximately fivefold with the mixed (RS-BG) growth substrates were obtained when compared to cellulose-grown extract (Table 1).

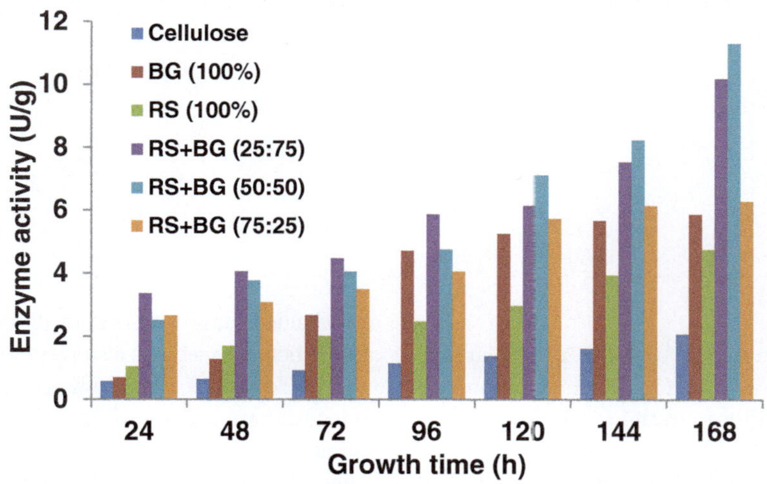

Figure 1 FPase activity from *A. terreus* during growth on rice straw (RS), bagasse (BG), and cellulose substrates.

The FPase activities in the culture extracts of the RS75: BG25-, RS50:BG50-, and RS25:BG75-mixed growth substrates were 6.3 ± 1.1, 11.34 ± 0.7, and 10.22 ± 0.7, respectively (Figure 1). Although the FPase activities were comparatively higher in the RS75:BG25 growth culture extract, higher BG content in the mixed growth substrate produced effective cellulase components in the enzyme cocktail that enhanced the total cellulase activity. Recently, it is reported that FPase activity of 1.02 and 3.2 U ml^{-1} was obtained with *A. fumigates* NIYDGPKA3 [23] and *A. heteromorphus* [24] using rice straw and wheat straw biomass residues as growth residues. The FPase activities presented in our study were significantly higher than other reports. A brief comparison was made with other reported cellulases and xylanases and their activities from *Aspergillus* strains and is given in Table 2.

Cellobiohydrolase activity

In similar to FPase, the exoglucanase (Ex) activities (U g^{-1}) were also found to be maximum in the BG-grown culture (10.5 ± 0.3) with a twofold increase in enzyme activity followed by RS- (6.86 ± 0.3) and cellulose-(5.32 ± 0.3) grown cultures, respectively (Figure 2). However, with the mixed RS-BG growth substrates, the exoglucanase activities were comparable with the BG-grown culture. Whereas, in contrary to the findings on enzyme activities in different growth substrates, zymogram studies showed that cellulose substrate induced more number of exoglucanase proteins compared to BG- and RS-biomass substrates (Figure 3). In cellulose-grown culture, two high molecular proteins (Ex1 and Ex2) and one low molecular protein (Ex3) with exoglucanase activity were detected (Figure 3C). Whereas, Ex2 protein band in RS-grown culture and Ex3 protein band in BG-grown culture were not detected. Similar to RS-grown culture, both Ex1 and Ex3 protein bands were observed in RSBG-grown culture (Figure 3D). Among these protein bands, Ex1 protein showed high intensity in RS- and BG-grown culture extracts, while Ex3 in cellulose-grown culture was highly intense. Correlating both zymogram and enzyme activity analysis, it is clearly evident that majority of the exoglucanase activity might be associated with these high intense bands of their respective growth substrate. Detailed molecular and biochemical investigation on characterization of these individual exoglucanase active protein bands will provide further insight into the functional properties of the enzyme isoforms.

Table 1 Comparative analysis of cellulase and hemicellulase enzyme activities from *A. terreus*

Enzyme activity (U/g biomass)	Growth carbon substrate[#]					
	Cellulose	Sugarcane baggage	Rice straw	RS:BG (25:75)	RS:BG (50:50)	RS:BG (75:25)
FPase	1.96 ± 0.3[d]	5.74 ± 0.3[b]	4.62 ± 0.4[c]	10.22 ± 0.7[a]	11.34 ± 0.7[a]	6.3 ± 1.1[b]
Exoglucanase	5.32 ± 0.3[h]	10.5 ± 0.3[e]	6.86 ± 0.3[g]	8.4 ± 0.8[f]	9.8 ± 0.4[e]	10.22 ± 0.5[e]
Endoglucanase	62.44 ± 1.3[n]	145.32 ± 2.5[l]	96.6 ± 1.1[m]	203.42 ± 2.2[j]	213.22 ± 0.3[i]	159.32 ± 1.7[k]
β-Glucosidase	13.72 ± 0.3[s]	75.18 ± 1.4[o]	74.9 ± 1.7[o]	61.74 ± 2.2[r]	69.14 ± 1.5[p]	65.94 ± 1.3[q]
Xylanase	1,763 ± 251[x]	8,472 ± 490[t]	3,921 ± 406[w]	7,280 ± 462[u]	6,959 ± 252[u]	6,050 ± 294[v]

Growth on rice straw (RS), sugarcane bagasse (BG), cellulose, and mixed RSBG carbon substrates.
[#]Each value represents the mean ± standard error; values with no common alphabetic letter are significantly different at $p < 0.05$.

Table 2 Comparison of cellulase and xylanase production by some of the *Aspergillus* strains

Microorganism	Substrate	Enzyme activity					Reference
		FPase	Avicelase	CMCase	β-glucosidase	Xylanase	
A. fumigatus NITDGPKA3	Rice straw	1.02 U ml^{-1}	–	6.53 U ml^{-1}	80.1 U ml^{-1}	193.58 U ml^{-1}	[23]
A. heteromorphus	Wheat straw	3.2 U ml^{-1}	–	83 U ml^{-1}	–	–	[24]
A. terreus strain EMOO 6-4	Rice straw	–	–	–	4,457 U g^{-1}	–	[28]
A. terreus RWY	Bagasse	11.3 U g^{-1}	10.3 U g^{-1}	103 U g^{-1}	122 U g^{-1}	872 U g^{-1}	[29]
A. niger	Bagasse	153 U g^{-1}	67 U g^{-1}	–	–	–	[30]

Endoglucanase activity

The endoglucanase activity (U g^{-1}) in BG-grown culture extract was significantly higher (145.32 ± 2.5) than the RS- (96.6 ± 1.1) and cellulose- (62.44 ± 1.3) grown cultures, respectively (Figure 4). The enzyme activities were approximately 1.5- to 2-fold higher in BG compared to other growth carbon substrates. Similar to FPase, the endoglucanase activities were also increased in the all the RSBG-mixed growth cultures. The enzyme activities were 159.32 ± 1.7, 213.22 ± 0.3, and 203.42 ± 2.2 in RS75: BG25-, RS50:BG50-, and RS25:BG75-grown culture extracts, respectively (Figure 4). Jecu [25] also used mixed substrates (wheat straw and wheat bran) for cellulase production by *A. niger*; however, the endoglucanase activity was only 11.25 U ml^{-1}, which is very low compared to our mixed growth substrates. Five endoglucanase proteins in BG- (Eg1, Eg2, Eg3, Eg5, and Eg6) and four each in cellulose- (Eg2, Eg3, Eg4, and Eg5) and RS- (Eg1, Eg2, Eg3 and Eg5) grown cultures, respectively, were detected by zymogram analyses (Figure 5). However, protein bands corresponding to Eg1 and Eg6 in cellulose-grown culture, Eg4 in BG- and RS-grown cultures and Eg4 and Eg5 in RSBG-grown cells, were not detected. These results indicate that BG induced production of additional endoglucanase proteins compared to RS-and cellulose-grown substrates. Sequential expression of multiple endoglucanase and β-glucosidase isoforms in different culture conditions and carbon sources was also recently reported in *A. terreus*; where, addition of fructose and cellobiose upregulated the endoglucanase activity while addition of mannitol, ethanol, and glycerol selectively repressed three endoglucanase isoforms [26]. Similarly, in *Hypocrea jecorina*, sophorose was reported to induce selective expression of endoglucanses and one xylanase isoform where xylobiose specifically induced xylanase I and xylanase II proteins [27].

β-Glucosidase activity

In contrary to the other cellulase activities, high β-glucosidase activity (U g^{-1}) was obtained in both RS- and BG-grown culture extracts (75.18 ± 1.4) (Figure 6). Whereas, in cellulose-grown culture extract, negligible amount of β-glucosidase activity (1.01 ± 0.02) was observed, indicating that cellulose repressed production of β-glucosidase. Unlike other cellulase activities, β-glucosidase activities were not affected with the type of growth substrate. In RSBG-grown culture extracts, the enzyme activity was nearly similar (69.14 ± 1.5) in all the tested substrate ratios (Figure 6). Our results were in close comparison to *A. fumigatus*, where 80.1 U ml^{-1} β-glucosidase was obtained [23]. Very recently, a newly isolated *A. terreus* strain EMOO 6-4 showed a maximum β-glucosidase activity of 4457.162 U g^{-1} using rice straw under

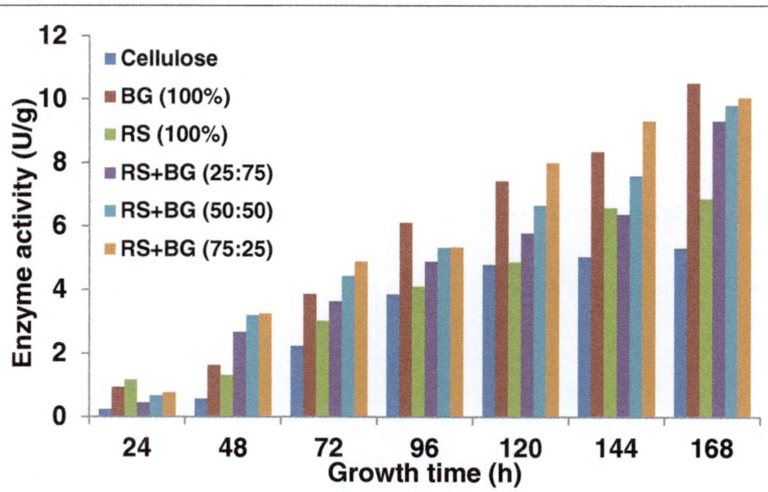

Figure 2 Exoglucanase activity from *A. terreus* during growth on rice straw (RS), bagasse (BG), and cellulose substrates.

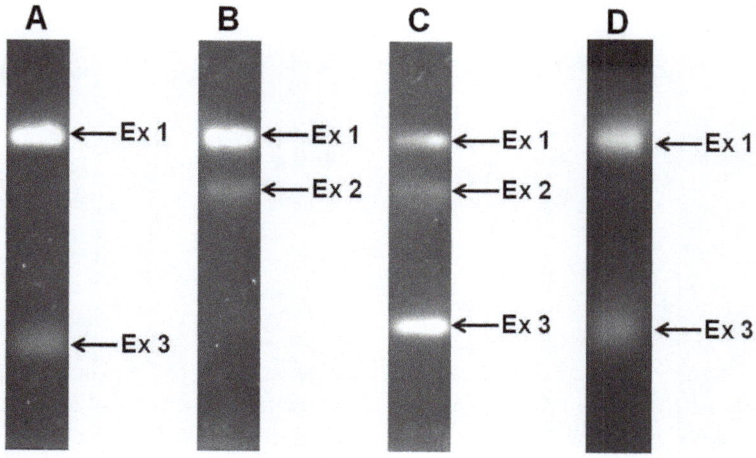

Figure 3 Detection of exoglucanases by zymogram analysis. Detection of exoglucanases by zymogram analysis using *A. terreus* crude cellulase cocktail produced from **(A)** rice straw-(RS), **(B)** sugarcane bagasse-(BG), **(C)** cellulose-, and **(D)** RS50:BG50-grown substrates.

statistically optimized growth conditions [28]. The enzyme activity results were also supported by the zymogram studies. A single broad protein band (Bg1) was detected in cellulose-grown cells, whereas, two additional proteins (Bg2 and Bg3) were detected in BG- and only Bg3 in RS-grown culture extracts (Figure 7). Similar to RS-grown culture, RSBG-grown extracts produced only Bg1 and Bg3 proteins. Apparently, the slightly higher β-glucosidase activity in BG-grown extract could be correlated with the additional Bg2 protein. In contrary to the induction of endoglucanase expression in *A. terreus*; glucose, fructose, sucrose, cellobiose, mannitol, and glycerol addition also resulted in the repression of few β-glucosidase isoforms [26].

Xylanase activity

Xylanase activities (U g^{-1}) in BG-grown culture extract (8,472 ± 490) was approximately twofold higher than the RS-grown extracts (3,921 ± 406) and approximately fivefold higher than cellulose-grown extracts (1,763 ± 251) (Figure 8). Interestingly, the xylanase activities were increased linearly with the amount of BG content. The enzyme activities in RSBG-grown culture extracts were 6,050 ± 294, 6,959 ± 252, and 7,280 ± 462 from RS75:BG25-, RS50:BG50-, and RS25:BG75-grown culture extracts, respectively (Figure 8). It is recently reported that a newly isolated *A. terreus* RWY produced xylanase activity of 872 U g^{-1} when sorghum bagasse was tested under solid-state conditions [29]. Significantly lower xylanase activity obtained from cellulose growth culture in correlation with the zymogram studies found that it could be due to the poor expression of xylanase active proteins. Only two xylanase proteins (Xy2 and Xy3) were detected in cellulose growth culture extract while additional proteins *viz.* Xy1, Xy4, and Xy5 in RS- and Xy1 and Xy5 in BG-grown culture extracts were detected

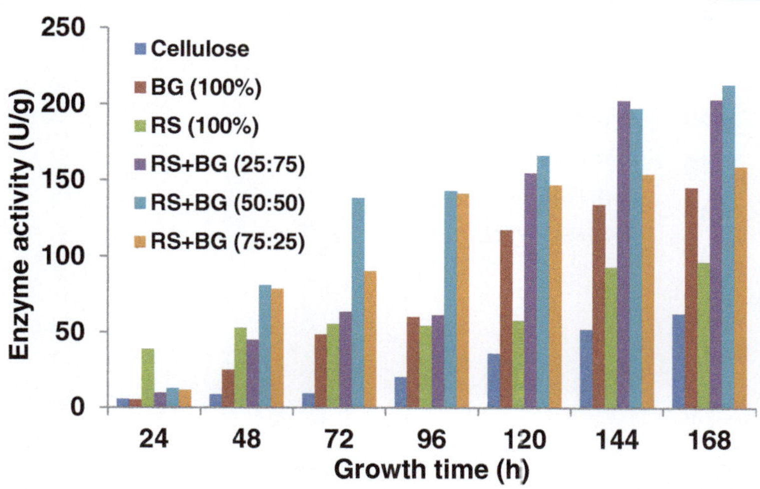

Figure 4 Endoglucanase activity from *A. terreus* during growth on rice straw (RS), bagasse (BG), and cellulose substrates.

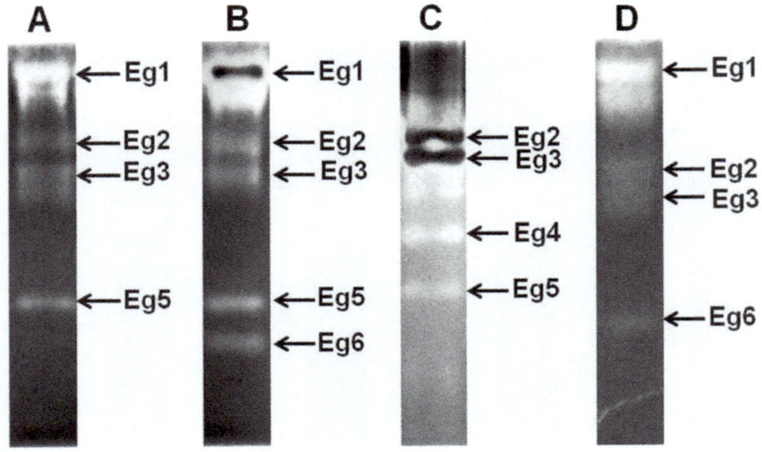

Figure 5 Detection of endoglucanases by zymogram analysis. Detection of endoglucanases by zymogram analysis using *A. terreus* crude cellulase cocktail produced from **(A)** rice straw-(RS), **(B)** sugarcane bagasse-(BG), **(C)** cellulose-, and **(D)** RS50:BG50-grown substrates.

(Figure 9). Although four xylanase proteins were detected in RS- and BG-grown extracts, however, a high molecular weight protein (Xy3) in RS-grown extract and a low molecular weight protein (Xy4) in BG-grown extract were not detected. In contrary to cellulase enzyme variants, all the xylanase enzymes were detected when RSBG-mixed growth culture extract was used in the growth medium (Figure 9D). From zymogram analysis and xylanase assay activity measurements, we could infer that the low intense protein (Xy3) might be crucial for enhancing the overall xylanase activity in the BG-grown culture extract. Our results on the identification of various xylanases from RS-, BG-, cellulose-, and RSBG-grown culture extracts also correlate with the earlier reports on the expression of distinct enzyme isoforms with the culturing conditions, where production of different multiple xylanases in *Myceliopthora* sps [31] and exoglucanases in *A. niger* [30] is detected.

From our results, it is clearly evident that the expression and repression of cellulase and hemicellulase genes in *A. terreus* D34 varied with the type of growth carbon source. This may be mostly due to the structural difference and architecture of the carbon source that play a crucial role in differential expression of the cellulolytic enzymes [32]. In addition to the culture conditions, small molecules and metabolites were also found to affect the production of cellulase enzymes in microbes like *Aspergillus* sps [25,32]. However, it is not clearly understood how the expression of these multiple proteins is regulated at gene level and further detailed molecular and proteomic analysis will provide the mechanism on the metabolic role of cell-bound regulatory proteins and the cell-substrate interactions. But to some extent, it was reported that the regulation of cellulases was mediated by induction and repression of the genes that control the expression of cellulose-degrading enzymes [33,34]. Such studies on

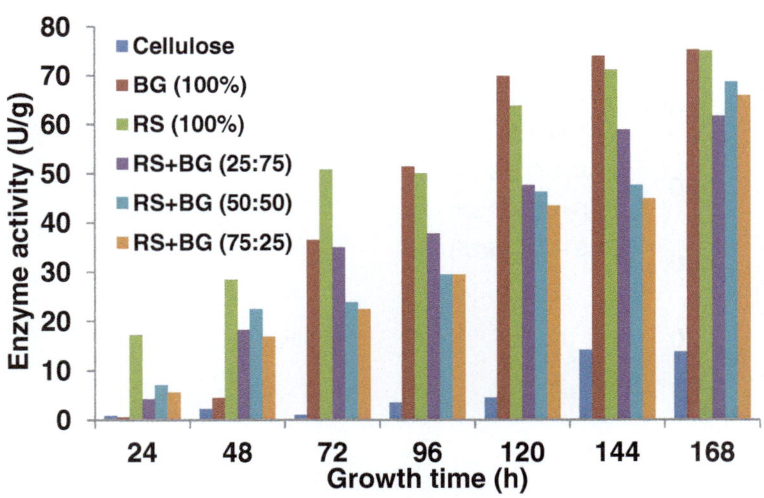

Figure 6 ***β*-Glucosidase activity from *A. terreus* during growth on rice straw (RS), bagasse (BG), and cellulose substrates.**

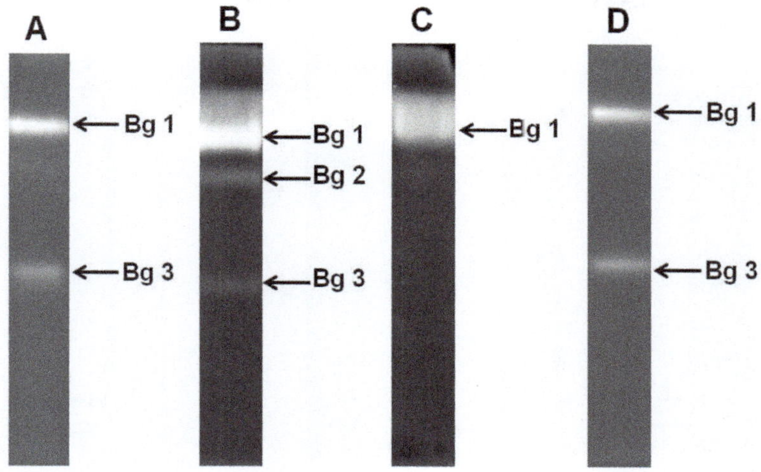

Figure 7 Detection of β-glucosidases by zymogram analysis. Detection of β-glucosidases by zymogram analysis using *A. terreus* crude cellulase cocktail produced from **(A)** rice straw-(RS), **(B)** sugarcane bagasse-(BG), **(C)** cellulose-, and **(D)** RS50:BG50-grown substrates.

multiple and differential expression of cellulase and hemicellulase enzymes and the corresponding enzyme activities in the culture extracts from different growth carbon substrates will produce an effective cellulase enzyme cocktail with all the cellulase enzyme components in a single mixture. The mechanism of expression and repression of each enzyme variant is under the control of a complex metabolic process, and this could be mostly due to the production and accumulation of various extracellular metabolites in the culture medium and their continuous interactions with the membrane transporter proteins [33]. Thus, it triggers the secretion of cellulose-degrading proteins into the extracellular medium during microbial growth on a specific lignocellulosic biomass residue [34]. For example, our experimental results showed nearly twofold higher xylanase activity in BG- than RS-biomass residues and may be correlated with high hemicellulose content present in the BG biomass that may induce the secretion of additional xylanase active proteins. From zymogram and enzyme activity studies, the induction and repression of each enzyme variant and their probable correlation with the enzyme activities as discussed in this report strongly suggest the significant effect of biomass on the corresponding enzyme activities and was clearly demonstrated in Table 1. As shown in the case of exoglucanase, although Ex3 was not produced in BG-grown culture extract, the exoglucanase activity was significantly higher than the RS- and cellulose-grown extracts, indicating that the total exoglucanase activity was primarily from the Ex1 protein with no major contribution from Ex3 protein and was also clearly evident from the detection of a high intense sharp exoglucanase active protein band in zymographic analysis. Similarly, the role of Eg4 protein in endoglucanase activity could also support this inference. Moreover, the presence of Bg3

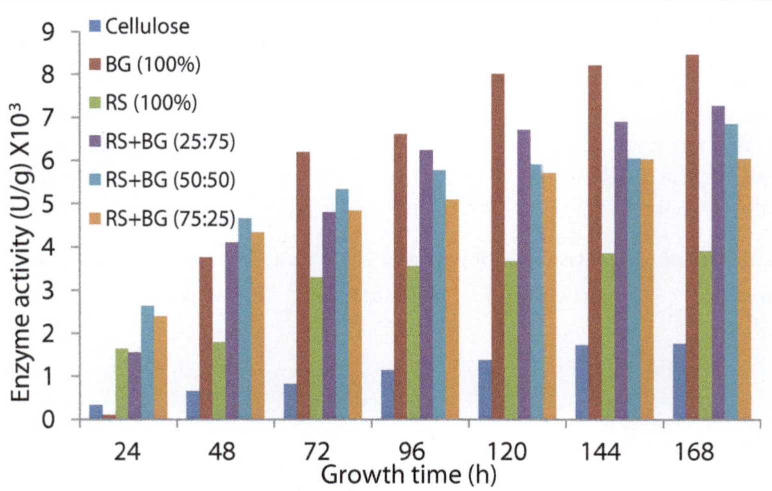

Figure 8 Xylanase activity from *A. terreus* during growth on rice straw (RS), bagasse (BG), and cellulose substrates.

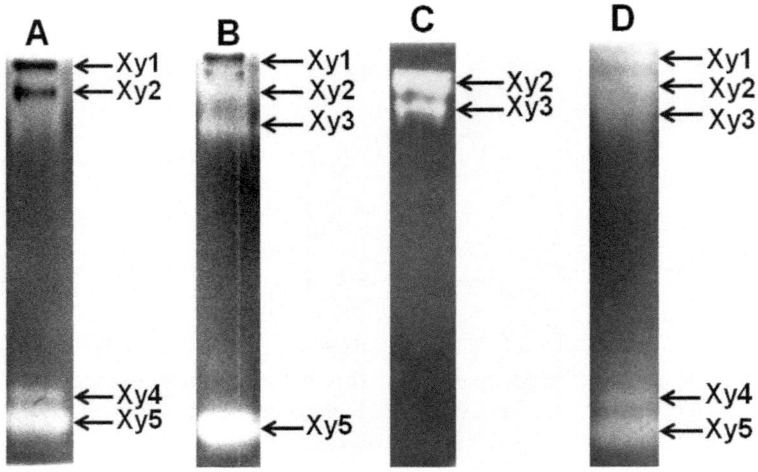

Figure 9 Detection of xylanases by zymogram analysis. Detection of xylanases by zymogram analysis using *A. terreus* crude cellulase cocktail produced from **(A)** rice straw-(RS), **(B)** sugarcane bagasse-(BG), **(C)** cellulose-, and **(D)** RS50:BG50-grown substrates.

protein is crucial for increased β-glucosidase activity. This conclusion was made on the observations with cellulose growth extract where Bg3 protein was not detected and a fourfold decrease in β-glucosidase activity was obtained. Also, the correlation data between xylanase activity and two low molecular xylanase isoforms (Xy4 and Xy5) strongly supports our results.

Pretreatment of lignocellulosic biomass residues

Delignification of RS biomass by mild-alkali pretreatment method increased the cellulose content from 40.0% ± 1.8% to 60.1% ± 1.2% with a marginal loss of hemicellulose content (28.0% ± 2.0% to 19.2% ± 1.6%). The cellulose availability was slightly higher after mild-alkali treatment. Interestingly, more than 70% of lignin was removed after mild-alkali pretreatment (Table 3). While after acid treatment, the holocellulose (mainly cellulose content) in the pretreated RS-biomass residue was increased to 76.8% ± 1.1% with nearly complete removal of hemicellulose (1.3% ± 0.23%). Moreover, similar results were also obtained with BG biomass after both pretreatment methods. The cellulose content of BG after mild-alkali treatment was increased from 35.1% ± 2.1% to 49.9% ± 1.2% with hemicellulose content of 25.2% ± 1.6%. Although the holocellulose content was comparatively lower after acid pretreatment (55.2% ± 1.4%), similar to RS, the hemicelluloses (1.6% ± 0.9%) in BG were completely removed. Comparative amount of lignin was removed by mild-alkali treatment in the biomass residues while majority of the lignin content was retained after dilute-acid pretreatment (Table 3).

Enzymatic hydrolysis of pretreated biomass

Saccharification efficiency (%) from mild-alkali pretreated RS-biomass residue was found to be 82.8 ± 1.0, 79.3 ± 1.2, and 73.0 ± 2.0 using crude concentrated cellulase mixture obtained from the RS-, BG-, and RSBG-grown culture extracts, respectively. With mild-alkali pretreated BG-biomass residue, the saccharification yields were comparatively lower (37.1 ± 2.1, 52.4 ± 1.9, and 59.3 ± 1.7), where approximately 59% higher saccharification yields were obtained with enzyme obtained from RSBG culture extract. Similarly, with acid treated RS biomass, maximum reducing sugars were obtained with RSBG-grown culture extracts followed by RS-grown extract and vice versa when acid treated BG biomass was used for saccharification. However, with Celluclast 1.5 L, the saccharification yields reached up to 93.3 ± 0.7 and 82.1 ± 0.4 with mild-alkali treated RS- and BG-biomass residues and 52.9 ± 0.5 and 49.0 ± 1.7 were

Table 3 Compositional analysis of untreated and pretreated rice straw and sugarcane bagasse

Biomass	Treatment	Residual straw (g)	Cellulose (%)	Hemicellulose (%)	Holocellulose (%)	Lignin (%)
Rice straw	Untreated	200	40.0 ± 1.8	28.0 ± 2.0	68.0 ± 1.9	9.0 ± 0.8
	Alkali treated	144	60.1 ± 1.2	19.2 ± 1.6	79.3 ± 1.3	3.5 ± 0.6
	Acid treated	110	76.8 ± 1.1	1.3 ± 0.23	78.1 ± 0.8	8.1 ± 0.2
Sugarcane bagasse	Untreated	200	35.1 ± 2.1	31.2 ± 1.7	67.3 ± 2.0	19.2 ± 0.4
	Alkali treated	152	49.9 ± 1.2	25.2 ± 1.6	75.1 ± 0.6	10.6 ± 1.3
	Acid treated	125	49.6 ± 2.5	1.6 ± 0.9	52.2 ± 1.4	18.1 ± 0.5

obtained with acid treated RS- and BG-biomass residues, respectively. While with mixed Celluclast 1.5 L and Novozyme 188, the saccharification yields reached maximum and were 96.5 ± 0.3 and 88.3 ± 0.4 with mild-alkali treated RS- and BG-biomass residues and 74.5 ± 0.9 and 54.2 ± 1.0 with acid treated RS- and BG-biomass residues, respectively. The synergetic affect of β-glucosidase in enhancing reducing sugar production from the pretreated biomass residues is clearly evident from these results. Moreover, the RS-grown culture extract enzyme produced reducing sugars which are nearly comparable with the commercial enzyme, Celluclast 1.5 L. Detailed reducing sugar concentrations and corresponding saccharification efficiencies of the respective pretreated biomasses and their enzyme sources were given in Table 4. Interestingly, saccharification studies with the formulated enzyme cocktails showed slightly lower efficiencies for both alkali and acid treated biomass residues when compared with the crude cellulase enzymes from individual growth carbon sources (Additional file 1 Supplementary Table S1). With alkali treated RS- and BG-biomass residues, the maximum saccharification yields obtained with the formulated enzymes were 57.0 ± 1.8 and 42.5 ± 1.7, respectively. Whereas, with acid treated RS- and BG-biomass residues, the maximum saccharification yields were 44.2 ± 1.4 and 21.5 ± 1.9, respectively. A comparison data of enzymatic hydrolysis of alkali and acid treated rice straw along with alkali and acid treated sugarcane bagasse was detailed in supplementary Table 2.

Beside zymogram studies of enzyme variants, interesting observations were made during this investigation and showed that nearly 1.25- and 2.5-fold increase in FPase, exoglucanase, endoglucanase, and xylanase activities were obtained with BG- and RS50:BG50-grown culture extracts when compared to RS-grown culture. Since these enzymes are known to be used in a wide variety of industrial applications, this may be an important factor to consider for large-scale production of enzymes and may significantly reduce the capital expenditures towards handling large capacity of biomass residues, reactor vessel volumes, and downstream processing costs, etc. [35]. However, for bioethanol production, the saccharification efficiencies with the isolated enzymes from RS-grown culture extracts were marginally higher

Table 4 Enzymatic saccharification of mild alkali and dilute acid pretreated lignocellulosic biomass residues

Enzyme	Pretreatment condition		Rice straw (RS)		Sugarcane bagasse (BG)	
			Reducing sugars (g g^{-1})	Saccharification (%)	Reducing sugars (g g^{-1})	Saccharification (%)
RS-grown cells	Alkali	24 h	0.621 ± 0.013^{c}	70.2 ± 1.5	0.249 ± 0.016^{e}	29.6 ± 1.9
		48 h	0.733 ± 0.009^{g}	82.8 ± 1.0	0.312 ± 0.018^{j}	37.1 ± 2.1
	Acid	24 h	0.356 ± 0.015^{k}	40.9 ± 1.7	0.179 ± 0.011^{k}	30.8 ± 1.9
		48 h	0.424 ± 0.012^{p}	48.8 ± 1.4	0.216 ± 0.015^{p}	37.2 ± 2.6
BG-grown cells	Alkali	24 h	0.648 ± 0.008^{b}	73.2 ± 0.9	0.336 ± 0.021^{d}	40.0 ± 2.5
		48 h	0.702 ± 0.011^{h}	79.3 ± 1.2	0.440 ± 0.016^{i}	52.4 ± 1.9
	Acid	24 h	0.337 ± 0.021^{l}	38.8 ± 2.4	0.059 ± 0.005^{m}	10.3 ± 0.9
		48 h	0.352 ± 0.013^{q}	40.5 ± 1.5	0.112 ± 0.009^{q}	19.3 ± 1.5
Mixed growth cells (RS50:BG 50)	Alkali	24 h	0.454 ± 0.019^{d}	51.3 ± 2.1	0.368 ± 0.018^{c}	43.8 ± 2.1
		48 h	0.646 ± 0.018^{i}	73.0 ± 2.0	0.498 ± 0.014^{h}	59.3 ± 1.7
	Acid	24 h	0.293 ± 0.012^{m}	33.7 ± 1.4	0.121 ± 0.011^{l}	20.8 ± 1.9
		48 h	0.451 ± 0.026^{o}	51.9 ± 3.0	0.201 ± 0.019^{p}	34.6 ± 3.3
Celluclast 1.5 L	Alkali	24 h	0.635 ± 0.008^{b}	71.8 ± 0.9	0.462 ± 0.006^{b}	55.0 ± 0.7
		48 h	0.826 ± 0.006^{f}	93.3 ± 0.7	0.690 ± 0.003^{g}	82.1 ± 0.4
	Acid	24 h	0.285 ± 0.010^{m}	32.8 ± 1.2	0.105 ± 0.004^{l}	18.1 ± 0.7
		48 h	0.460 ± 0.004^{o}	52.9 ± 0.5	0.285 ± 0.010^{o}	49.0 ± 1.7
Celluclast 1.5 L + Novozyme 188	Alkali	24 h	0.714 ± 0.008^{a}	80.7 ± 0.9	0.513 ± 0.008^{a}	61.0 ± 1.0
		48 h	0.854 ± 0.003^{e}	96.5 ± 0.3	0.742 ± 0.003^{f}	88.3 ± 0.4
	Acid	24 h	0.420 ± 0.012^{j}	48.3 ± 1.4	0.168 ± 0.011^{k}	28.9 ± 1.9
		48 h	0.648 ± 0.008^{n}	74.5 ± 0.9	0.315 ± 0.006^{n}	54.2 ± 1.0

At 10% solid loading and enzyme loadings[#] of 9 FPU/g substrate using crude cellulase enzymes and formulated enzyme cocktails.
[#]Enzyme loadings of fixed concentration (9FPU/g substrate) for respective biomass were obtained directly from the crude enzyme from the respective biomass sources. In each case, the final enzyme loadings were calculated based on the enzyme activities shown in Table 1. Commercial enzymes (Celluclast 1.5 L and Novozyme 188) were also tested for comparison. Each value represents the mean ± standard error; values with no common letter in a column (calculated separately for respective time, biomass, and pretreatment) are significantly different at $p < 0.05$.

compared to the RSBG-mixed growth culture extracts but at a fixed enzyme loading of 9 FPU/g biomass; the volume of protein required was significantly low (on the basis of respective enzyme activities shown in Table 1), where, in order to obtain the same enzyme loading, nearly twofold of the enzyme was required for RS-grown culture extract.

Similarly, most of the commercially available cellulase enzymes are formulated cocktails and there is a continuous thrust towards the development of efficient cellulase enzyme mixtures prepared by physical mixing of subtle amounts of other cellulose-degrading enzyme components for synergistic action and high product yields at low enzyme loadings. Formulation of cellulase enzyme activities for saccharification process will not only reduce the total enzyme loadings and but also decrease the capital expenditure costs (as described above) in a cellulosic biofuel production. Moreover, adding requisite amounts of enzyme components that were deficient in the cocktail mixture may further improve the overall process efficiency. It was clearly evident from the high saccharification yields of pretreated biomasses that using commercial enzymes were mainly attributed due to the external supplementation of β-glucosidase. All together, optimization of each enzyme component is attractive for lowering enzyme loadings; preparing such type of complex mixtures is quite challenging. Because it generally requires the separation of a particular enzyme component from the broth and re-addition into the cocktail which is considered to be one of the costly process [36], attempts were made to overcome such challenges in the preparation of formulated cellulase cocktails by physical mixing of crude enzymes and optimizing the growth carbon substrates, separately resulted in the production of an efficient enzyme mixture with at least twofold higher cellulase activities. Thus, it was concluded that crude cellulose-degrading enzymes produced by the *A. terreus* D34 with mixed growth substrates produced efficient crude enzyme mixture with the required enzyme components in appropriate proportions to yield high reducing sugars. These studies provide a clear evidence of the production of multiple cellulase proteins and their correlation with the corresponding enzyme activities with a complex synergetic mechanism for combined action between the enzyme components for effective lignocellulosic breakdown.

Conclusion

In the present study, we have demonstrated that during microbial growth, addition of BG-biomass residue to the RS-biomass residue enhanced the production of multiple cellulases and hemicellulases. Whereas, in RS-biomass residue and cellulose substrates, *A. terreus* was deficient in producing some cellulose-degrading enzymes that were found to be critical in increasing the overall enzyme activity. We further suggest that RSBG-mixed growth substrate could be a potential growth carbon source for cost-effective production of cellulases at a larger scale. Moreover, using RS-grown culture extract enzymes, saccharification yields of up to 82.8% ± 1.0% were obtained from mild-alkali treated RS biomass and with mild-alkali treated BG biomass, up to 59.3% ± 1.7% were obtained using RS50:BG50-grown culture extract enzymes.

Additional file

Additional file 1: Table S1. Enzymatic saccharification of mild-alkali and dilute-acid pretreated lignocellulosic biomass residues at 10% solid loading and enzyme loadings# of 9 FPU/g substrate using crude cellulase enzymes and formulated enzyme cocktails.

Competing interests

The authors declare that they have no competing interests.

Authors' contributions

BSP performed the research experiments; AKK designed the research and wrote the paper. Both authors read and approved the final manuscript.

Authors' information

AKK, is a Senior Scientist and BSP is a Technical Assistant at the Sardar Patel Renewable Energy Research Institute, Bioconversion Technology Division, near BVM Engineering College, Vallabh Vidyanagar, Gujarat, India.

Acknowledgements

The authors are thankful to the Director, Sardar Patel Renewable Energy Research Institute, Gujarat, India, for the support of this research. The research work is financially supported by DBT, Govt. of India, through grant BT/PR10103/PBD/26/131/2007 and ICAR, Govt. of India. The authors acknowledge helpful discussions with Dr. S.P. Singh, SPRERI, and Prof. Datta Madambar, BRD School of Biosciences, Sardar Patel University, Gujarat, India.

References

1. Ray LG, Pal A, Ghosh A, Chattopadhyay P (1993) Cellulases and β-glucosidase from *Aspergillus niger* and saccharification of some cellulosic wastes. World J Micorbiol Biotechnol 8:85–94
2. Kim S, Dale BE (2004) Global potential bioethanol production from wasted crops and crop residues. Biomass Bioenergy 26:361–375
3. Yoswathana N, Phuriphipat P, Treyawutthiwat P, Eshtiaghi MN (2010) Bioethanol production from rice straw. Energy Res J 1:26–31
4. Margeot A, Hagerdal BH, Edlund M, Slade R, Monot F (2009) New improvements for lignocellulosic ethanol. Curr Opin Biotechnol 3:372–380
5. Sims RE, Mabee W, Saddler JN, Taylor M (2010) An overview of second generation biofuel technologies. Bioresour Technol 101:1570–80
6. Van Dyk JS, Pletchke BI (2012) A review of lignocelluloses bioconversion using enzymatic hydrolysis and synergistic cooperation between enzymes-factors affecting enzymes, conversion and synergy. Biotechnol Adv 30:1458–1480
7. Gusakov AV (2011) Alternatives to *Trichoderma reesei* in biofuel production. Cell 29:419–425
8. Zhang L, Wang X, Ruan Z, Liu Y, Niu X, Yue Z, Li Z, Liao W, Liu Y (2014) Fungal cellulase/xylanase production and corresponding hydrolysis using pretreated corn stover as substrates. Appl Biochem Biotechnol 172:1045–1054
9. Ouyang J, Yan M, Kong D, Xu L (2006) A complete protein pattern of cellulase and hemicellulase genes in the filamentous fungus *Trichoderma reesei*. Biotechnol J 1:1–9
10. Foreman PK, Brown D, Dankmeyer L, Dean R, Diener S, Dunn-Coleman NS, Goedegebuur F, Houfek TD, England GJ, Kelley AS, Meerman HJ, Mitchell T, Mitchinson C, Olivares HA, Teunissen PJM, Yao J, Ward M (2003) Transcriptional regulation of biomass degrading enzymes in the filamentous fungus *Trichoderma reesei*. J Biol Chem 278:31988–31997

11. De Vries RP, Visser J (2001) *Aspergillus* enzymes involved in degradation of plant cell wall polysaccharides. Microb Mol Biol Rev 65:497–522
12. Narra M, Dixit G, Divecha J, Kumar AK, Madamwar D, Shah AR (2012) Production of cellulases by solid state fermentation with *Aspergillus terreus* and enzymatic hydrolysis of mild alkali-treated rice straw. Biores Technol 121:355–361
13. Narra M, Dixit G, Divecha J, Madamwar D, Shah AR (2013) Production, purification and characterization of a novel GH12 family endoglucanase from *Aspergillus terreus* and its application in enzymatic degradation of delignified rice straw. Int Biodeterior Biodegrad 88:150–161
14. Pietrobon VC, Monteiro RTR, Pompeu GB, Borges EP, Lopes ML, Amorim HV, Cruz SH, Viegas EKD (2011) Enzymatic hydrolysis of sugarcane bagasse pretreated with acid or alkali. Braz Arch Biol Technol 54:229–233
15. Chaturvedi V, Verma P (2013) An overview of key pretreatment processes employed for bioconversion of lignocellulosic biomass into biofuels and value added products. 3. Biotech 3:415–431
16. Mood SH, Golfeshan AH, Tabatabaei M, Jouzani GS, Najafi GH, Gholami M, Ardjmand M (2013) Lignocellulosic biomass to ethanol, a comprehensive review with a focus on pretreatment. Renew Sustain Energy Rev 27:77–93
17. Ghose T (1994) Measurement of cellulase activities. In: Commission on Biotechnology. IUPAC, New York, pp 1–12
18. Dashtban M, Maki M, Leung KT, Mao C, Qin W (2010) Cellulase activities in biomass conversion: measurement methods and comparison. Crit Rev Biotechnol 30:302–309
19. Kubicek CP (1982) Beta-glucosidase excretion by *Trichoderma pseudokoningii*: correlation with cell wall bound beta-1,3-glucanase activities. Arch Microbiol 132:349–354
20. Taniguchi M, Suzuki H, Watanabe D, Sakai K, Hoshino K, Tanaka T (2005) Evaluation of pretreatment with *Pleurotus ostreatus* for enzymatic hydrolysis of rice straw. J Biosci Bioeng 100:637–643
21. Bradford MM (1976) A rapid and sensitive method for the quantitation of microgram quantities of protein utilizing the principle of protein-dye binding. Anal Biochem 72:248–254
22. Goering HK, Van Soest PJ (1970) Forage fiber analysis (apparatus, reagents, procedures and some applications). In: USDA Agricultural Handbook No. 379, Washington
23. Sarkar N, Aikat K (2014) *Aspergillus fumigatus* NITDGPKA3 provides for increased cellulase production. Int J Chem Engg 2014:1–9
24. Singh A, Singh N, Bishnoi NR (2009) Production of cellulases by *Aspergillus heteromorphus* from wheat straw under submerged fermentation. Int J Environ Sci Engg 1:23–26
25. Jecu L (2000) Solid state fermentation of agricultural wastes for endoglucanase production. Ind Crop Products 11:1–5
26. Nazir A, Soni R, Saini HS, Kaur A, Chadha BS (2010) Profiling differential expression of cellulases and metabolic footprints in *Aspergillus terreus*. Appl Biochem Biotechnol 162:538–547
27. Rauscher AJ, Wurleitner E, Wacenovsky C, Aro N, Stricker AR (2006) Transcriptional regulation of xyn1, encoding xylanase I in *Hypocrea jecorina*. Eukaryotic Cell 5:447–456
28. Ahmady NE, Naggar E (2015) Optimization of β-glucosidase production by *Aspergillus terreus* strain EMOO 6-4 using response surface methodology under solid state fermentation. Prep Biochem Biotechnol 45:568–587
29. Sharma R, Kocher GS, Bhogal RS, Oberoi HS (2014) Cellulolytic and xylanolytic enzymes from thermophilic *Aspergillus terreus* RWY. J Basic Microbiol 54:1367–1377
30. Hanif A, Yasmeen A, Rajoka MI (2004) Induction, production, repression and de-repression of exoglucanase synthesis in *Aspergillus niger*. Biores Technol 94:311–319
31. Badhan AK, Chadha BS, Kaur J, Saini HS, Bhat MK (2007) Production of multiple xylanolytic and cellulolytic enzymes by thermotolerant fungus *Myceliophthora sp.* IMI 387099. Biores Technol 98:504–510
32. Resch MG, Donohoe BS, Baker JO, Decker SR, Bayer EA, Beckham GT, Himmel ME (2013) Fungal cellulases and complex cellulosomal enzymes exhibit synergistic mechanism in cellulose deconstruction. Energy Environ Sci 6:1858–1867
33. Sanchez-Herrera LM, Ramos-Valdivia AC, Torrre M, Salgado LM, Ponce-Noyola T (2007) Differential expression of cellulases and xylanases by *Cellulomonas flavigena* grown on different carbon sources. Appl Microbiol Biotechnol 77:589–595
34. Messner R, Gruber F, Kubicek CP (1998) Differential regulation of synthesis of multiple forms of specific endoglucanases by *Trichoderma reesei* QM9414. J Bacteriol 170:3689–3693
35. Kuhad RC, Gupta R, Singh A (2011) Microbial cellulases and their industrial applications. Enz Res. doi:10.4061/2011/280696
36. Falls M, Shi J, Ebrik MA, Redmond T, Yang B, Wyman CE, Garlock R, Balan V, Dale BE, Pallapolu VR, Lee YY, Kim Y, Mosier NS, Ladisch MR, Hames B, Thomas S, Donohoe BS, Vinzant TB, Elander RT, Warner RE, Ramirez RS, Holtzapple MT (2011) Investigation of enzyme formulation on pretreated switchgrass. Biores Technol 102:11072–11079

Disposal of municipal solid waste with *in situ* termireactors: proof-of-concept

SA Abbasi* and S Gajalakshmi

Abstract

Background: Termites, which are among the nature's most effective scavengers and earthmovers, are distinguished by their ability to process lignin. In this report, arguably the first of its type, we present studies in which termites have been utilized for solid waste disposal. Twenty-five different types of solid wastes were positioned by us in *in situ* termireactors, at different locations and at varying distances away from the mounds of different termite species. The rate of consumption of the substrates was monitored as a function of time.

Results: The results reveal that substrates such as cotton waste, coconut shells, and torn jute bags, which resist composting, vermicomposting, or anaerobic digestion, are successfully decomposed by termites. Different termite species were able to bypass some substrates kept nearer their mounds and go to some other substrates kept farther away, showing their preference for different wastes as well as their ability to discern one waste type from the other.

Conclusions: The work provides proof-of-concept that termites can be used for the assimilative disposal of MSW.

Keywords: Termites; MSW; Termigradation; Termireactors; Anaerobic digestion; Vermicomposting

Background

Anthropogenic processes for the treatment of biodegradable solid waste revolve round the use of aerobic, anaerobic, and facultative bacteria [1,2]. Be it a sanitary landfill, a composting system, a solid-feed anaerobic digester, or a bioprocess of some other kind, bacterial digestion has been central to the treatment of biodegradable solid wastes [3-7]. The only exception to this general rule has been vermicomposting wherein the action of bacteria and enzymes on solid waste is mediated (and controlled) by earthworms. The animal ingests solid waste along with soil and deposits the digested material in the form of seed-like vermicast. During the passage through the worm gut, the feed is acted upon by the gut microflora and gets significantly stabilized. The resulting vermicast is a good soil conditioner and fertiliser [8-13].

But neither vermicomposting nor direct bacterial action during any of the economically viable solid waste degradation processes can handle lignin ([4,5] a, b). 'Hard' biowastes such as coconut shells and woody biomass also defy swift biodegradation.

* Correspondence: abbasi.cpee@gmail.com
Center for Pollution Control and Environmental Engineering, Pondicherry University, R.Venkataraman Nagar, Kalapet, Puducherry 605 014, India

In an attempt to find a quicker and more widely applicable way to dispose large volumes of biowaste, especially the type of biowaste - mentioned above - which resists treatment methods currently in general use, we have begun exploring a new frontier: termigradation. This is one of the first reports, perhaps the first ever, on the use of termites in disposing solid waste. Termites are among the nature's most powerful scavengers and earth movers, alongside earthworms and ants [14-16]. But unlike the other two, termites harbour in their midst microflora which have the special ability not possessed by other animals: ability to digest lignin [17]. In case of lower termites, ligneous material is masticated and ingested which is then digested by microflora present in certain species of protozoa living symbiotically in the termite gut [18,19]. In case of higher termites, the microflora capable of digesting lignin is present directly in the animal gut [18,19].

Other characteristics of termites which make them potential candidates for bioprocessing of solid waste are [20]: a) their voracious appetite; b) their ability to consume a wide variety of wastes; c) diversity of their habitat preference which makes it possible to always find one or other species suitable for a given geo-climatic situation [21-23]; d) their very fast rate of population growth; and e) good quality of protein represented by termite bodies, making them ideal as poultry feed or source of chemicals such as biofuel [24,20].

But any endeavour to develop bioreactors based on termites has to overcome a unique challenge. It lies in the fact that termites are 'eu-social' animals with well-defined and uncompromising social hierarchy [25]. Unlike earthworms, of which every single individual has the potential to reproduce sexually while it is feeding upon the waste in a bioreactor, the worker termites cannot breed. It is, therefore, not possible to inoculate a pile of waste with worker termites and expect that the workers would feed and breed till the entire waste is consumed. For any termireactor to function sustainably, it has to be ensured that the workers keep coming from termite nests where the workers are being born and reared continuously (along with, of course, other termites of higher caste *viz* soldiers). To achieve this objective, we have explored the concept of *in situ* termireactor wherein chambers containing the waste are placed near pre-existing termite mounds. The chambers have sufficient openings to allow access to termites but are otherwise closed from all sides.

Methods

Substrate

Different constituents of municipal solid waste (MSW) were assessed including cardboard, paper waste, pieces of jute and cotton, coconut shells, leaf litter, etc. Paper waste, tattered jute bags, pieces of cardboard, and coconut shells were collected from the piles of municipal solid waste generated at the residential quarters of the Pondicherry University campus. Scraps and pieces of wood were obtained from a saw mill. Fresh cow dung and paddy straw were collected from a cattle farm. Tissue paper rolls made out of recycled paper, and cosmetic grade cotton were purchased from the market.

Leaf litter from 13 different types of trees: asoka (*Polyalthia longifolia*), pencil tree (*Acacia auriculiformis*), coconut (*Cocos nucifera*), cashew (*Anacardium occidentale*), palm (*Borassus flabellifer*), guava (*Psidium guajava*), mango (*Mangifera indica*), teak (*Tectona grandis*), banyan (*Ficus benghalensis*), jack (*Artocarpus heterophyllus*), tamarind (*Tamarindus indica*), peepal (*Ficus religiosa*), and pungam (*Pongamia pinnata*) was assayed in the experiments. The litter was collected from below the tree canopies. After removing the debris and leaves of other species, dry weight of the leaf litter of each tree species was estimated by oven drying weighed samples at 105°C to constant weight.

In situ termireactors

To assess the consumption of different MSW constituents by termites, each constituent was kept in an *in situ* termireactor of rectangular shape, made up from 3-mm-thick aluminium sheets. Each of the termireactors was placed in a shallow pit (45 × 30 × 2 cm), dug by us in the ground, at a measured distance away from the termite mound. All termireactors had holes of 1 cm diameter at the sides and the bottom to allow termites to access the waste.

Assessment of termigradation

The termireactors were located in such a way that the minimum distance between any two reactors as well as the distance between the reactors and the mound was 30 cm. This setup helped to spatially demarcate one substrate from the other as well as to keep the distance travelled by the termites from the nest to the pit almost equal. In each of the termireactors, 100 g (fresh weight) of the substrate was placed. The equivalent dry weight of each substrate was concurrently determined by oven-drying a known mass to a constant weight at 105°C. The substrates were moistened with tap water, to the extent that the contents become damp but not soggy. To protect the substrates from rainfall, direct sunlight, and disturbance by other animals, the pits were covered with polythene sheets reinforced with aluminium mesh. The covers were removed once in 10 days, to assess the termite action, to maintain adequate moisture content and to see if any interference of other soil macro fauna like ants, was occurring. After 60 days, the substrate remaining unconsumed by the termites was carefully removed from the termireactors and air-dried. Care was taken to brush off any soil particles adhering to the surface of the residue. Then, the residue was oven dried at 105°C to constant weight.

Experiments were also conduced to see whether the termites go to the nearest available substrate or go further in search of more favoured substrate if the one available at shorter distance is less agreeable. Termireactors of the type identical to ones described above were set up at varying distances away from the mounds. The substrates were kept in the termireactors in such a way that different substrates were located at different distances from the mound. The first substrate was kept 1 ft away from the mound, the second 2 ft away and the third 3 ft away. In all, 14 different substrates were placed at distances progressively away from the mound. The farthest a substrate was placed was 14 ft away from the mound.

Results and discussion

The 'termigradability' of 25 different substrates studied by us, in terms of fraction of the substrate consumed (dry weight basis), is presented in Figure 1.

The results (Figure 2) show that not only ligninous material but other tough-to-degrade substrates like cardboard were also consumed by termites (up to 41.65%). Moreover, substrates such as cotton waste and torn jute

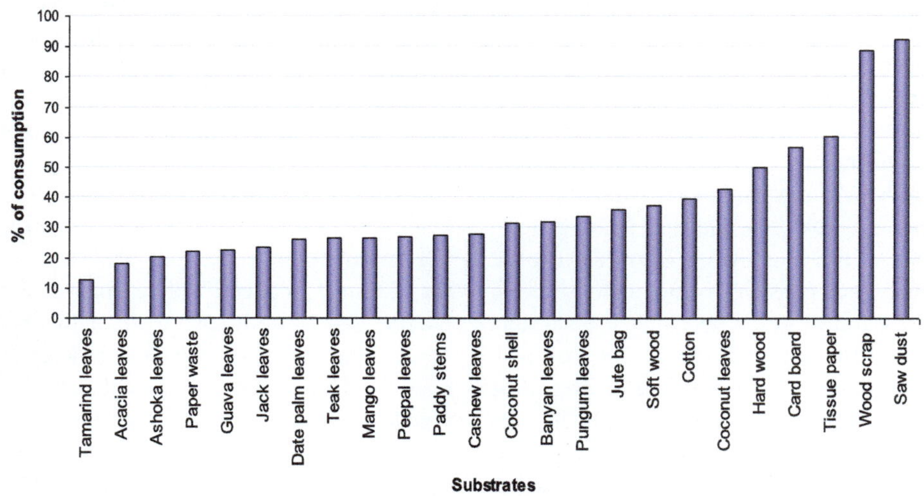

Figure 1 Average consumption of quadruplicate samples of the substrates by the termites.

bags, which resist bioprocessing during composting or vermicomposting, were also successfully consumed by termites. The studies thus confirm the potential of termites in processing ligninous and other 'hard' substances which defy composting, vermicomposting, and other forms of bacterial action in conventional solid waste management systems. Of special significance is our finding that pernicious weeds, like ipomoea, can be termidegraded. This is a finding of high promise. It is also significant that termites were able to degrade about 40% of cardboard within 60 days because otherwise cardboard is resistant to biodegradation [26]. Indeed, due to their resistance to biodegradation, cardboard pieces are often used as bulking agents in composting/vermicomposting systems [27].

Several species of white rot fungi (WRF) notably *Phanerochaete chrysosporium, Ceriporiopsis subvermispora, Pycnoporus cinnabarinus, Fuscoporia ferrea, Trametes pubescens, T. multicolor, Coriolus versicolor,* and *Pleurotus sajar-caju* are known to degrade lignin [28,29]. Attempts to utilize this ability in treating ligninous wastes have been made since long, with increasing intensity since early 1970s [30-34]; But the rate at which WRF acts is very very slow

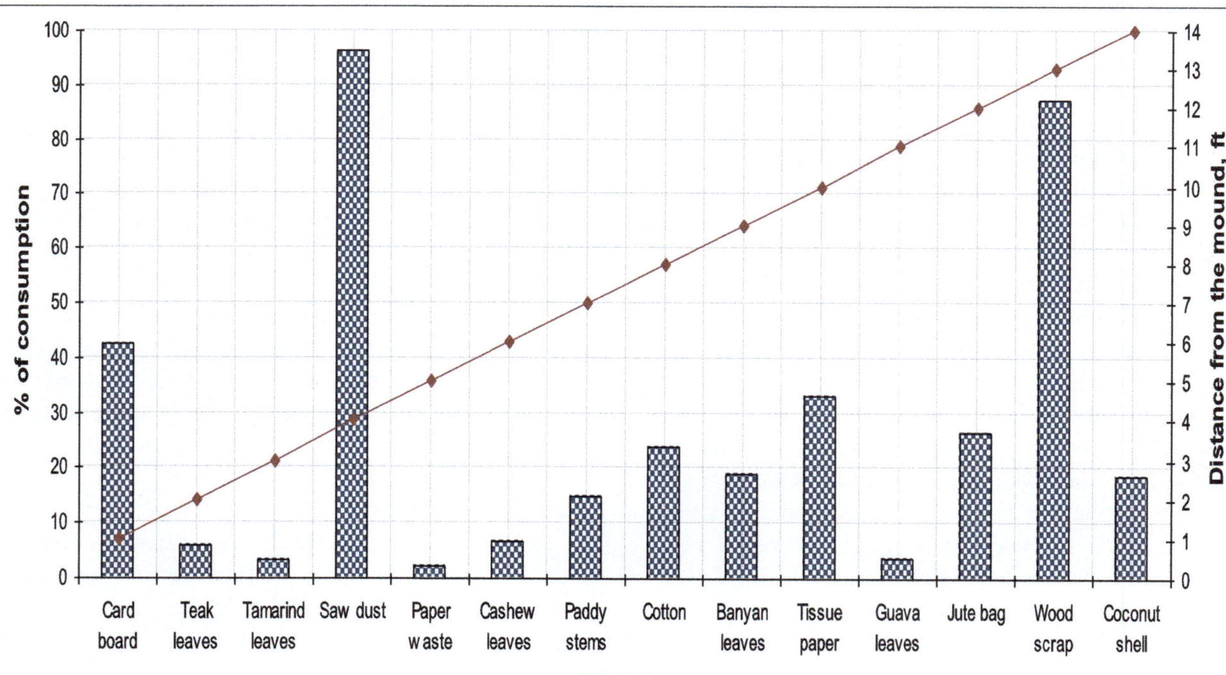

Figure 2 Consumption of the substrates by termites in the distance-based experiment.

[35,32] and, due to this, not a single process has been developed till now which makes use of WRF in degrading lignin on a commercial scale [36]. The present work suggests that termites may be able to process lignin at rates that are faster than possible with WRF but further work is needed to get a confirmation of this indication.

Figure 2, which indicates how far will termites venture out from some substrates in preference to others, indicates that the two substrates most voraciously consumed - sawdust and wood scrap (Figure 1) - are also the ones to which termites will go even if they lie a little farther away than the other substrates.

The preceding discussion is based on the extent of substrate consumed and does not give any information on the quantity or the characteristics of termicast. The reason is that in contrast to vermicomposting - which happens to be the only pre-existing bioprocess used in pollution control that is mediated by a multi-cellular animal - 50% to 60% of the substrate is converted to vermicast, termites produce no 'termicast'. This is because termites are extremely efficient utilizers of food due to the staggering diversity of the microflora present in their gut [37]. Up to 93% of the ingested food is assimilated by termites [38,39]. Very little of what termites ingest is excreted; the excreta are made up of organomineral aggregates which are carried away by the worker termites to line the termite nests [40,16]. As a result, no 'termicast' is obtained in the termireactors and the extent of substrate consumption is quantified on the basis of what is left off [10].

Moreover, whereas in a vermireactor, we can control the number of the feeders (adult earthworm) and can work out per capita vermicast production, per unit time, such assessments are not possible in termite-based systems - for reasons that have been explained in the paper (last para of the 'Background' section). Hence, the rate of disposal of the waste is assessed on the basis of the fraction consumed rather than on the basis of cast produced (as is done in vermireactors).

In summary, there is really no way to precisely determine how much of the ingested lignin is decomposed in the 'termireactors' but given the fact that termites leave little residue, it can be easily surmised that the bulk of the ligneous material that termites eat is converted to termite zoomass. Also, evidence of termite action is very stark in all these reactors while no other organisms are seen who could have contributed to the consumption.

Conclusions

The studies provide a fairly convincing 'proof-of-concept', and we are now aiming to translate these findings into practicable waste treatment systems. At present, attempts are underway to develop captive colonies of some of the termite species within the confines of laboratories so that they can be used in a controlled fashion to treat specific components, as well as unsorted forms, of municipal solid waste.

We hope that further work on this new frontier would lead to new technology which could have great potential in handling 'difficult' organic solid waste - including that which is generated in several industries.

Competing interests

The authors declare that they have no competing interests.

Authors' contributions

The idea of using termites in treating ligninous waste was conceived and developed by SAA. It was tested and refined with the assistance of SG.

Acknowledgements

The authors would thank the Department of Biotechnology, Government of India for support in the form of R&D project. We also thank our research students, Mr. Ramachander, Mr. Anbalagan, and Mr. Santhosh: especially Dr. P. Sankar Ganesh and Ms. Gurjeet Kaur for their help from time to time.

References

1. Kreith F, Tchobanoglous G (2002) Handbook of solid waste management, 2nd edn. McGraw-Hill Professional, USA
2. Dhamija U (2006) Sustainable solid waste management: issues policies and structures. Academic Foundation, New Delhi, 264 pages
3. Gajalakshmi S, Abbasi SA (2008) Solid waste management by composting: state of the art. Crit Rev Environ Sci Technol 38:311–400
4. Abbasi T, Abbasi SA (2010) Biomass energy and the environmental impacts associated with its production and utilization. Renew Sust Energ Rev 14:919–937
5. Abbasi T, Abbasi SA (2010) Production of clean energy by anaerobic digestion of phytomass - new prospects for a global warming amelioration technology. Renew Sust Energ Rev 14:1653–1659
6. Abbasi T, Tauseef SM, Abbasi SA (2012) Anaerobic digestion for global warming control and energy generation: an overview. Renew Sust Energ Rev 16:3228–3242
7. Ramasamy EV, Abbasi SA (2001) Solid waste management with earthworms. Discovery Publishing House, New Delhi
8. Abbasi T, Gajalakshmi S, Abbasi SA (2009) Towards modeling and design of vermicomposting systems: mechanisms of composting/vermicomposting and their implications. Indian J Biotechnol 8:177–182
9. Abbasi SA, Nayeem-Shah M, Abbasi T (2014) Vermicomposting of phytomass: limitations of the past approaches and the promise of the clean and efficient high-rate vermicomposting technology. Journal of Cleaner Production. doi:10.1186/s40643-014-0026-4
10. Karthikeyan M, Gajalakshmi S, Abbasi SA (2014a) Effect of storage on the properties of vermicompost generated from paper waste: with focus on pre-drying and extent of sealing. Int J Energy Environ Eng 5(135):1–11
11. Karthikeyan M, Hussain N, Gajalakshmi S, Abbasi SA (2014) Effect of vermicast generation form an allelopathic weed lantana *(Lantana camara)* on seed germination, plant growth, and yield of cluster bean *(Cyamopsis tetragonoloba)*. Environ Sci Pollut Res. doi:10.1007/s11356-014-3103-5
12. Karthikeyan M, Gajalakshmi S, Abbasi SA (2014) Ingestion of sand and soil by phytophagous earthworm *Eudrilus eugeniae*: a finding of relevance to earthworm ecology as well as vermitechnology. Arch Agron Soil Sci 60(12):1795–1804
13. Ganeshkumar T, Premalatha M, Gajalakshmi S, Abbasi SA (2014) A new process for the rapid and direct vermicomposting of the aquatic weed salvinia (*Salvinia molesta)*. Bioresources Bioprocessing 1:26
14. Abe T, Bignell DE, Higashi M (2000) Termites: evolution. Symbiosis, Ecology, Kluwer Academic Publishers, Dordrecht, Sociality
15. Konig H, Varma A (2006) Intestinal microorganisms of termites and other invertebrates. Springer-Verlag Berlin, Heidelberg
16. Bignell DE, Roisin Y, Lo N (2011) Biology of termites: a modern synthesis. Springer, London

17. Ohkuma M (2003) Termite symbiotic systems: efficient bio-recycling of lignocelluloses. Appl Microbiol Biotechnol 61:1–9
18. Butler JHA, Buckerfield JC (1979) Digestion of lignin by termites. Soil Biol Biochem 11:507–513
19. Breznak JA, Brune A (1994) Role of microorganisms in the digestion of lignocellulose by termites. Annu Rev Entomol 39:453–487
20. Premalatha M, Tauseef SM, Abbasi T, Abbasi SA (2013) A website to showcase the human-friendly aspects of termites. Curr Sci 105(11):1459–1460
21. Harit A, Gajalakshmi S, Abbasi SA (2011) Species composition and duration of termite swarms at a wooded location in Puducherry- II: The pattern in the year 2010, ECOCORPS-2011, National Seminar on Environmental Challenges in conservation of Resources of 'ONE' Planet for Sustainable Development. Department of Environmental Sciences, Bishop Heber College, Tiruchirappalli, India
22. Harit AK, Gajalakshmi S, Abbasi SA (2013) Swarming of the termite *Coptotermes gestroi* in north-eastern Puducherry. Zool Ecol 24:62–69
23. Anantharaju T, Kaur G, Gajalakshmi S, Abbasi SA (2014) Sampling and identification of termites in Northeastern Puducherry. J Entomol Zool Studies 2(3):225–230
24. Premalatha M, Abbasi T, Abbasi T, Abbasi SA (2011) Energy-efficient food production to reduce global warming and ecodegradation: the use of edible insects. Renew Sust Energ Rev 15:4357–4360
25. Krishna and Weesner (1969) Biology of termites. Vol. 1. Academic Press, New York
26. Ruiz M, Pastor K, Acevedo A (2013) Biodegradability of disposable products in a vermicomposting system. Informacion Tecnologica 24(2):47–56
27. Dominguez J, Edwards CA, Webster M (2000) Vermicomposting of sewage sludge: effect of bulking materials on the growth and reproduction of the earthworm *Eisenia andrei*. Pedobiologia 44(1):24–32
28. Giles RL, Zackeru JC, Elliott GD, Parrow MW (2012) Fungal growth necessary but not sufficient for effective biopulping of wood for lignocellulosic ethanol applications. International Biodeterioration & Biodegradation, 67, pp 1–7.
29. Hatakka A, Hammel KE (2012) Fungal bioremediation of lignocelluloses. In: Industrial applications, 2nd Edition, The Mycota X; M.Hofrichter (ed.), Springer-Verlag Berlin Heidelberg, 319–340.
30. Abbasi SA (1985) Occurrence, toxicity and treatment of lignin in pulp and paper effluents-state of the art. J S Afr Inst Civ Eng 65:1–8
31. Abbasi SA, Bhatia KKS, Nandeshwar MD, Nipaney PC (1991) Environmental impact of pulpmill effluents on an estuary on the Malabar coast. J Inst Publ Health Eng 4:28–37
32. Li P, Lee C, Lin J (2012) Using biological process to destroy lignocellulosic structure for energy conversion. Adv Mater Res 343–344:273–277
33. Narayanaswamy N, Dheeran P, Verma S and Kumar S (2013) Biological pretreatment of lignocellulosic biomass for enzymatic saccharification. In: Fang Z (Ed.) Pretreatment Techniques for Biofuels and Biorefineries. Green Energy and Technology. Springer-Verlag Berlin Heidelberg, pp. 3–34.
34. Malakani M, Khademieslam H, Hosseinihashemi SH, Zeninaly F (2014) Influence of fungal decay on chemi-mechanical properties of beech wood (*Fagus orientalis*). Cellul Chem Technol 48(1-2):97–103
35. Gupta R, Mehta G, Khasa YP, Kuhad RC (2011) Fungal delignification of lignocellulosic biomass improves the saccharification of cellulosics. Biodegradation 22(4):797–804
36. Kaur G (2014) Sampling and identification of termites in Northeastern Puducherry and exploration of their use in treating ligninous solid waste, PhD thesis, Pondicherry University
37. Warnecke F et al (2007) Metagenomic and functional analysis of hindgut microbiota of a wood feeding higher termite. Nature 450(7169):560–565
38. Wood TG (1978) Food and feeding habits of termites. In: Brian MV (ed) Production ecology of ants and termites. Cambridge University Press, London
39. Lavelle P (1997) Faunal activities and soil processes: adaptivestrategies that determine ecosystem function. Adv Ecol Res 27:93–132
40. Grube S (2001) Soil modification by the harvester termite *Hodotermes mossambicus* (Isoptera: Hodotermitidae) in a semi-arid savanna glass land of Namibia. Sociobiology 37:757–767

Chemoenzymatic construction of chiral alkenyl acetylenic alcohol, a key building block to access diastereoisomers of polyacetylenes

Sha-Hua Huang[1,2,3*], Wenhua Li[2], Long Chen[2], Jianhe Xu[3] and Ran Hong[2*]

Abstract

Background: Enzymatic kinetic resolution is proved as an efficient strategy of accessing chiral secondary alcohols in organic synthesis. Although several synthetic methods have been developed for the preparation of chiral acetylenic alcohol, biotransformation remains as a direct approach in the synthesis of polyacetylene lipids, which represent an intriguing class of marine natural products featuring interesting biological profiles.

Results: Novozym 435, a commercial lipase immobilized on macroporous acrylic resin, is utilized in the kinetic resolution of 1-yn-3-ol-4-(*E*)-ene (alkenyl acetylenic alcohol), which exists as terminus in recently isolated isofulvinol (**1**) both from the mollusk *Peltodoris atromaculata* and the sponge *Haliclona fulva*. The kinetic resolution enabled by Novozym 435 resulted in acetylenic alcohol and the corresponding acetate both in excellent enantiomeric excess and high isolated yield. The optimized reaction conditions allow us to realize the reaction at room temperature in toluene. The reaction is readily scaled up for synthetic use.

Conclusions: The current investigation builds up a platform for future exploration on stereoisomers of isofulvinol, synthetic intermediates with different chain length, and natural product analogs.

Keywords: Enzymatic catalysis; Novozym 435; Kinetic resolution; Polyacetylene; Total synthesis

Introduction

Over 100 polyacetylenes known so far are characterized from the extract of marine sponges of the genera *Petrosia*, *Xestospongia*, *Strongylophora*, etc. [1]. Characteristic of these natural products is a long chain that ranged from C44 to C47 that are bearing acetylenic, hydroxyl or keto, and isolated *cis*/*trans*-alkenes. Most polyacetylenic lipids have shown promising biological activities, such as antitumor, antimicrobial, and antiviral [2,3]. Very recently, Ciavatta and coworkers isolated several long-carbon-chain (C46) and fulvinol-like polyacetylenes from the ether extract of both the mollusk *Peltodoris atromaculata* and the sponge *Haliclona fulva* [4]. Among them, two compounds, isofulvinol (**1**) and hydroxydehydroisofulvinol (**2**) (Figure 1), showed *in vitro* promising growth inhibitory activity against a panel of six human cancer cell lines with the IC_{50} values that ranged from 3 to 36 μM. It may also be essential for understanding some basic biological phenomena, including chemical defense mechanisms since *H. fulva* was reported as the nudibranch feeds for the mollusk *P. atromaculata* [5].

The rare resources of marine sponges as well as diverse biological profiles attract several total syntheses of selected targets [6-11]. These syntheses focused on the construction of acetylenic alcohols in a highly enantioselective manner through hydroboration, catalytic asymmetric hydrogenation, as well as enzymatic resolution. The direct acetylene addition of enal with chiral alkynylzinc complex was documented to be highly substrate dependent [8,12,13]. The enzymatic kinetic resolution in literature precedents revealed the power of achieving high enantioselectivity of the recovered acetylenic alcohols as well as the esterication product [7-11,14]. Lipase AK, Lipase PS, and Novozym 435 were proved as efficient biocatalysts for this type of substrates. However,

* Correspondence: shahua@sit.edu.cn; rhong@sioc.ac.cn
[1]School of Chemical and Environmental Engineering, Shanghai Institute of Technology, Shanghai 201418, China
[2]CAS Key Laboratory of Synthetic Chemistry of Natural Substances, Shanghai Institute of Organic Chemistry, CAS, 345 Lingling Road, Shanghai 200032, China
Full list of author information is available at the end of the article

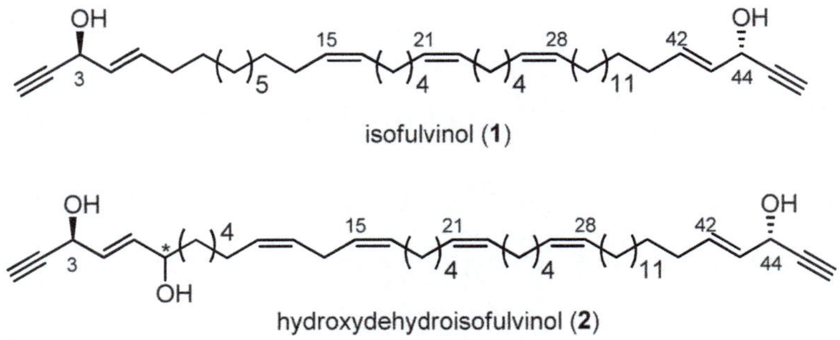

Figure 1 Isofulvinol (1) and hydroxydehydroisofulvinol (2).

target-oriented synthesis (TOS) is limited to a specific target instead of developing a general approach to diversify structural features of polyacetylenes. Only two recent reports undertook this task and disclosed interesting biological profiles [13,14]. However, the chain length and levels of desaturation (numbers and geometry) as well as the chirality for adjusting biological activity remain elusive. Guided under this line, we initiated a synthetic effort to devise a unified and practical strategy for the future evaluation of the structure-and-activity relationship in terms of carbon chain length and chirality of acetylenic alcohols.

With the consideration of isolated *cis*-alkenes, the long carbon chain, and two identical chiral 1-yn-3-ol-4-(*E*)-ene terminuses, isofulvinol (**4**) is denoted as a good starting point to develop a general approach which may eventually guide us to synthesize its stereoisomers with different chain length using a flexible synthetic tactics. Retrosynthetically, *cis*-alkenyl bisaldehyde **5** can be derived from partial hydrogenation (Lindlar catalyst) of symmetric internal alkyne **6** and stepwise Wittig olefination (Scheme 1). In order to collect both enantiomers of two 1-yn-3-ol-4-(*E*)-ene units **3** and **4**, an enzymatic kinetic resolution is envisioned to be an ideal option since both isomers are expected to achieve in high enantiomeric excess (e.e.). Moreover, this flexible tactics may offer an opportunity to establish the unknown chiral center in hydroxydehydroisofulvinol (**2**) (Figure 1) via introducing a hydroxyl group at C6 in **3** or its derivative.

Results and discussion

The synthesis of acetylenic alcohol *rac*-**3** was begun from the commercially available brominated alcohol **7** (Scheme 2). After standard Swern oxidation [15] and Wittig chain elongation, the corresponding ester **9** was further reduced by diisobutylaluminum hydride (DIBAL-H) and subsequent MnO_2 oxidation [16] to deliver enal **10**, which smoothly reacted with an acetylenic Grignard reagent to give acetylenic alcohol **3** in an impressively overall 76% yield for five straightforward steps. The corresponding

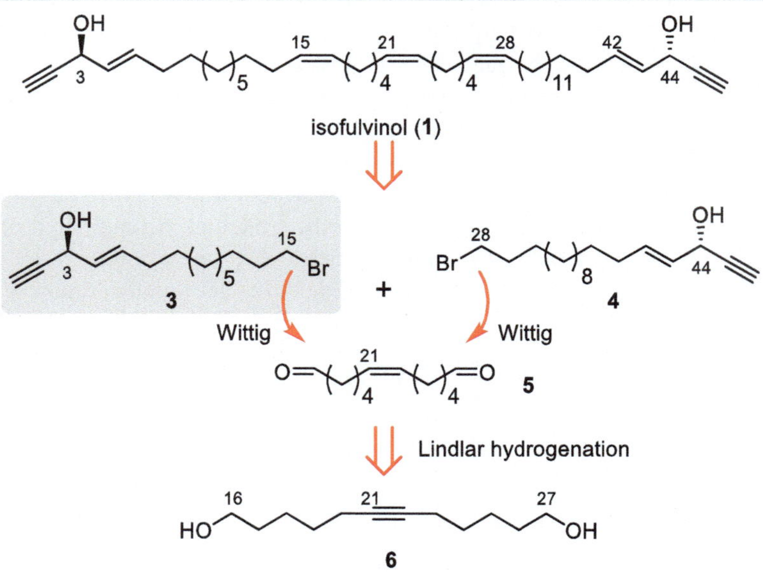

Scheme 1 Synthesis design of isofulvinol (1) (chiral moiety 3 was highlighted as the focus of the current research).

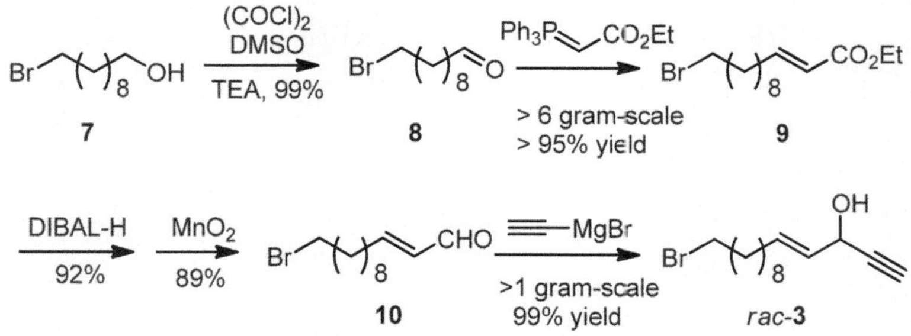

Scheme 2 Gram-scale synthesis of racemic acetylenic alcohol 3.

substance was readily cumulated in a gram scale. Three literature-known biocatalysts were screened, and only Novozym 435 gave a controllable conversion in 97% e.e. of the recovered alcohol (entry 2 in Table 1). For Lipase AK, the recovered alcohol was confirmed without a considerable enantiomeric excess during the reaction process (entry 1). Lipase PS gave a sluggish reaction and low selectivity ($E = 7$) (entry 3).

We then focused on solvent screening with the combination of Novozym 435. As shown in Table 2, enzymatic kinetic resolution was performed with a slow rate in polar solvents (entries 1 to 4). The activity of enzyme is correlated with the hydrophobicity of the solvent (Log P, P is the partition coefficient between 1-octanol and water) [17] (entries 1 to 9). In hexane and octane, however, the reaction was rather fast and not easy to maintain the alcohol **3** and the corresponding acetate **11** both in high enantioselectivities ($E = 17$ in entry 10 vs $E = 340$ in entry 8). Toluene was identified as the solvent of choice both for the reaction time and enantioselectivity (entry 7 vs 14). The resolution efficiency (E) for a synthetically useful scale (>200 mg) remains excellent (entry 15). The dramatic deterioration of activity in n-hexane during the increase of conversion[a] may be due to the conformational adjustment of biocatalyst [17] (entry 10 vs 8). Although the E number remains excellent when the reaction was performed in octane (entry 11 vs 9), overesterification was found when the reaction was scaled up. The absolute configuration of the recovered chiral acetylenic alcohol **3** was determined to be R by the modified Mosher ester [18] (Additional file 1). Since the corresponding ester (S)-**11** was not able to achieve a baseline separation on chiral HPLC columns, hydrolysis with K_2CO_3 in MeOH was performed and (S)-**3** was obtained in good yield (Scheme 3). The corresponding acetylic alcohol was assigned as S by the comparison with the HPLC profile of the recovered (R)-**3**. Its enantiomeric excess is also impressively high as 97%. That means all diastereoisomers of isofulvinol (**1**) can be accessed in excellent enantiomeric purity by utilizing Novozym 435. Enzymatic resolution provided one pair of enantiomers in high e.e. which is suitable for further synthetic usage and derivatization[b] [19].

Conclusion

In summary, we devised an efficient protocol to access both enantiomers of the chiral acetylenic alcohol motif in high enantiomeric excess (both >97% e.e.). The synthesis design discussed in this work would allow us to access a series of isofulvinol stereoisomers and derivatives through the mutation of chain length and the alternation of the stereochemistry of secondary alcohols,

Table 1 Identification of biocatalyst for the resolution of *rac*-3

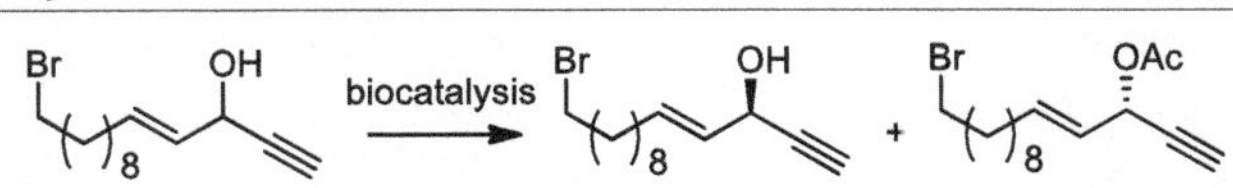

rac-3 (*R*)-3 (*S*)-11

Entry	Biocatalyst	Time (h)	Conversion (%)[a]	(*R*)-3[b] % e.e.	(*S*)-11[b] % e.e.	E^c
1	Lipase AK	0.5	>99	0	0	–
2	Novozym 435	0.5	56	96	97	340
3	Lipase PS	27.5	76	98	32	7

Reaction conditions: *rac*-**3** (151 mg, 0.5 mmol), biocatalyst (250 mg, 165% *w/w*), vinyl acetate (307 μL, 6.6 equiv), n-hexane (5 mL), R.T. [a]Conversion was determined by HPLC (Chiralcel AS-H; flow rate: 0.50 mL/min; ^{i}PrOH/n-hexane: 1/99; temperature: 25°C; UV: λ = 199 nm); [b]the determination of enantiomeric excess (HPLC chiralcel AS-H) and the absolute configuration (Mosher ester) of **3** are illustrated in Additional file 1; $^cE = \ln[(1 - e.e._S)(1 - C)]/\ln[(1 + e.e._S)(1 - C)]$, where $e.e._S$ represents the enantiomeric excess of the recovered alcohol.

Table 2 Solvent effect on the optimization of enzymatic kinetic resolution of acetylenic alcohol 3

Entry	Solvent	Log *P*	*t* (h)	Conversion (%)[a]	(*R*)-3[b]	(*S*)-11[b]	*E*[c]
					% e.e.	% e.e.	
1	Dioxane	−1.10	1.5	4	2.2	58.0	4
2	CH_3CN	−0.33	1.5	22	23.3	95.2	51
3	Acetone	−0.23	1.5	15	20.6	80[e]	11
4	THF	0.49	1.5	13	10.0	84.5	13
5	Et_2O	0.85	1.5	46.6	78.7	95.0	94
6	^{t}BuOMe	0.96	1.5	45	76.4	97.0	152
7	Toluene	2.50	1.5	49.3	91.1	96.6	185
8	n-Hexane	3.50	0.5	56	96.0	97.7	340
9	n-Octane	4.50	0.5	51.1	89.0	97.9	284
10	n-Hexane	3.50	6	60.0	96.0	64.2	17
11	n-Octane	4.50	6	54.3	98.4	96.2	300
12	Et_2O	0.85	6	52.9	97.7	89.2	78
13	^{t}BuOMe	0.96	6	52.4	98.4	91.6	109
14	Toluene	2.50	6	50.3	98.4	97.3	354
15[d]	Toluene	2.50	6	50.3	97.4	97.3	319

Reaction conditions: *rac*-**3** (30 mg, 0.1 mmol), Novozym 435 (20 mg, 67% *w/w*, 1,000 U/mmol), vinyl acetate (62 μL, 6.6 equiv), solvent (1 mL), R.T. [a]Conversion (*C*) was determined by HPLC (Chiralcel AS-H; flow rate: 0.50 mL/min; ^{i}PrOH/n-hexane: 1/99; temperature: 25°C; UV: $\lambda = 199$ nm); [b]the determination of enantioselectivity (HPLC Chiralcel AS-H) and the absolute configuration (Mosher's method) of **3** are illustrated in Additional file 1; [c]$E = \ln[(1 - e.e._S)(1 - C)]/\ln[(1 + e.e._S)(1 - C)]$, where $e.e._S$ represents the enantiomeric excess of the recovered alcohol; [d]257 mg of *rac*-**3** was applied, see the Experimental section for details; [e]since considerable side products were contaminated, the HPLC determination (80% e.e.) of (*R*)-**3** was not calibrated.

which may further provide a comprehensive investigation of biological evaluation. Current investigation of total synthesis of the target molecule as well as its derivatization is underway in this laboratory and will be reported in due course.

Experimental section

General procedure

All the reactions were carried out under N_2 atmosphere unless otherwise stated. Solvents utilized for reactions were dried using the standard procedure and distilled before use. All reagents were used after receiving albeit special treatment as indicated.

^{1}H NMR and ^{13}C NMR were recorded on Bruker AM-400 (Bruker AXS, Inc., Madison, WI, USA) and Agilent NMR-vnmrs 400 (Agilent Technologies, Inc., Santa Clara, CA, USA) spectrometers in $CDCl_3$. ^{1}H NMR data were recorded as follows: multiplicity (s = singlet, d = doublet, t = triplet, m = multiplet or unresolved, coupling constant in Hz, integration). IR spectra (cm^{-1}) were recorded on a PerkinElmer 983 Digital FT-IR spectrometer (PerkinElmer, Waltham, MA, USA) or Bruker Tensor 27 (Bruker AXS, Inc., Madison, WI, USA). Mass spectra were determined on a Shimadzu LCMS-2010EV (ESI) mass spectrometer (Shimadzu Corporation, Kyoto, Japan) or Agilent G6100 LC/MSD (ESI) single Quand mass spectrometer (Agilent Technologies, Inc., Santa Clara, CA, USA). High-resolution mass spectra were recorded on Bruker Daltonics, Inc. APEXIII 7.0 TESLA FTMS (ESI) mass spectrometers (Bruker Daltonics, Inc., Billerica, MA, USA). Optical rotation was taken on JASCO P1030 (JASCO Analytical Instruments, Easton, MD, USA). Enantiomeric excess was determined by DIONEX UltiMate 3000 chiral HPLC (Thermo Fisher Scientific, Inc., Waltham, MA, USA). Lipase from *Candida antarctica* (5,000 U/g) was used (Novozym 435, Sigma-Aldrich, Saint Louis, USA). The melting point was not calibrated.

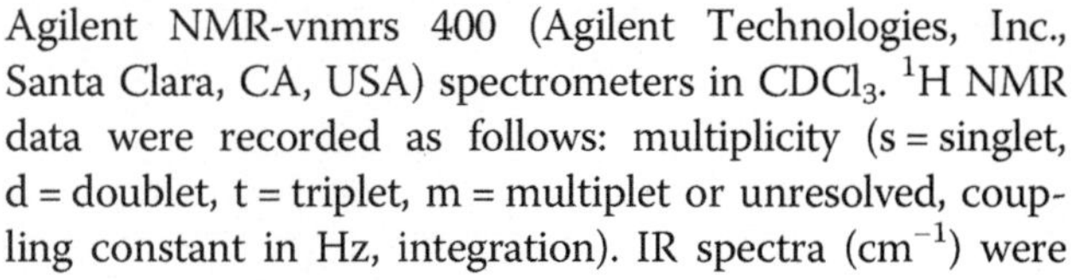

Scheme 3 Preparation of (*S*)-3 from hydrolysis of (*S*)-11.

Preparation of (E)-15-bromopentadec-4-en-1-yn-3-ol (rac-3)

11-Bromoundecanal (**8**): A solution of CH_2Cl_2 (25 mL) and oxalyl chloride (1.0 mL, 11 mmol) was added a solution of DMSO (1.7 mL, 22 mmol) in dry CH_2Cl_2 (5 mL) at −60°C. The corresponding mixture was stirred for 15 min, and then a solution of 11-bromo-1-undecanol (7) (2.512 g, 10 mmol) in CH_2Cl_2 (10 mL) was added dropwise over 5 min. After being stirred for 15 min, Et_3N (7.0 mL, 50 mmol) was added and then the reaction mixture was gradually warmed up to room temperature. The reaction was quenched with water (50 mL), and the aqueous layer was extracted with CH_2Cl_2 (2 × 25 mL). The combined organic layers were washed with HCl (aq., 1% *v*/*v*) and brine (3 × 40 mL) and then dried over anhydrous Na_2SO_4. After the removal of the organic solvent, the residue was subjected to purification by flash column chromatography (PE/EA = 50/1) to give **8** as colorless oil (2.473 g, 99% yield). ^{1}H NMR (400 MHz, $CDCl_3$): δ 9.76 (t, J = 1.4 Hz, 1H), 3.41 (t, J = 6.8 Hz, 2H), 2.42 (td, J = 7.3, 1.6 Hz, 2H), 1.81 to 1.89 (m, 2H), 1.59 to 1.65 (m, 2H), 1.38 to 1.44 (m, 2H), 1.27 to 1.34 (m, 10H).

(*E*)-ethyl 13-bromotridec-2-enoate (**9**): A solution of 11-bromoundecanal (**8**) (6.2 g, 25 mmol) and Ph_3P = CHCOOEt (9.135 g, 26.25 mmol) in toluene (125 mL) was refluxed about 3 h until the aldehyde was completely consumed. The reaction mixture was cooled to room temperature and diluted with petroleum ether (250 mL). A white precipitate was removed through a pad of Celite® (Shanghai Dahe Chemicals, Inc., Shanghai, China) and thoroughly washed with petroleum ether. The combined filtrate was concentrated under reduced pressure, and the residue was subjected to purification by flash column chromatography (PE/EA = 50/1) to afford **9** as yellow oil (7.15 g, 89% yield). ^{1}H NMR (400 MHz, $CDCl_3$): δ 6.96 (dt, J = 15.6, 7.0 Hz, 1H), 5.81 (dt, J = 15.6, 1.4 Hz, 1H), 4.18 (q, J = 7.1 Hz, 2H), 3.41 (t, J = 6.9 Hz, 2H), 2.16 to 2.22 (m, 2H), 1.81 to 1.89 (m, 2H), 1.38 to 1.47 (m, 4H), 1.26 to 1.33 (m, 13H); LRMS-ESI: calcd. for $C_{15}H_{27}BrNaO_2$ $[M + Na]^+$: 341.1, found: 341.0.

(*E*)-13-bromotridec-2-enal (**10**): A solution of (*E*)-ethyl 13-bromotridec-2-enoate (**9**) (1.877 g, 5.88 mmol) in toluene (30 mL) was stirred at 0°C for 10 min. Then DIBAL-H (10.3 mL, 1.2 M in toluene) was added dropwise within 15 min, and the resulting mixture was stirred for additional 30 min. The reaction was quenched by H_2O (0.65 mL), and the resulting reaction mixture formed into gel; vigorous stirring was needed before the gel ages into a precipitate. The suspension was filtered by Celite® and washed with ethyl acetate. The filtrate was concentrated under reduced pressure, and the corresponding residue was subjected to purification by flash column chromatography (PE/EA = 10/1) to deliver (*E*)-13-bromotridec-2-en-1-ol as colorless oil (1.505 g, 92% yield). ^{1}H NMR (400 MHz, $CDCl_3$): δ 5.59 to 5.73 (m, 2H), 4.08 (m, 2H), 3.41 (t, J = 6.9 Hz, 2H), 2.03 (q, J = 7.2 Hz, 2H), 1.81 to 1.89 (m, 2H), 1.33 to 1.44 (m, 4H), 1.25 to 1.32 (m, 10H); ^{13}C NMR (100 MHz, $CDCl_3$): δ 133.36, 128.81, 63.68, 34.01, 32.79, 32.17, 29.43, 29.40, 29.38, 29.12, 29.09, 28.72, 28.13; FT-IR: ν (cm^{-1}) 3,334 (br), 2,925, 2,853, 1,464, 1,438, 1,254, 1,089, 1,002, 969, 722, 645, 562; HRMS-ESI: calcd. for $C_{13}H_{25}BrNaO$ $[M + Na]^+$: 299.0986, found: 299.0981.

A suspension of above (*E*)-13-bromotridec-2-en-1-ol (1.45 g, 5.23 mmol) and MnO_2 (4.47 g, 52.3 mmol) in CH_2Cl_2 (20 mL) was stirred at room temperature. The reaction process was monitored by TLC. After completion, the mixture was filtered by Celite® and washed with CH_2Cl_2. The filtrate was concentrated under reduced pressure, and the residue was subjected to purification by flash column chromatography (PE/EA = 50/1) to give **10** as yellow oil (1.277 g, 89% yield). ^{1}H NMR (400 MHz, $CDCl_3$): δ 9.51 (d, J = 7.9 Hz, 1H), 6.85 (dt, J = 15.6, 6.8 Hz, 1H), 6.12 (dd, J = 15.8 Hz, J = 7.8 Hz, 1H), 3.41 (t, J = 6.8 Hz, 2H), 2.34 (q, J = 6.8 Hz, 2H), 1.81 to 1.89 (m, 2H), 1.47 to 1.54 (m, 2H), 1.39 to 1.44 (m, 2H), 1.27 to 1.36 (m, 10H); ^{13}C NMR (100 MHz, $CDCl_3$): δ 194.06, 158.98, 132.90, 33.98, 32.76, 32.67, 29.31, 29.24, 29.06, 28.67, 28.09, 27.77; FT-IR: ν (cm^{-1}) 2,927, 2,854, 2,731, 1,692, 1,637, 1,460, 1,438, 1,256, 1,153, 1,009, 974, 722, 644, 561; HRMS-ESI: calcd. for $C_{13}H_{23}BrNaO$ $[M + Na]^+$: 297.0830, found: 297.0824.

(*E*)-15-bromopentadec-4-en-1-yn-3-ol (*rac*-**3**): A solution of (*E*)-13-bromotridec-2-enal (**10**) (1.164 g, 4.23 mmol) in tetrahydrofuran (THF) (30 mL) was stirred at 0°C for 10 min. Then ethynylmagnesium bromide (12.7 mL, 0.5 M in THF) was added dropwise, and the resulting reaction mixture was stirred for 3 h. The reaction was diluted with ether (50 mL) and quenched by addition of NH_4Cl (aq., 5 mL). The organic layer was separated, and the aqueous layer was extracted with Et_2O (2 × 15 mL). The combined organic layers were washed successively by $NaHCO_3$ (aq.), water, and brine and dried over Na_2SO_4. After removal of the solvent, the residue was subjected to purification by flash column chromatography (PE/EA = 10/1) to deliver the corresponding alcohol (*rac*-**3**) as radial crystal (1.3 g, 99% yield). mp: 27°C to 28°C; ^{1}H NMR (400 MHz, $CDCl_3$): δ 5.92 (dt, J = 15.2, 7.0 Hz, 1H), 5.60 (dd, J = 15.2, 6.4 Hz, 1H), 4.84 (t, J = 5.6 Hz, 1H), 3.41 (t, J = 6.9 Hz, 2H), 2.56 (d, J = 2.2 Hz, 1H), 2.06 (q, J = 7.6 Hz, 2H), 1.88 (d, J = 6.0 Hz, 1H), 1.81 to 1.89 (m, 2H), 1.36 to 1.46 (m, 4H), 1.26 to 1.33 (m, 10H); ^{13}C NMR (100 MHz, $CDCl_3$): δ 134.36, 128.38, 83.40, 73.87, 62.65, 34.02, 32.79, 31.89, 29.40, 29.36, 29.11, 28.78, 28.71, 28.12; FT-IR: ν (cm^{-1}) 3371 (br), 3,305, 2,926, 2,854, 1,459, 1,438, 1,255, 1,089, 1,012, 968, 721, 647, 561; HRMS-ESI: calcd. for $C_{15}H_{25}BrNaO$ $[M + Na]^+$: 323.0986, found: 323.0981.

Enzymatic resolution of (E)-15-bromopentadec-4-en-1-yn-3-ol (rac-3)

(*R*)-15-bromopentadec-4-en-1-yn-3-ol ((*R*)-**3**) and (*S*)-15-bromopentadec-4-en-1-yn-3-yl acetate ((*S*)-**11**): A suspension of alcohol (*rac*-**3**) (257 mg, 0.853 mmol), Novozym 435 (172 mg, 5,000 U/g, recombinant, expressed in *Aspergillus niger*) and vinyl acetate (524 μL, 8.53 mmol) in toluene (8.5 mL) was stirred at room temperature. The reaction process was monitored by chiral HPLC. After completion (6 h), the mixture was filtered by Celite®. The filtrate was concentrated under reduced pressure, and the residue was subjected to purification by flash column chromatography (PE/EA = 50/1 to 10/1) to give (*R*)-**3** (125 mg, 48% yield) and (*S*)-**11** (150 mg, 51% yield).

(*R*)-**3**: $[\alpha]_D^{25} = -20.8$ (c 1.00, $CHCl_3$); 97.4% e.e., HPLC (Chiralcel AS-H; flow rate: 0.50 mL/min; ^{i}PrOH/n-hexane: 1/99; temperature: 25°C; UV: $\lambda = 199$ nm); mp: 31°C to 32°C; ^{1}H NMR (400 MHz, $CDCl_3$): δ 5.92 (dt, $J = 15.2$, 7.0 Hz, 1H), 5.60 (dd, $J = 15.2$, 6.4 Hz, 1H), 4.84 (t, $J = 5.6$ Hz, 1H), 3.41 (t, $J = 6.9$ Hz, 2H), 2.56 (d, $J = 2.2$ Hz, 1H), 2.06 (q, $J = 7.6$ Hz, 2H), 1.88 (d, $J = 6.0$ Hz, 1H), 1.81 to 1.89 (m, 2H), 1.36 to 1.46 (m, 4H), 1.26 to 1.33 (m, 10H); ^{13}C NMR (100 MHz, $CDCl_3$): δ 134.36, 128.38, 83.40, 73.87, 62.65, 34.02, 32.79, 31.89, 29.40, 29.36, 29.11, 28.78, 28.71, 28.12; FT-IR: ν (cm^{-1}) 3,371 (br), 3,305, 2,926, 2,854, 1,459, 1,438, 1,255, 1,089, 1,012, 968, 721, 647, 561.

(*S*)-**11**: $[\alpha]_D^{25} = +22.3$ (c 1.00, $CHCl_3$); 96.5% e.e., HPLC (Chiralcel AS-H; flow rate: 0.50 mL/min; ^{i}PrOH/n-hexane: 1/99; temperature: 25°C; UV: $\lambda = 199$ nm); ^{1}H NMR (400 MHz, $CDCl_3$): δ 6.02 (dtd, $J = 15.4$, 6.8, 0.8 Hz, 1H), 5.83 (d, $J = 5.6$ Hz, 1H), 5.54 (ddt, $J = 15.2$, 6.8, 1.4 Hz, 1H), 3.41 (t, $J = 6.8$ Hz, 2H), 2.56 (d, $J = 2.2$ Hz, 1H), 2.09 (s, 3H), 2.06 (m, 2H, overlap), 1.81 to 1.89 (m, 2H), 1.36 to 1.44 (m, 4H), 1.25 to 1.32 (m, 10H); ^{13}C NMR (100 MHz, $CDCl_3$): δ 169.67, 137.15, 124.24, 79.85, 74.67, 64.05, 34.01, 32.79, 31.94, 29.39, 29.36, 29.33, 29.08, 28.72, 28.58, 28.13, 21.06; FT-IR: ν (cm^{-1}) 3,296, 2,927, 2,854, 1,743, 1,458, 1,437, 1,370, 1,228, 1,015, 968, 642; HRMS-ESI: calcd. for $C_{17}H_{27}BrNaO_2$ $[M + Na]^+$: 365.1092, found: 365.1087.

Hydrolysis of acetate to prepare (S)-15-bromopentadec-4-en-1-yn-3-ol ((S)-3)

A suspension of (*S*)-**11** (68.5 mg, 0.2 mmol) and anhydrous K_2CO_3 (276 mg, 2 mmol) in CH_3OH (1 mL) was stirred at room temperature. The reaction process was monitored by TLC. After completion, the mixture was diluted by Et_2O. The organic layer was washed successively by water and brine and dried over Na_2SO_4. After removal of the organic solvent, the residue was subjected to purification by flash column chromatography (PE/EA = 10/1) to deliver the corresponding alcohol (*S*)-**3** as radial crystal (48 mg, 81% yield). $[\alpha]_D^{25} = +24.0$ (c 1.00, $CHCl_3$); 97.3% e.e., HPLC (Chiralcel AS-H; flow rate: 0.50 mL/min; ^{i}PrOH/n-hexane: 1/99; temperature: 25°C; UV: $\lambda = 199$ nm); mp: 27°C to 28°C; ^{1}H NMR (400 MHz, $CDCl_3$): δ 5.92 (dt, $J = 15.2$, 7.0 Hz, 1H), 5.60 (dd, $J = 15.2$, 6.4 Hz, 1H), 4.84 (t, $J = 5.6$ Hz, 1H), 3.41 (t, $J = 6.9$ Hz, 2H), 2.56 (d, $J = 2.2$ Hz, 1H), 2.06 (q, $J = 7.6$ Hz, 2H), 1.88 (d, $J = 6.0$ Hz, 1H), 1.81 to 1.89 (m, 2H), 1.36 to 1.46 (m, 4H), 1.26 to 1.33 (m, 10H); ^{13}C NMR (100 MHz, $CDCl_3$): δ 134.36, 128.38, 83.40, 73.87, 62.65, 34.02, 32.79, 31.89, 29.40, 29.36, 29.11, 28.78, 28.71, 28.12; FT-IR: ν (cm^{-1}) 3,371 (br), 3,305, 2,926, 2,854, 1,459, 1,438, 1,255, 1,089, 1,012, 968, 721, 647, 561.

Endnotes

[a]The conversions in Tables 1 and 2 were directly taken from the HPLC integration. This number should be adjusted since there are different absorption of alcohol (**3**) and the resulting acetate (**11**). This system error will not affect the qualitative comparison of enzymatic efficiency in different organic solvents.

[b]A recent elegant dynamic kinetic resolution of 1-yn-3-ol-4-ene utilizing Novozym 435 with acid resin as a racemization reagent was reported by Xu and coworkers (see [19]).

Additional file

Additional file 1: Supporting information available. The detailed information of the Mosher method for the absolute configuration identification of (*R*)-**3**, ^{1}H and ^{13}C NMR spectra of all unknown compounds, and HPLC profiles for compounds **3** and **11** are available free of charge via the Internet or directly from the authors.

Competing interests

The authors declare that they have no competing interests.

Authors' contributions

SHH and RH conceived the project and designed the strategy of synthesis. WL and LC performed the experiments. SHH, JX and RH wrote the manuscript. All authors discussed results and proofread the manuscript.

Acknowledgements

Financial supports from the National Natural Science Foundation of China (No. 21402121 to S.-H. Huang) and The MOST (2011CB710800 to R. Hong) are highly appreciated. SH Huang also thanks the financial support by the Open Funding Project of the State Key Laboratory of Bioreactor Engineering at ECUST.

Author details

[1]School of Chemical and Environmental Engineering, Shanghai Institute of Technology, Shanghai 201418, China. [2]CAS Key Laboratory of Synthetic Chemistry of Natural Substances, Shanghai Institute of Organic Chemistry, CAS, 345 Lingling Road, Shanghai 200032, China. [3]State Key Laboratory of Bioreactor Engineering, East China University of Science and Technology, 130 Meilong Road, Shanghai 200237, China.

References

1. Dembitsky VM (2006) Anticancer activity of natural and synthetic acetylenic lipids. Lipids 41:883–924
2. Siddiq A, Dembitsky V (2008) Acetylenic anticancer agents. Anti-Cancer Agents Med Chem 8:132–170
3. Minto RE, Blacklock BJ (2008) Biosynthesis and function of polyacetylenes and allied natural products. Prog Lipid Res 47:233–306
4. Ciavatta ML, Nuzzo G, Takada K, Mathieu V, Kiss R, Villani G, Gavagnin M (2014) Sequestered fulvinol-related polyacetylenes in Peltodoris atromaculata. J Nat Prod 77:1678–1684
5. Gemballa S, Schermutzki F (2004) Cytotoxic haplosclerid sponges preferred: a field study on the diet of the dotted sea slug *Peltodoris atromaculata* (Doridoidea: Nudibranchia). Mar Biol 144:1213–1222
6. Morishita K, Kamezawa M, Ohtani T, Tachibana H, Kawase M, Kishimoto M, Naoshima Y (1999) Chemoenzymatic synthesis of (1)-docosa-4,15-dien-1-yn-3-ol,

a component of the marine sponge Cribrochalina vasculum, and confirmation of the structure and absolute configuration of the acetylenic alcohol, by lipase-catalysed biotransformations. J Chem Soc, Perkin Trans 1:513–518
7. Gung BW, Dickson H, Shockley S (2001) A concise synthesis of (+)- and (−)-adociacetylene B. Tetrahedron Lett 42:4761–4763
8. Gung BW, Dickson HD, Seggerson S, Bluhm K (2002) A short synthesis of an acetylenic alcohols from the sponge cribrochalina vasculum. Synth Commun 32:2733–2740
9. Gung BW, Omollo AO (2008) First total synthesis of the potent anticancer natural product dideoxypetrosynol A: preparation of the "skipped" (*Z*)-enediyne moiety by oxidative coupling of homopropargylphosphonium ylide. Eur J Org Chem 4790–4795
10. Gung BW, Omollo AO (2008) Total synthesis of (+)- and (−)-duryne: a potent anticancer agent from the marine sponge Cribrochalina dura. Establishment of the central double bond geometry and the absolute configuration of the chiral centers. J Org Chem 73:1067–1070
11. Mori K, Akasaka K, Matsunaga S (2014) Chemoenzymatic synthesis and HPLC analysis of the stereoisomers of miyakosyne A [(4*E*,24*E*)-14-methyloctacosa-4,24-diene-1,27-diyne-3,26-diol], a cytotoxic metabolite of a marine sponge Petrosia sp., to determine the absolute configuration of its major component as 3*R*,14*R*,26*R*. Tetrahedron 70:392–401
12. Kirkham JED, Courtney TDL, Lee V, Baldwin JE (2005) Asymmetric synthesis of cytotoxic sponge metabolites R-strongylodiols A and B and an analogue. Tetrahedron 61:7219–7232
13. El Arfaoui El, Listunov D, Fabing I, Oukessou M, Frongia C, Lobjois V, Samson A, Ausseil F, Ben-Tama A, El Hadrami El H, Chauvin R, Génisson Y (2013) Identification of chiral alkenyl- and alkynylcarbinols as pharmacophores for potent cytotoxicity. ChemMedChem 8:1779–1786
14. Shirouzu T, Watari T, Ono M, Koizumi K, Saiki I, Tanaka C, Rob WM, Soest V, Miyamoto T (2013) Structure, synthesis, and biological activity of a C-20 bisacetylenic alcohol from a marine sponge *Callyspongia* sp. J Nat Prod 76:1337–1342
15. Mancuso AJ, Huang SL, Swern D (1978) Oxidation of long-chain and related alcohols to carbonyls by dimethyl sulfoxide "activated" by oxalyl chloride. J Org Chem 43:2480–2482
16. Paquette LA, Heidelbaugh TM (1996) (4S)-(−)-tert-Butyldimethylsiloxy-2-cyclopen-1-one. Org Synth 73:44–47
17. Carrea G, Ottolina G, Riva S (1995) Role of solvents in the control of enzyme selectivity in organic media. Trends Biotech 13:6–70
18. Dale JA, Mosher HS (1973) Nuclear magnetic resonance enantiomer reagents. Configurational correlations *via* nuclear magnetic resonance chemical shifts of diastereomeric mandelate, *O*-methylmandelate, and *α*-methoxy-*α*-trifluoromethylphenylacetate (MTPA) esters. J Am Chem Soc 95:512–519
19. Wu JP, Meng X, Wang L, Xu G, Yang LR (2014) Highly efficient dynamic kinetic resolution of secondary aromatic alcohols at low temperature using a low-cost sulfonated sepiolite as racemization catalyst. Tetrahedron Lett 55:5129–5232

Permissions

All chapters in this book were first published in BB, by Springer; hereby published with permission under the Creative Commons Attribution License or equivalent. Every chapter published in this book has been scrutinized by our experts. Their significance has been extensively debated. The topics covered herein carry significant findings which will fuel the growth of the discipline. They may even be implemented as practical applications or may be referred to as a beginning point for another development.

The contributors of this book come from diverse backgrounds, making this book a truly international effort. This book will bring forth new frontiers with its revolutionizing research information and detailed analysis of the nascent developments around the world.

We would like to thank all the contributing authors for lending their expertise to make the book truly unique. They have played a crucial role in the development of this book. Without their invaluable contributions this book wouldn't have been possible. They have made vital efforts to compile up to date information on the varied aspects of this subject to make this book a valuable addition to the collection of many professionals and students.

This book was conceptualized with the vision of imparting up-to-date information and advanced data in this field. To ensure the same, a matchless editorial board was set up. Every individual on the board went through rigorous rounds of assessment to prove their worth. After which they invested a large part of their time researching and compiling the most relevant data for our readers.

The editorial board has been involved in producing this book since its inception. They have spent rigorous hours researching and exploring the diverse topics which have resulted in the successful publishing of this book. They have passed on their knowledge of decades through this book. To expedite this challenging task, the publisher supported the team at every step. A small team of assistant editors was also appointed to further simplify the editing procedure and attain best results for the readers.

Apart from the editorial board, the designing team has also invested a significant amount of their time in understanding the subject and creating the most relevant covers. They scrutinized every image to scout for the most suitable representation of the subject and create an appropriate cover for the book.

The publishing team has been an ardent support to the editorial, designing and production team. Their endless efforts to recruit the best for this project, has resulted in the accomplishment of this book. They are a veteran in the field of academics and their pool of knowledge is as vast as their experience in printing. Their expertise and guidance has proved useful at every step. Their uncompromising quality standards have made this book an exceptional effort. Their encouragement from time to time has been an inspiration for everyone.

The publisher and the editorial board hope that this book will prove to be a valuable piece of knowledge for researchers, students, practitioners and scholars across the globe.

List of Contributors

Dharmesh N Adhyaru
P. G. Department of Microbiology, Gujarat Vidyapeeth, M. D. Gramseva Mahavidhyalaya, Talim Kendra Campus, Sadra 382 320, Gujarat, India

Nikhil S Bhatt
P. G. Department of Microbiology, Gujarat Vidyapeeth, M. D. Gramseva Mahavidhyalaya, Talim Kendra Campus, Sadra 382 320, Gujarat, India

Hasmukh A Modi
Department of Life Sciences, School of Sciences, Gujarat University, Navrangpura, Ahmedabad, Ahmedabad 380009, Gujarat, India

Mrinalini J Singh
Department of Botany, Nirmala college for Women, Red fields, Coimbatore 641018, Tamilnadu, India

Padmavathy Sedhuraman
Department of Botany, Nirmala college for Women, Red fields, Coimbatore 641018, Tamilnadu, India

Franck Junior Anta Akouan Ekorong
Department of Process Engineering, National School of Agro-Industrial Sciences (ENSAI), University of Ngaoundere, P.O. Box 455, Ngaoundere, Cameroon

Gaston Zomegni
Department of Textile and Leather Engineering, The Higher Institute of the Sahel, University of Maroua (ISS), P.O. Box 46 Maroua, Cameroon

Steve Carly Zangué Desobgo
Department of Food Processing and Quality Control, University Institute of Technology (UIT) of The University of Ngaoundere, P.O. Box 455, Ngaoundere, Cameroon

Robert Ndjouenkeu
Department of Food Science and Nutrition, National School of Agro-Industrial Sciences (ENSAI) of The University of Ngaoundere, P.O. Box 455, Ngaoundere, Cameroon

Yu Matsuoka
Kyushu Institute of Technology, Iizuka, Fukuoka 820-8502, Japan

Kazuyuki Shimizu
Kyushu Institute of Technology, Iizuka, Fukuoka 820-8502, Japan
Institute of Advanced Biosciences, Keio University, Tsuruoka, Yamagata 997-0017, Japan

Jing Wang
State Key Laboratory of Bioreactor Engineering, Department of Bioengineering, East China University of Science and Technology, 130 Meilong Rd., Shanghai 200237, China
Shanghai Collaborative Innovation Center for Biomanufacturing Technology, 130 Meilong Road, Shanghai 200237, People's Republic of China

Jie Qian
School of Life Science and Technology, Tongji University, Shanghai 200092, China

Lingyun Yao
State Key Laboratory of Bioreactor Engineering, Department of Bioengineering, East China University of Science and Technology, 130 Meilong Rd., Shanghai 200237, China
Shanghai Collaborative Innovation Center for Biomanufacturing Technology, 130 Meilong Road, Shanghai 200237, People's Republic of China

Yanhua Lu
State Key Laboratory of Bioreactor Engineering, Department of Bioengineering, East China University of Science and Technology, 130 Meilong Rd., Shanghai 200237, China
Shanghai Collaborative Innovation Center for Biomanufacturing Technology, 130 Meilong Road, Shanghai 200237, People's Republic of China

Mehrdad Khatami
Department of Biotechnology, Shahid Bahonar University of Kerman, End of 22-Bahman Blvd, 76169–133, Kerman 76189-18951, Iran
Department of Enviroment, The Enviromental Researches Center, Havaniroz Street, Jomhori Blvd, Kerman 76189-18951, Iran

Shahram Pourseyedi
Department of Biotechnology, Shahid Bahonar University of Kerman, End of 22-Bahman Blvd, 76169–133, Kerman 76189-18951, Iran

Mansour Khatami
Department of Enviroment, The Enviromental Researches Center, Havaniroz Street, Jomhori Blvd, Kerman 76189-18951, Iran

Hadi Hamidi
Department of Biotechnology, Shahid Bahonar University of Kerman, End of 22-Bahman Blvd, 76169–133, Kerman 76189-18951, Iran

Mehrnaz Zaeifi
Department of Plant Protection, Shahid Bahonar University of Kerman, End of 22-Bahman Blvd, 76169-133, Kerman 76169-14111, Iran

Lida Soltani
Department of Plant Protection, Shahid Bahonar University of Kerman, End of 22-Bahman Blvd, 76169-133, Kerman 76169-14111, Iran

Liqun Wang
School of Pharmaceutical Engineering and Life Sciences, Changzhou University, 1 Ge Hu Road, Jiangsu 213164, China

Junjie Miao
School of Pharmaceutical Engineering and Life Sciences, Changzhou University, 1 Ge Hu Road, Jiangsu 213164, China

Zhongqiang Wang
William G. Lowrie Department of Chemical and Biomolecular Engineering, The Ohio State University, 140 W 19th Ave, Columbus, OH 43210, USA

Lijuan Wang
School of Pharmaceutical Engineering and Life Sciences, Changzhou University, 1 Ge Hu Road, Jiangsu 213164, China

Qing Qing
School of Pharmaceutical Engineering and Life Sciences, Changzhou University, 1 Ge Hu Road, Jiangsu 213164, China

Shang-Tian Yang
William G. Lowrie Department of Chemical and Biomolecular Engineering, The Ohio State University, 140 W 19th Ave, Columbus, OH 43210, USA

Shubo Li
State Key Laboratory of Food Science and Technology, Jiangnan University, Wuxi 214122, China
College of Light Industry and Food Engineering, Guangxi University, Nanning 530004, China
The Key Laboratory of Industrial Biotechnology, Ministry of Education, Jiangnan University, Wuxi 214122, China
Laboratory of Food Microbial-Manufacturing Engineering, Jiangnan University, Wuxi 214122, China

Liming Liu
State Key Laboratory of Food Science and Technology, Jiangnan University, Wuxi 214122, China
The Key Laboratory of Industrial Biotechnology, Ministry of Education, Jiangnan University, Wuxi 214122, China
Laboratory of Food Microbial-Manufacturing Engineering, Jiangnan University, Wuxi 214122, China

Jian Chen
State Key Laboratory of Food Science and Technology, Jiangnan University, Wuxi 214122, China
The Key Laboratory of Industrial Biotechnology, Ministry of Education, Jiangnan University, Wuxi 214122, China

Guo-Chao Xu
The Key Laboratory of Industrial Biotechnology, Ministry of Education, School of Biotechnology, Jiangnan University, Wuxi 214122, China

Ye Ni
The Key Laboratory of Industrial Biotechnology, Ministry of Education, School of Biotechnology, Jiangnan University, Wuxi 214122, China

Hesong Liu
State Key Laboratory of Bioreactor Engineering, East China University of Science and Technology, 130 Meilong Road, Shanghai 200237, China

Yue Wang
State Key Laboratory of Bioreactor Engineering, East China University of Science and Technology, 130 Meilong Road, Shanghai 200237, China

Jingfan Xiao
State Key Laboratory of Bioreactor Engineering, East China University of Science and Technology, 130 Meilong Road, Shanghai 200237, China

Qiyao Wang
State Key Laboratory of Bioreactor Engineering, East China University of Science and Technology, 130 Meilong Road, Shanghai 200237, China

Qin Liu
State Key Laboratory of Bioreactor Engineering, East China University of Science and Technology, 130 Meilong Road, Shanghai 200237, China
Shanghai Collaborative Innovation Center for Biomanufacturing Technology (SCICBT), East China University of Science and Technology, 130 Meilong Road, Shanghai 200237, China

Yuanxing Zhang
State Key Laboratory of Bioreactor Engineering, East China University of Science and Technology, 130 Meilong Road, Shanghai 200237, China
Shanghai Collaborative Innovation Center for Biomanufacturing Technology (SCICBT), East China University of Science and Technology, 130 Meilong Road, Shanghai 200237, China

Jersson Plácido
Department of Biological and Agricultural Engineering, Texas A&M University, 201 Scoates hall, 2117 TAMU, College Station, TX 77843, USA

Sergio Capareda
Department of Biological and Agricultural Engineering, Texas A&M University, 201 Scoates hall, 2117 TAMU, College Station, TX 77843, USA

Ana L Gonçalves
LEPABE, Departamento de Engenharia Química, Faculdade de Engenharia, Universidade do Porto, Rua Dr. Roberto Frias, 4200-465 Porto, Portugal

Carla Ferreira
LEPABE, Departamento de Engenharia Química, Faculdade de Engenharia, Universidade do Porto, Rua Dr. Roberto Frias, 4200-465 Porto, Portugal

Joana A Loureiro
LEPABE, Departamento de Engenharia Química, Faculdade de Engenharia, Universidade do Porto, Rua Dr. Roberto Frias, 4200-465 Porto, Portugal

José CM Pires
LEPABE, Departamento de Engenharia Química, Faculdade de Engenharia, Universidade do Porto, Rua Dr. Roberto Frias, 4200-465 Porto, Portugal

Manuel Simões
LEPABE, Departamento de Engenharia Química, Faculdade de Engenharia, Universidade do Porto, Rua Dr. Roberto Frias, 4200-465 Porto, Portugal

Adepu K. Kumar
Bioconversion Technology Division, Sardar Patel Renewable Energy Research Institute, Near BVM Engineering College, Post Box No. 2, Vallabh Vidyanagar 388120, Gujarat, India

Dayanidhi Sarkar
Department of Microbiology, Primeasia University, Banani, Dhaka 1213, Bangladesh

Kazuyuki Shimizu
Kyushu Institute of Technology, Iizuka, Fukuoka 820-8502, Japan
Institute of Advanced Bioscience, Keio University, Tsuruoka, Yamagata 997-0017, Japan

Diogo Portugal-Nunes
CICECO-Aveiro Institute of Materials and Chemical Department, University of Aveiro, Campus Universitário de Santiago, 3810-193 Aveiro, Portugal
Division of Applied Microbiology, Department of Chemistry, Lund University, PO Box 124, SE-221 00 Lund, Sweden

Violeta Sànchez i Nogué
Division of Applied Microbiology, Department of Chemistry, Lund University, PO Box 124, SE-221 00 Lund, Sweden

Susana R Pereira
CICECO-Aveiro Institute of Materials and Chemical Department, University of Aveiro, Campus Universitário de Santiago, 3810-193 Aveiro, Portugal

Sandra C Craveiro
Department of Biology and GeoBioTec Research Unit, University of Aveiro, Campus Universitário de Santiago, 3810-193 Aveiro, Portugal

António J Calado
Department of Biology and GeoBioTec Research Unit, University of Aveiro, Campus Universitário de Santiago, 3810-193 Aveiro, Portugal

Ana MRB Xavier
CICECO-Aveiro Institute of Materials and Chemical Department, University of Aveiro, Campus Universitário de Santiago, 3810-193 Aveiro, Portugal

Junjie Yang
Key Laboratory of Synthetic Biology, Institute of Plant Physiology and Ecology, Shanghai Institutes for Biological Sciences, Chinese Academy of Sciences, 300 Fenglin Road, Shanghai 200032, China
Shanghai Research Center of Industrial Biotechnology, Shanghai 201201, China
Shanghai Collaborative Innovation Center for Biomanufacturing Technology, Shanghai 200237, China

Yimeng Kong
Key Laboratory of Synthetic Biology, Institute of Plant Physiology and Ecology, Shanghai Institutes for Biological Sciences, Chinese Academy of Sciences, 300 Fenglin Road, Shanghai 200032, China
University of the Chinese Academy of Sciences, Beijing 100080, China

Sheng Yang
Key Laboratory of Synthetic Biology, Institute of Plant Physiology and Ecology, Shanghai Institutes for Biological Sciences, Chinese Academy of Sciences, 300 Fenglin Road, Shanghai 200032, China
Shanghai Research Center of Industrial Biotechnology, Shanghai 201201, China
Shanghai Collaborative Innovation Center for Biomanufacturing Technology, Shanghai 200237, China

Prosun Tribedi
Department of Microbiology, University of Calcutta, 35 B.C. Road, Kolkata 700019, India

Anirban Das Gupta
Department of Microbiology, University of Calcutta, 35 B.C. Road, Kolkata 700019, India

Alok K Sil
Department of Microbiology, University of Calcutta, 35 B.C. Road, Kolkata 700019, India

Harshvadan Patel
BRD School of Biosciences, Sardar Patel Maidan Satellite Campus, Sardar Patel University, PO Box No. 39, Vallabh Vidyanagar 388120, Gujarat, India

Digantkumar Chapla
Department of Microbiology, Shree Ramkrishna Institute of Computer Education and Applied Sciences - MTB College Campus, Athwalines, Surat 395001, Gujarat, India

Jyoti Divecha
Department of Statistics, Sardar Patel University, Vallabh Vidyanagar 388120, Gujarat, India

Amita Shah
BRD School of Biosciences, Sardar Patel Maidan Satellite Campus, Sardar Patel University, PO Box No. 39, Vallabh Vidyanagar 388120, Gujarat, India

Venkatesh Chaturvedi
School of Biotechnology, Banaras Hindu University, Varanasi, UP 221005, India

Pradeep Verma
Department of Microbiology, Central University of Rajasthan, N.H. 8 Bandarsindri, Kishangarh, Ajmer, Rajasthan 305801, India

Yan Cao
National University of Singapore (Suzhou) Research Institute, 377 Linquan Street, Suzhou, Jiangsu Province, China

Zhen-ni He
School of Biotechnology, Jiangnan University, 1800 Lihu Road, Wuxi, Jiangsu Province, China

Zhong-ping Shi
School of Biotechnology, Jiangnan University, 1800 Lihu Road, Wuxi, Jiangsu Province, China

Mpofu Enock
Department of Food Processing Technology, Harare Institute of Technology, 1505 Ganges Road, P.O. Box BE 277, Belvedere, Harare, Zimbabwe

Ting Yu
State Key Laboratory of Bioreactor Engineering, East China University of Science and Technology, 130 Meilong Rd, Shanghai 200237, People's Republic of China

Xin Gao
State Key Laboratory of Bioreactor Engineering, East China University of Science and Technology, 130 Meilong Rd, Shanghai 200237, People's Republic of China

Yuhong Ren
State Key Laboratory of Bioreactor Engineering, East China University of Science and Technology, 130 Meilong Rd, Shanghai 200237, People's Republic of China

Dongzhi Wei
State Key Laboratory of Bioreactor Engineering, East China University of Science and Technology, 130 Meilong Rd, Shanghai 200237, People's Republic of China

Adepu K Kumar
Bioconversion Technology Division, Sardar Patel Renewable Energy Research Institute, Near BVM Engineering College, Vallabh Vidyanagar, Post box No. 2, 388 120 Gujarat, India

Bhumika S Parikh
Bioconversion Technology Division, Sardar Patel Renewable Energy Research Institute, Near BVM Engineering College, Vallabh Vidyanagar, Post box No. 2, 388 120 Gujarat, India

SA Abbasi
Center for Pollution Control and Environmental Engineering, Pondicherry University, R.Venkataraman Nagar, Kalapet, Puducherry 605 014, India

S Gajalakshmi
Center for Pollution Control and Environmental Engineering, Pondicherry University, R.Venkataraman Nagar, Kalapet, Puducherry 605 014, India

Sha-Hua Huang
School of Chemical and Environmental Engineering, Shanghai Institute of Technology, Shanghai 201418, China
CAS Key Laboratory of Synthetic Chemistry of Natural Substances, Shanghai Institute of Organic Chemistry, CAS, 345 Lingling Road, Shanghai 200032, China
State Key Laboratory of Bioreactor Engineering, East China University of Science and Technology, 130 Meilong Road, Shanghai 200237, China

Wenhua Li
School of Chemical and Environmental Engineering, Shanghai Institute of Technology, Shanghai 201418, China

Long Chen
CAS Key Laboratory of Synthetic Chemistry of Natural Substances, Shanghai Institute of Organic Chemistry, CAS, 345 Lingling Road, Shanghai 200032, China

Jianhe Xu
State Key Laboratory of Bioreactor Engineering, East China University of Science and Technology, 130 Meilong Road, Shanghai 200237, China

Ran Hong
CAS Key Laboratory of Synthetic Chemistry of Natural Substances, Shanghai Institute of Organic Chemistry, CAS, 345 Lingling Road, Shanghai 200032, China